Exploring Poisonous Plants

Poisonous plants are used in traditional medicine systems in various healing therapies. They are a rich resource of ingredients used in herbal drug formulations that are also used in the development of synthetic drugs. They are recognized for their antioxidant, anti-inflammation, anti-cancer, and anti-diabetic activities and for many other health benefits. *Exploring Poisonous Plants: Medicinal Values, Toxicity Responses, and Therapeutic Uses* provides an analysis of the most important poisonous herbs, shrubs, and trees, detailing poisonous plants while demonstrating endorsements for their potential therapeutic values.

Features:

Presents therapeutic potentials on various poisonous herbs, shrubs, and trees.

Provides descriptions of notable toxic compounds and discusses their adverse effects when consumed by animals or people.

Gives practical guidance for botanical description, distribution, phytochemical constituents, pharmacological studies, and traditional and other potential uses of selected poisonous plants.

This volume in the *Exploring Medicinal Plants* series is appropriate for scientists, researchers, and students working with poisonous plants, as well as in areas of economic botany, plant biochemistry, biotechnology, pharmacognosy, pharmaceuticals, industrial chemistry, and nanomedicine.

Exploring Medicinal Plants

Series Editor:
Azamal Husen
Wolaita Sodo University, Ethiopia

Medicinal plants render a rich source of bioactive compounds used in drug formulation and development; they play a key role in traditional or indigenous health systems. As the demand for herbal medicines increases worldwide, their supply is declining as most of the harvest is derived from naturally growing vegetation. Considering global interests and covering several important aspects associated with medicinal plants, the Exploring Medicinal Plants series comprises volumes valuable to academia, practitioners, and researchers interested in medicinal plants. Topics provide information on a range of subjects including diversity, conservation, propagation, cultivation, physiology, molecular biology, growth response under extreme environment, handling, storage, bioactive compounds, secondary metabolites, extraction, therapeutics, mode of action, and healthcare practices.

Led by Azamal Husen, PhD, this series is directed to a broad range of researchers and professionals consisting of topical books exploring information related to medicinal plants. It includes edited volumes, references, and textbooks available for individual print and electronic purchases.

Sustainable Uses of Medicinal Plants
Learnmore Kambizi and Callistus Bvenura

Omics Studies of Medicinal Plants
Ahmad Altaf

Traditional Herbal Therapy for the Human Immune System
Azamal Husen

Environmental Pollution and Medicinal Plants
Azamal Husen

Herbs, Shrubs, and Trees of Potential Medicinal Benefits
Azamal Husen

Phytopharmaceuticals and Biotechnology of Herbal Plants
Sachidanand Singh, Rahul Datta, Parul Johri, and Mala Trivedi

Exploring Poisonous Plants: Medicinal Values, Toxicity Responses, and Therapeutic Uses
Azamal Husen

Plants as Medicine and Aromatics: Conservation, Ecology, and Pharmacognosy
Mohd Kafeel Ahmad Ansari, Bengu Turkyilmaz Unal, Munir Ozturk, and Gary Owens

Medicinal Plant Responses to Stressful Conditions
Arafat Abdel Hamed Abdel Latef

Exploring Poisonous Plants

Medicinal Values, Toxicity Responses, and Therapeutic Uses

Edited by

Azamal Husen

CRC Press
Taylor & Francis Group
Boca Raton London New York

CRC Press is an imprint of the
Taylor & Francis Group, an **informa** business

Designed cover image: © Shutterstock

First edition published 2023
by CRC Press
6000 Broken Sound Parkway NW, Suite 300, Boca Raton, FL 33487-2742

and by CRC Press
4 Park Square, Milton Park, Abingdon, Oxon, OX14 4RN

CRC Press is an imprint of Taylor & Francis Group, LLC

Library of Congress Cataloging-in-Publication Data
Names: Husen, Azamal, editor.
Title: Exploring poisonous plants : medicinal values, toxicity responses, and therapeutic uses / edited by Azamal Husen.
Description: First edition. | Boca Raton, FL: CRC Press, 2023. | Includes bibliographical references and index.
Identifiers: LCCN 2022038657 (print) | LCCN 2022038658 (ebook) | ISBN 9781032311371 (hardback) | ISBN 9781032311388 (paperback) | ISBN 9781003308249 (ebook)
Subjects: LCSH: Poisonous plants—Therapeutic use. | Medicinal plants. | Materia medica, Vegetable.
Classification: LCC QK100.A1 E97 2023 (print) | LCC QK100.A1 (ebook) | DDC 581.6/59—dc23/eng/20220825
LC record available at https://lccn.loc.gov/2022038657
LC ebook record available at https://lccn.loc.gov/2022038658

ISBN: 9781032311371 (hbk)
ISBN: 9781032311388 (pbk)
ISBN: 9781003308249 (ebk)

DOI: 10.1201/b23017

Typeset in Times
by codeMantra

To my sister, Kaiser Jahan

Contents

About the Editor

Azamal Husen served as Professor & Head, Department of Biology, University of Gondar, Ethiopia and is a Foreign Delegate at Wolaita Sodo University, Wolaita, Ethiopia. Earlier, he was a Visiting Faculty of the Forest Research Institute, and the Doon College of Agriculture and Forest at Dehra Dun, India. His research and teaching experience of 20 years involves studies of biogenic nanomaterial fabrication and application, plant responses to environmental stresses and nanomaterials at the physiological, biochemical and molecular levels, herbal medicine, and clonal propagation for improvement of tree species.

He has conducted several research projects sponsored by various funding agencies, including the World Bank (FREEP), the National Agricultural Technology Project (NATP), the Indian Council of Agriculture Research (ICAR), the Indian Council of Forest Research Education (ICFRE); and the Japan Bank for International Cooperation (JBIC). He received four fellowships from India and a recognition award from the University of Gondar, Ethiopia, for excellent teaching, research, and community service. Husen has been on the Editorial board and the panel of reviewers of several reputed journals published by Elsevier, Frontiers Media, Taylor & Francis, Springer Nature, RSC, Oxford University Press, Sciendo, The Royal Society, CSIRO, PLOS, MDPI, John Wiley & Sons and UPM Journals. He is on the advisory board of Cambridge Scholars Publishing, UK. He is a Fellow of the Plantae group of the American Society of Plant Biologists, and a Member of the International Society of Root Research, Asian Council of Science Editors, and INPST. To his credit are over 200 publications; and he is Editor-in-Chief of the *American Journal of Plant Physiology.*

He is also working as Series Editor of '*Exploring Medicinal Plants*', published by Taylor & Francis Group, USA; '*Plant Biology, Sustainability, and Climate Change*', published by Elsevier, USA; and '*Smart Nanomaterials Technology*', published by Springer Nature Singapore Pte Ltd. Singapore.

Preface

Plants, including the poisonous ones, are primary producers and therefore important for healthy ecosystems. Poisonous plants (and their parts such as seeds, fruits, flowers, leaves, stems, and roots) are used in traditional medicine system as a natural source in various healing therapies. They are rich resource of important ingredients, the bioactive compounds, that are used in various herbal drug formulations. Additionally, these are also used in the development of synthetic drugs. These plant-based products are also considered as significant source of nutrition and are endorsed for their therapeutic value. For instance, some plants are rich in terms of phytochemicals, namely flavonoids, anthocyanins, phenolic acids, carotenoids, and tannins. Thus, they are at the same time used for potential health benefits such as antioxidant activity, anti-inflammation, anti-cancer activity, anti-diabetic activity, and hepatoprotective activity. However, quite often, recipes for the treatment of common ailments such as diarrhea, constipation, hypertension, dysentery, low sperm count, weak penile erection, piles, coated tongue, menstrual disorders, bronchial asthma, leucorrhea, and fevers are given by the traditional healers very effectively.

Poisonous plants, including numerous herbs, shrubs, and trees, cause adverse effects to animals or people when consumed. These plants, such as *Abrus precatorius*, *Aconitum napellus*, *Acorus calamus*, *Agapanthus orientalis*, *Antiaris toxicaria*, *Aristolochia indica*, *Atractylis gummifera*, *Atropa belladonna*, *Calotropis gigantea*, *Colchicum autumnale*, *Conium maculatum*, *Datura stramonium*, *Digitalis purpurea*, *Gloriosa superba*, *Nerium oleander*, *Ricinus communis*, *Thevetia peruviana*, and *Toxicodendron succedaneum*, contain specific notable toxic chemical compounds which cause toxicity. For instance, toxic alkaloids include piperidine, pyridine, indole, quinolizidine, and other alkaloids. Similarly, toxic phenolics, toxic glycosides, etc. constitute another category of plant toxins. In general, dogs, cats, sheep, goats, horses, and people, especially children, are often exposed to toxic plants. Moreover, pets are usually poisoned by these plants as a result of accidental ingestion of house or garden plants. Some plants exhibit mild effect, and thus no therapeutic intervention is required. However, some are extremely toxic, and ingestion of small amounts leads to rapid death. It has been noticed that the diagnosis of plant intoxication is quite challenging. However, many intoxicated pets and other patients are recovered under suitable symptomatic and supportive care.

Thus, the aim of this book is to provide an overview of the most important and selected poisonous herbs, shrubs, and trees. It must be beneficial to the scientists, researchers, and students working specially on poisonous plants, as well as in areas of economic botany, plant biochemistry, biotechnology, pharmacognosy, pharmaceuticals, industrial chemistry, nanomedicine, and other allied subjects.

I am grateful to all contributors for readily accepting my invitation, sharing their knowledge in specialized areas of research, and readily adjusting the suggestions for improving the shape of their contributions. With great pleasure, I also extend my sincere thanks to Randy Brehm, Tom Connelly, and all associates at Taylor & Francis Group, LLC/CRC Press for their active cooperation. Finally, my special thanks go to Shagufta, Zaara, Mehwish, and Huzaifa for providing their time and extended support to put everything together.

Azamal Husen

Contributors

Shreyshi Aggarwal
Department of Life Sciences
Graphic Era Deemed to be University
Dehradun, India

Mahendra Laxman Ahire
Department of Botany
Yashavantrao Chavan Institute of Science
Life Science Laboratory
Rayat Institute of Research and Development
Satara, India

Gulwaiz Akhter
Department of Unani Medicine
Central Council for Research in Unani Medicine Ministry of AYUSH
Govt. of India
New Delhi, India

Ahmad Ali
Department of Life Sciences
University of Mumbai
Mumbai, India

Hina Alim
Department of Life Sciences
University of Mumbai
Mumbai, India

Godwin Anywar
Department of Plant Sciences, Microbiology & Biotechnology
Makerere University
Kampala, Uganda

Lubna Azmi
Institute of Pharmaceutical Sciences
University of Lucknow
Lucknow, India

Senjuti Banerjee
Department of Microbiology
St. Xavier's College
Kolkata, India

Subhrajit Banerjee
Department of Physiology
Surendranath College
Kolkata, India

Ambreen Bano
Department of Biosciences
Faculty of Sciences
Integral University
Lucknow, India

Ankita Bhardwaj
Department of Silviculture and Agroforestry
Kerala Agricultural University
Thrissur, India

Noureddine Chaachouay
Agri-Food and Health Laboratory (AFHL), Faculty of Sciences and Techniques of Settat
Hassan First University
Settat, Morocco

Kolagani Chandramohan
National Forest Inventory, Forest Survey of India
Ministry of Environment, Central Zone, Seminary Hills
Nagpur, India

Umakanta Dash
Department of Silviculture and Agroforestry
Kerala Agricultural University
Thrissur, India

Abhijit Dey
Department of Life Sciences
Presidency University
Kolkata, India

Indu Anna George
Department of Life Sciences
University of Mumbai
Mumbai, India

Upagya Gyaneshwari
Departments of Biotechnology, School of Life Sciences
Mahatma Gandhi Central University
Motihari, India

Adria Hasan
Department of Bioengineering
Faculty of Engineering
Integral University
Lucknow, India

Azamal Husen
Wolaita Sodo University
Wolaita, Ethiopia

Tenzin Jamtsho
Yangchenphugh Higher Secondary School
Ministry of Education
Thimphu, Bhutan

Meenakshi K
Department of Life Sciences
University of Mumbai
Mumbai, India

Prabhjot Kaur
CSIR-Research Associate
Department of Biotechnology
Kurukshetra University
Kurukshetra, India

Johra Khan
Department of Medical Laboratory Sciences
College of Applied Medical Sciences
Majmaah University
Al Majmaah, Saudi Arabia

Salman Khan
National Forest Inventory, Forest Survey of India
Ministry of Environment, Central Zone, Seminary Hills
Nagpur, India

Rakesh Kumar
Department of Botany
Vijay Singh Pathik Govt. (P.G.) College
Kairana, India

Ram Kumar
Department of Botany
Vijay Singh Pathik Govt. (P.G.) College
Kairana, India

Ashwath M.N.
Department of Forest Biology and Tree Improvement
Kerala Agricultural University
Thrissur, India

Snober S. Mir
Department of Biosciences
Faculty of Sciences
Integral University
Lucknow, India

Sourav Ranjan Mohapatra
Department of Forest Biology and Tree Improvement
Kerala Agricultural University
Thrissur, India

Pankaj Shivnarayan Mundada
Department of Biotechnology
Yashavantrao Chavan Institute of Science
Satara, India

Mustehasan
Department of Unani Medicine
Central Council for Research in Unani Medicine Ministry of AYUSH
Govt. of India
New Delhi, India

Sudeshna Nandi
Molecular and Applied Mycology and Plant Pathology Laboratory
Department of Botany
University of Calcutta
Kolkata, India

Priti Pal
Department of Civil Engineering
Shri Ramswaroop Memorial College of Engineering & Management
Lucknow, India

Vedanshi Pal
Department of Life Sciences
Graphic Era Deemed to be University
Dehradun, India

Brijesh Pandey
Departments of Biotechnology, School of Life Sciences
Mahatma Gandhi Central University
Motihari, India

Manu Pant
Department of Life Sciences
Graphic Era Deemed to be University
Dehradun, India

Apekshakumari Patel
Department of Life Sciences
University of Mumbai
Mumbai, India

Nimisha Patel
Department of Life Sciences
J.C. Bose University of Science & Technology, YMCA
Faridabad, India

Yashashree Pradhan
Department of Life Sciences
University of Mumbai
Mumbai, India

Manjula Rai
Department of Botany
St. Joseph's College
Darjeeling, India

Kajal Samantara
Department of Genetics and Plant Breeding
Institute of Agricultural Sciences
Siksha 'O' Anusandhan (Deemed to be University)
Bhubaneswar, India

Rupa Sanyal
Department of Botany
Bhairab Ganguly College
Kolkata, India

Kasturi Sarkar
Department of Microbiology
St. Xavier's College
Kolkata, India

Surjit Sen
Department of Botany
Fakir Chand College
Diamond Harbour, India

Mansi Shah
Department of Life Sciences
Centre for Excellence in Marine Studies
University of Mumbai
Mumbai, India

Jitender Sharma
Department of Biotechnology
Kurukshetra University
Kurukshetra, India

Neha Sharma
Amity Institute of Biotechnology
Amity University
Noida, India

Swati Sharma
Department of Biosciences
Faculty of Sciences
Integral University
Lucknow, India

Parames C. Sil
Department of Molecular Medicine
Bose Institute
Kolkata, India

Akhilesh Kumar Singh
Departments of Biotechnology, School of Life Sciences
Mahatma Gandhi Central University
Motihari, India

Amarjit Singh
Department of Botany
Guru Nanak Khalsa College
Yamuna Nagar, India

Arundhati Singh
School of Agriculture and Environment
The University of Western Australia
Perth, Australia

Riddhima Singh
Amity Institute of Biotechnology
Amity University
Noida, India

Tina
Department of Life Sciences
Graphic Era Deemed to be University
Dehradun, India

Patience Tugume
Department of Plant Sciences, Microbiology & Biotechnology
Makerere University
Kampala, Uganda

Ugyen
Jigme Khesar Strict Nature Reserve
Department of Forests & Park Services
MoAF
Haa, Bhutan

Suraj Dhanyakumar Umdale
Department of Botany
Jaysingpur College, Jaysingpur (Affiliated to Shivaji University)
Kolhapur, India

Preeti Vats
Department of Forest Products and Utilization
Kerala Agricultural University
Thrissur, India

Phurpa Wangchuk
Center for Molecular Therapeutics
Australian Institute of Tropical Health and Medicine (AITHM)
James Cook University
Smithfield, Australia

Karma Yeshi
Center for Molecular Therapeutics
Australian Institute of Tropical Health and Medicine (AITHM)
James Cook University
Smithfield, Australia

Belkıs Muca Yiğit
Vocational School of Technical Sciences
Department of Forestry
Igdir University
Igdir, Turkey

Kamal Fatima Zahra
Faculty of Sciences and Techniques
Laboratory of Physical Chemistry of Processes and Materials/Agri-Food and Health
Hassan First University
Settat, Morocco

Lahcen Zidane
Plant, Animal Productions and Agro-Industry Laboratory
Department of Biology, Faculty of Sciences
Ibn Tofail University
Kenitra, Morocco

Section A

General Information

1 Poisonous Plants
An Overview

Mustehasan, Gulwaiz Akhter, and Azamal Husen

CONTENTS

1.1 INTRODUCTION

Plants are beneficial for human life in every aspect. Life on earth can't be expected without plants. Plant and their parts are used for nutrition, ornamental and medicinal purposes (Husen 2021, 2022; Husen et al., 2021). Some plants are poisonous in nature (Bailey and Connor, 2005; Njugi, 2018). The history of poisonous plants is quiet old and can be traced from antiquity to present day. The execution of Socrates (399 BC) is a historical event. He died by consuming extract of a poisonous plant known as hemlock (*Conium maculatum*). The progressive centripetal paralysis leading to respiratory failure is said to be the cause of death (Dyan, 2009). It is believed that Egyptian queen Cleopatra VII (69 BC–30 BC) had good knowledge of poisons, and her suicide is attributed to snake venom. However, some scholars are of the view that she might have consumed a cocktail of drugs made of opium, aconitum and hemlock, to end her life in few hours, or applied a toxic ointment (Singh et al., 2011; Ana María, 2021).

In general, it has been noticed that whether someone is speaking of people, pets or livestock, the situation surrounding plant intoxication varies. Pets and children, on the other hand, usually share the same habitat, rendering the same plants exposed to both. Plants involved in exposures or poisonings are comparable for children and dogs, with a few significant differences, and include typical household plants such as those possessing insoluble calcium oxalates (Buck, 1992; Bronstein et al., 2008). Further, due to the prevalence of plants in pastures and rangelands, livestock are sometimes intoxicated by plants to which people and pets are not exposed. People who are unaware of plants' toxicity unknowingly provide clippings from toxic plants such as *Taxus* spp. or *Nerium oleander* to their livestock and thus intoxicate them (Alden et al., 1977; Galey et al., 1996). When it is realized that the presence of meadow saffron or water hemlock in a meadow may occasion the loss of valuable animals or that the ingestion of certain wild berries by a child may result in death, it will be clearly seen that some knowledge as to which plants are poisonous is desirable not only on the part of farmers and others but also of all dwellers in the countryside.

The intoxication of animals is caused by a variety of reasons. As many hazardous plants are unpleasant to eat, cattle avoid them as there are other high-quality feeds available (James, 1988). Toxic plant ingestion is substantially more frequent at times of forage scarcity, especially in droughts.

DOI: 10.1201/b23017-2

Toxic plants can flourish to the point where non-toxic fodder is sparse, compelling animals to eat the toxic plants if pastures and ranges are not properly managed (Panter et al., 2007). Many plants that are harmful to animals have different levels of toxicity depending on their stage of development. As a result, cattle can safely use pastures or rangelands at particular periods of the year when plant toxicity is minimal (Panter et al., 2007). The ancient men were not aware of agriculture, and they used plants and animals for their food. Investigation showed that a field associated with the losses contained a large amount of meadow saffron and water hemlock, and that these plants were the cause of the loss of stock. Human knowledge of poisonous plant is accidental. The plants exerting harmful effects to human or animal in any manner are called poisonous plants. Humans applied certain plant extracts on their arrows for hunting animals (Thothathari et al., 1985; Singh et al., 2011). Laburnum, yew, and other noxious plants may be browsed upon by stock in fields near large gardens; others, such as box, rhododendron, monkshood, hellebore, and larkspur, may be present in clippings and other rubbish from gardens and shrubberies.

When it comes to susceptibility to intoxication, there might be major differences between species, particularly when it comes to cattle. Sheep, for example, can tolerate intake of larger pyrrolizidine alkaloids (PAs) than cattle or horses (Duringer et al., 2004). This is attributable to sheep rumen microorganisms' ability to break down PAs before absorption, as well as a reduction in the synthesis of reactive pyrroles in the liver from absorbed PAs (Bythe and Craig, 1994). There are some prominent small animal species differences in terms of intoxication susceptibility (e.g., cats experience acute renal failure after eating lilies, while dogs are unaffected). Alternatively, a variety of poisonous plants have the same effect on humans and animals. When it comes to the distribution of dangerous plants, there are regional disparities. Plants related to intoxication in one geographical region are absent in other regions. However, a number of poisonous plants have been introduced into non-native places and have thrived, posing serious poisoning dangers to humans and animals (Knight, 1988; Panter et al., 2007). Herbal plant intoxication has become more common as a result of widespread usage of herbal remedies, particularly among people (Hung and Lewin, 2001). Herbal intoxication is typically affected by misidentification or overuse of a specific herb. For example, intake of a weight-loss product containing *Aristolochia fangchi* instead of *Stephania tetrandra* was shown to be the cause of a series of cases of progressive renal interstitial fibrosis (Debelle et al., 2008). Despite the fact that herbal preparations are used in veterinary treatment, there have been no reports of widespread animal intoxication as a result of their use. Intoxication by the volatile oils such as pennyroyal oil and melaleuca oil, as well as the dietary supplement 5-hydroxytryptophan, has been reported in the veterinary literature (Bischoff and Guale, 1998; Sudekum et al., 1992; Gwaltney-Brant et al., 2000).

Apart from indigenous plants of the farm that are well known to be poisonous, there are quite a few others that are suspected of being poisonous or handful. In addition, many plants, though perhaps not really poisonous, do much damage by tainting the milk of dairy cows and the products made from it, as well as the flesh of animals. It would not be possible in this little volume to exhaust the whole subject, but some information is given about plants that might well be the cause of loss, together with notes on minor and suspected species, plants that taint the milk of cows, and plants that cause mechanical injury to stock.

The main aim of this chapter is to highlight the current available information and understanding of plants which are most typically implicated in plant intoxication and how to diagnose and treat intoxications. The effects of poisonous plants on livestock and how to minimize or prevent the livestock poisoning are discussed. Utility of poisonous plant in some drug preparations such as those for neurological disorder and cancer treatments has also been included. Information on plant-based poison commonly used as suicidal poison along with toxin responsible for death has also been summarized. Additionally, information on the clinical effects of a variety of poisonous plants with their family and other details are mentioned in Table 1.1.

TABLE 1.1
Poisonous Plants and Their Phytochemicals

Family Name	Botanical Name	Common Names	Phytochemicals
Acoraceae	*Acorus calamus* L.	Sweet-flag	Soluble oxalates
Adoxaceae	*Sambucus ebulus* L.	Elder, red berried elder, dwarf elder, danewort	Resinous substances, soluble oxalates; falcarinol; known for alkylating agent and irritant
Alstroemeriaceae	*Alstroemeria aurantiaca* Muñoz Schick & Brinck	Peruvian lily, lily of the Incas	Tulipalin A
Amaryllidaceae	*Clivia miniata* (Lindl.) Bosse	Kaffir lily	Amaryllidaceae alkaloid lycorine, galanthamine; few grams fatal; insoluble calcium oxalates throughout plant
Amaryllidaceae	*Galanthus nivalis* L.	Snowdrop	Lycorine (Amaryllidaceae alkaloid) and galanthamine
Amaryllidaceae	*Haemanthus multiflorus* Martyn	Blood lily, powder puff lily	Lycorine, chidanthine, hippeastrine, haemanthidine, and haemultine
Amaryllidaceae	*Hippeastrum aulicum* (Ker Gawl.) Herb.	Lily of the palace, naked lady, amaryllis	Lycorine (phenanthridine alkaloid)
Apiaceae	*Ammi majus* L.	Bishop's weed, queen Anne's lace	Furocoumarins
Apiaceae	*Conium maculatum* L.	Hemlock	Coniine (piperidine alkaloid); 0.25%–4% of body weight in green material can be fatal
Apocynaceae	*Allamanda cathartica* L.	Yellow allamanda	Unknown strong irritant
Apocynaceae	*Apocynum album* Greene	Dogbane, Indian hemp	Apocynamarin (cardioactive glycoside)
Apocynaceae	*Catharanthus coriaceus* Markgr.	Madagascar periwinkle	Alkaloids throughout plant
Apocynaceae	*Nerium oleander* L.	Oleander	Oleandrin (cardenolide cardiac glycoside), as little as 0.005% of animal's body weight in plant material can be fatal
Apocynaceae	*Periploca graeca* L.	Silk vine	Cardioactive glycosides
Araceae	*Alocasia acuminata* Schott	Elephant's ear	Calcium oxalate raphides and idioblasts; additional unknown toxin; small amounts of root or leaf fatal to children
Araceae	*Arisaema triphyllum* (L.) Schott	Jack in the pulpit	Insoluble calcium oxalates
Araceae	*Arum maculatum* L.	Lords-and-ladies, cuckoo pint, Adam-and-Eve	Calcium oxalate raphides and idioblasts, and also, soluble oxalates
Araceae	*Epipremnum aureum* (Linden & André) G.S. Bunting	Devil's ivy, ivy arum	Insoluble calcium oxalates

(*Continued*)

TABLE 1.1 (*Continued*)
Poisonous Plants and Their Phytochemicals

Family Name	Botanical Name	Common Names	Phytochemicals
Araliaceae	*Hedera helix* L.	English ivy, common ivy, Irish ivy	Hedera saponins B, C; rutin, caffeic acid, chlorogenic acid, α-hederin, emetine; falcarinol, falcarinone, and didehydrofalcarinol (alkylating agents, irritants)
Arecaceae	*Caryota mitis* Lour.	Fishtail palm	Calcium oxalate raphides
Asparagaceae	*Bowiea volubilis* Harv.	Zulu potato, climbing onion	Bufadienolide cardiotoxins: banocide A, B, C and D, hellebrin
Asparagaceae	*Hyacinthoides non-scripta* (L.) Chouard ex Rothm	Bluebell	Glucosidase inhibitors: 2,5-dihydroxy-methyl-3,4-dihydroxypyrrolidine (DMDP); and 1,4-dideoxy-1,4-imino-D-arabinitol (DAB)
Asparagaceae	*Hyacinthus orientalis* L.	Hyacinth, garden hyacinth, Dutch hyacinth	Lycorine (phenanthridine alkaloids), bulb, less toxic than Narcissus
Asparagaceae	*Maianthemum bifolium* (L.) F.W. Schmidt	May lily	Saponins
Asparagaceae	*Urginea maritima* (L.) Baker	Sea onion	Proscillaridin A (bufadienolide cardiotoxins)
Begoniaceae	*Begonia abbottii* Urb.	Wax begonia	Insoluble calcium oxalates and soluble oxalates
Berberidaceae	*Caulophyllum thalictroides* (L.) Michx.	Blue cohosh	Lupin alkaloids (quinolizidine alkaloids) and N-methylcytisine
Berberidaceae	*Nandina domestica* Thunb.	Nandina, heavenly bamboo	Cyanogenic glycosides
Boraginaceae	*Borago officinalis* L.	Borage	Pyrrolizidine alkaloids
Boraginaceae	*Cynoglossum officinale* L.	Hound's tongue	Pyrrolizidine alkaloids
Boraginaceae	*Symphytum officinale* L.	Comfrey	Pyrrolizidine alkaloids
Brassicaceae	*Armoracia rusticana* P. Gaertn., B. Mey. & Scherb.	Horseradish, red color	Sinigrin (allylglucosinolate)
Bromeliaceae	*Aechmea fasciata* (Lindl.) Baker	Urn plant	Insoluble calcium oxalate and proteolytic enzymes
Bromeliaceae	*Ananas comosus* (L.) Merr.	Pineapple, pineapple tree	Bromelin (digestive enzyme), calcium oxalate raphides, and ethyl acrylate (skin sensitizer)
Bromeliaceae	*Conium maculatum* L.	Lily-of-the-valley	Cardenolide cardiotoxins: primarily convallotoxin and convallamarin
Bromeliaceae	*Cryptanthus acaulis* (Lindl.) Beer	Green earth star	Insoluble calcium oxalates and proteolytic enzymes
Bromeliaceae	*Tillandsia cyanea* Linden ex K. Koch	Pink quill	Insoluble calcium oxalates and proteolytic enzymes
Bromeliaceae	*Vriesea fenestralis* Linden & André	Netted vriesea	Insoluble calcium oxalate and proteolytic enzymes

(*Continued*)

TABLE 1.1 (*Continued*)
Poisonous Plants and Their Phytochemicals

Family Name	Botanical Name	Common Names	Phytochemicals
Buxaceae	*Buxus sempervirens* L.	Box, boxwood, common box	Buxine (cyclobuxine, pregnane alkaloids); 0.1 g/kg fatal to dogs; also, tannins
Cactaceae	*Lophophora williamsii* (Lem. ex Salm-Dyck) J.M. Coult.	Peyote, mescal, mescal button	Mescaline
Cannabaceae	*Cannabis sativa* L.	Marijuana	Tetrahydrocannabinol
Colchicaceae	*Gloriosa superba* L.	Gloriosa lily, glory lily, climbing lily	Gloriosine and colchicine, both antimitotic; 7–11 mg fatal to adult human
Commelinaceae	*Rhoeo spathacea* (Sw.) Stearn	Oyster plant, boat lily, Moses in a boat	Unidentified irritant
Compositae	*Ageratum conyzoides* L.	Ageratum, floss flower	Precocenes I and II, and hepatotoxic chromenes
Compositae	*Arnica montana* L.	Arnica	Helenalinester (sesquiterpene lactone with strong sulfhydryl cross-linking ability)
Compositae	*Eupatorium rugosum* Houtt.	White snakeroot	Tremetol/tremetone; 0.5%–1.5% of body weight fatal; recovery rare, slow, can be incomplete
Compositae	*Atractylis gummifera* Salzm. ex L.	Mediterranean thistle	Atractyligenin, atractyloside, and carboxyatractyloside
Compositae	*Senecio jacobaea* L.	Common ragwort, tansy ragwort	Jacobine plus five other pyrrolizidine alkaloids
Compositae	*Artemisia absinthium* L.	Wormwood, sagewort, sagebrush	Thujone (monoterpene) in essential oil
Cupressaceae	*Juniperus sabina* Sm.	Savin	Podophyllotoxins; sabinene and sabinyl acetate (terpene derivatives)
Cupressaceae	*Thuja occidentalis* L.	White cedar	Thujone (monoterpene)
Dioscoreaceae	*Tamus communis* L.	Black bryony	Calcium oxalate raphides and idioblasts; traces of alkaloids; saponins; and photosensitive phenanthrene derivatives
Dryopteridaceae	*Dryopteris filix-mas* (L.) Schott	Male-fern	Filicin
Ericaceae	*Ledum palustre* L.	Wild rosemary	Grayanotoxin I (cardioactive diterpenoid); ledol and palustrol (sesquiterpene alcohols)
Ericaceae	*Leucothoe davisiae* Torr. ex A. Gray	Black laurel, mountain laurel	Grayanotoxin I (cardioactive diterpenoid); ledol and palustrol (sesquiterpene alcohols)
Ericaceae	*Menziesia ferruginea* Sm.	Mock azalea, rusty leaf	Grayanotoxin I (cardioactive diterpenoid); ledol and palustrol (sesquiterpene alcohols)
Ericaceae	*Pieris japonica* (Thunb.) D. Don ex G. Don	Pieris	Grayanotoxin I (cardioactive diterpenoid)

(*Continued*)

TABLE 1.1 (*Continued*)
Poisonous Plants and Their Phytochemicals

Family Name	Botanical Name	Common Names	Phytochemicals
Euphorbiaceae	*Acalypha hispida* Burm.f.	Chenille plant, red-hot cattail	Euphorb latex (diterpene esters)
Euphorbiaceae	*Codiaeum variegatum* (L.) Rumph. ex A.Juss.	Croton (ornamental)	Low-level phorbol diterpene esters
Euphorbiaceae	*Croton tiglium* L.		Crotin (toxalbumin); tigliane phorbol esters; few drops of sap can cause a reaction
Euphorbiaceae	*Euphorbia lacteal* Haw.	Candelabra cactus, false cactus, mottled spurge, dragon bones	Diterpene esters in milky sap (type not specified)
Euphorbiaceae	*Euphorbia milii* Des Moul.	Crown of thorns	Hydroxyphorbol diterpene esters in milky sap (milliamines A-I); also, triterpenes and flavonoids
Euphorbiaceae	*Euphorbia aaronrossii* A.H. Holmgren & N.H. Holmgren	Spurge, creeping spurge, donkey tail	Ingenane diterpene esters
Euphorbiaceae	*Euphorbia pulcherrima* Willd. ex Klotzsch	Poinsettia, Christmas star	Very little diterpene ester and considerable natural variability
Euphorbiaceae	*Euphorbia resinifera* O.Berg	Euphorbium	Daphnane diterpene esters (resiniferatoxin); one of the most irritating euphorbs
Euphorbiaceae	*Euphorbia tirucalli* L.	Pencil tree, milkbush, Indian tree, rubber euphorbia, finger tree, naked lady	Phorbol diterpene esters
Euphorbiaceae	*Ricinus communis* L.	Castor bean plant	Ricin (toxalbumin, 2 chain); one seed can be fatal; 1 mg ricin/g seed; LD mouse, rat, dog ~1 g/kg; rabbit, 0.1 g/kg
Euphorbiaceae	*Hura crepitans* L.	Sandbox tree, monkey pistol	Hurin (toxalbumin) and daphnane diterpene esters
Euphorbiaceae	*Synadenium grantii* Hook.f.	African milk bush	Tigliane-type diterpene esters (primary irritants and cocarcinogens)
Garryaceae	*Aucuba japonica* Thunb	Aucuba, Japanese aucuba, Japanese laurel, spotted laurel	Aucubin (triterpenoid saponins of the α-amyrin group)
Gelsemiaceae	*Gelsemium sempervirens* (L.) J.St.-Hil.	Yellow jessamine, Carolina jessamine	Indolizidine alkaloids
Hydrangeaceae	*Hydrangea macrophylla* (Thunb.) Ser.	Hydrangea, hills of snow, French hydrangea, hortensia	Hydrangin (cyanogenic glycoside)
Lauraceae	*Laurus nobilis* L.	Laurel, bay laurel, sweet bay	Unknown
Lauraceae	*Persea americana* Mill.	Avocado	Persin [(Z, Z)-1-(acetyloxy)-2-hydroxy-12,15-heneicosadien-4-one]

(*Continued*)

TABLE 1.1 (*Continued*)
Poisonous Plants and Their Phytochemicals

Family Name	Botanical Name	Common Names	Phytochemicals
Leguminosae	*Abrus precatorius* L.	Rosary pea, prayer bean, jequerity, precatory bean	Abrin (toxalbumin/lectin, 2 chains) in seeds; 0.5–2 seeds fatal to adult human
Leguminosae	*Albizia julibrissin* Durazz.	Mimosa, silk-tree	5-Acetoxymethyl-3-hydroxy-4-methoxymethyl-2-methylpyridine (alkaloid)
Leguminosae	*Glycyrrhiza glabra* L.	Licorice	Glycyrrhizin (triterpenoid saponin)
Leguminosae	*Laburnum anagyroides* Medik.	Laburnum, golden chain tree	Cytisine (quinolizidine alkaloid); 23 pods fatal to adult human
Leguminosae	*Robinia acacia* L.	Black locust	Robin (toxalbumin/lectin); less toxic than abrin, ricin, and phasin
Leguminosae	*Wisteria sinensis* (Sims) Sweet	Wisteria, Chinese kidney bean	Wisterin "sapotoxin"; lectins; 2 seeds cause vomiting/ gastroenteritis in a child
Liliaceae	*Lilium abchasicum* Baker	Easter lily, star-gazer lily, tiger lily	Unknown toxin; only known to affect cats
Malvaceae	*Abutilon hybridum* Voss	Chinese bellflower, flowering maple, parlor maple	Unknown
Melanthiaceae	*Paris quadrifolia* L.	Herb-Paris	Steroidal saponins throughout plant
Moraceae	*Ficus benjamina* L.	Weeping fig, Java willow, Benjamin tree, small-leaved rubber plant	Allergen; irritant white sap
Nyctaginaceae	*Mirabilis jalapa* L.	Four o'clock	Unknown toxin in rootstocks and seeds
Papaveraceae	Chelidonium majus L.	Greater celandine (bright yellow sap)	21 Alkaloids (benzophenanthridine and protoberberine) with chelidonic acid
Papaveraceae	*Eschscholzia californica* Cham.	California poppy	Rhoeadine-type alkaloids, thebaine, papaverrubines, papaverine; small amounts compared to opium poppy
Papaveraceae	*Papaver somniferum* L.	Opium poppy	Morphine, narcotine, codeine, papaverine, thebaine
Phytolaccaceae	*Phytolacca Americana* L.	Pokeweed, pokeberry, poke salad	Phytolaccatoxin, triterpenoids
Plantaginaceae	*Digitalis purpurea* L.	Foxglove, common foxglove, long purples, dead men's fingers	Cardenolide cardiotoxins; desacetyl-digilanids A, B; glucogitaloxin; 10%–20% hydrolyzed to digitoxin, gitoxin, gitaloxin; therapeutic index is steep; it also contains saponins digitonin, gitonin, tigonin

(*Continued*)

TABLE 1.1 (*Continued*)
Poisonous Plants and Their Phytochemicals

Family Name	Botanical Name	Common Names	Phytochemicals
Polygonaceae	*Rheum rhabarbarum* L.	Rhubarb, garden rhubarb, pie plant, water plant, wine plant	Anthraquinone cathartics; low level
Primulaceae	*Cyclamen persicum* Mill.	Persian violet, alpine violet, sowbread	Cyclamen (triterpenoid saponin); small pieces of tuber may cause convulsions
Ranunculaceae	*Aconitum abietetorum* W.T. Wang & L.Q. Li	Monk's-hood	Aconitine, others (diterpene/nor-diterpene alkaloids with alkylated amine); LD adult human is a few grams plant material
Ranunculaceae	*Caltha palustris* L.	Marsh-marigold, kingcup	Alkaloids, saponins, and very low level of protoanemonin
Rhamnaceae	*Karwinskia humboldtiana* (Schult.) Zucc	Coyotilla, tullidora	Karwinol A (neurotoxic polyphenol)
Rosaceae	*Prunus dulcis* (Mill.) D.A.Webb	Bitter almond	Amygdalin
Rosaceae	*Prunus laurocerasus* L.	Cherry laurel	Prunasin, amygdalin
Rosaceae	*Prunus serotina* Ehrh.	Wild black cherry	Amygdalin Glucosinolates Glucofrangulin (anthracene glycoside)
Rutaceae	*Thamnosma texana* Torr.	Dutchman's breeches	Furocoumarins
Santalaceae	*Viscum album* L.	European mistletoe	Viscotoxins; viscumin (toxalbumin)
Saxifragaceae	*Tolmiea menziesii* (Pursh) Torr. & A.Gray	Piggyback plant, pickaback plant, thousand-mothers, youth-on-age	Unknown
Solanaceae	*Atropa belladonna* L.	Deadly nightshade	L-Hyoscyamine, scopolamine, atropine; 2–5 berries kill a child
Solanaceae	*Brunfelsia calycina* Benth.	Yesterday, today, and tomorrow	Unknown
Solanaceae	*Capsicum annuum* L.	Ornamental and cultivated peppers	Capsaicin (70% of irritant capacity), less of dihydrocapsaicin, nordihydrocapsaicin, homocapsaicin, homodihydrocapsaicin
Solanaceae	*Cestrum diurnum* L.	Day-blooming jessamine	1,25-Dihydrodroxycholecalciferol
Solanaceae	*Datura stramonium* L.	Jimsonweed, thornapple, moonflower	L-Hyoscyamine, scopolamine
Solanaceae	Hyoscyamus niger	Henbane	L-Hyoscyamine, scopolamine
Solanaceae	*Physalis acutifolia* (Miers) Sandwith	Chinese lantern, winter cherry, Cape gooseberry	Unknown
Solanaceae	*Solanum dulcamara* L.	Bittersweet, woody nightshade, wild nightshade	Variable major alkaloids: solanine, tomatine, solamarine, atropine, solasadine; 10 berries needed to produce signs and 200 to kill an adult human

(*Continued*)

TABLE 1.1 (*Continued*)
Poisonous Plants and Their Phytochemicals

Family Name	Botanical Name	Common Names	Phytochemicals
Solanaceae	*Solanum nigrum* L.	Black nightshade	Solasadine; solanine; only a few berries needed to produce clinical signs
Solanaceae	*Solanum tuberosum* L.	Potato	Solanine; 0.035% critical level for human poisoning
Thymelaeaceae	*Daphne mezereum* L.	Mezereon, spurge olive	Mezerein (daphnane diterpene ester); 2–3 fruits fatal; several flowers: week-long hospitalization
Verbenaceae	*Lantana camara* L.	Lantana, shrub verbena, yellow sage, bunchberry	Pentacyclic triterpenes lantadene A and B; unripe berries are most toxic

1.2 PHYTOCHEMISTRY OF POISONOUS PLANTS AND THEIR CLASSIFICATION BY CHEMICAL STRUCTURE

There are literally thousands of distinct compounds in a single plant. The majority of them aren't hazardous. It is true that many plant compounds are extremely important, and some are required for the survival of all forms of life on earth, including plant and animal life. We know that life is built of carbon and that simple sugar molecules (such as glucose) are generated by sunlight interacting with green plant cells in photosynthesis power practically every food chain in the universe. These simple sugars are frequently converted into other essential compounds, such as amino acids, through biochemical pathways. Amino acids are the building blocks of proteins, and they also serve as the structure of protoplasm (the living substance); about 20 amino acids are required for life and the metabolic processes that follow it (metabolism). Primary metabolism refers to the metabolic pathways that are required for life to exist, and primary chemicals or primary metabolites are the compounds or metabolites that are directly involved in these pathways (such as glucose and essential amino acids). Despite this, plants manufacture a range of different chemicals through secondary routes, including primary metabolites like glucose (as components of certain types of glucosides, for example). Secondary chemicals (alkaloids; glucosides; or more generally, glycosides, phenolic compounds, and so on) are less well understood, and not all of them appear to be required for life to exist (Audi et al., 2005). Toxic plant chemicals fall into this large group of secondary compounds with a hazy definition; toxins are considered a subset of secondary compounds (that is, those secondary compounds which are toxic). In some cases, this is a reasonable argument: some of these deadly chemicals, but not all, appear to have no clear role in plants other than survival adaptations (defense mechanisms). It's still up for dispute whether some of these molecules originated simply as "evolutionary noise," with no discernible effect on the plant. Toxic plants, no matter what, are mostly limited to a few broad types of chemicals:

1. **Alkaloids:** Alkaloids ("alkali-like") are basic chemicals that include nitrogen, usually in the form of a heterocyclic ring. They're usually bitter, and the majority of them are toxic. They can be divided into subcategories based on the chemical nature of the nitrogen-containing heterocyclic ring:
   A. **Pyrrolizidine Alkaloids:** Plants containing PA can be found all over the world and are potentially harmful to a variety of animals, including horses, cattle, poultry, and humans (Burrows and Tyrl, 2001). The nucleus consists of two five-membered rings. PAs are responsible for the toxicity of various *Senecio* spp. (tansy particularly ragwort,

common groundsel), crotalaria, *Echium plantagineum* (Paterson's curse), and *Heliotropium europaeum.* The latter two are important in Australia. PAs cause irreversible liver damage. Examples are monocrotaline, heliotrine, lasiocarpine, and senecionine (Stewart and Steenkamp, 2001).

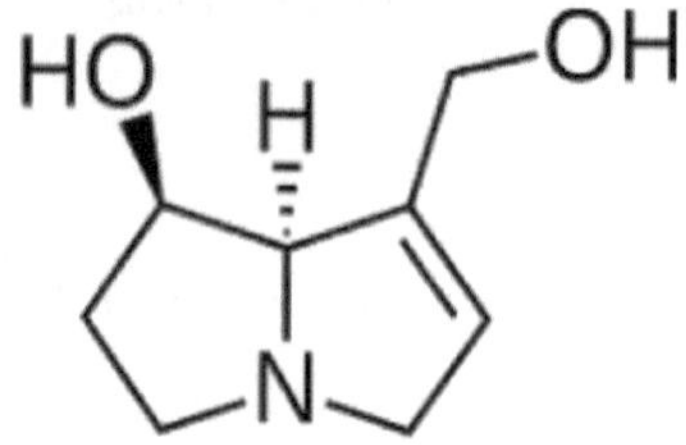

B. **Piperidine Alkaloids:** The most important piperidine alkaloids in animal production are coniine and related alkaloids found in *Conium maculatum* (poison hemlock; Panter et al., 2007). These alkaloids affect the central nervous system and are also teratogens.

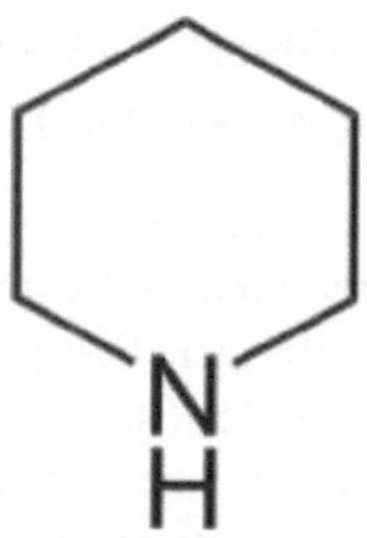

C. **Pyridine Alkaloids:** Pyridine alkaloids found in *Nicotiana* spp. (wild or cultivated tobacco, tree tobacco), indole alkaloids found in *Phalaris* spp. grasses and grasses infested with ergot (ergot alkaloids), quinolizidine alkaloids found in *Lupinus* spp. (lupines), and indolizidine alkaloids found in many *Astragalus* spp. (locoweeds) and Swain (larkspurs).

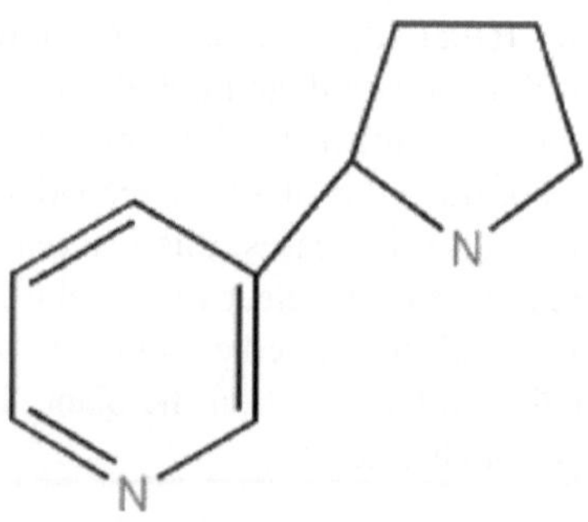

D. **Indole Alkaloids:** Indoles are derivatives of amino acid tryptophan. Examples are ergot alkaloids, alkaloids such as perloline in tall fescue, and 3-methylindole, an alkaloid implicated in bovine pulmonary emphysema.

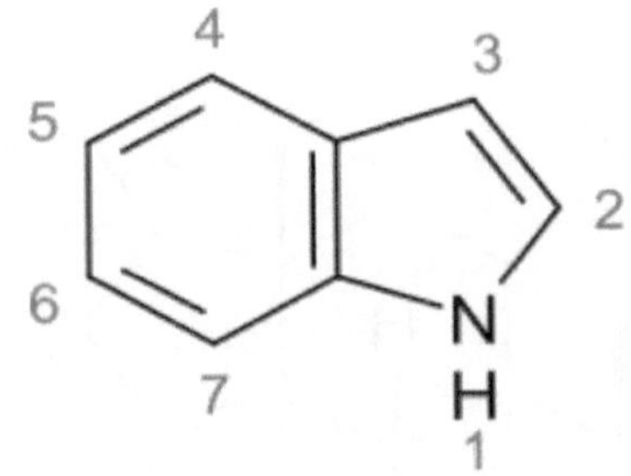

E. **Quinolizidine Alkaloids:** The quinolizidine nucleus consists of two six-membered rings. Lu pines contain these alkaloids, which cause acute poisoning in sheep and teratogenic effects in calves (crooked calf disease).

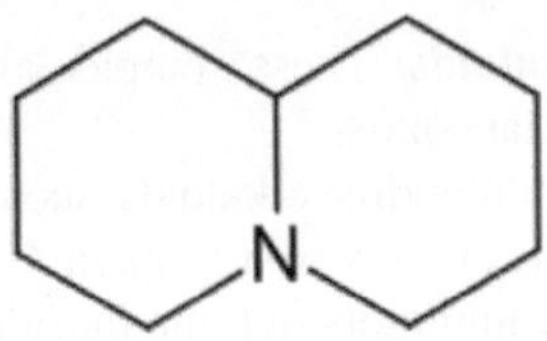

F. **Steroid Alkaloids**

i. ***Solanum* Type:** These are found in green potatoes, tomatoes, and nightshade. They are central nervous system poisons and cholinesterase inhibitors. Solanidine in potatoes has human health implications; an upper limit for alkaloid in new potato cultivars in the U.S. has been established by regulatory agencies.

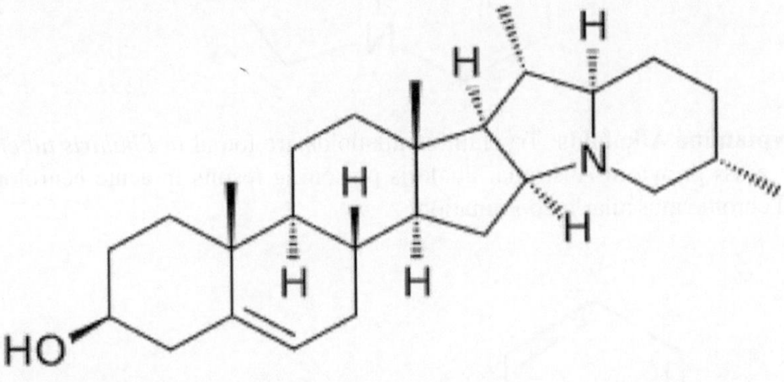

ii. ***Veratrum* Type:** False hellebore (*Veratrum californicum*) contains veratrum alkaloids that produce teratogenic effects in lambs (cyclops lamb) and prolonged gestation in ewes. Death camas, a source of extensive sheep losses in the American West in the past, contains alkaloids of this type.

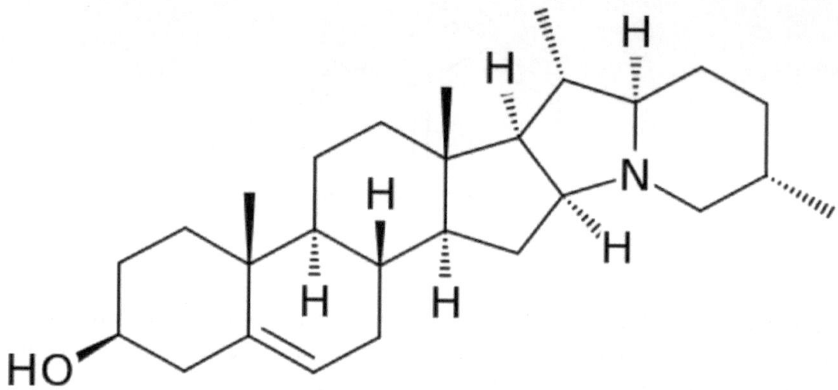

G. **Polycyclic Diterpene Alkaloids:** These complex alkaloids are found in *Delphinium* spp. Commonly known as larkspurs.

H. **Indolizidine Alkaloids:** Indolizidine alkaloids, such as swainsonine, have been identified as the toxic components of *Swainsona* spp. in Australia and *Astragalus* spp. (locoweed) in the U.S. They are inhibitors of a-mannosidase, resulting in the accumulation of mannose deposits in lysosomes of nerve cells. Slaframine is an alkaloid produced on red clover by a fungus; it causes profuse salivation.

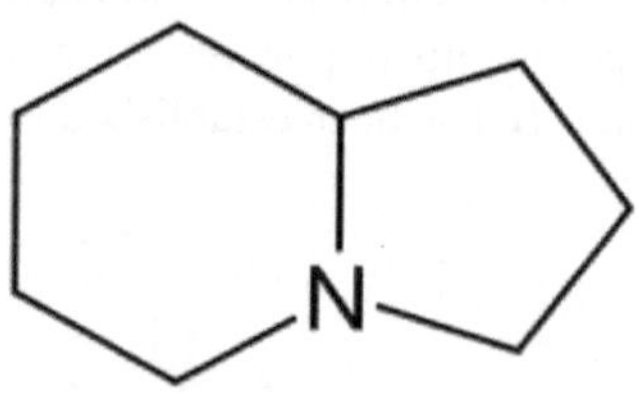

I. **Tryptamine Alkaloids:** Tryptamine alkaloids are found in *Phalaris tuberosa*, a forage grass grown in Australia. Phalaris poisoning results in acute neurological signs and chronic muscular incoordination.

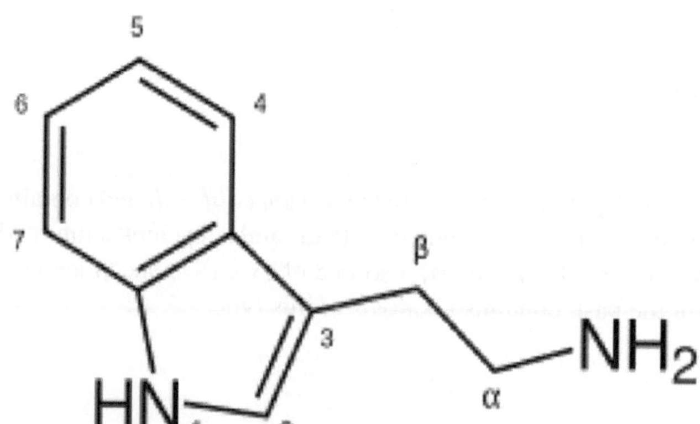

J. **Tropane Alkaloids:** Atropine, found in *Datura* spp. (Jimsonweed), is an example of tropane alkaloid. It has pronounced effects on the central nervous system.

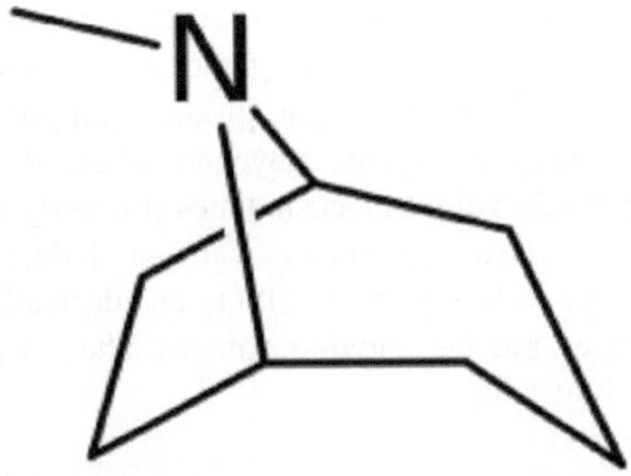

2. **Glycosides:** Glycosides are ethers containing a carbohydrate moiety and a non-carbohydrate moiety (aglycone) joined with an ether bond. They are usually bitter substances. Often the aglycone is released by an enzymatic action when the plant tissue is damaged, such as by wilting, freezing, mastication, or trampling. They are classified on the basis of the structure and/or properties of the aglycone.

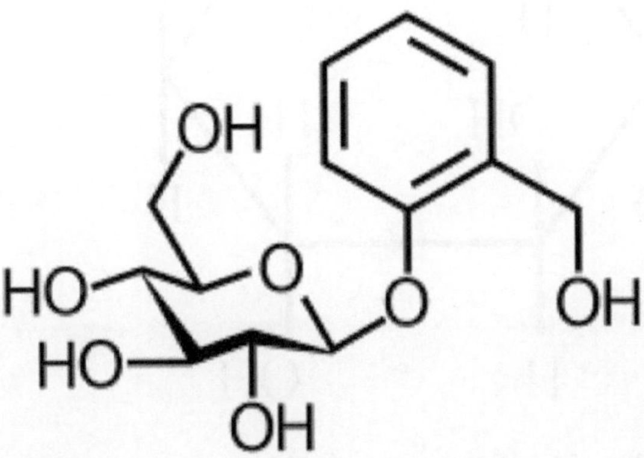

A. **Cyanogenic Glycosides:** They produce a toxic substance namely hydrogen cyanide upon hydrolysis which occurs during chewing or digestion. An example is laetrile (amygdalin) found in the kernels of almonds, apricots, peaches, apples, and the leaves of chokecherries. It is hydrolyzed by B glycosidase to release hydrogen cyanide, glucose, and benzaldehyde as follows:

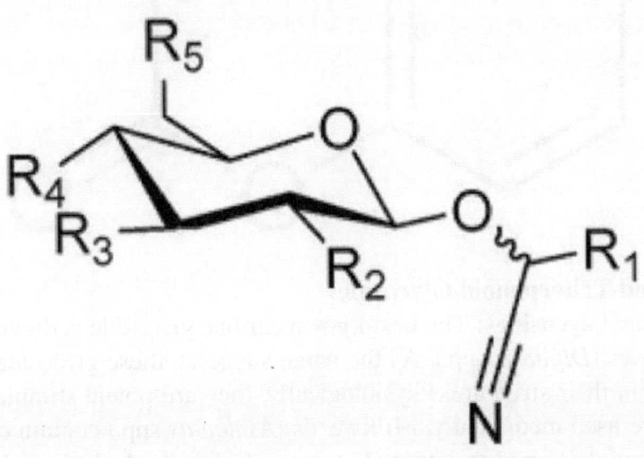

B. **Goitrogenic Glycosides:** Goitrogens decrease production of the thyroid hormones by inhibiting their synthesis by the thyroid gland. As a result, the thyroid enlarges to compensate for reduced thyroxine output, producing a goiter. Goitrogenic glycosides are thioethers, containing an organic aglycone, where R (the aglycone) is an alkyl group. These compounds, called glucosinolates (formerly called thioglucosides), are commonly found in *Brassica* spp. such as cabbage, kale, and rape. They are hydrolyzed by glucosinolases to B-D-glucose, HSO_4, and derivatives of the aglycone. These include isothiocyanates, nitriles, and thiocyanates. The glucosinolates are anions, usually found as potassium salts.

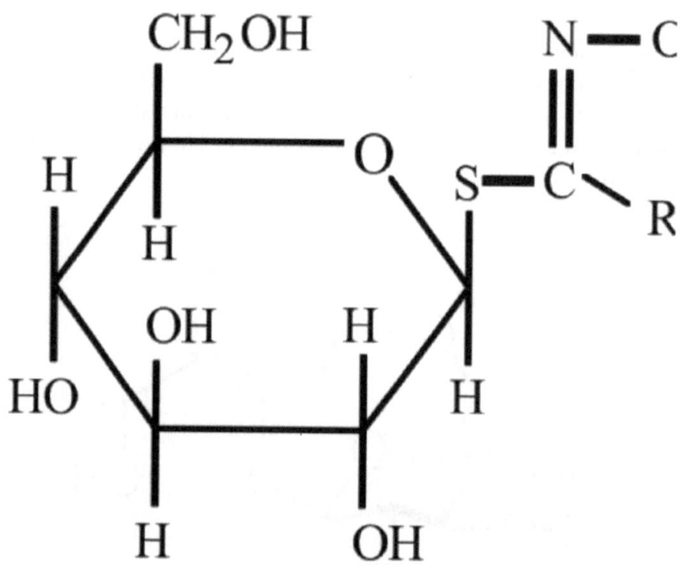

C. **Coumarin Glycosides:** Coumarin is found in sweet clover (*Melilotus* spp.) as melilotoside. Coumarin is converted by mold growth to dicoumarol, an antagonist of vitamin K. Sweet clover poisoning, caused by feeding moldy sweet clover hay, is therefore an induced vitamin K deficiency.

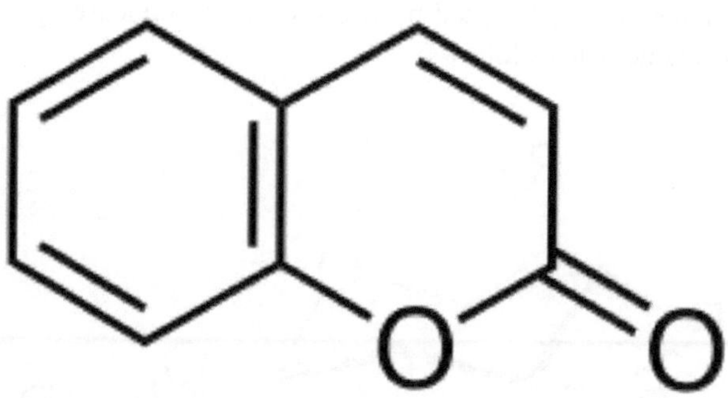

D. **Steroid and Triterpenoid Glycoside:**

i. **Cardiac Glycosides:** The best-known cardiac glycoside is digitonin, contained in foxgloves (*Digitalis* spp.). As the name suggests, these glycosides contain a sterol group in their structure. Physiologically, they are potent stimulators of heart rate and are used medicinally. Milkweeds (*Asclepias* spp.) contain cardiac glycosides (cardenolides) and are extremely toxic to livestock. As little as 100 g of milkweed can kill a sheep.

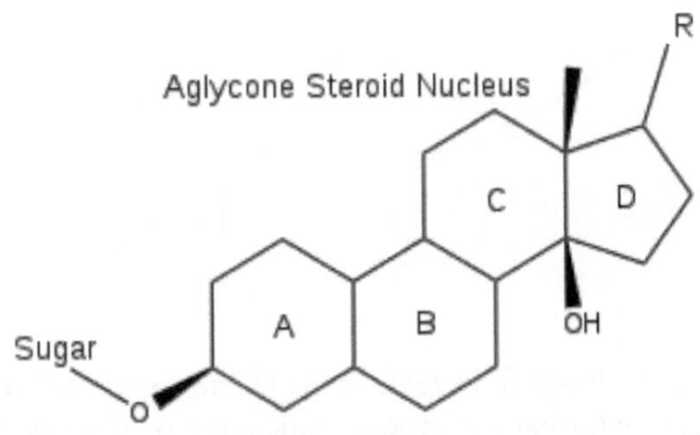

ii. **Saponins:** Saponins are glycosides which have profuse foaming properties, producing a distinctive honeycombed stable foam when an aqueous solution is shaken. They are widely distributed in plants and, in animal nutrition, are particularly important in temperate legume forages. They are bitter compounds, affecting palatability and feed intake. They have growth-depressing properties in poultry and swine, and they have been implicated in bloat in ruminants. Saponins contain a polycyclic aglycone (steroid or triterpenoid) and a side chain of sugars attached by an ether bond to Ca as follows:

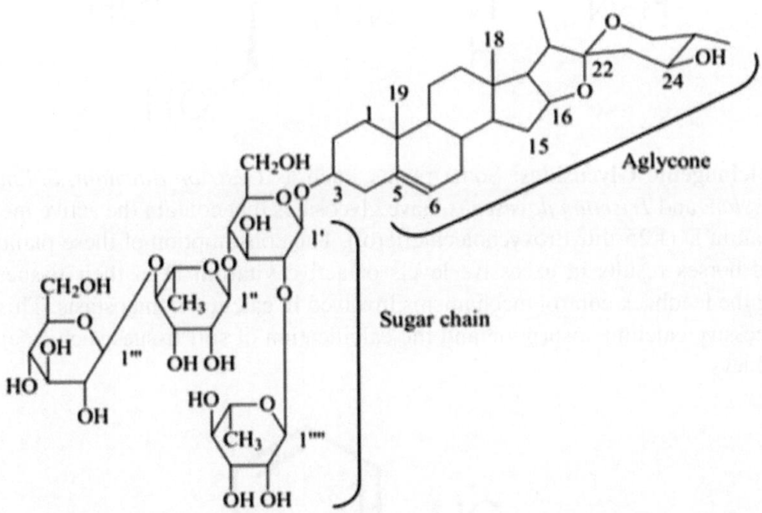

E. **Nitropropanol Glycosides:** Many *Astragalus* species, such as timber milk vetch (*Astragalus miser*), owe their toxicity to this class of glycoside. They are metabolized to 3-nitro-1-propanol (3NPOH) in ruminants and 3-nitropropionic acid in nonruminants. These compounds, especially 3NPOH, are acutely toxic, producing methemoglobinemia. They also produce chronic toxicity symptoms, involving permanent nerve damage.

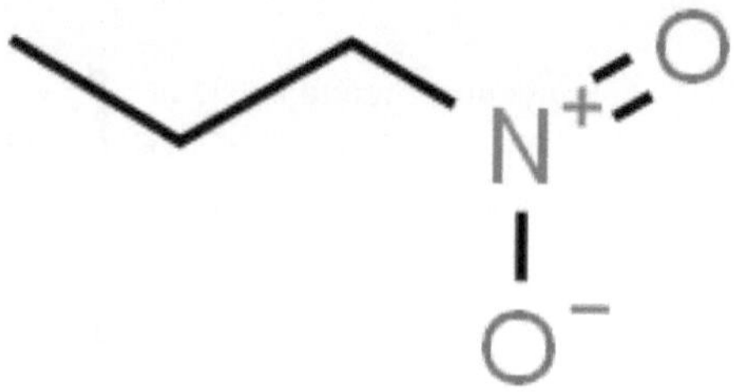

F. **Vicine:** Vicine is a glycoside in fava beans (*Vicia faba*). It causes hemolytic anemia (favism) in people who have a genetic deficiency of glucose-6-phosphate dehydrogenase activity in their red blood cells. Fava beans are being utilized, especially in Canada, as a protein supplement for livestock.

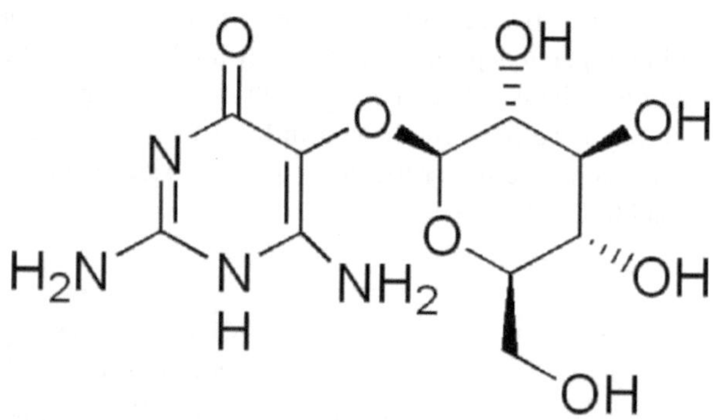

G. **Calcinogenic Glycosides:** Some plants, such as *Cestrum diurnum*, *Solanum malacoxylon*, and *Trisetum flavescens*, have glycosides that contain the active metabolite of vitamin D (1,25-dihydroxycholecalciferol). The consumption of these plants by cattle and horses results in excessive levels of active vitamin D in their tissues, overriding the feedback control mechanisms involved in calcium homeostasis. This results in excessive calcium absorption and the calcification of soft tissues such as arteries and kidneys.

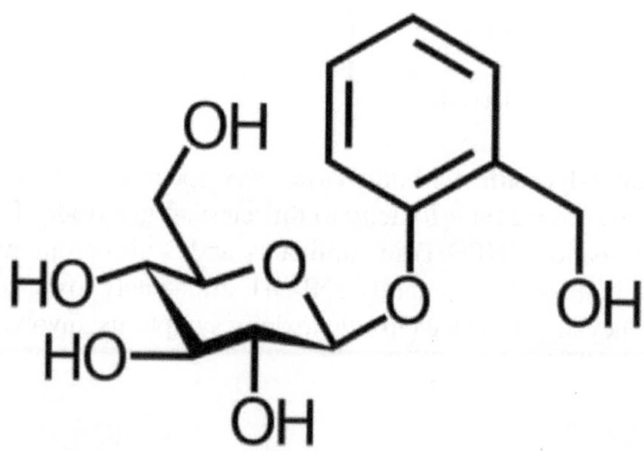

H. **Carboxyatractyloside:** A glycoside called carboxyatractyloside has been identified as the toxicant in cocklebur (*Xanthium strumarium*). It produces hepatic lesions, convulsions, and severe hypoglycemia.

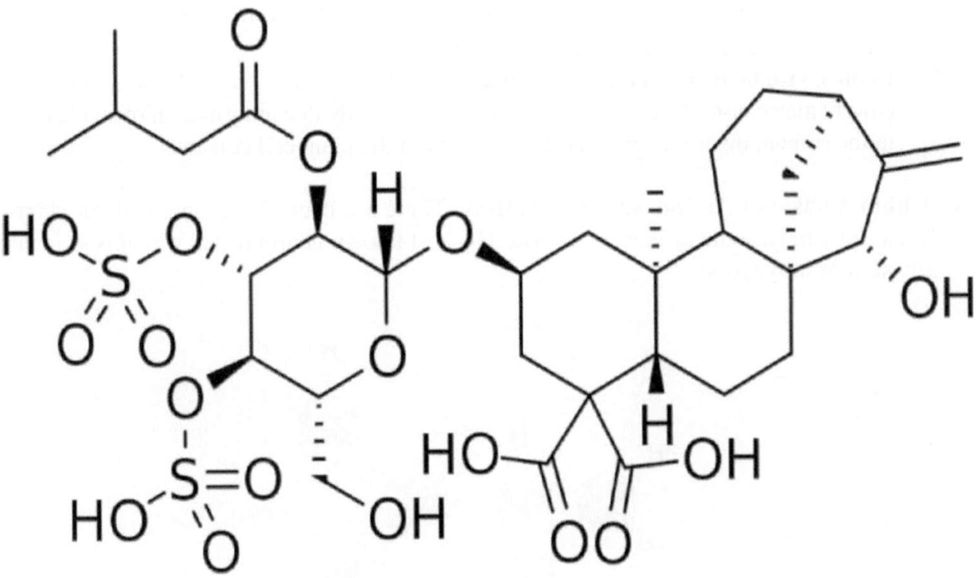

I. **Isoflavone:** These compounds, which are called phytoestrogens, contain a flavone nucleus. Examples of isoflavones are genistein, formononetin, and coumestrol. Isoflavones are particularly important in subterranean clover and have caused extensive reproductive problems in sheep in Australia.

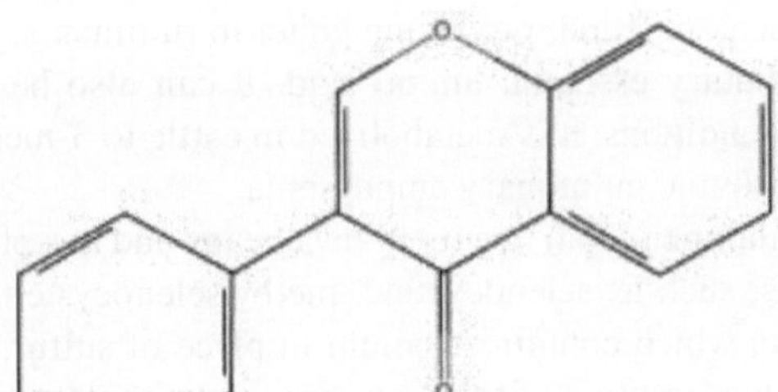

3. **Proteins:** Several important inhibitors in plants are proteins. It's interesting to note that in some cases, the effect of these is to inhibit the utilization of other proteins by animals.
   A. **Protease (Trypsin) and Amylase Inhibitors:** Soybeans, most other legume seeds, and some grains (e.g., rye and triticale) contain trypsin inhibitors. These are small protein molecules which combine with and inactivate the digestive enzyme trypsin in the small intestine. They cause reduced growth and pancreatic hypertrophy. Amylase inhibitors occur in beans and have been commercialized as "starch blockers" to reduce obesity in humans.
   B. **Lectins (Hemagglutinins):** These are glycoproteins of 60,000–100,000 MW that cause agglutination (clumping) of red blood cells in vitro. They are found in most types of beans, including soybeans. They cause reduced growth and diarrhea, and they interfere with nutrient absorption.
   C. **Enzymes:** An example of an enzyme toxin is thiaminase, found in bracken fern (*Pteridium aquilinum*) and certain fish such as carp. The enzyme cleaves the B vitamin thiamine, thereby inactivating it. Consumption of bracken fern causes thiamine deficiency in some animals. Use of carp and other types of thiaminase-containing fish

in mink diets has caused thiamine deficiency (Chastek paralysis). Other enzymes in feeds produce deleterious effects in livestock.

D. **Plant Cytoplasm Protein:** Many leguminous forages such as alfalfa and numerous clovers may cause bloat in ruminants. This is primarily due to formation of stable foam in the rumen, involving cytoplasmic proteins in the plant cell contents.

4. **Amino Acids and Amino Acid Derivatives:** There are over 300 amino acids in plants. Not surprisingly, some of these are toxic. The best known is mimosine, which is structurally similar to tyrosine.

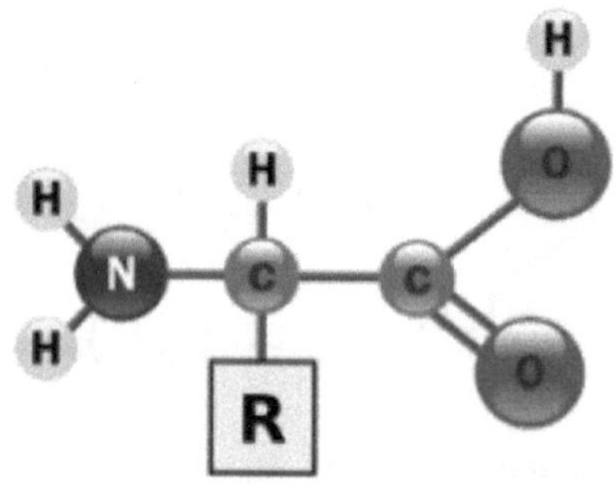

Mimosine occurs in the tropical forage *Leucaena leucocephala*, which produces protein-rich leaves that have considerable potential as livestock feed. Mimosine causes reduced growth and alopecia (loss of hair) in nonruminants and is metabolized in the rumen to a goitrogenic compound, producing goiter in ruminants.

Tryptophan is a dietary essential amino acid. It can also be regarded as a toxicant because under some conditions, it is metabolized in cattle to 3-methylindole, a compound responsible for acute bovine pulmonary emphysema.

Dihydroxyphenylalanine (dopa) occurs in fava beans and has pharmacological effects.

Selenoamino acids, such as selenocystine, methylselenocysteine, selenocystathionine, and selenomethionine, which contain selenium in place of sulfur, are implicated in selenium toxicity due to consumption of selenium accumulator plants such as *Astragalus* spp.

There are several lathyrogenic amino acids, including β-cyano-L alanine in *Vicia* spp. and β-amino propionitrile in *Lathyrus* spp. Lathyrus seeds are consumed as food in India, causing a major public health problem because it may result in permanent paralysis and skeletal deformity. In livestock, consumption of *Lathyrus odoratus* seeds by cattle and horses causes paralysis, aortic aneurysm, and skeletal deformity. Lathyrism is the term used to describe the symptoms, and both neurolathyrism (paralysis) and osteolathyrism (skeletal deformity) are observed.

Another toxic amino acid is 1-amino-D-proline. It is produced from linatine, found in linseed meal. 1-Amino-D-proline is an antagonist of pyridoxine (vitamin B_6). Indospecine is a toxic amino acid found in *Indigofera spicata*, a potentially useful tropical pasture legume. It is an antagonist of arginine, resulting in depressed incorporation of arginine into liver-synthesized proteins.

The brassica anemia factor, S-methylcysteine sulfoxide, is an amino acid derivative. It leads to red blood cell hemolysis and anemia. It is found in forage brassicas such as kale and turnips.

A. **Polypeptides:**

Amanita mushrooms are extremely poisonous and owe their toxicity to cyclic polypeptides. An unusual livestock poisoning problem in Australia is caused by the larvae of the sawfly, which contain an octapeptide called lophyrotomin. The sawfly larvae congregate in mounds beneath the silver-leaf ironbark tree (*Eucalyptus melanophloia*) and are avidly consumed by cattle. Affected animals show incoordination, trembling, and liver damage.

5. **Carbohydrates:** There are few toxicity problems due to carbohydrates. Xylose, a hexose sugar, causes reduced growth and eye cataracts in pigs and poultry. As certain oligosaccharides, such as raffinose, are not digested in the small intestine, they promote bacterial growth in the hindgut. These are the flatulence factors in beans. The β-glucans in certain barley varieties sometimes cause nutritional problems in poultry.
6. **Lipids:** Several fatty acids are toxic. These include erucic acid in rapeseed, which may result in myocardial lesions in rats. Cyclopropenoid fatty acids, such as sterculic and malvalic acids in cottonseed, have toxic properties and cause pink albumins to develop in stored eggs. They are also cocarcinogens, leading to increase in the carcinogenicity of aflatoxins.
7. **Glycoproteins:** Lectins (discussed under Proteins) are glycoproteins. Avidin, a glycoprotein in egg albumin, is an antagonist of the B vitamin biotin. Raw eggs can be used to induce biotin deficiency in experimental animals. Biotin deficiency has occurred in fur animals (foxes and mink) fed raw eggs or poultry-processing plant wastes containing unlaid eggs.
8. **Glycolipids:** The cause of annual ryegrass toxicity (ARGT) has been identified as a glycolipid(s) called corynetoxin. This toxin is synthesized by a *Corynebacterium* which colonizes galls, produced by a nematode, in the ryegrass seed head. The toxin affects the brain, leading to incoordination and staggering.
9. **Metal-Binding Substance:**
   A. **Oxalates:** Oxalic acid is a chelating agent which chelates calcium very effectively. Plants with a high oxalate content, such as the U.S. range weed *Halogeton glomeratus*, may produce acute metabolic calcium deficiency (hypocalcemia) when consumed by livestock.
   B. **Phytates:** Phytic acid in cereal grains and soybean meal causes reduced mineral availability, particularly of zinc, through the formation of absorbable phytates. Organic phosphorus (phytin phosphorus) is of low availability to nonruminant animals.
   C. **Mimosine:** Mimosine, the toxic amino acid in *L. leucocephala*, is reputed to have metal-binding properties.
10. **Resin:** These compounds do not share any particular structural features, but have certain common physical characteristics. They are soluble in a number of organic solvents, are insoluble in water, and do not contain nitrogen. An example of a resin is cicutoxin, the poisonous principle of *Cicuta* spp. (water hemlock). It is one of the most spectacular poisons known, acting directly on the central nervous system to produce violent convulsions.
11. **Phenolic Compound:** Phenolics contain an aromatic ring(s) with one or more hydroxyl groups.
    A. **Hypericin:** Hypericin is a phenolic compound in *Hypericum perforatum* (St. John's wort). It is a primary photosensitizing agent, producing dermatitis and skin lesions in light-skinned animals by reacting with ultraviolet light at the skin surface to produce a photodynamic reaction.
    B. **Gossypol:** Cottonseed meal contains a toxic polyphenol called gossypol. Gossypol causes reduced growth and feed intake, cardiac lesions, and male infertility.
    C. **Tannins:** Tannins are phenolic compounds that react with proteins. The term was originally used for plant extracts that were used in tanning leather. They are astringent and adversely affect feed intake. Tannins are important in oak poisoning and in nutritional problems with sorghum grain (milo).

12. **Sesquiterpene Lactones:** Sesquiterpene lactones are derivatives of the germacranolide nucleus. These compounds are the toxic constituents of sneezeweeds (*Helenium* spp.) and bitterweeds (*Hymenoxys* spp.). They cause severe irritation of nasal and intestinal membranes.
13. **Mycotoxins:** Mycotoxins are toxic substances released by fungi (molds). A total of 25 specific diseases in livestock can be attributed to mycotoxins. Some of the mycotoxins involve aflatoxins, phomopsin, tremorgens, T-2 toxin, citrinin, ochre sporidesmin, and zearalenone. They are implicated in diverse conditions such as acute death of poultry (turkey X disease), liver care trout, lupinosis, fescue foot of cattle, sweet clover poisoning, and eczema of sheep, as well as ryegrass staggers and ergotism.

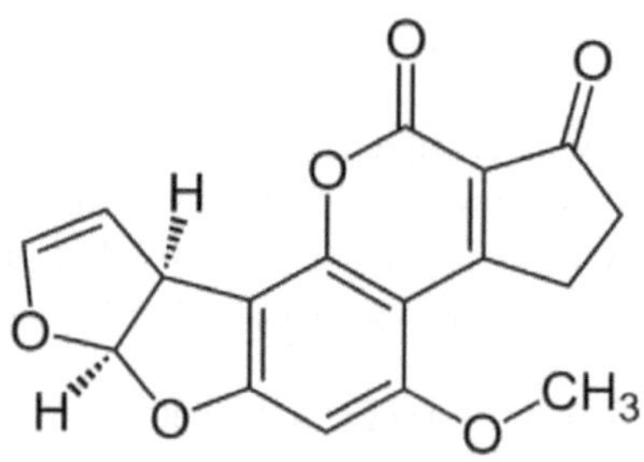

1.3 DIAGNOSIS AND TREATMENT OF INTOXICATION IN SOME RUMINANTS

Animals under nutritional stress are most susceptible to plant poisoning. The diagnosis of plant intoxication is a challenging task for veterinarian as history of ingestion or exposure of toxic plants is often not noted by animal owners or affected individuals. Only a few specific tests are available to confirm exposure to a specific plant, while the number of toxic plants is much more. However, exposure to a plant containing a cyanogenic glycoside can potentially be confirmed by measurement of blood cyanide concentration. Consumption of some plants even at small quantities can result in rapid death. The most common plant species that are responsible for plant toxicity in ruminants include *Abrus precatorius, Datura stramonium, Lantana camara, Ricinus communis, Nerium oleander*, and *Pteridium aquilinum*. Here is a summary of diagnosis and management of intoxication caused by the abovementioned toxic plants:

***Abrus precatorius* L.:** It is commonly known as rosary pea, bead vine, coral bead plant, crab's eyes, jumbo beads, and love bead. It is generally grown as an ornamental plant, and its seeds are used in jewelry and handicrafts. Cattle are more prone to abrin poisoning in comparison to other animal species.

Toxin: A lectin namely abrin present in seeds is the toxic principle of the plant.

Mechanism of Toxicity: It inhibits protein synthesis mediated through ribosomal subunit inactivation. Toxicity of abrin is similar to ricin poisoning.

Sign and Symptoms: Oral ingestion leads to gastroenteritis with vomiting and diarrhea, which then leads to circulatory collapse. Other local signs such as conjunctivitis and dermatitis are also seen.

Management: Symptomatic treatment along with fluid therapy is given to facilitate faster recovery. Activated charcoal is administered at a total dose of 250–500 g in large ruminants. Anti-abrin serum as an antidote is also given. For the excretion of absorbed toxins and alkalinization of urine, saline purgatives at the rate of 100–200 g are also administered (Rajeev, 2012).

***Datura stramonium* L.:** It is commonly known as devil's trumpet, mad apple, jimsonweed, and thornapple. It is an annual ornamental plant with white funnel-shaped flowers. Due to its strong smell and unpleasant taste, it is not preferred by animals. The whole plant including nectar is considered as toxic, and seeds are the most toxic ones (Glen, 2008).

Toxin: Atropine, hyoscyamine, scopolamine, and other anticholinergic alkaloids are toxic principles.

Sign and Symptoms of Toxicity: Initially, rapid pulse with increased heart rhythm, pupil dilatation, dry mouth, and vision impairment are seen (Everest et al., 2005). Later, decreased body temperature, nausea, loss of muscle coordination, violent tremors, aggressive behavior, slow breathing, and rapid and weak pulse are observed.

Management: The supportive and symptomatic care including administration of an antidote, physostigmine, should be done in severe intoxication. Several other topics related to *D. stramonium* have been discussed in Chapter 16 of this book.

***Lantana camara* L.:** It is commonly known as Lantana weed, Wild sage, and red sage. It is the most popular ornamental garden plant. Some varieties of *L. camara* complex are known to cause toxicity particularly in sheep.

Toxin: The triterpene ester metabolites, lantadene A and B, are main toxic principles of the plant. The leaves and immature berries are more toxic to livestock.

Mechanism of Toxicity: Metabolism of lantadene A in the liver yields more polar compounds which are then excreted into the bile. The lantana poisoning shows cholestasis and hepatotoxicity due to the continuous absorption of toxins from the rumen. Other related consequences of cholestasis are jaundice and photosensitization. Experimental studies show that lantana-poisoned sheep develops dehydration and hypokalemia.

Management: Administration of laxatives is done to remove toxins from the body (Blood et al., 1983). To enhance the recovery, manual removal of the toxic rumen contents is done. To adsorb the toxins in the rumen and to prevent further absorption, administration of activated charcoal can be done. Antihistaminic and antibiotics should also be administered. Moving the animal into the shade helps to prevent the development of photosensitive dermatitis (Blood et al., 1983).

***Ricinus communis* L.:** It is commonly known as castor oil plant or castor bean. Horses are the most susceptible and ruminants have intermediate susceptibility, whereas the chicken is most resistant animal to the poisoning of castor bean (Aslani et al., 2007).

Toxins: Its seeds contain a water-soluble, heat-sensitive toxic principle namely ricin, while the leaves and fruits contain an alkaloid known as ricinine (Audi et al., 2005).

Mechanism of Toxicity: Toxicity may be due to the inhibition of protein synthesis, direct damage to cell membrane, and release of cytokine inflammatory mediators (Day et al., 2002).

Sign and Symptoms of Toxicity: The initial toxicity signs include nausea, gastrointestinal irritation, abdominal pain, diarrhea, tenesmus, dehydration, cessation of rumination, muscle twitching, dullness of vision, convulsions, weakness, profuse watery diarrhea, and dehydration within 6–24 hours.

Diagnosis: It can be made from case history and clinical signs exhibited by the animal. Due to the non-specific nature of the clinical manifestations in plant toxicity, one cannot confirm the intoxication merely based on the clinical signs unless the suspicious particles are not found in vomitus, feces, or in the intestine. For confirmatory diagnosis, detection and quantification of toxin is made by antibody-based immunoassay (Roy et al., 2003; He et al., 2010). Ricinine can be detected successfully instead of ricin using paper chromatography, UV detection, and liquid chromatography (Knight and Walter, 2001).

Management: Its poisoning is managed by symptomatic therapy as there is no antidote against ricin. Administration of activated charcoal to adsorb toxins would be helpful.

Intravenous hydration and fluid replacement is necessary to combat dehydration and fluid loss in affected animals. Additional detail information on this plant can be seen in Chapter 20 of this book.

***Nerium oleander* L.:** It is commonly known as oleander, oleana, and rose bay. It is an ornamental flowering shrub commonly grown in gardens and parks. It is widely distributed in tropical and subtropical regions and known to be poisonous to animals (Knight and Walter, 2001; Frohn and Pfander, 1983).

Toxins: Cardiotoxicity exhibited by the cardiac glycosides, i.e., oleandrin, folinerin, and digitoxigenin, is the main toxic principle.

Mechanism of Toxicity: Cardiac glycosides inhibit Na^+/K^+-ATPase pump resulting in electrolyte disturbance, and thus, affecting the electrical conductivity of the heart by increasing the intracellular calcium ion concentration (Joubert, 1989).

Sign and Symptoms of Toxicity: In initial stages of intoxication, there is a ruminal atony with moderate tympany. Other symptoms include abdominal pain, frequent urination, bradycardia, tachycardia, depression, and weakness. The toxic effects exhibited by oleander include ventricular arrhythmia which leads to ventricular fibrillation and finally death of the animal.

Management: As there is no specific remedy, the general symptomatic and supportive care including administration of activated charcoal to eliminate or reduce toxins from the gastrointestinal tract is done. To relieve cardiac symptoms, i.e., tachycardia and atrioventricular block, adrenergic blockers along with atropine are given. Rehydration of animal is also taken care of (Knight and Walter, 2001). Administration of anti-digoxin antibodies is considered as the first line of treatment as it is highly effective but restricted due to high cost and limited availability (Joubert, 1989). Several other topics related to *N. oleander* have been discussed in Chapter 19 of this book.

Pteridium aquilinum: *Pteridium aquilinum* var. *pubescens* is commonly known as bracken fern, Brake, and Hog Pasture Break. It is a poisonous weed widely distributed in many places around the world where it grows in woodlands and other shaded places on hillsides and open pastures (Stegelmeier et al., 1999). All parts of the fern are poisonous.

Toxin: Thiaminase inhibitors present in the plant cause thiamine deficiency. Norsesquiterpene glycoside called ptaquiloside present in plant causes aplastic anemia in cattle. Consumption of milk from poisoned cows is harmful to humans.

Sign and Symptoms: In cattle and sheep, high fever, loss of appetite, depression, dyspnea, excessive salivation, nasal and rectal bleeding, hematuria, hemorrhages on mucous membranes, thrombocytopenia, anemia, leukopenia, plastic bone marrow, and bladder tumors are seen (Davis et al., 2011).

Management: Administration of thiamine hydrochloride, batyl alcohol, activated charcoal, and saline cathartics is recommended to counter toxicity.

1.4 EFFECT OF POISONOUS PLANTS ON ANIMALS (LIVESTOCK) AND PREVENTION OF POISONING

Poisonous plants exert major impact on livestock industry in various ways. Livestock loss increases in certain conditions: when animals graze ranges infested with poisonous plants or when animals are unloaded from trucks onto areas where poisonous plants are abundant. When the animals are under stress, they tend to graze less selectively and are more likely to consume poisonous plants; they are more likely to do so when they are not given regular access to water, when they are allowed to become hungry, when they are more likely to eat lethal quantities of poisonous plants, and when they are allowed to graze on lands early in the spring when there is no other green vegetation except poisonous plants. Poisonous plants cause major economic losses to livestock industry due to improper management. The economic losses can be classified into two categories: direct losses and indirect losses (James, 1978). In the direct loss, the animals are directly affected, i.e., deaths of livestock, abortions, birth defects, weight loss (due to illness or decreased feed intake), lengthened calving interval, decreased fertility, decreased immune response, decreased function (due to damage of organs such as the nervous system, lungs, and liver), and loss of breeding stock (due to deaths, functional inefficiency, etc.). However, indirect loss includes the management costs namely for building and maintaining fences, increased feed requirements, increased medical expenses, decreased forage availability, decreased land values, loss of time in management, and stress to management. Thus, to avoid livestock poisoning, the person involved in management should have following abilities: must know to identify the poisonous plants that grow in a particular localities, must know the conditions under which these plants can be dangerous, must learn to develop a grazing plan to improve local areas and prevent poisoning, must feed the animal timely to avoid hunger stress, must provide adequate water to livestock, must be careful when grazing newly acquired livestock in particular localities, must provide adequate salt and other supplements as needed, must avoid placing them in an area where poisonous plants are growing, must control poisonous plants where feasible, and if animals get sick, must consult local veterinarian to ensure proper diagnosis and treatment. Moreover, if a poisonous plant is involved, identification of the plant by an experienced botanist or poisonous plant expert is essential for any corrective and timely action (Panter et al., 2011).

1.5 CATEGORIES OF PLANT TOXINS

Plant toxins can broadly be categorized as toxins as drugs and toxins as suicidal poisons. The phytochemicals or compounds presently used in clinics for disease management or as research tools to probe biological functions are derived from natural sources, i.e., plants. Since ancient times, venoms and other toxic substances are included in numerous systems of traditional healing. In the 1940s, tubocurarine, a vegetal compound, was used as an anesthetic agent which is one of the key active ingredients in curare, a South American arrow poison. Plant-derived secondary metabolites such as alkaloids, terpenes, steroids, and phenolic compounds exhibit toxicity to humans and animals. These metabolites are good source for providing templates for drug design. Many of the plant-based molecules have significant effect on neural transmission and cell division processes. Such molecules have given rise to drugs for treating central nervous disorders and cancer (Rates et al., 2015). Various dugs have been isolated from plants (*Papaver somniferum*, *Cannabis sativa*, *Erythroxylum coca*, *Atropa belladonna*, *D. stramonium*, *Hyoscyamus niger*, and *Hyoscyamus muticus*) which are being used to treat nervous disorders. Some important plant toxins such as podophyllotoxin, taxanes, vincristine, and vinblastine are reported to have antitumor activity and are supposed to be effective in cancer treatment (Rates et al., 2015). Further, there is a lot of information available on various bioactive compounds, which is summarized in the following section and has also been discussed in detail in the chapters that follow.

Opioids: The alkaloids isolated from opium poppy plant fall under the category of opioid drugs or narcotics, namely, morphine, codeine, and papaverine. Morphine sulfate that is approved by the FDA is still accepted as gold standard for pain management in comparison to other analgesic. Codeine works as an analgesic and antitussive drug. Papaverine is the basis for developing an antihypertensive drug verapamil, a calcium channel blocker. The most common side effects of opioids are respiratory depression, antinociceptive tolerance, physical dependence, and constipation.

Cannabinoids: Terpenophenolic compounds isolated from *C. sativa* fall under the category of cannabinoids. For recreational purpose, this plant is smoked to get pleasure and relaxing perception that may be followed by dysphoria, anxiety, panic episodes, impairment of memory, and reductions in psychomotor and cognitive performance. Active compounds derived from *Cannabis* extracts are tetrahydrocannabinol (THC, a strong psychotropically active component) followed by cannabidiol (CBD), cannabigerol, and cannabichromene.

Pharmacologically, cannabinoids are found effective in spasticity, pain, nausea, vomiting, loss of appetite, neuronal injury, neurodegeneration, and psychiatric disorders, such as schizophrenia.

Cocaine: A tropane alkaloid obtained as white powder from the leaves of the South American shrub *E. coca* is called cocaine (benzoylmethylecgonine). Cocaine is one of the widely used worldwide illegal recreational drugs. Cocaine is the only local anesthetic with vasoconstrictor properties. The local anesthetic effect occurs as the action of axonal membrane sodium channels is blocked. Cocaine's derived formulations, including co-phenylcaine forte (5% lignocaine and 0.5% phenylephrine), lignocaine alone, or 4% lignocaine with 1:1,000 adrenaline, amethocaine, and oxymetazoline, are still in use. The most important adverse effects are addiction and cardiotoxicity.

Ergot Alkaloids: The ergot alkaloids are mainly produced by fungi from the Ascomycota phylum, such as *Claviceps*, *Epichloë*, *Penicillium*, and *Aspergillus* spp. Ergot alkaloids are used in the management of a broad array of diseases, especially as psychoactive and vasoconstricting agents in migraine. Nasal spray and injection of dihydroergotamine (an ergotamine analog) is available for use. Ergotamine is available in oral and sublingual tablet formulations and rectal suppositories.

Atropine, Scopolamine, and Hyoscyamine: These alkaloids are found in *A. belladonna*, *D. stramonium*, *H. niger*, and *H. muticus*. Atropine is used in ophthalmic treatment for pupil dilatation or even for the induction of complete cycloplegia; it is also used in anesthesia procedures. The other belladonna alkaloids and muscarinic receptors can also be employed in the respiratory, gastrointestinal, and cardiovascular systems. Toxic effects of atropine, scopolamine, hyoscyamine, and other belladonna alkaloids may be exhibited such as skin rash, skin flushing, mouth dryness, and eye perturbations (Rates et al., 2015).

Podophyllotoxin (Etoposide): It is isolated from a resin namely podophyllin which is produced by *Podophyllum emodi* and *Podophyllum peltatum*. Due to its cytotoxic activity, it is topically applied in the treatment of genital warts and condylomata. Etoposide, a semisynthetic derivative of podophyllotoxin, is used in combination with other chemotherapeutic agents for the treatment of refractory testicular tumors, small-cell lung cancer, lymphoma, nonlymphocytic leukemia, and glioblastoma multiforme. Podophyllotoxin toxicity is exhibited by vomiting, diarrhea, abdominal pain, abnormal hepatic functions, and neurological disturbance in humans.

Taxanes: They are modified diterpenes initially produced by a small slow-growing evergreen coniferous tree known as yew tree, *Taxus brevifolia*. Later, due to scarcity of *T. brevifolia*, it was derived from *Taxus baccata*. They are discovered in 1979 and got approval for clinical use against ovarian cancer in 1992 and against breast cancer in 1994. In the present time, paclitaxel and docetaxel are widely used as antineoplastic agents for a broad range of malignant solid tumors. Toxic effects include neutropenia, hypersensitivity reactions, hematological toxicity, neurotoxicity, and cardiac effects.

Vincristine and Vinblastine: The vinca alkaloids are derived from *Catharanthus roseus*, known as the vinca plant. These are currently used in the treatment of Hodgkin's lymphoma, Kaposi's sarcoma, ovarian cancer, and testicular and infant acute lymphoblastic leukemia. Peripheral neuropathy and neutropenia are toxic effects of vincristine and vinblastine (Rates et al., 2015).

1.6 PLANT TOXINS AS SUICIDAL POISON

Toxic plants are found throughout the world. Due to public awareness and accessibility to toxic plants, suicidal deaths are reported in many parts of the world. Suicide is the second leading cause of death in the age range of 15–29 years. There are some toxic plants commonly used for suicidal purpose, e.g., hemlock (*Conium maculatum*), nicotine (in *Nicotiana* spp. and Solanaceae), monkshood (*Aconitum napellus*), yellow oleander (*Thevetia peruviana*), rosary pea (*A. precatorius*), and castor oil plants (*Ricinus communis*; Fernando, 2016). Some of the basic information related to these selected plants is summarized in the following section and has also been discussed in detail in the chapters that follow.

Hemlock (*Conium maculatum*): It is a highly poisonous perennial herbaceous flowering plant native to Europe and North Africa. It was used as poison to execute Greek philosopher Socrates. It contains a potent toxin known as coniine. Ingestion of even small amounts of coniine (6–8 leaves, or an even smaller dose of the seeds or roots) causes death by disrupting the body's neuromuscular junctions, resulting in "ascending muscular paralysis." The paralysis typically begins in a person's legs and ascends up the body until it reaches the respiratory muscles, resulting in death. Many other information and explanations related to this plant are discussed in Chapter 15 of this book.

Nicotine (in *Nicotiana* spp. and Solanaceae): It is one of the commonly abused and easily accessible drugs found worldwide. Nicotine naturally occurs in large amounts in the leaves of tobacco plants. The lethal dose of nicotine has been reported to be between 40 and 60 mg. Exposures to lethal dose of nicotine cause autonomic ganglionic blockade leading to hypotension, bradycardia, dyspnea, and, eventually, coma and respiratory failure followed by death.

Monkshood (*Aconitum napellus*): The roots and seeds are freely available in open market. It contains aconitine which has neurotoxic and cardiotoxic properties. The lethal dose in adults is 2–6 mg. Clinical features of poisoning appear about 30 minutes after herb intake. Symptoms of poisoning include muscle weakness, palpitations, chest tightness, nausea and vomiting, hypotension, arrhythmias including polymorphic ventricular tachycardia, and finally, ventricular fibrillation leading to death. Many other information and explanations related to this plant are discussed in Chapter 6 of this book.

Yellow Oleander (*Thevetia peruviana*): It is an ornamental tree that is common throughout the tropics. It contains thevetin A and B and neriifolin that are toxic to cardiac myocytes and autonomic nervous system. Symptoms of poisoning developed after 3–4 hours of oral intake of its seeds include vomiting, abdominal pain, diarrhea, and neurological manifestations. Resistant ventricular fibrillation leads to death. Many other information and explanations related to this plant are discussed in Chapter 21 of this book.

Rosary Pea (*A. precatorius*): It is cultivated in many subtropical areas. It contains a toxin namely abrin, an enzyme which inhibits ribosomal function, halting protein synthesis and leading to cellular death. Symptoms of poisoning include vomiting and watery diarrhea which can result in hypovolemia and life-threatening electrolyte imbalance.

Castor Oil Plants (*Ricinus communis*): It is noticed worldwide in houses/gardens and considered as the world's most poisonous plant. The seeds have a toxin called ricin or ricinine. Lethal dose is 4–8 seeds. Ricin, similar to abrin, is an enzyme which inhibits ribosomal function, halting protein synthesis and leading to cellular death. Oral intake of the seeds can lead to burning sensations in the mouth and throat, intense abdominal pain, and bloody diarrhea within 36 hours and can lead to dehydration, shock, respiratory failure, and death within 3–5 days if left untreated (Fernando, 2016). Several other information related to this plant is presented in Chapter 20 of this book.

1.7 CONCLUSION

The current chapter discusses some of the often observed potentially poisonous plants. Toxic plants can harm every organ system and put animal health and production at danger. However, plant toxicity depends on many factors, ranging from phytochemical features to human or animal or pet age, size, weight and so on. Because of their widespread distribution, it is critical and necessary to examine dangerous plants and their toxic chemical features. To minimize poisoning in livestock, various measures such as removing the poisonous weeds from the grazing area, providing good-quality forages, adapting good practices of grazing, and proper observation of grazing animals are required. The intense knowledge about poisonous plants and strategic approaches to therapy helps to resolve the problem. For prevention of suicides through plant poisons, an effective strategy to restrict access to the most common means, which include pesticides, household chemicals, firearms, medications, and plants, is also required. At the same time, such poisonous plants possess various medicinal values; therefore, all aspects required to be carefully examined. New methods and instrumentation techniques should be adopted to understand for safety and their identification.

NOTES

Mustehasan, Department of Unani Medicine, Central Council for Research in Unani Medicine Ministry of AYUSH, Govt. of India, New Delhi, India
Gulwaiz Akhter, Department of Unani Medicine, Central Council for Research in Unani Medicine Ministry of AYUSH, Govt. of India, New Delhi, India
Azamal Husen, Wolaita Sodo University, Wolaita, Ethiopia

REFERENCES

Alden, C. L., Fosnaugh, C. J., Smith, J. B., & Mohan, R. 1977. Japanese yew poisoning of large domestic animals in the Midwest. *American Veterinary Medical Association* 170: 314–316.

Ana María R. (2021). Toxicology and snakes in ptolemaic Egyptian dynasty: The suicide of Cleopatra. *Toxicology Reports*, 8: 676–695. https://doi.org/10.1016/j.toxrep.2021.03.004

Aslani, M. R., Maleki, M., Mohri, M., Sharifi, K., Najjar-Nezhad, V., & Afshari, E. 2007. Castor bean (*Ricinus communis*) toxicosis in a sheep flock. *Toxicon* 49(3): 400–406.

Audi, J., Belson, M., Patel, M., Schier, J., & Osterloh, J. 2005. Ricin poisoning: A comprehensive review. *Journal of the American Medical Association* 294: 2342–2351.

Bailey, C. P., & Connor, M. 2005. Opioids: Cellular mechanisms of tolerance and physical dependence. *Current Opinion in Pharmacology* 5: 60–68.

Bischoff, K., & Guale, F. 1998. Australian tea tree (*Melaleuca alternifolia*) oil poisoning in three purebred cats. *Journal of Veterinary Diagnostic Investigation* 10: 208–210.

Blood, D. C., Radostits, O. M., & Henderson, J. A. 1983. *Veterinary Medicine: A Textbook of the Diseases of Cattle, Sheep, Pigs, Goats and Horses*. Veterinary Medicine, Bailliere Tindall, London.

Bronstein, A. C., Spyker, D. A., Cantilena, L. R. Jr., Green, J. L., Rumack, B. H., & Heard, S. E. 2008. Annual report of the American Association of Poison Control Centers' data system (NPDS): 25th annual report. *Clinical Toxicology* 46: 927–1057.

Buck, W. B. 1992. Top 25 generic agents involving dogs and cats managed by the National Animal Poison Control Center in 1992. In: Bonagura, J. D., & Kirk, R. W. (eds.): *Kirk's Current Veterinary Therapy XII: Small Animal Practice*. WB Saunders, Philadelphia, PA, 210.

Burrows, G. E., & Tyrl, R. J. 2001. *Toxic Plants of North America*. Iowa State University Press, Ames, IA.

Bythe, L. L., & Craig, A. M. 1994. Role of the liver in detoxification of poisonous plants. *Veterinary and Human Toxicology* 36: 564–566.

Davis, T. Z., Stegelmeier, B. L., Green, B. T., Welch, K. D., Panter, K. E., & Hall, J. O. 2011. Acute toxicity of selenium compounds commonly found in selenium-accumulator plants. Poisoning by plants, mycotoxins and related toxins. In: Riet-Correa, F., Pfister, J., Schils, A. L., & Wierenga, T. (eds.): *Poisoning by Plants, Mycotoxins, and Related Toxins*. CAB International, Cambridge, MA, 525–531.

Day, P. J., Pinheiro, T. J., Roberts, L. M., & Lord, J. M. 2002. Binding of ricin A-chain to negatively charged phospholipid vesicles leads to protein structural changes and destabilizes the lipid bilayer. *Biochemistry* 41(8): 2836–2843.

Dayan A. D. (2009). What killed Socrates? Toxicological considerations and questions. *Postgraduate Medical Journal*, 85(999): 34–37. https://doi.org/10.1136/pgmj.2008.074922

Debelle, F. D., Vanherweghem, J. L., & Nortier, J. L. 2008. Aristolochic acid nephropathy: A worldwide problem. *Kidney International* 74: 158–169.

Duringer, J. M., Buhler, D. R., & Craig, A. M. (2004). Comparison of hepatic in vitro metabolism of the pyrrolizidine alkaloid senecionine in sheep and cattle. *American journal of veterinary research*, 65(11): 1563–1572. https://doi.org/10.2460/ajvr.2004.65.1563

Everest, J. W., Powe, T. A. Jr., & Freeman, J. D. 2005. Poisonous plants of the Southeastern United States. Alabama Cooperative Extension. ANR-975. https://giles.tennessee.edu/wp-content/uploads/sites/194/2020/10/Ag-Poisonous-Plants-of-the-Southeastern-United-States.pdf (Accessed 03 Oct. 2022).

Fernando, R. 2016. Suicidal plant poisoning. In: Gopalakrishnakone, P., Carlini, C., & Ligabue-Braun, R. (eds.): *Plant Toxins. Toxinology*. Springer, Dordrecht, the Netherlands. https://doi.org/10.1007/978-94-007-6728-7_20-1

Frohn, D., & Pfander, H. J. 1983. *A Colour Atlas of Poisonous Plants*. 1st ed. A Wolf Science Book, London, England, 47–48.

Galey, F. D., Holstedge, D. M., Plumlee, K. H., Tor, E., Johnson, B., Anderson, M. L., Blanchard, P. C., & Brow, F. 1996. Diagnosis of oleander poisoning in livestock. *Journal of Veterinary Diagnostic Investigation* 8: 358–364.

Glen, N. 2008. Guide to toxic plants in forages. https://www.extension.purdue.edu/extmedia/ws/ws_37_toxicplants08.pdf (Accessed 03 Oct. 2022).

Gwaltney-Brant, S. M., Albretsen, J. C., & Khan, S. A. 2000. 5-Hydroxytryptophan toxicosis in dogs: 21 cases (1989–1999). *Journal of the American Veterinary Medical Association* 216: 1937–1940.

He, X., McMahon, S., Henderson, T. D. II, Griffey, S. M., & Cheng, L. W. 2010. Ricin toxicokinetics and its sensitive detection in mouse sera or feces using immuno-PCR. *PloS One*, 5(9): e12858.

Hung, O. L., & Lewin, N. A. 2001. Herbal preparations. In: Flomenbaum, N. E., Howland, M. A., Goldfrank, L. R., Lewin, N. A., Hoffman, R. S., & Nelson, L. S. (eds.): *Goldfrank's Toxicologic Emergencies*. McGraw Hill, New York, NY, 664–684.

Husen, A. 2021. *Traditional Herbal Therapy for the Human Immune System*. CRC Press, Boca Raton, FL. https://doi.org/10.1201/9781003137955

Husen, A. 2022. *Herbs, Shrubs and Trees of Potential Medicinal Benefits*. Taylor & Francis Group, LLC, Boca Raton, FL. https://doi.org/10.1201/9781003205067

Husen, A., Bachheti, R. K., & Bachheti, A. 2021. *Non-Timber Forest Products (Food, Healthcare and Industrial Applications)*. Springer Nature Switzerland AG, Cham, Switzerland. https://doi.org/10.1007/978-3-030-73077-2

James, L. F. 1978. Overview of poisonous plant problems in the United States. In: Keeler, R. F., Van Kampen, K. R., & James, L. F. (eds.): *Effects of Poisonous Plants on Livestock*. Academic Press, New York, NY, 3–5.

James, L. F. 1988. Introduction. In: James, L. F., Ralphs, M. H., & Nielson, D. B. (eds.): *The Ecology and Economic Impact of Poisonous Plants on Livestock Production*. Westview Press, Boulder, CO, 1–4.

Joubert, J. P. J. 1989. Cardiac glycosides. In: Cheeke, P. R. (ed.): *Toxicants of Plant Origin*, vol. II. CRC Press, Boca Raton, FL, 61–96.

Knight, A. P. 1988. Poisonous plants: Oleander poisoning. *Compendium Food Animal* 10: 262–263.

Knight, A. P., & Walter, R. G. 2001. *A Guide to Plant Poisoning of Animals in North America*. Teton New Media, Jackson, WY.

Njugi, W. 2018. A constructive approach on lethal plants for medicinal use. *Journal of Medical Toxicology and Clinical Forensic Medicine 4*(1): 2.

Panter, K. E., Gardner, D. R., Lee, S. T., Pfister, J. A., Ralphs, M. H., Stegelmeier, B. L., & James, L. F. 2007. Important poisonous plants of the United States. In: Gupta, R. C. (ed.): *Veterinary Toxicology: Basic and Applied Principles*. Elsevier Academic Press, Amsterdam, the Netherlands, 825–872.

Panter, K. E., Ralphs, M. H., Pfister, J. A., Gardner, D. R., Stegelmeier, B. L., Lee, S. T., Welch, K. D., Green, B. T., Davis, T. Z., & Cook, D. 2011. *Plants Poisonous to Livestock in the Western States, Agriculture Information Bulletin Number 415*. U.S. Department of Agriculture, Agricultural Research Service, Poisonous Plant Research Laboratory, Logan, Utah.

Rajeev, N. 2012. Heamatopathological changes in rats following experimental feeding of Abrus precatorius seeds. *International Journal of Pharmacology and Phytochemical Research* 1(5): 283–286.

Rates, S. M. K., Betti, A. H., Müller, L. G., & Nunes, J. M. 2015. Plant toxins as sources of drugs. In: Gopalakrishnakone, P., Carlini, C., & Ligabue-Braun, R. (eds.): *Plant Toxins. Toxinology*. Springer, Dordrecht, the Netherlands. https://doi.org/10.1007/978-94-007-6728-7_5-1

Roy, C. J., Hale, M., Hartings, J. M., Pitt, L., & Duniho, S. 2003. Impact of inhalation exposure modality and particle size on the respiratory deposition of ricin in BALB/c mice. *Inhalation Toxicology* 15(6): 619–638.

Singh, R., Sharma, P. K., & Malviya, R. 2011. Pharmacological properties and ayurvedic value of Indian buch plant (Acorus calamus): A short review. *Advances in Biological Research* 5(3): 145–154.

Stegelmeier, B. L., Edgar, J. A., Colegate, S. M., Gardner, D. R., Schoch, T. K., Coulombe, R. A., & Molyneux, R. J. 1999. Pyrrolizidine alkaloid plants, metabolism and toxicity. *Journal of Natural Toxins* 8(1): 95–116.
Stewart, M. J., & Steenkamp, V. 2001. Pyrrolizidine poisoning: A neglected area in human toxicology. *Therapeutic Drug Monitoring* 23: 698–708.
Sudekum, M., Poppenga, R. H., Raju, N., & Braselton, W. E. Jr. 1992. Pennyroyal oil toxicosis in a dog. *Journal of the American Veterinary Medical Association* 200: 817–818.
Thothathari, K., Sen, R., Pal, D. C., & Molla, H. A. 1985. *Selected Poisonous Plants from the Tribal Areas of India*. Botanical Survey of India, Kolkata, India.

2 Biotechnological Approaches in Poisonous Plants

Lubna Azmi

CONTENTS

2.1 INTRODUCTION

The world's livestock industries suffer significant financial losses as a result of poisonous plants and the poisons they release. Toxic plants are thought to be responsible for annual losses of more than $340 million in the 17 western states of the USA alone. The USDA-ARS Poisonous Plant Research Laboratory in Logan, Utah has conducted research on poisonous plants for the past 20 years, and the outcomes have provided knowledge to land managers and livestock farmers that they may utilise to lessen losses from toxic plants and increase economic viability in rural regions. This study places attention on identifying poisoning biomarkers for diagnoses and research in various places. The updated knowledge provided here will aid medical professionals and diagnostic facilities in the identification of harmful plant toxicoses, even if much more study is still needed (Gupta, 2018). The relationship between plants and animals, including humans, has long been a difficult topic to discuss due to the great diversity of organism interactions, particularly their reciprocal effects on individual metabolisms. Even before domestication and the development of agriculture, plants have always been the foundation of human nourishment. Fewer than 20 species and associated cultivars and variants currently meet 90% of the world's food requirement, despite the fact that at least 7,000 plant species are thought to have been consumed by humans throughout history. Based on the LD50 measurement in rats, the World Health Organization (WHO) distinguishes four kinds of toxicity: Class Ia is the most hazardous category, Class Ib is very dangerous (5–50 mg/kg bodyweight), Class II is moderately hazardous (50–500 mg/kg bodyweight), and Class III is mildly harmful (>500 mg/kg bodyweight). Class III plants are the least dangerous, followed by class II plants, which

DOI: 10.1201/b23017-3

are classified as poisonous. Plants classified as classes Ia and Ib are thought to be very poisonous (Mezzasalma et al., 2017). Worldwide, plant poisoning is a problem that is regularly documented, and there are many different species that are affected. Severe occurrences of intoxication or death brought on by exposure to plants are, however, uncommon in industrialised nations. In 24,950 cases of plant poisoning that the Swiss Toxicological Information Centre reported between 1966 and 1994, only 152 of them resulted in significant poisoning and only 5 of them resulted in fatal poisoning. These cases were caused by the consumption of Colchicum (in two cases), *Oenanthe crocata*, *Taxus baccata*, and *Narcissus pseudonarcissus* (Eddleston and Persson, 2003). One of the most frequent hazardous plant exposures documented in the USA is exposure to plants that produce oxalate crystals, such *Philodendron* and *Dieffenbachia* (Tyrl, 2006). To deliver the most reliable therapy, instances of intoxication require a quick diagnosis. It can be difficult to accurately taxonomically characterise the species that have been swallowed, though. In this study, we give a general overview of the new DNA-based methods and technologies being developed to solve the problem of identifying dangerous plants (Mezzasalma et al., 2017). Particularly, traditional DNA barcoding and its implementations employing Great Resolution Melting (Bar-HRM) assure high universality and quick response, whilst high-throughput sequencing (HTS) approaches offer a thorough characterisation of plant residues in complicated matrices.

2.2 ASPECTS OF POISONOUS PLANT HISTORY

The majority of kids have heard tales of sadhus who regularly ate Datura seeds, which were thought to be toxic. The majority of individuals have also read fiction about *Vishkanyas* and nonfiction on poison arrows. The interesting tale of the *Vishkanyas* is said to have been introduced in Chanakya's Arthashastra, which he wrote as Chandragupta's advisor and prime minister. *Vishkanyas*, also known as poison damsels, were young women whose blood were poisonous and were used to kill enemies (Chaurasia, 2002). According to archaeological evidence, early man looked for more effective weapons to defend themselves against predators or foes. Poisonous substances were regarded as enigmatic things in antiquity. Around 300 BC, Egyptian monarchs researched toxic plants. Claudius, the Roman emperor, was assassinated by his wife Agrippina using a death cap mushroom administered with skill. The famous Greek philosopher Socrates was killed by water hemlock containing coniine, another infamous toxic substance. Ancient India employed the use of poisoned weapons in their military strategies. Sanskrit poem "*Jalam visravayet sarmavamavisravyam ca dusayet*" refers to contaminated well water that has been tainted with poison. The Pontus king Mithridates used to frequently eat little doses of poison to fortify his body against deadly substances (Hashim et al., 2016). The toxic qualities of mistletoe (*Viscum album*) have been recognised since antiquity. In paganism, especially among the Celts, it was regarded as a sacred plant. According to Nordic texts, a mistletoe-tipped arrow was the reason for the well-known demise of the deity Baldr. Many harmful substances, including deadly lectins and viscumins, are found in mistletoe. Mistletoe extracts have proved effective in treating a variety of illnesses and are still widely used in alternative medicine, especially for the treatment of cancer (Dubey et al., 2018).

2.3 PLANT POISONING CAUSES

Poisonous instances can result from a variety of factors and situations. The majority of these can be classified into the following categories: unintentional plant misidentification; use of botanical products whose taxonomy and safety have not yet been characterised; and trade in phytochemical products for cosmetic, food, and medical uses without establishing appropriate *control* and regulations (Mezzasalma et al., 2017). The collection and use of wild plants by individuals with little to no understanding of systematic botany is now a favourable trend. Typically, the word "wild plant" refers to plants that are growing naturally (Evans-Brown et al., 2015) This meaning is not always accurate since in certain instances, the phrase is misused to promote a particular species. Producers frequently back the assertion that wild plants have more nutritional benefits and are healthier than cultivated ones in the context of the food supply chain. *Juglans regia*, *Corylus avellana*, *Prunus avium*, *Ficus carica*, and *Sambucus nigra* are only a few examples of plants that are frequently misidentified as being in the wild. The same species, nevertheless, can exist in cultivated, wild, or semi-wild conditions depending on the circumstances. There are several case reports reporting persons who were searching for wild plants that resembled salad or young shoots that resembled asparagus and found toxic species that were morphologically similar. In other instances, dangerous *Colchicum* spp. was used in place of *Allium ursinum*, according to Colombo and colleagues (2010). The primary harmful components in plants include organic molecules such as tannins, flavonoids, diterpenes, cardiac and cyanogenic glycosides, proanthocyanidins, phenylpropanoids, lignans, alkaloids, nitrogen-containing compounds, resins, oxalates, and certain proteins or amino acids (Bernhoft, 2010). Numerous metabolic processes that are interrelated and reliant on a variety of biotic and abiotic elements of the environments where the plants thrive are used to synthesise harmful secondary metabolites. As a result, these compounds have different properties depending on their location. This is also demonstrated by the amount of dangerous piperidine alkaloids from *Conium maculatum*, which is positively connected with herbivory in the oecophorid caterpillar known as *Agonopterix alstroemeriana*, which is located in New York and Washington, USA (Shivkar and Kumar, 2003). An effective traceability method is hampered by the large variety of species utilised as botanical superfoods, as well as the lack of specific trade regulations. Restani and colleagues revealed a list of more than 70 widely used plants that are harmful to consumers' health, despite the fact that practically all of these items are put through toxicological testing. The gastrointestinal system was affected by poisoning the most, followed by issues with the brain system (such as sleeplessness) and the cardiovascular system (e.g. tachycardia). Unexpectedly, this was evident in the consumer survey conducted by Restani et al. (2016).

2.4 ACTIVE COMPOUND AND PHARMACOLOGICAL ACTIVITY

Numerous poisons have been employed in suicide attempts in the recent years. Many toxic or harmful plant species are regularly found in our gardens or are purposefully or unintentionally planted as roadside trees by the forest department without first researching their effects on people (Suhag, 2001). Different forms of poisoning exist, including those caused by touch (skin irritation), ingestion (internal poisoning), absorption (by the dermal layer), and inhalation (in the respiratory system). Animals should generally avoid contact with any part of the plant. The fruits, which are extremely deadly to people, are devoured by birds to the point of "intoxication" despite their great poisonousness to humans. Since the beginning of time, people have been aware of the toxicity of the seeds of the castor bean (*Ricinus communis*) and the *jequirity bean* (*Abrus precatorius*; Dubey et al., 2018). The castor bean was first domesticated in Asia and Africa, although it is now found in all temperate and subtropical climates. More than 700 cases of ricin in poisoning in humans have been reported. Ricin is the main toxin from castor beans with a wide spectrum of biological effects on higher species. Ricin is largely degraded in the colon since it is a protein. However, when a poison is given as an injection, its potency multiplies by 100. *Adenia digitata*, a plant native to Southern Africa, contains modeccin in its roots, which, because of how closely they resemble the roots of edible plants, can seriously intoxicate users (Sahoo et al., 2008).

It has been suggested to use immunochemical screening, such as the assays used in dipstick testing, to quickly find target compounds in a given matrix. On the basis of a visual evaluation of the changes (such as colour) on a functionalised strip, dipsticks are simple-to-use, inexpensive lateral flow devices (LFDs) that can offer a quick response. The DNA-based diagnostic molecular tools and, in particular, the DNA barcoding method (and its technological variations), have emerged as the most promising diagnostic procedures in this area (Figure 2.1; Galimberti et al., 2015b).

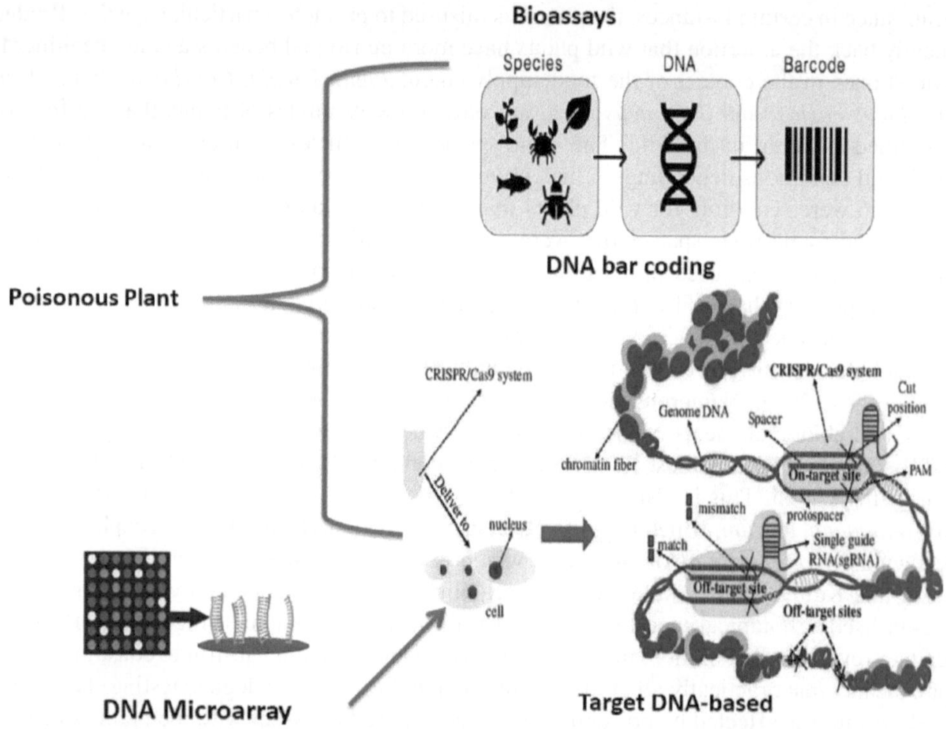

FIGURE 2.1 Biotechnological approaches or techniques for poisonous plants studies.

2.5 BIOASSAYS APPROACH

Breeding programmes have employed a variety of biotechnology techniques over the past few decades to promote agricultural performance and yield. A variety of crucial methods are available in bioinformatics for evaluating and deciphering the vast amounts of data produced by molecular biology-based approaches. These methods are crucial for evaluating and integrating data to infer information from a whole systems perspective as a result of the development of high-throughput techniques (Sharma and Sarkar, 2013). An in-depth investigation of genomic, proteomic, and metabolomic data is needed to improve our understanding of the biological processes connected to plants. Identification of genes and pathways that may be linked to significant bioactive secondary metabolites from poisonous plant is made possible by bioinformatics techniques, which provide crucial tools. The use of such an approach is examined in several fields where it can produce worthwhile results quickly and cheaply. The integration of disparate information, molecular data analysis, drug development and design, authentication, and toxicity are covered with an emphasis on computational approaches.

The standardisation and quality assurance of diverse botanical products might benefit particularly from bioassays. Due to the existence of bioactive component mixes from the same or purposely mixed botanical sources, such products may be referred to as "heterogeneous." Since they frequently lack sensitivity to the chemical complexity present in unprocessed plant extracts, physical analytical procedures like chromatography are used less frequently for this purpose. The majority of the time, a desirable biological reaction is caused by a combination of bioactive plant components, and the amounts of individual bioactive chemicals can change from batch to batch whilst the bioactivity still stays within acceptable bounds. As a result, a component's physical or chemical examination in such mixes is not always adequate. Four "bench top" bioassays have been used by the authors to help with the "drug discovery" process for botanicals (McLaughlin et al., 1998).

2.6 DNA BARCODING APPROACH

The morphological study of plant pieces found in the stomach contents is often used to make the clinical diagnosis of inebriated people. This approach is exceedingly time-consuming to use, needs extensive training, and typically leaves some plant bits unidentified. Furthermore, without residuals displaying discrete taxonomic components, plant species identification might be challenging. DNA barcoding is a recent method for plant identification based on the examination of a brief, standardised DNA region; the fundamental tenet of the discriminating system is straightforward: in order to distinguish one species from another, there must be a significant enough change in the DNA sequence between them (Hebert et al., 2003). In actuality, a DNA sequence from such a standardised DNA area may be produced from a tiny tissue sample obtained from an unidentified creature and compared to a library of reference sequences from documented species. A quick and accurate identification can be made by comparing the sequence from the unknown organism to one of the reference sequences. The trade-off that emerges between the necessity for universal application and the highest rates of sequence divergence complicates the choice of a barcode locus, though (Hollingsworth et al., 2009). For animal identification, a fragment of the mitochondrial coxI (cytochrome c oxidase subunit 1) gene has been used extensively as a DNA barcode sequence; but in terrestrial plants, this sequence is very invariant and is hence not suited for use as a DNA barcode. Many different alternative areas have been suggested, including the plastidial conserved genes rpoB, rpoC1, rbcL, and matK, as well as many intergenic plastidial spacers (trnH-psbA, atpF-atpH, and psbK-psbI) and nuclear internal transcribed spacers. Numerous research teams have suggested various pairings of these loci to improve the accuracy of species identification, particularly for the crucial groups. But there is disagreement on the identity and quantity of the best areas for plant DNA barcoding (Kress and Erickson, 2007).

When food cannot be identified by morphological criteria, DNA barcoding promises to provide a useful, standardised, species-level identification tool that may be utilised for many studies, including forensic examinations and determining animal diets. Similar circumstances may be seen in poison centres, where dangerous plants should be recognised beginning with various plant parts (fruits, seed, leaves, etc.). The sole way to identify a species is typically by bits of the most appealing fruits and seeds that kids mistakenly eat (Valentini et al., 2009). Due to these factors, our research was done using a variety of plant components, such as appealing fruits and seeds. Without any special preservation techniques, DNA was isolated from plant samples that were gathered in the field or inside and kept at room temperature. The utility of the DNA barcoding technique as a realistic identification tool in the toxic centres is supported by the findings of DNA amplification and sequence analysis, which show that plant materials used for DNA extraction have no impact on the results of DNA barcoding analysis (Bruni et al., 2010).

One important potential benefit of DNA barcoding is its ability to detect cryptic species as well as several related species, even from degraded biological material or minuscule fragments of organisms. This is possible by determining the minimum genetic distances that two individuals must meet in order to be designated members of the same species or members of distinct species. Among closely related species, the trnH-psbA intergenic spacer is readily amplified and has substantial genetic diversity (De Mattia et al., 2012). Due to its faster pace of development, the nuclear ITS region—more particularly, the ITS2 portion—was also suggested as an additional DNA barcode area (China Plant BOL Group et al., 2011). The strength of DNA barcoding depends on the availability of well-populated, high-quality reference datasets that are provided by worldwide platforms in addition to the universality, resolution, and standardisation of the selected marker regions. All plastid DNA regions and nuclear ITS results are accepted in a dedicated DNA barcoding database created exclusively for medicinal plant resources. Medical Materials DNA Barcode is a website that includes key references to the American Herbal Pharmacopoeia, the People's Republic of China's Medical Register, and DNA sequences together with their information and other pertinent references (Chen et al., 2010). This database also contains information on distinguishing therapeutic materials (including plant, animal, and fungal species) from their common adulterants and replacements. The DNA barcoding strategy can be more or less ineffective as a diagnostic tool; thus, it is important to remember that flaws in the process more often than not have to do with biological species in general. DNA barcoding can also be a trustworthy method for identifying poisonous plants, but its use in the production of processed plant-based dietary supplements needs to be carefully adjusted to account for the potential lower quality of the total DNA derived from these products.

Targeted techniques should be adopted to obtain an advantage in terms of the amount of time and money spent on analysis when the species that cause poisoning are well recognised. TaqMan probes, SCARs, and the recently discovered PCR-HRM are among the target techniques that are most suited for finding toxic plants (Lou et al., 2010).

2.6.1 Target DNA-Based Methods

Studies on the veracity of poisonous plants have included a variety of analytical techniques. Even if traditional techniques like liquid chromatography, electrophoretic profiling, and sensory analysis are still often used, the addition of cutting-edge DNA-related technologies to the roster of already available food testing addresses the drawbacks of those traditional techniques. DNA is the ideal target molecule for determining the authenticity of food, despite the fact that enzyme-linked immunosorbent tests (ELISAs) are still one of the most used procedures. This is because PCR, which amplifies DNA, has superior specificity, sensitivity, and molecular stability. This makes it possible to find plant fraud and mislabelling even at low levels and in processed plants.

From DNA variation in nucleotide sequences (including the DNA barcoding regions) between poisonous and palatable plants, it may be possible to find species-specific genomic regions to give an easy and clever detection method. There are several approaches from those based on species-specific primers to those that employ probes to target certain DNA sections.

2.6.1.1 DNA Microarray

DNA microarray uses species-specific probes. A target species' numerous loci may be simultaneously analysed using this high-throughput method. The method is based on first determining the DNA oligonucleotide sequences that are specific to each species, and then creating matching probes that are immobilised in a regular pattern on an impermeable solid substrate (glass, silicon chips, or nylon membrane). The microarray DNA is then hybridised with DNA that has been isolated from the samples and marked with a particular fluorescent molecule. Fluorescence scanning or imaging technology is used to identify and see a positive hybridisation. The identification of poisonous plants among traditional Chinese medicines has also been done using this method. The biggest barriers to the array system's general implementation in the context of plant exposure diagnostics are its lack of universality and the high development costs.

2.6.1.2 PCR Primers

Sequence Characterised Amplified Regions (SCARs) analysis is the basis for species-specific PCR primers. The main goal of this approach is to locate polymorphic DNA in dangerous plant sections so that a particular PCR primer may be created. Such primers only anneal when the template contains DNA from the target species. Firstly, using discontinuous molecular markers like random amplified polymorphic DNA (RAPD) and amplified fragment length polymorphism (AFLP) to create polymorphic fragments, the SCAR method was devised.

2.6.1.3 Species-Specific PCR Primers

Analysis of SCARs involves the use of species-specific PCR primers. Identifying polymorphic DNA regions of dangerous plant species allows for the construction of specialised PCR primers, which is the main component of this approach. Such primers only anneal when the target species' DNA is present in the template. Firstly, the SCAR method was created starting with polymorphic fragments generated using discontinuous molecular markers such as RAPD and AFLP. At any locus inside or surrounding the RAPD/AFLP amplicon, specific SCAR sequence primers can be positioned. The most important thing to remember is that polymorphisms in the annealing region(s) must be retained within a species and extremely variable between congenerics or often replaced plant species. Because of this, a single SNP is insufficient to determine a suitable SCAR area. It is easy to discover in silico SCARs and develop particular primers targeting hypervariable DNA regions since plant genomic regions data, including DNA barcoding sequences, are widely available in public databases.

2.6.1.4 Loop-Mediated Isothermal Amplification

Notomi et al. published the initial description of the loop-mediated isothermal amplification (LAMP) in 2000 (Notomi et al., 2000). Several LAMP-based techniques have been published so far for the detection of harmful bacteria, genetically modified components, tumour tissues, and the sexing of bovine embryos (Fu et al., 2011). Using stem-loop DNA constructions, the LAMP method amplifies DNA (DNA hairpins). This technique is based on the employment of a collection of four specifically created primers that can identify a total of six different DNA target sequences. There is no need for thermal cycling or any heat-related denaturation of the template DNA during the isothermal amplification, which is based on an autocycling strand-displacement synthesis and is catalysed by the thermostable Bst DNA polymerase (Chaudhary et al., 2012). LAMP findings might be seen by colour change or by looking at the presence or absence of bands on an agarose gel electrophoresis. LAMP is a highly sensitive approach that doesn't require pricey equipment to operate. LAMP has been used to identify many species, including plants, and especially, adulterants in conventional herbal medicines (Lai et al., 2015).

2.7 HIGH-THROUGHPUT SEQUENCING (HTS)

A rapidly evolving method called high-throughput sequencing (HTS) offers significant potential for epidemiology and diagnostics. This method provides the rapid sequencing of millions of nucleotides, allowing for the identification of the majority of viral pathogens in the sample. Furthermore, before sequencing, HTS does not require any prior knowledge of pathogens (Villamor et al., 2019). It can give a worldwide spectrum of naturally existing disease strains or species because of their potential in sequencing millions of nucleotide sequences. It enables the detection of all known pathogens and the discovery of novel ones from symptomatic or asymptomatic plants, as well as substrates, such as water or soil (Thorburn et al., 2015). This is done in conjunction with the bioinformatic analysis of the collected raw data.

2.8 DNA BARCODING COMBINED WITH HIGH-RESOLUTION MELTING (BAR-HRM) ANALYSIS

The target area is amplified using a typical PCR procedure in the first phase whilst being bound to double-stranded DNA (dsDNA) by a specific binding dye. dsDNA intercalating dyes come in a variety of forms, including SYBR Green, LC Green PLUS, Eva Green, SYTO9, and Reso Light. The most popular dsDNA intercalating non-saturating dye is SYBR Green. Due to its ability to block PCR at high concentrations and its potential to redistribute from melted areas back into the dsDNA amplicon, it is typically inappropriate for most HRM applications (Herrmann et al., 2006). Non-saturating doses can be utilised with "release-on-demand" colours like EvaGreen. When the dye is free in solution, the fluorescent signal is muted due to its innovative fluorescence emission mechanism. Instead, when the dye binds to duplex DNA, a strong fluorescence signal is produced. In comparison to conventional approaches, HRM analysis provides a number of benefits for locating dangerous plants in various matrices. The first benefit of this method is that it may be used for high-throughput amplification without the need for amplicon purification, making it more cost-effective than other methods of a similar kind. These benefits have led to an increase in the application of HRM in various scientific fields, including forensics, healthcare contexts (such as poison control centres), and the verification of plant-derived food products (Graham et al., 2005).

Using this technique, it is possible to accurately track how the release of an intercalating DNA dye from a DNA duplex during gradual heating causes the fluorescence to alter. The melting profile of the amplification result is dependent on its length, strand complementarity, and GC content; hence, Bar-HRM analysis opens the door to the potential of identifying single-base variations or species-specific variations in a small section of DNA (Graham et al., 2005). By analysing the disassociation behaviour of nucleic acid samples, High-Resolution Melting technique uses direct melting to find minute sequence variations in PCR-amplified sequences. High-end instruments, specialised analytic tools, and DNA-specific dyes are used to find these changes. The composition, length, GC content, and strand complementarity of the samples are used to make distinctions (Bag et al., 2015). Furthermore, recent research has validated the use of this method to find species substitutions and admixtures in herbal goods with unambiguous results in the context of ensuring customer safety because poisonous plants may be present in these situations. Singtonat and Osathanunkul specifically showed how Bar-HRM might be used to find the harmful adulterant *Crotalaria spectabilis* in herbal items made from *Thunbergia laurifolia* at concentrations as low as 1%. The contamination of *Phyllanthus amarus* with other *Phyllanthus* species. However, by confirming the presence of the pollutant within 2 hours, the scientists also emphasised the quick detection power of this approach. Due to its inexpensive cost and quick analysis time, Bar-HRM is a very promising method (Su et al., 2015). The study of processed meals and complicated matrices is also made possible by the use of tiny amplicons. One small drawback is the requirement to get a melting curve database to use as a reference when attempting to identify unknown samples (Galimberti et al., 2015a). Development of techniques that allow the curves to be shared among several laboratories is therefore crucial.

2.9 BIOMARKERS APPROACH FOR POISONOUS PLANT

The biomarkers, with the exception of pathway specific metabolites, exhibit varying sensitivity and predictive capacity that is sometimes complicated by fluctuations in growth circumstances, making them unsuitable as standalone indicators of environmental stress. Due to the lack of understanding of predictive power, sensitivity, and specificity, the application of gene expression to identify pollution has been and remains immature. Additionally, it turns out that utilising gene expression to identify the method of action of unknown toxicants is not as simple as first thought. The main patterns in gene expression are not as well defined as first thought. Instead of responding to chemicals with recognised mechanisms of action, the main patterns in gene expression are often obtained from genes that are triggered by stress (Baerson et al., 2005). Technological and scientific advancements over the past 50 years have led to the identification of tens of thousands of phytochemicals, the characterisation of their biological effects, and the creation of management techniques or therapies to lessen toxicity. For doctors, diagnostic labs, and scientists, the identification of biomarkers to enhance plant toxicosis diagnosis has always been crucial. There are, however, not many biomarkers specifically for plant toxicoses. Development of biomarkers specifically for toxic plant diagnosis is one area of study that is prioritised at the USDA Poisonous Plant Research Lab. This comprises histological description, biochemical alterations linked to certain harmful plant toxicoses, and chemical detection of poisons in particular tissues (Panter et al., 2019).

2.10 CONCLUSION

When considering complex matrices made up of several known and/or undiscovered plant species, the pipeline has become considerably more complex in terms of the required technologies and resources. When using PCR to analyse universal markers (including DNA barcodes), which yield multiple amplicons for each species present in the sample, the HTS technologies provide the best possibility to screen the whole composition of matrices. The presence of recognised and undiscovered poisonous species may be discovered by bioinformatics analysis of the collected data. HTS data are useful information to create target PCR methods in this situation as well. It should be highlighted that HTS procedures also require additional expertise and lab resources, particularly when it comes to bioinformatics analysis and the creation of DNA reference databases. The diagnosis of plant poisoning cases will be more quick and portable thanks to the ability to link these equipment to rapid devices and have direct access to online reference sequence matching systems.

NOTE

Lubna Azmi, Institute of Pharmaceutical Sciences, University of Lucknow, Lucknow, India

REFERENCES

Baerson, S. R., Sánchez-Moreiras, A., Pedrol-Bonjoch, N., Schulz, M., Kagan, I. A., Agarwal, A. K., Reigosa, M. J., & Duke, S. O. 2005. Detoxification and transcriptome response in Arabidopsis seedlings exposed to the allelochemical benzoxazolin-2 (3H)-one. *J. Biol. Chem.* 280, 21867–21881.

Bag, S., Al Rwahnih, M., Li, A., Gonzalez, A., Rowhani, A., Uyemoto, J. K., & Sudarshana, M. R. 2015. Detection of a new luteovirus in imported nectarine trees: A case study to propose adoption of metagenomics in post-entry quarantine. *Phytopathology.* 105, 840–846.

Bernhoft, A. 2010. A brief review on bioactive compounds in plants. *Bioact. Compd. Plants-Benefits Risks Man Anim.* 50, 11–17.

Bruni, I., De Mattia, F., Galimberti, A., Galasso, G., Banfi, E., Casiraghi, M., & Labra, M. 2010. Identification of poisonous plants by DNA barcoding approach. *Int. J. Legal Med.* 124, 595–603.

Chaudhary, A. A., Mohsin, M., & Ahmad, A. 2012. Application of loop-mediated isothermal amplification (LAMP)-based technology for authentication of Catharanthus roseus (L.) *G. Don. Protoplasma.* 249, 417–422.

Chaurasia, R. S. 2002. *History of Ancient India: Earliest Times to 1000 AD.* Atlantic Publishers & Distributors, New Delhi, India.

Chen, S., Yao, H., Han, J., Liu, C., Song, J., Shi, L., Zhu, Y., Ma, X., Gao, T., Pang, X., et al. 2010. Validation of the ITS2 region as a novel DNA barcode for identifying medicinal plant species. *PLoS One.* 5, e8613.

China Plant BOL Group, Li, D.-Z., Gao, L.-M., Li, H.-T., Wang, H., Ge, X.-J., Liu, J.-Q., Chen, Z.-D., Zhou, S.-L., Chen, S.-L., et al. 2011. Comparative analysis of a large dataset indicates that internal transcribed spacer (ITS) should be incorporated into the core barcode for seed plants. *Proc. Natl. Acad. Sci.* 108, 19641–19646.

Colombo, M. L., Assisi, F., Della Puppa, T., Moro, P., Sesana, F. M., Bissoli, M., Borghini, R., Perego, S., Galasso, G., Banfi, E., et al. 2010. Most commonly plant exposures and intoxications from outdoor toxic plants. *J. Pharm. Sci. Res.* 2, 417.

De Mattia, F., Gentili, R., Bruni, I., Galimberti, A., Sgorbati, S., Casiraghi, M., & Labra, M. 2012. A multi-marker DNA barcoding approach to save time and resources in vegetation surveys. *Bot. J. Linn. Soc.* 169, 518–529.

Dubey, N. K., Dwivedy, A. K., Chaudhari, A. K., & Das, S. 2018. Common toxic plants and their forensic significance. In: Subhash C. Mandal, Vivekananda Mandal, Tetsuya Konishi (eds.), *Natural Products and Drug Discovery.* Elsevier, Amsterdam, the Netherlands, pp. 349–374.

Eddleston, M., & Persson, H. 2003. Acute plant poisoning and antitoxin antibodies: Antivenoms. *J. Toxicol. Clin. Toxicol.* 41, 309–315.

Evans-Brown, M., Gallegos, A., Francis, W., Christie, R., Cunningham, A., Sekula, J., Almeida, A., & Sedefov, R., 2015. New psychoactive substances in Europe. An update from the EU early warning system. *Eur. Monitor. Centre Drugs Drug Addict.* 34, 12.

Fu, S., Qu, G., Guo, S., Ma, L., Zhang, N., Zhang, S., Gao, S., & Shen, Z. 2011. Applications of loop-mediated isothermal DNA amplification. *Appl. Biochem. Biotechnol.* 163, 845–850.

Galimberti, A., Bruno, A., Mezzasalma, V., De Mattia, F., Bruni, I., & Labra, M. 2015a. Emerging DNA-based technologies to characterize food ecosystems. *Food Res. Int.* 69, 424–433.

Galimberti, A., Sandionigi, A., Bruno, A., Bruni, I., Barbuto, M., Casiraghi, M., & Labra, M. 2015b. Towards a universal molecular approach for the quality control of new foodstuffs. *Adv. Food Biotechnol.* https://doi.org/10.1002/9781118864463.ch04

Graham, R., Liew, M., Meadows, C., Lyon, E., & Wittwer, C. T. 2005. Distinguishing different DNA heterozygotes by high-resolution melting. *Clin. Chem.* 51, 1295–1298.
Gupta, P. K. 2018. *Illustrated Toxicology: With Study Questions*. Academic Press, London, England.
Hashim, A., Vaswani, V., Pramod, K. L., & Rasheed, S. 2016. Homicidal poisons-past, present and future. *Indian J. Forensic Med. Toxicol.* 10(2), 245–249.
Hebert, P. D. N., Cywinska, A., Ball, S. L., & DeWaard, J. R. 2003. Biological identifications through DNA barcodes. *Proc. R. Soc. London. Ser. B Biol. Sci.* 270, 313–321.
Herrmann, M. G., Durtschi, J. D., Bromley, L. K., Wittwer, C. T., & Voelkerding, K. V. 2006. Amplicon DNA melting analysis for mutation scanning and genotyping: Cross-platform comparison of instruments and dyes. *Clin. Chem.* 52, 494–503.
Hollingsworth, M. L., Andra Clark, A., Forrest, L. L., Richardson, J., Pennington, R. T., Long, D. G., Cowan, R., Chase, M. W., Gaudeul, M., & Hollingsworth, P. M. 2009. Selecting barcoding loci for plants: Evaluation of seven candidate loci with species-level sampling in three divergent groups of land plants. *Mol. Ecol. Resour.* 9, 439–457.
Kress, W. J., & Erickson, D. L. 2007. A two-locus global DNA barcode for land plants: The coding rbcL gene complements the non-coding trnH-psbA spacer region. *PLoS One*. 2, e508.
Lai, G.-H., Chao, J., Lin, M.-K., Chang, W.-T., Peng, W.-H., Sun, F.-C., Lee, Meng-Shiunn, L., & Meng-Shiou, L. 2015. Rapid and sensitive identification of the herbal tea ingredient Taraxacum formosanum using loop-mediated isothermal amplification. *Int. J. Mol. Sci.* 16, 1562–1575.
Lou, S.-K., Wong, K.-L., Li, M., But, P.P.-H., Tsui, S.K.-W., & Shaw, P.-C. 2010. An integrated web medicinal materials DNA database: MMDBD (Medicinal Materials DNA Barcode Database). *BMC Genom.* 11, 1–8.
McLaughlin, J. L., Rogers, L. L., & Anderson, J. E. 1998. The use of biological assays to evaluate botanicals. *Drug Inf. J.* 32, 513–524.
Mezzasalma, V., Ganopoulos, I., Galimberti, A., Cornara, L., Ferri, E., & Labra, M. 2017. Poisonous or non-poisonous plants? DNA-based tools and applications for accurate identification. *Int. J. Legal Med.* 131, 1–19.
Notomi, T., Okayama, H., Masubuchi, H., Yonekawa, T., Watanabe, K., Amino, N., & Hase, T. 2000. Loop-mediated isothermal amplification of DNA. *Nucleic Acids Res.* 28(12), e63.
Panter, K. E., Welch, K. D., & Gardner, D. R. 2019. Poisonous plants: Biomarkers for diagnosis. *Biomarkers Toxicol.* 627–652. https://doi.org/10.1016/B978-0-12-814655-2.00037-2
Restani, P., Di Lorenzo, C., Garcia-Alvarez, A., Badea, M., Ceschi, A., Egan, B., Dima, L., Lüde, S., Maggi, F. M., Marculescu, A., et al. 2016. Adverse effects of plant food supplements self-reported by consumers in the PlantLIBRA survey involving six European countries. *PLoS One*. 11, e0150089.
Sahoo, R., Hamide, A., Amalnath, S. D., & Narayana, B. S. 2008. Acute demyelinating encephalitis due to Abrus precatorius poisoning—Complete recovery after steroid therapy. *Clin. Toxicol.* 46, 1071–1073.
Sharma, V., & Sarkar, I. N. 2013. Bioinformatics opportunities for identification and study of medicinal plants. *Brief. Bioinform.* 14, 238–250.
Shivkar, Y. M., & Kumar, V. L. 2003. Anthelmintic activity of latex of Calotropis procera. *Pharm. Biol.* 41, 263–265.
Su, L., Zhou, F., Ding, Z., Gao, Z., Wen, J., Wei, W., Wang, Q., Wang, W., & Liu, H. 2015. Transcriptional variants of Dmrt1 and expression of four Dmrt genes in the blunt snout bream, Megalobrama amblycephala. *Gene*. 573, 205–215.
Suhag, P. 2001. *Phytochemical Investigation on Melia Azedarach L. and Its Bioefficacy against Some Lepidopterous Pests. Chemistry*. CCSHAU, Hisar, India.
Thorburn, F., Bennett, S., Modha, S., Murdoch, D., Gunson, R., & Murcia, P. R. 2015. The use of next generation sequencing in the diagnosis and typing of respiratory infections. *J. Clin. Virol.* 69, 96–100.
Tyrl, R. J. 2006. *Handbook of Toxic Plants of North America*. Blackwell Pub, Ames, Iowa.
Valentini, A., Miquel, C., Nawaz, M. A., Bellemain, E. V. A., Coissac, E., Pompanon, F., Gielly, L., Cruaud, C., Nascetti, G., Wincker, P., et al. 2009. New perspectives in diet analysis based on DNA barcoding and parallel pyrosequencing: The trnL approach. *Mol. Ecol. Resour.* 9, 51–60.
Villamor, D. E. V, Ho, T., Al Rwahnih, M., Martin, R. R., & Tzanetakis, I. E. 2019. High throughput sequencing for plant virus detection and discovery. *Phytopathology*. 109, 716–725.

3 Uses of Poisonous Plants in Nanoscience

Apekshakumari Patel, Nimisha Patel, Ahmad Ali, and Hina Alim

CONTENTS

DOI: 10.1201/b23017-4

3.1 INTRODUCTION

In most animals to protect them from predator's attack, evolution in their defense mechanism takes place, such as motion, calcified shell, camouflage, claws, echolocation and various other features to protect them from predator's attack. To encounter any foreign particles or bodies, some higher animals adopted immune response for survival. When considering slow motile animals and non-motile plants, they have naturally developed a prodigious evolution of production of poisonous compounds. Such compounds have low molecular weight (<1,000 Da), including alkaloids, glycosides, tannins, terpenoids, proteins, peptides and other secondary metabolites (Wink and Wyk 2008). Main elements of poisoning are the poisonous compound, the poisoned organism, the degree of damage caused to the organism, and signs and symptoms observed. These elements depict the cause, effects and consequences of poisoning in the subject. Poisoning initiates when the subject is exposed to the poisonous chemical. Upon entry into the subject, the compound accumulates in the targeted cells, tissues or organ. As a consequence, it causes abruption of cell structure or function. Observable signs and symptoms develop as an onset of poisoning, and if toxicity passes a certain threshold, it can lead to detrimental situations, resulting in death of the subject (Klaassen et al. 2020).

Poisonous plants produce diversified secondary metabolites that result in large economic loss due to its effect majorly on the global livestock industries. Ideally, the poisonous plants are naturally unpalatable, indicating that animals avoid consumption of such plants. However, in arid regions and areas where the lands are desertified, the limitations of forage compel them to graze upon poisonous plants. Such a situation may also arise due to the seasonal changes in rangelands and pastures, causing scarce non-toxic forage (Panter et al. 2007). The common poisonous plants that endanger the production chain of livestock are locoweeds (*Astragalus* spp. and *Oxytropis* spp.), *Lupinus* spp., *Aconitum* spp., *Cynanchum* spp. and *Conium maculatum* (Lu et al. 2012; Welch et al. 2012). Livestock grazing upon such toxic forage suffers severely from neurotoxicity, hepatotoxicity and reproductive toxicity (Keeler et al. 1978). In humans, poisonous plants affect all age groups, but the most vulnerable category is the children, as they are prone to be sensitive to acute poisoning of plants. A study on pediatric patients showed that the majority of cases of intoxication by poisonous plant occurred in children under the age of 20 years in the USA (<1 year, 57.8% cases), Botswana and South Africa (<20 years, 60% cases) and Thailand (<13 years, 69.8% cases; Krenzelok and Mrvos 2011; Malangu 2014; Sriapha et al. 2015). Certain household plants can impose intoxication to the majority of children and pets due to the presence of insoluble compounds such as calcium oxalates, a toxic compound present in some common household plants (Poppenga 2010). The most common causative poisonous plants for the average age group of 16 years were oleander, lantana, Bougainvillea, datura, milkweed and morning glory (Enfield et al. 2018).

Despite its adverse effect, under controlled conditions, these secondary metabolites can facilitate the advancement of health care for the livestock as well as humans (Panter et al. 2019). The practice of using poisonous plants as a form of medicine dates back to the Siddha medicinal system, the most ancient medicinal system descended from the Dravidians of peninsular South India (Thas 2008). The most essential criterion for the use of any compound in the field of medicine is to elucidate the prescribed dose, which will act as potent therapeutic agents. Toxins are potentially detrimental to humans depending on the concentration at which the body is exposed to. In the sixteenth century, this was adroitly quoted by Paracelsus, "dose makes the poison" (Tamilselvan et al. 2014).

A novel field of nanotechnology is augmented for various real-life applications (Husen and Siddiqi 2014; Husen and Iqbal 2019a,b; Husen and Jawaid 2020; Jin-Chul et al. 2021; Sharma et al. 2021; Kumar et al. 2021, 2022; Husen 2022) in which nanoparticles are the focal point of many substantial developments. For the synthesis of such an accelerating large-scale production line, a greener and cleaner approach is essential to achieve a verdant future. This can be accomplished with the progression of nanoparticles' green synthesis. Green synthesis is a momentous procedure to obtain desired nanoparticles and/or nanomaterials with the absence of any toxic by-products (Husen and Siddiqi 2014). The main aim of this chapter is to insight the biosynthesis of some important metal and metal oxide nanoparticles from poisonous plant extract. Additionally, this chapter also explores some common poisonous plants and their bioactive compounds.

3.2 COMMON POISONOUS PLANTS

Plants appear to be harmless enough to misunderstand the potential lethal effect it can possess on humans and animals. Most commonly found poisonous plants do not cause adverse effects; however, some may be lethal resulting in death. Due to this, the children and pets are the most vulnerable category frequently and unintentionally affected by such plants. Moreover, some of the antidotes against plant poison are still not available, but for mild toxic plant poisoning, proper coherent care the pain may subside (Poppenga 2010). Even though this category is varied, the research conducted on such plants is scarce, and therefore entitled as least explored field.

The common poisonous plants are majorly applied in areas for enhancing productivity of life through food and medicines, while also in dire fields such as fishing poison, forensics and bioterrorism. The modern medicines have utilized these bioactive compounds of poisonous plants to assist in drug development and therapeutics (Anywar 2020).

Overview of some common poisonous plants found worldwide with their route of administration and toxic effects is given in Tables 3.1 and 3.2. Certain physiochemical activities performed by the poisonous plants' bioactive compounds are given in Table 3.3.

3.3 BIOACTIVE COMPOUNDS (PHYTOCHEMICALS)

A bioactive compound is produced by plants as a secondary metabolite, possessing a defense mechanism against animals upon its ingestion above specific dosage. Apart from the compound's toxicological effects, it also elicits pharmacological effects on humans and animals (Bernhoft 2010).

Classification of bioactive compounds is still inconsistent; the classification is mostly based on similar characteristics. The three broad classes of bioactive compounds are (I) alkaloids, (II) terpenes and terpenoids and (III) phenolic compounds (Azmir et al. 2013).

I. **Alkaloids:**

These bioactive compounds are named because of their alkali-like nature and ability to produce salts upon reacting with acids. There are more than 5,000 alkaloid compounds identified in plants and animals (Odeh et al. 2014), and they are classified into three main classes depending upon their molecular precursor: true alkaloids, protoalkaloids and pseudoalkaloids (Dey et al. 2020); most common alkaloid backbone structures are given in Figure 3.1. Except pseudoalkaloids, every alkaloid is made from precursor amino acids such as aliphatic amino acids originating from the TCA cycle and aromatic amino acids originating from the shikimic acid pathway (Azmir et al. 2013). They have therapeutic potential due to their antiviral, anticancer, antibacterial, analgesic and antioxidant potential, which is preferable for drug development (Qiu et al. 2014). However, natural occurrence of such compounds in various foods appears to be present as a contaminant. High levels of alkaloids found in coffee seeds, cacao seeds, and tea leaves prove to be toxic causing severe poisoning upon its consumption (Crews 2014).

II. **Terpenes and Terpenoids:**

These compounds are comprised mainly of five carbon isoprene arranged in various ways. Terpenes are made up of simple hydrocarbons; upon modification of terpenes with oxidized methyl groups and diverse functional groups, terpenoids are obtained (Perveen 2018) and they share terpenes' carbon backbones. Few examples of terpenoids are shown in Figure 3.2.

There are more than 40,000 compounds identified, representing as the most widely discovered bioactive compound (Odeh et al. 2014). Terpenes are synthesized via mevalonic

TABLE 3.1
Brief Information of Common Poisonous Plants' Toxin Effects on Ingestion

Plant Name	Location	Toxin/Bioactive	Toxic Effects	Reference
Lathyrus sativus (grass pea)	Asia, Northern Africa, Europe, North America	β-Aminopropionitrile	Pain in spine, legs weaken and lead to paralysis; causes death in children	Klaassen et al. (2020)
Abrus precatorius (rosary pea, prayer bean or jequirity pea)	Tropical regions	Abrin (*N*-Methyltryptophan) and abric acid	Hours to 2 days delay in onset: chills, acute gastroenteritis, convulsions diarrhea, vomiting, heart failure and death; child upon ingestion of seed may prove to be fatal	Poppenga (2010)
Kalmia latifolia (mountain laurel)	North America	Andromedotoxin	Tears, hyper-salivation, visual impairment, paresthesia of skin, lethargy, vomiting, muscle paralysis, seizures, coma, death; children are vulnerable to the leaves ingestion	Klaassen et al. (2020)
Atropa belladonna (belladonna)	United States, Europe, Asia	Atropine, hyoscine, hyoscyamine and other complexes of alkaloids	Xeroderma, xerostomia, dysphagia, difficulty in face flushing, cyanosis, slurred speech, nausea, vomiting, coma, death; animals and children poisoned frequently due to ingestion of fruit	Klaassen et al. (2020)
Melia azedarach (chinaberry)	Asia, Southern Africa, North America	Azedarin	Stomatitis and hematemesis, body paralysis	Klaassen et al. (2020)
Cannabis sativa (marijuana)	United States, Mexico, tropical America	Cannabidiol, cannabinol and derived compounds	Exaltation, inebriety, hallucination, CNS depression; long-term heavy consumption may result in cognitive impairment or psychosis; larger amount of consumption may lead to death by cardiac depression	Volkow et al. (2014)
Cicuta maculata (water hemlock)	Northern temperate regions	Cicutoxin	One of the extreme poisonous plants; respiratory distress, enteralgia, nausea, diarrhea, vomiting, convulsions, hyper-salivation and death	Klaassen et al. (2020)
Erythroxylon coca (huanuco cocaine)	Tropics of both hemispheres	Cocaine and other alkaloids	Indians in Bolivia and Peru often chew on these leaves as a analeptic drug; CNS stimulation and depression, respiratory distress, paresthesia of tongue, cyanosis, respiratory centers paralysis and death	Klaassen et al. (2020)
Papaver somniferum (opium poppy)	Europe, Asia, tropics	Codeine, papaverine, morphine, thebaine, narcotine	CNS depression, meiosis, cyanosis, respiration depression, coma, death	Klaassen et al. (2020)
Conium maculatum (poison hemlock)	Asia, Northern Africa, Temperate United States, South America	Coniine, coniceine, conhydrine, *N*-Methyleonine and other alkaloids	Asthenia, blindness, paralysis of extremities, respiratory failure, death; causes high number of fatalities in humans; on flowering, the leaves are the most toxic ones	Klaassen et al. (2020)
Croton tiglium (purging croton)	Asia, Africa, Pacific Islands	Croton, ricinine and croton resin	Purging disorder, vomiting, collapse, death; croton oil acts as an irritant to the skin, causing reddening, swelling and pustules	Klaassen et al. (2020)

(*Continued*)

TABLE 3.1 (*Continued*)
Brief Information of Common Poisonous Plants' Toxin Effects on Ingestion

Plant Name	Location	Toxin/Bioactive	Toxic Effects	Reference
Jatropha curcas (Barbados nut)	Tropics	Curcin	Bloating, burning throat, vertigo, vomiting, diarrhea, lethargy, violent purgative action, dysuria, leg in cramps; may lead to fatality in children	Klaassen et al. (2020)
Delphinium species (larkspur)	Northern temperate regions	Delphinine, methyllycaconitine, deltaline and other alkaloids	Most extreme poison for fatality of grazing livestock; dysesthesia of mouth, vomiting, nausea, itching, respiratory distress, cyanosis	Welch et al. (2012)
Dioscorea hispida (wild yam)	Southern Asia, Pacific Islands	Dioscorine	Frequent consumption of raw tubers led to death in the Philippines; irritation, burning sensation in throat, dizziness, hematemesis, respiratory distress, lethargic, fatigue, nervous system paralysis and death	Klaassen et al. (2020)
Datura stramonium (jimsonweed or thornapple)	Temperate and tropical regions	Hyoscine, hyoscyamine, atropine	Headache, dizziness, vomiting, nausea, dryness, thirst and paresthesia of skin, delirium, psychosis, amnesia, convulsions and death; often children are subjected to poisoning by sucking flowers or ingestion of seeds	Klaassen et al. (2020)
Blighia sapida (akee)	West Africa, Caribbean	Hypoglycin A, B	Drowsiness, sudden vomiting, prostration, nervous and muscular exhaustion, coma, death	Emanuel and Benkeblia (2011)
Brassica napus (rape)	Western Europe, Canada, China and India	Isothiocyanates	Anemia, respiratory distress, pulmonary emphysema, constipation, irritability	Woodfield and Harwood (2017)
Phytolacca americana (pokeberry)	North America, Europe, southern Africa	Phytolaccine	Bitterness and burning sensation in mouth, vomiting, spasms, purging, convulsions, death	Klaassen et al. (2020)
Apocynaceae (Dogbane family—oleander, milkweed, periwinkle)	Tropical, subtropical areas	Reserpine, ajmaline, neriine, deserpidine, apocynin, oleandrin, cyamarin	Tachycardia, vomiting, blue coloration of mucous membrane, convulsion, dilation of pupils, coma, death	Britannica (2020); Dey et al. (2017)

(*Continued*)

TABLE 3.1 (*Continued*)
Brief Information of Common Poisonous Plants' Toxin Effects on Ingestion

Plant Name	Location	Toxin/Bioactive	Toxic Effects	Reference
Ricinus communis (castor bean)	United States, tropics	Ricin, A-Toxalbumin	Paresthesia of stomach, throat and mouth, diarrhea, vomiting, cramps in the abdomen, blurred vision, paralysis, convulsions and death; children consuming of more than 2–3 seeds may be fatal	Klaassen et al. (2020)
Solanum nigrum (black nightshade)	North America, Europe	Solanine, a glycoalkaloid	Vomiting, nausea, diarrhea, trembling, abdominal pain, coma, death	Klaassen et al. (2020)
Astragalus and Oxytropis (locoweed, vetches, selenium accumulator plants)	Northern Hemisphere	Locoine, swainsonine, 3-Nitropropanol, selenium	Weakness, behavioral abnormality, vision impairment, periorbital cellulitis, ataxia, loss of appetite, starvation, emaciation; causes death in cattle, horses and sheep	Welch et al. (2012)

TABLE 3.2
Brief Information of Common Poisonous Plants' Toxic Effects on Contact

Plant Name	Location	Toxin/Bioactive	Toxic Effects	Reference
Euphorbia species (euphorbia, spurge)	Worldwide	Alkaloid derivatives, glycoside derivatives and other complexes	Ocular pruritis, blindness, swelling around the mouth, paresthesia of the mouth, blistering of the skin, unconsciousness, death; arrow poison is prepared from milky sap	Klaassen et al. (2020)
Jatropha urens (spurge nettle)	North America, Europe, Asia	Bioactive unknown	Contact with toxic substance induces intense itching and stinging; stinging hairs facilitate injection into the skin resulting in skin rashes lasting for 30 minutes (red papular rash) and a remain of purplish discoloration of the skin for weeks	Klaassen et al. (2020)
Metopium toxiferum (poisonwood)	West Indies, Florida	Similar to poison ivy	Skin turns black on contact with the any part of the plant, particularly the sap, resulting in blisters and rashes; burning tree smoke is highly infuriating, leading to temporary blindness and illness	Klaassen et al. (2020)
Strychnos toxifera (curare)	Central, Northern, South America	Toxiferines, caracurines and other alkaloids	Haziness vision, unable to raise head, facial palsy, loss control over legs, arms, respiratory muscles, may lead to death; applied as blowgun dart and arrows poison	Klaassen et al. (2020)
Strophanthus species (strophanthus)	Florida, tropical America, Africa	Trigonelline alkaloid, numerous cardiac aglycones and glycosides	Bradycardia, heart arrhythmia, delirium, blurred vision, respiratory failure, death; toxin applied as arrow poison	Klaassen et al. (2020)
Toxicodendron radicans (poison ivy)	North America	Urushiol	Edema, rashes, itching, blistering; fatal for children; toxic smoke from plant burning	Klaassen et al. (2020)

TABLE 3.3
Brief Information of Common Poisonous Plants That Release Toxic Compounds

Plant Name	Location	Toxin/Bioactive	Toxic Effects	Reference
Plants Which Produce Photosensitizing Compounds				
Froelichia humboldtiana (Snake cotton)	Brazilian northeastern region	Bioactive unknown	Maximum toxicity in rainy season; lesions, edema, hyperemia, necrosis	Santos et al. (2017)
Hypericum perforatum (St. John's wort)	North America, Europe	Hypericin, a naphthodianthrone derivative	Liver dysfunction, results in photosensitizing pigment deposition in the skin; sunlight then causes redness of the skin, nervousness, blepharitis, seizures	Klaassen et al. (2020)
Fagopyrum sagittatum (buckwheat)	North America, Europe	Fagopyrin, a naphthodianthrone derivative	Leaves consumption by animals leads to liver dysfunction, results in photosensitizing pigment deposition in the skin; sunlight then causes redness of the skin, nervousness, blepharitis, seizures and prostration in animals of farm	Klaassen et al. (2020)
Ammi majus (Bishop's flower)	West and Central Asia, South Europe, North America	Furocoumarins	Eye injuries, dermatitis, erythema, rash, skin tumor; high concentration in fruits and seeds	Knupp et al. (2016)
Plants Which Produce Airborne Allergies				
Salsola tragus (Tumbleweed or Russian thistle)	Southwest America, Asia	Sethoxydim, fluazifop, clethodim	Irritant dermatitis, forearms show pruritic erythematous papules; pollen causing allergic rhinitis	Burgdorf et al. (2015)
Acer negundo (Box elder)	Northern Hemisphere	Oleoresin and a water-soluble antigen	Hay fever, exposure of body part is susceptible to eczematous dermatitis	Klaassen et al. (2020)

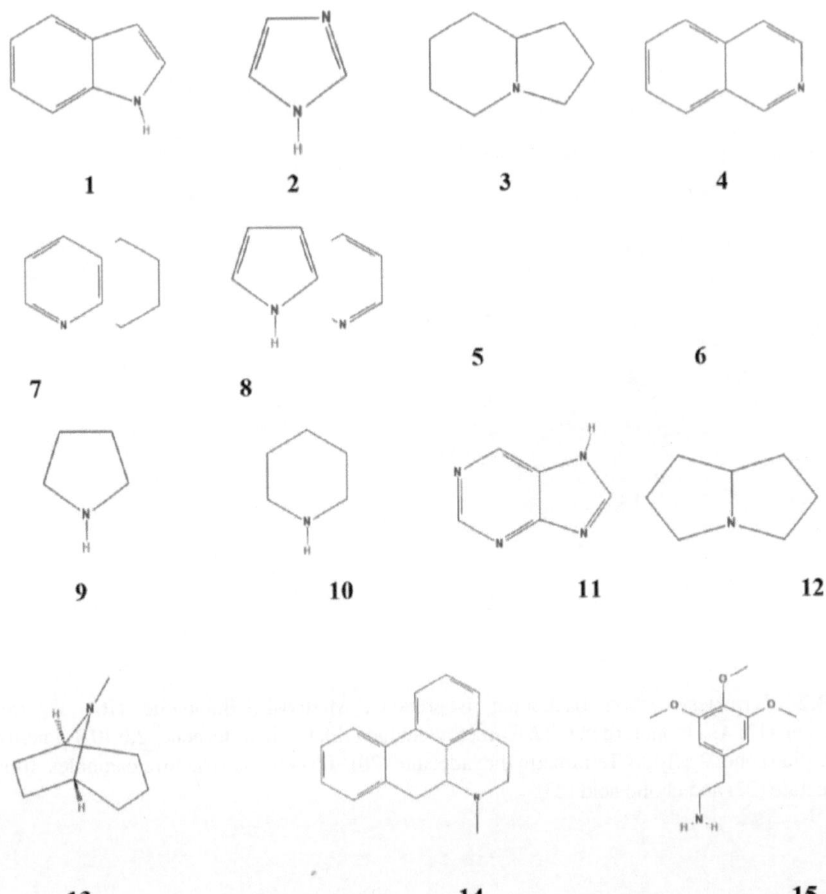

FIGURE 3.1 Alkaloid backbone structures: indole (1), imidazole (2), indolizidine (3), isoquinoline (4), quinolizidine (5), quinoline (6), pyridine (7), pyrrole (8), pyrrolidine (9), piperidine (10), purine (11), pyrrolizidine (12), tropane (13), aporphine (14) and mescaline (15).

acid pathway and non-mevalonate pathway (Azmir et al. 2013). Although widely explored in cancer biology, inflammation and malaria, it has toxic effects on the human gastrointestinal and central nervous system (Mbaveng et al. 2014).

III. **Phenolic:**

These are the compound which comprises hydroxylated aromatic rings, where the hydroxyl group attaches to aryl, phenyl or substituted phenyl group; the most commonly found phenolic compounds are shown in Figure 3.3. There are more than 8,000 compounds identified, including mostly flavonoids, and other classes are phenolic quinones, coumarins, lignans, polymeric lignins and tannins (Odeh et al. 2014). Their intake in diet is mainly due to the presence of hydroxyl group, which plays a role of antioxidant rather than a nutrient. They are bioactive compounds ubiquitously present in most plant tissues, synthesized via the non-mevalonate pathway and shikimic acid pathway (De la Rosa et al. 2019).

The other methods of classification of the bioactive compounds are based on various criteria, such as chemical properties, effective organs and nature of compound, as shown in Figure 3.4.

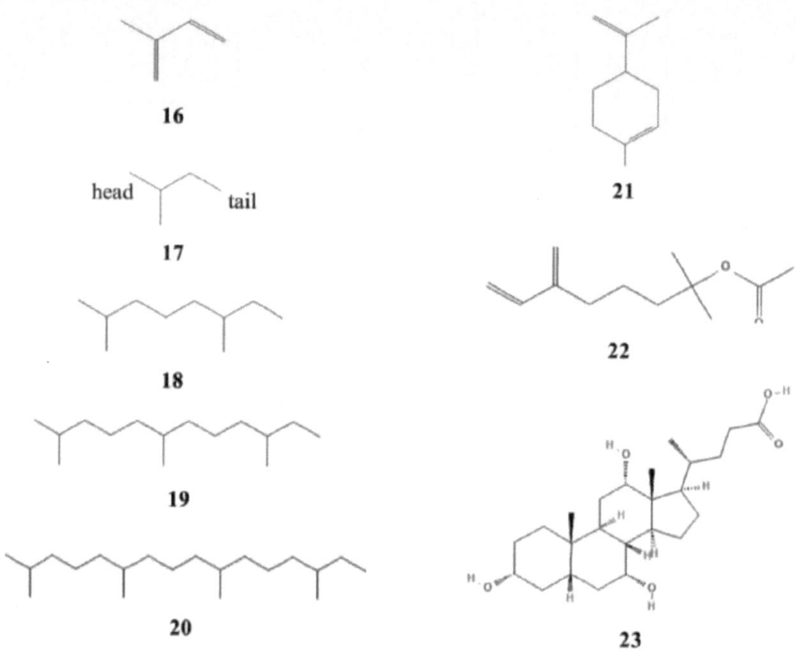

FIGURE 3.2 Terpenes carbon backbone: isoprene: 2-Methyl-1,3-Butadiene (16), C_5 hemiterpene: 2-Methylbutane (17), C_{10} monoterpene: 2,6-Dimethyloctane (18), C_{15} hemiterpene: 2,6,10-Trimethyldodecane (19) and C_{20} diterpene: 2,6,10,14-Tetramethylhexadecane (20). Terpenoids structure examples: limonene (21), myrcenyl acetate (22) and cholic acid (23).

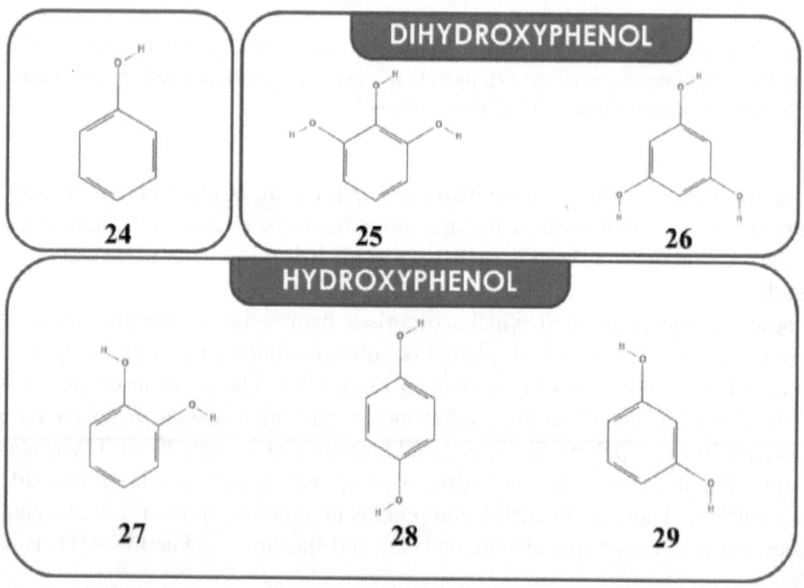

FIGURE 3.3 Structures of various phenolic compounds: phenol (24), gallocatechol (25), phloroglucinol (26), catechol (27), hydroquinone (28) and resorcinol (29).

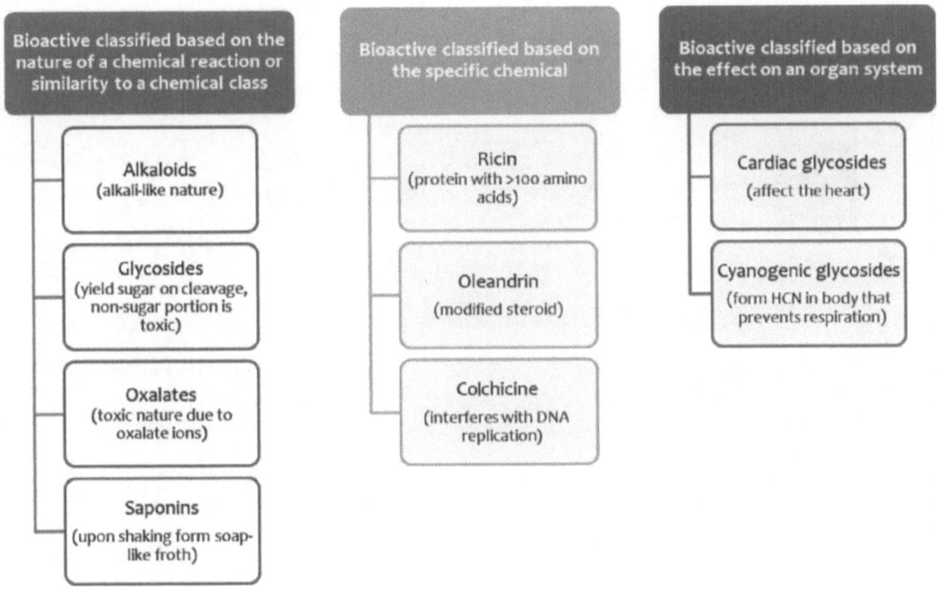

FIGURE 3.4 Other methods for bioactive compounds classification.

3.4 NANOTECHNOLOGY

Nanotechnology is the application part of nanoscience; as a result of such research, one can attain a novel nanomaterial. It is the most fascinating and advancing interdisciplinary field of research that has been emerging over the past two decades (Kurath and Maasen 2006) and has significance in fields of science and technology. The field majorly deals with the laws of chemistry, physics, molecular biology and material sciences at the nanoscale (i.e. 10^{-9}m). The most stable molecule at the visible spectrum is seen to be thermodynamically unstable at nanoscale. To visualize such a dynamic world, one requires the use of electron microscopes or microscopes with built-in lasers. The most wide application of nanotechnology is the manufacturing of nanostructured materials such as nanoparticles, nanocomposites, nanofibers, nanocapsules, nanowires, nanoporous materials, single-walled or multi-walled carbon nanotubes, dendrimers, fullerenes, quantum dots, thin films and molecular electronics (Willems and Wildenberg 2005).

Nanoparticles are the type of nanomaterial with controlled size and composition, which have the potential to serve as the building block for developing complex and dynamic nanostructures. These particles exhibit remarkable characteristics on account of the large surface area to volume ratio decreasing in its size. Upon increasing the surface energy, their biological effectiveness can simultaneously increase (Willems and Wildenberg 2005). Nanoparticles occur in various morphological shapes and sizes, including spherical, cubic, prism, helical, oval and many other shapes (Benelmekki 2015). There are two major approaches to produce nanoparticles: top-down and bottom-up approach (Figure 3.5; Husen and Siddiqi 2014). Ideally, the top-down approach executes dry methodology, while bottom-up approach is often employing wet methodology. However, as these particles are unstable at nanoscale, there is a possibility of agglomeration. Therefore, techniques need to be employed to prevent such phenomena. Ideally, the synthesis of nanoparticles has been accomplished via three major approaches: physical, chemical and green syntheses. In physical approach, evaporation–condensation and laser ablation transpire to be the chief techniques. The extensive energy required for maintaining high temperature and pressure can increase the production cost. In chemical and green syntheses, a commonly adapted principle for the synthesis of nanoparticles includes the reduction of metal or metal oxide into nanoparticles using organic and inorganic reducing agents. Additionally, following the formation of nanoparticles, the immediate

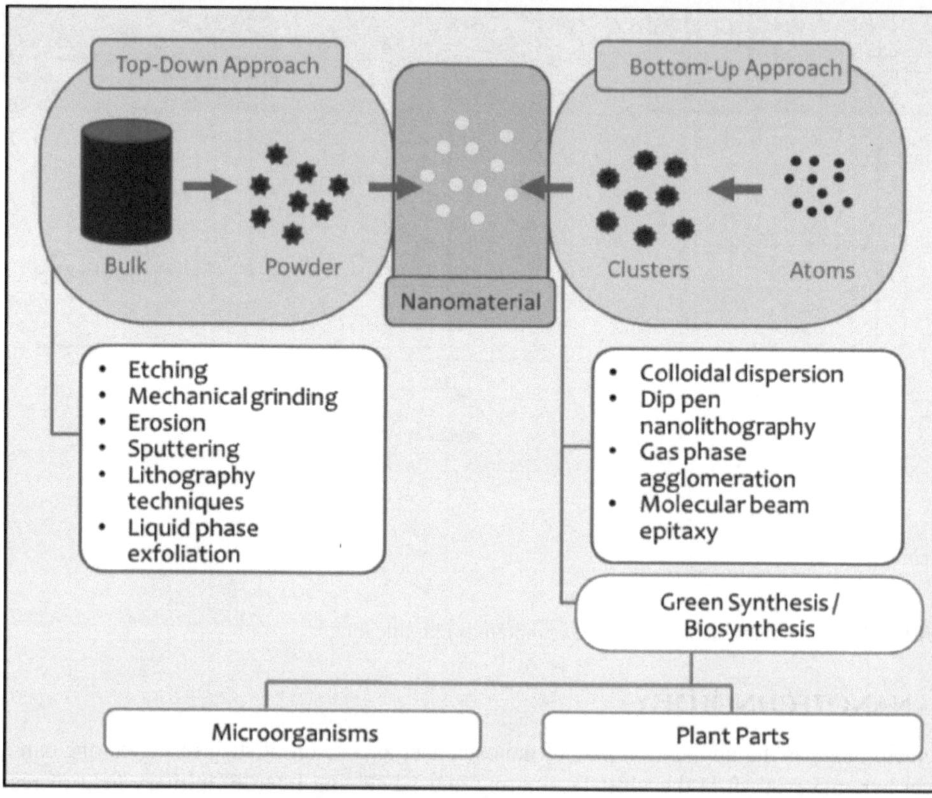

FIGURE 3.5 Illustration of the most common approaches used for the production of nanoparticles.

capping of these is essential to prevent agglomeration of the nanoparticle. However, some of the chemicals used in chemical synthesis are non-biodegradable and some may produce toxic by-products (Song and Kim 2008). Out of these, green synthesis substantiates to be the most cost-effective, eco-friendly, and efficiently scalable to large-scale nanoparticles production. The major advantage is that it is a single step process in which reduction and capping of metal and metal oxides occurs, as a result of the attainment of utmost stable nanoparticles. The scope of green synthesis will be explored further in this review with the aid of poisonous plant extracts.

3.5 GREEN SYNTHESIS OF NANOPARTICLES

The attainment of metal nanoparticles with the assist of organic solvent system from biological entity (e.g. bacteria, fungi and plant extract), whose by-product is non-toxic and eco-friendly, the products are collectively called biogenic nanoparticles, while such method is called "green synthesis" (Singh et al. 2018). Majority of the nano-scientists are recently attracted toward the biological system due to the increasing demand for environmentally cordial procedures (Husen and Siddiqi 2014; Husen and Iqbal 2019a, b; Husen 2019; Husen and Jawaid 2020). Especially in the field of pharmaceutical industries, accomplishment of manufacturing nanodevices for medical purposes and drug delivery technology is achievable through green synthesis (Shafey 2020). The biosynthesized nanomaterials are of enterprising nature as they eliminate or reduce by-product pollution and attain versatile properties such as optical, magnetic, electric and catalytic potentials (Yeary et al. 2005; Shafey 2020). The properties exhibited by the metal nanoparticles differ majorly from the bulk metallic materials by various unique properties. These nanoparticles are applicable in engender augmented contrast agent, active material for food packaging, nano-biosensors, nano-electronics, disinfectant, antibiotics and antiseptic.

In the green synthesis of various nanoparticles, plant extract is adopted as the most cogent source to synthesize metal or metal oxide nanoparticles at larger scale, as it eliminates the sophisticated steps involved in cell culture and avoidance of contamination (Singh et al. 2018; Husen and Siddiqi 2014; Husen 2019). Plant extract as a source had advanced with the achievements in biotechnology and genetics (Shafey 2020). Diverse parts of the plants are utilized for the green synthesis of metal or metal oxide nanoparticles including seeds, leaves, roots, fruits and barks. The nanoparticles obtained by such a procedure exhibit a wide range of different properties and prevent further aggregation or agglomeration of nanoparticles. The properties of the nanoparticles can be enhanced by applying certain conditions such as temperature, pH and the concentration of reagent (Shafey 2020).

3.6 METAL AND METAL OXIDE NANOPARTICLES

The metal and metal oxide nanoparticles are applied in multidisciplinary areas. Recent era of nanotechnology is immensely incorporated in our daily lifestyle products, medicines, cosmetics, clothing and electronic devices. The industrial sectors that are commercializing such nanotechnological products are prospering continuously. The synthesis of nanoparticles can be accomplished through conventional physical or chemical synthesis, as well as through green synthesis methods. Since the green synthesis is advantageous at various commercial levels, such methods are given priority to achieve products with desired properties. In green synthesis, plants are preferred over microorganisms and other biological entities due to their sustainability, and ability to utilize and transform 75% light energy to chemical energy. Thereby, plants contain certain chemicals such as sugars, antioxidants and other bioactive compounds, which execute an essential role in nanoparticle formation. Accordingly, plants are scrutinized for green synthesis (Shafey 2020).

Diverse plant species are studied and utilized for nanoparticle biosynthesis. However, it is necessary to consider the favorable conditions to optimize the process of synthesis. The leaf extracts of plants play a key role in the nanoparticle biosynthesis, as it contains the most antioxidant and reducing agents. To achieve metal nanoparticles from metallic ions, reduction of metal ions to metal nanoparticle through addition of strong base (e.g. sodium hydroxide), followed by the immediate capping of the nanoparticles are two essential steps (Figure 3.6; Bystrzejewska-Piotrowska et al. 2009). Plant executes the procedure in one step solely: reduction and capping in a single

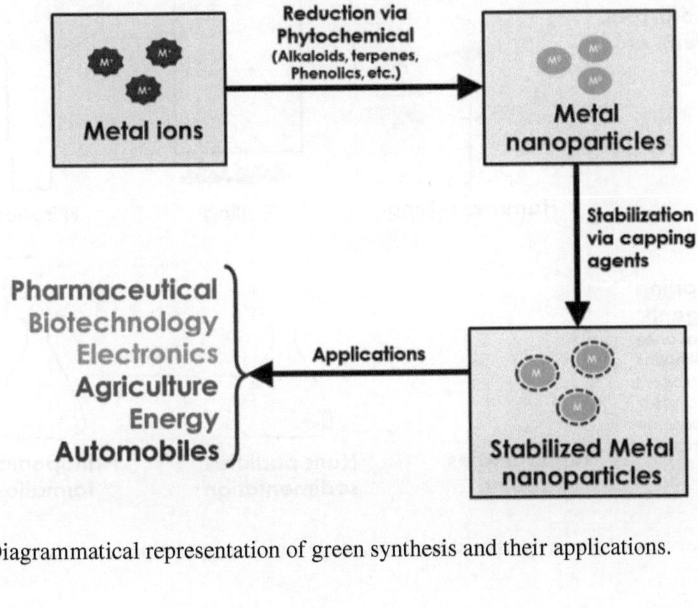

FIGURE 3.6 Diagrammatical representation of green synthesis and their applications.

step. Therefore, plants containing such phytochemicals are preferential for synthesis of noble metal nanoparticles under the absence of interfering chemicals. These phytochemicals are bioactive compounds present in the plant for the purpose of producing physiological activity (Cazarin et al. 2022). Various factors of bioactive compounds affect the nanoparticles yield and physicochemical properties. The factors include temperature, pH, salt content, contact time and phytochemical properties of the extract. Plants consume less time to reduce metal ions and microorganism, owing to the long incubation due to the presence of phytochemicals that are water soluble. The reducing capacities of different plant species vary as the composition of extract is different, affecting the nanoparticle synthesis to an extent (Shafey 2020). The two processes for attaining the nanoparticles via phytochemical (Figure 3.7) are milling (for dry sample) and blending (for wet sample).

The metal and metal oxide nanoparticles consist of greater surface area to volume ratio, and such properties of nanoparticles intrigue researchers to extensively study their unique physicochemical characteristics, optical properties, electronic properties, magnetic properties and antimicrobial properties. Various metal and metal oxide nanoparticles are discussed further in this chapter.

3.6.1 Silver Nanoparticles

Silver nanoparticles are extensively used in many fields. Due to their rising demand, its large-scale production through conventional means causes extensive pollution of harmful chemicals. To counter this demand, plant-based green synthesis seems to be the most preferential option for production at larger scale (Husen 2019; Siddiqi and Husen 2020). Various plant extracts are examined to achieve desired properties of silver nanoparticles. The most recent research was conducted by using leaf extract of neem (*Azadirachta indica*), lemongrass (*Cymbopogon citratus*) and green tea (*Camellia sinensis*), while plant extract of *Aloe vera*, starch and natural rubber can also be used (Rupiasih et al. 2013). The poisonous plants utilized for production of various nanoparticles and their properties are listed as follows:

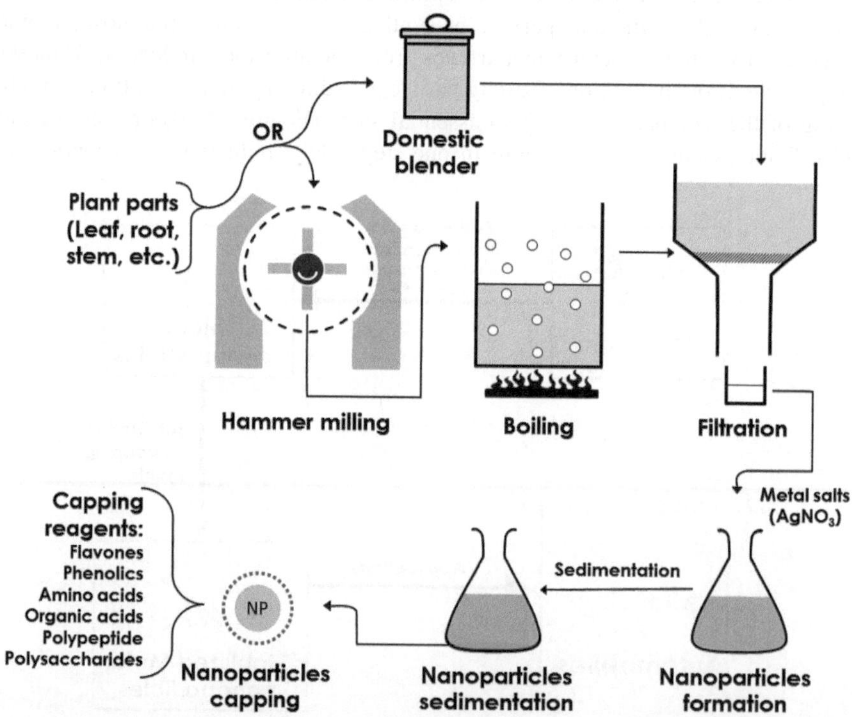

FIGURE 3.7 Overview of green synthesis of nanoparticles utilizing plant parts as a source.

3.6.1.1 *Thevetia peruviana*

Yellow oleander is an ornamental dicotyledonous plant belonging to the family Apocynaceae. It is typically located in tropical and subtropical areas, containing white latex (milky juice), called thevetin. The compound thevetin is extremely poisonous due to the presence of alkaloids and cardiac glycosides, but used in pharmaceuticals as heart stimulants. They are found in high concentration in the plant's flowering period. Tincture of bark has been utilized to reduce fever even though it is emetic in nature (Rupiasih et al. 2013). Thevetin is a glycosidic toxin containing thevetin A and B, while other cardiac glycosides includes cerberin, digoxin, neriine, oleandrin, and perusitin (Dear 2014; Oluwaniyi and Samuel 2007). Additionally, the compound is proven to be inactive against avian malaria and used as fish poison in Indonesia (Rupiasih et al. 2013). The fresh flowers consist of kaempferol, quercetin and its derived polyphenol quercetin 7-o-galactoside. Leaf extracts have been deduced to contain proteins, phenols and flavonol glycosides (Oluwaniyi et al. 2016).

To obtain glycoside from *T. peruviana* seed extract, alcoholic extraction was employed by scientists Oluwaniyi and Samuel (2007). In the experiment, the seeds' mesocarp and pericarp were removed and the remaining was crushed. Thenceforth, defatting of the extract was carried out through mechanical pressing, and then the solvent extraction of the paste was performed using distilled hexane. The results showed that seed extract treated with 70% aqueous alcohol was extracted with the highest protein concentration (Oluwaniyi and Samuel 2007).

In an experiment, Rupiasih et al. (2013) used the leaf extract of Thevetia peruviana. They prepared the leaf extract by collecting of the white latex of plant green stem and filtering the extract through Whatman No. 1 filter paper, and it was stored at 4°C. The reaction mixture was 50 μL of 1 M silver nitrate (aqueous solution) with 10 mL of extract, resulting in the reduction of Ag ions. After 6 hours of incubation period at room temperature, color changes to light purple and then slowly to purplish brown, showing the formation of silver nanoparticles. The Ag nanoparticles TEM (transmission electron microscopy) micrograph indicated the shape to be primarily spherical with average size around 10–20 and 20–30 nm. The characterization studies showed that the major component present in the plant extract was thevetin, whereas amino acid carbonyl group and polypeptides acted as encapsulating agents (Rupiasih et al. 2013).

3.6.1.2 *Catharanthus roseus*

Catharanthus roseus formerly known as periwinkle is an extremely poisonous plant belonging to the family Apocynaceae (Nejat et al. 2015). All the parts of this plant are toxic for human and animal consumption, due to the presence of terpenoid indole alkaloid, which is of greater medical significance (Almagro et al. 2015). Silver nanoparticles formed using the leaf extract can be able to perform optimum antiplasmodial activity (Ponarulselvam et al. 2012). Ponarulselvam et al. (2012) prepared leaf extract of *C. roseus* by boiling it and filtering it through Whatman No.1 filter paper. After 1 week of storage at –15°C, it was incubated in the Erlenmeyer flask with $AgNO_3$ and brownish-yellow coloration of solution develops indicating the formation of silver nanoparticles. SEM (scanning electron microscope) imaging revealed the size of silver nanoparticles to be 30–35 nm in diameter.

3.6.1.3 *Datura metel*

Datura belongs to the family Solanaceae, containing anticholinergic activity due to the presence of atropine, scopolamine and hyoscyamine (Trancă et al. 2017). Kesharwani et al. (2009) experimented on producing silver nanoparticles from leaf extract of *Datura metel*. The results showed the changes in coloration from yellow to dark brown after incubation with 1 mM silver nitrate, indicating colloidal silver nanoparticle formation. TEM showed 16–40 nm of diameter nanoparticles that exist in a quasilinear superstructure. The research also concluded the possibility of quinol compound from leaf extract to be responsible for silver ion reduction, while chlorophyll pigment serves as a stabilizing agent between the electron donor (silver ions) and acceptor (plastoquinone) molecules.

3.6.1.4 *Convolvulus arvensis*

Convolvulus arvensis, commonly known as bindweed, is from the family Convolvulaceae containing various alkaloids such as meso-cuscohygrine tropinone and tropine. *C. arvensis* toxicity in mice's feed causes hepatitis and gastritis histologic lesions (Schultheiss et al. 1995). The reduction of silver ions to silver nanoparticles was indicated by the color change of solution to dark brown. Rasheed et al. (2018) investigated *C. arvensis* dry plant extract for silver nanoparticle synthesis and examined its potential to degrade dyes. The silver nanoparticles formed had intrinsic properties including an 18.5 surface charge and a 90.9 nm size distribution which was concluded through zeta potential and dynamic light scattering technique. The catalytic reducing activity of the silver nanoparticles on azo dye was depicted, indicating the potency to degrade dye pollutants.

3.6.1.5 *Portulaca oleracea*

The herb *Portulaca oleracea*, commonly referred to as Purslane, contains high level of L-norepinephrine and can be used as a natural herb to reduce hemorrhage in tissues (Firdhouse and Lalitha 2012). The leaves of *P. oleracea* contain cardiac glycoside and oxalic acid which are extremely toxic to the humans (Heydari et al. 2019). In Firdhouse and Lalitha's (2012) research paper, plant extract prepared using fresh leaves of *P. oleracea* was used for the synthesis of gold nanoparticles. To 1 mL of aqueous plant extract, silver nitrate solution was added; after 1 hour of incubation at room temperature, the coloration of the solution changed from yellow to reddish brown. Then, it is preceded with water bath at 70°C–75°C and sonicated. These synthesized silver nanoparticles were separated using centrifugation at 13,000 rpm for 15 minutes, and the supernatant was collected and stored at −4°C for further analysis. Through SEM and XRD analysis, the gold nanoparticle size was elucidated to be less than 60 nm calculated using Debye–Scherrer's equation.

3.6.2 Gold Nanoparticles

Gold nanoparticles have exceptional application in tumor cells such as bioimaging of tumor cells, dark-field microscopy for microorganism and their metabolites detection, surface receptor detection and analyses of endocytosis. In therapeutics, studying the efficacy of cells labeled with conjugates is executed as qualitative estimation. For the quantitative estimation for examining the efficacy of labeled cells, gold nanoparticles are labeled, for example, pig embryo kidney cells labeled with gold nano-shell (Dykman and Khlebtsov 2011). The gold nanoparticles have various pharmaceutical applications due to their antibacterial, antioxidant and anticancerous properties; they are also broadly applied in aquaculture and biomedicine (Babu et al. 2020).

3.6.2.1 *Gymnocladus assamicus*

Gymnocladus assamicus is a Northeast India leguminous tree and is a critically endangered species. The seed pods are rich in saponin, which are used to treat dermatological disorders and remove hair leaches. As the pods are reservoir of bioactive compound saponin, Tamuly et al. (2013) demonstrated the green synthesis of gold nanoparticles using pod extract of *G. assamicus*. Pod extract was continuously stirred with 0.001 M of chloroauric acid solution for 4 hours at 30°C, resulting in the development of ruby red colored solution which indicated the synthesis of gold nanoparticles. The HRTEM imaging indicated the shape to be triangular, pentagonal and hexagonal with size ranging from 4.5 ± 70.23 to 22.5 ± 1.24 nm. The synthesized gold nanoparticles showed potential catalytic activity in reducing certain dyes such as 4-nitophenol to 4-aminophenol.

3.6.2.2 *Nerium oleander*

Nerium oleander is extremely poisonous ornamental shrub belonging to the Apocynaceae family. The most common bioactive compounds present in all parts of the plants are nondigitalis cardiac glycosides, which include oleandrin found in oleander (Khan et al. 2010). Tahir et al.'s (2015) research article showed the successful synthesis of gold nanoparticles from leaf aqueous extract of *N. oleander*. The procedure includes continuously stirring 10 mL of leaf extract with 50 mL of 0.03 M of chloroauric acid solution. As a result, yellow solution changes into black after 2 hours of incubation at room temperature. The synthesized gold nanoparticles were examined to be in nearly spherical shape and showed potential antioxidant activity.

3.6.2.3 *Plumeria alba*

Plumeria alba belongs to the family Apocynaceae, and it is widely used in traditional medicine for the treatment of type 2 diabetes, diuretic and skin disease (Tessou et al. 2013). *P. alba* flower aqueous extract was prepared for the synthesis of nanoparticles, which showed that by mixing 10 mL of flower extract with 25 mL 1 mM chloroauric acid and incubating the mixture at room temperature for 2 minutes, a color change was observed. The reaction mixture color changed from yellow to blue or pink, depending on the flower extract concentration. The nanoparticles were analyzed for their catalytic activity and antibacterial activity. Results showed strong potential of gold nanoparticles to degrade organic dye and significantly inhibit the growth of *Escherichia coli* (Mata et al. 2016).

3.6.2.4 *Terminalia catappa*

Terminalia catappa, commonly known as almond, belongs to the Combretaceae family; its fruit consists of one-third pulp and two-third seed (edible). The leaves are used as an anticancer, antioxidant and anti-inflammatory agent, whereas fruits and leaves of this plant have antidiabetic activity and anthelmintic activity, respectively. Researcher Ankamwar (2010) experimented on gold nanoparticle synthesis. The tetrachloroauric acid was mixed with aqueous extract of plant and incubated for approximately 50 minutes, after which red coloration of solution is observed. The color change indicates the formation of gold nanoparticles through reduction reaction and encapsulation by *T. catappa*. The size was observed to be in the range of 10–35 nm, elucidating the antibacterial, antioxidant and anticancerous properties. Further enhancement of the gold nanoparticle is pursued to perform as a biogenic nanoparticle for cancer therapy.

3.6.3 Platinum Nanoparticles

Platinum nanoparticles are extensively used in nanocarriers, nanosensors and nanozymes owing to their inert properties. Recently, the field of theranostics (which is combination of therapeutic and diagnostic field) is including platinum nanoparticles, which are multi-functionalized. Industrial application of platinum nanoparticle includes their role in fuel cells and catalysis. Their application in pharmaceutical and drug discovery seems propitious (Bloch et al. 2021).

3.6.3.1 *Bacopa monnieri*

Bacopa monnieri is commonly known as water hyssop, belonging to the Scrophulariaceae family (Ray et al. 2021). Consumption of the plant extracts shows reduction in the rate of memory loss till a certain dose. The extract has side effects in the human body causing deterioration of the gastrointestinal tract; main symptoms are abdominal cramps, increased frequency of stool and nausea. Adverse toxicity was reported in a women's body after the consumption of homeopathic herbs which include *B. monnieri*. Moreover, upon pausing the consumption of such herbs, women's liver function was normalized (Sireeratawong et al. 2016). This was due to the presence of saponins in the *B. monnieri* herbs (Ray et al. 2021). At room temperature, platinum (IV) ions are mixed with the leaf extract of *B. monnieri* in the ratio of 1:4, resulting in the formation of platinum nanoparticles having size of 5–20 nm and spherical shape. The platinum nanoparticles formed showed antioxidant and neuroprotective properties (Fahmy et al. 2020).

3.6.3.2 *Prunus* x *yedoensis*

Prunus x *yedoensis*, commonly called as Somei-yoshinio in Japan, is a blooming tree with soft pink flowers having a hybrid origin with maternal descendant from *Prunus spachiana* f. and paternal descendant from *Prunus serrulata* var. species (Cho and Kim 2019). *Prunus* species are well known through two economically essential plants, sweet cherry and almond. However, the species produce bioactive compounds such as amygdalin and prunasin, which are hydrogen cyanide containing cyanogenic glucosides (Del Cueto et al. 2017). Velmurugan et al. (2016) examined the synthesis of platinum nanoparticles using tree gum of *P. yedoensis*. Tree exudate (gum) was collected, dried and powdered by mortar and pestle, then sieved to obtain desired particle size and autoclaved for 20 minutes for homogenization of extract. The gum extract is then reacted with 0.1 M chloroplatinic acid. As a result, color change in solution is observed, and the solution turns brown in color, which is a confirmation of obtaining platinum nanoparticles. By analyzing the platinum nanoparticles characteristics using TEM, spherical shaped 10–50 nm size nanoparticles were obtained. The nanoparticles were also analyzed for their antifungal activities, showing inhibition against *Cladosporium fulvum* and *Colletotrichum acutatum* (Velmurugan et al. 2016).

3.6.4 Copper Nanoparticles

Copper nanoparticles are widely applied therapeutics due to their antimicrobial, antioxidant, catalytic and photocatalytic properties and their ability to degrade organic dye (Murthy et al. 2018). Researchers are attracted toward application of copper nanoparticles in sensor technology, as a nanocomposite, and are used instead of metal-based sensor in perovskite or brittle oxidic spinel (Athanassiou et al. 2006).

3.6.4.1 *Hagenia abyssinica*

Belonging to the Rosaceae family, *Hagenia abyssinica* is commonly used in Ethiopian medicine to treat diarrhea; in East Africa, it is commonly known as African redwood (Kifle et al. 2021; Assefa et al. 2010). Dried leaves of the plant were used for preparing the extract; these leaves were grinded and packed in a brown bottle. Later on, the powdered extract was dissolved in deionized water and filtered using Whatman No.1 filter paper, and the filtrate was used for further analyses. The extract was mixed with 0.2 M copper (II) nitrate trihydrate solution, and then light brown coloration of solution develops which indicates the formation of copper nanoparticles. Subsequent wash with deionized water and ethanol eliminates remaining impurities in the solution. The copper nanoparticles obtained were analyzed through various techniques, such as SEM and TEM, elucidating the possible size to be an average of 34.76 nm and shape is to be spherical, cylindrical, triangular, and hexagonal. Antimicrobial activity of the copper nanoparticles was carried out, resulting in inhibition of the common pathogen under examination such as *E. coli*, *Bacillus subtilis*, *Staphylococcus aureus*, and *Pseudomonas aeruginosa* (Murthy et al. 2020).

3.6.5 Palladium Nanoparticles

Palladium nanoparticles have various features due to their diverse shapes and sizes that allure researchers to evaluate its mechanical, physical, optical, electronic and catalytic properties. It has wide application as photoacoustic and photothermal agent, antimicrobial and anticancerous agent, and nanocarrier and in other therapeutic techniques (Phan et al. 2020).

3.6.5.1 *Gymnema sylvestre*

This poisonous plant is commonly found in India, well known as "gurmar", belonging to the family Apocynaceae. Due to the presence of certain bioactive compounds, the leaves exhibit antidiabetic, antimicrobial and anti-inflammatory activities (Tiwari et al. 2014). The leaves of *G. sylvestre* were collected and ground in mortar and pestle, dissolved into deionized water, heated, filtered and stored at 3°C–5°C. The extract was mixed with palladium chloride and microwaved for 5 minutes; the yellow solution turns into dark brown solution with an appearance of precipitation indicating the formation of palladium nanoparticles. The SEM revealed the shape of the nanoparticle to be quasi-sphere, with the capacity to reduce chromium (III) (Cr^{+3}) to chromium (VI) (Cr^{+6}) (Kadam et al. 2020).

3.6.6 Nickel Nanoparticles

Nickel nanoparticles have an intrinsic property of ferromagnetism, including chemical stability, high coercive force and magnetocrystalline anisotropy. Broader applications are in areas of electronics (as supercapacitors), pharmaceuticals, catalysis and dye-sensitized solar panel (Jaji et al. 2020).

3.6.6.1 *Citrullus colocynthis*

Citrullus colocynthis is a plant belonging to the family Cucurbitaceae; it is a bitter fruit and poisonous plant found in desert regions of Asia, tropical regions of Africa and Mediterranean regions (Borhade et al. 2013). The plant has many bioactive compounds such as cucurbitacins, cucurbitacin glycosides, colocynthetin, colocynthin and flavonoids (Kapoor et al. 2020). The stem was collected, washed, dried, ground by mortar and pestle to obtain fine powder, and sieved. The powdered extract obtained was made into an aqueous solution using distilled water and heated for 20 minutes, later centrifuged, supernatant was removed and the extract is stored at 4°C. The extract was mixed with 0.3 M nickel nitrate, the flask was incubated at RT for 24 hours and the color changed from green to pale yellow indicating nickel nanoparticles formation. The nanoparticle was examined against poisonous Yellow 160 dye, and the results showed the degradation of dye with the production of non-toxic by-products (Kiran et al. 2020).

3.6.7 Zinc Oxide Nanoparticles

Zinc oxide nanoparticles have intrinsic features such as physical, optical and antimicrobial. Tremendous work has been executed for zinc nanoparticles applications in UV light emitters, piezoelectric devices chemical sensors and transparent electronics. It also performs high catalytic activity and has a high potential in photocatalytic degradation of pollutant properties which infers its utilization in environmental sustainability. As recommended by the FDA (Food and Drug Administration), using zinc oxide nanoparticle as a food additive is safe. Other applications include preservative, drug delivery, gas sensor, cosmetics, biosensors and optical devices (Sabir et al. 2014).

3.6.7.1 *Cassia fistula*

This poisonous plant was kept under the family Caesalpinioideae, and it is a widely used medicinal plant, commonly denoted as "golden shower" or "amaltas". Plant leaves contain various flavonoid compounds such as amentoflavone, and it is frequently applied in the treatment of liver disorders and inflammatory disease (Srividhya et al. 2017; Kaur et al. 2019). Plant leaves have other antioxidant compounds, for instance, terpenoids, alkaloids, phenolic, tannins, saponins and other bioactive compounds. The powdered extract of *C. fistula* was prepared from fresh leaves through washing, drying and then grinding into fine powder. The aqueous extract was carried out by rehydrating the fine powdered extract with distilled water, then heated at 70°C for 30 minutes, filtered using muslin cloth and then Whatman No.1 filter paper and stored at 4°C. For the synthesis of nanoparticles, aqueous extract was mixed with 0.01 M zinc acetate dihydrate, then the mixture was heated at 70°C for 1 hour and color changed from yellow to light yellow or translucent solution with white precipitate formed at the bottom of the flask and these precipitates indicate the formation of zinc oxide nanoparticles. Supernatant was removed and precipitate was further analyzed for the nanoparticle characteristics. The zinc oxide nanoparticles characterized by DLS (dynamic light scattering) evaluated the approximate size of zinc oxide nanoparticles to be 68.1 nm, the SEM imaging showed aggregated nanoparticles having uneven surfaces. Nanoparticles antimicrobial activity was examined, showing inhibition against *E. coli* and *S. aureus* (Naseer et al. 2020).

3.6.8 Iron Oxides Nanoparticle

Iron oxide nanoparticles have a wide range of applications in therapeutics due to its physical property, superparamagnetic property and biocompatibility. Hence, iron oxide nanoparticles are extensively used in cancer treatment, gene therapy, drug delivery and MRI (magnetic resonance imaging) scanning. The multi-functionalized form of iron oxide nanoparticle is applied in theranostics (Ali et al. 2016).

3.6.8.1 *Calotropis procera*

This plant comes under the family Asclepiadaceae and distributed in subtropical and tropical regions. The plant contains many bioactive compounds including terpenoids, flavonoids, polysaccharides, phenolic and other proteins. The glycosides found in the plant are calotropin, calactin, calotoxin, calotropogenin and uscharin (Farooq et al. 2017). Use of *Calotropis procera* in the synthesis of magnetite iron oxide nanoparticles initiates with the extract preparation: leaves of *C. procera* were collected and made into powdered form, then they were dissolved in distilled water, boiled, filtered and the filtrate collected was used in the further procedure. The extract was mixed with 1 mM iron chloride at 1:2 ratio and then heated for 2 hours; color change was observed indicating the calcination of copper nanoparticles. The iron oxide nanoparticles obtained were examined for its antifungal activities showing inhibition against *Alternaria alternata*, and the results indicate that smaller sized nanoparticles can efficiently inhibit the fungal growth (Ali et al. 2020).

3.6.9 Titanium Oxide Nanoparticles

Titanium oxide nanoparticles are essentially used in the degradation of toxic dyes and waste water treatment. Titanium oxide nanoparticles are used in biological applications such as food industries, silkworm reproduction and drug discovery. Their unique properties such as durability, corrosive resistance, light weight and compatibility with CFRP (carbon fiber reinforced polymer) enable its application in space technology. Especially, the alloy of titanium exhibits great strength-to-weight ratio. Routine application of titanium oxide nanoparticles in household owing to their anti-fogging and self-cleaning properties; they are also used as sanitizing agent (Seydi et al. 2019).

3.6.9.1 *Jatropha curcas*

This poisonous plant belongs to large family Euphorbiaceae, formally called Jangli erandi or Ratanjyot in southern parts of India. Ideally, it was used in traditional allopathic medicine, but recently it is applied in biodiesel production. The toxicity of *Jatropha curcas* is found in all parts of this plant with the seeds containing the highest toxin concentration (Singh et al. 2010; Abdelgadir and Staden 2013). The leaves contain phenolic compounds such as apigenin, its glycoside isovitexin and vitexin, as well as alkaloids that have anticancerous properties (Oskoueian et al. 2011). Titanium oxide nanoparticles from latex *J. curcas* L. using the milky white latex were extracted from the green stem of the plant. Latex was mixed with 2.5 mM dihydroxy(oxo)titanium [$TiO(OH)_2$] and the mixture was kept in a water bath for 10 minutes. After 12 hours, white cluster deposition was seen at the flask bottom, depicting the formation of titanium oxide nanoparticles. The nanoparticles obtained were of 25–50 nm in size and semi-crystalline sphere in structure (Hudlikar et al. 2012).

3.7 CONCLUSION AND FUTURE PERSPECTIVE

Poisonous plants have greater medicinal value due to the presence of bioactive compounds or phytochemical, and they have various pharmaceutical and therapeutic properties. Therefore, process of nanoparticle formation from its ion form was mediated through reduction and stabilization, facilitated by the bioactive compounds of poisonous plants, i.e. green synthesis. Plants are reservoirs of essential natural products that have antioxidant, antimicrobial, anti-inflammatory and many more pharmacological properties. With that, the nanoparticles formed through such process have versatile properties, which are more advantageous than traditional synthesis of nanoparticles. Therefore, metal and metal oxide nanoparticles formed from such procedures are a great asset when examining necessary factors. Such plants are considered as a nano-factory for the production of nanoparticles, which is cheaper, eco-friendly and faster than chemical synthesis of nanoparticles. The major drawback of the procedure is the lack of research on the poisonous plants for evaluating its uses in the field of nanoscience. Additionally, the primary bioactive compound, which is responsible for the reduction and stabilization, is not examined efficiently. Due to the great diversity in the bioactive compounds, finding the suitable solvent and condition for the extraction demands effective research and development. Especially, the poisonous plant potential in nanoscience is still largely unexplored by researcher. Poisonous plants can lead to a larger prospect in the industrial production of nanoparticles.

NOTES

Apekshakumari Patel, Department of Life Sciences, University of Mumbai, Mumbai, India
Nimisha Patel, Department of Life Sciences, J.C. Bose University of Science & Technology, YMCA, Faridabad, India
Ahmad Ali, Department of Life Sciences, University of Mumbai, Mumbai, India
Hina Alim, Department of Life Sciences, University of Mumbai, Mumbai, India

REFERENCES

Abdelgadir, H. A., & Staden, J. V. 2013. Ethnobotany, ethnopharmacology and toxicity of Jatropha curcas L. (Euphorbiaceae): A review. *South African Journal of Botany*, 88: 204–218.

Ali, A., Zafar, H., Zia, M., Ul Haq, I., Phull, A. R., Ali, J. S., & Hussain, A. 2016. Synthesis, characterization, applications, and challenges of iron oxide nanoparticles. *Nanotechnology, Science and Applications*, 9: 49–67.

Ali, M., Haroon, U., Khizar, M., Chaudhary, H. J., & Munis, M. F. H. 2020. Facile single step preparations of phyto-nanoparticles of iron in Calotropis procera leaf extract to evaluate their antifungal potential against Alternaria alternata. *Current Plant Biology,* 23: 100157. DOI: 10.1016/j.cpb.2020.100157

Almagro, L., Fernández-Pérez, F., & Pedreño, M. A. 2015. Indole alkaloids from Catharanthus roseus: Bioproduction and their effect on human health. *Molecules (Basel, Switzerland)*, 20 (2): 2973–3000.

Ankamwar, B. 2010. Biosynthesis of gold nanoparticles (Green-gold) using leaf extract of Terminalia Catappa. *E-Journal of Chemistry*, 7 (4): 1334–1339.

Anywar, G. 2020. Historical use of toxic plants. In Mtewa, A. G., Egbuna, C., & Rao, N. (Eds.), *Poisonous Plants and Phytochemicals in Drug Discovery*, Wiley, Hoboken, New Jersey, USA. 1–17.

Assefa, B., Glatzel, G., & Buchmann, C. 2010. Ethnomedicinal uses of Hagenia abyssinica (Bruce) J.F.Gmel. among rural communities of Ethiopia. *Journal of Ethnobiology and Ethnomedicine*, 6: 20.

Athanassiou, E. K., Grass, R. N., & Stark, W. J. 2006. Large-scale production of carbon-coated copper nanoparticles for sensor applications. *Nanotechnology*, 17 (6): 1668–1673.

Azmir, J., Zaidul, I. S. M., Rahman, M. M., Sharif, K. M., Mohamed, A., Sahena, F., & Omar, A. K. M. 2013. Techniques for extraction of bioactive compounds from plant materials: A review. *Journal of Food Engineering*, 117 (4): 426–436.

Babu, B., Palanisamy, S., Vinosha, M., Anjali, R., Kumar, P., Pandi, B., Tabarsa, M., You, S., & Prabhu, N. M. 2020. Bioengineered gold nanoparticles from marine seaweed Acanthophora spicifera for pharmaceutical uses: Antioxidant, antibacterial, and anticancer activities. *Bioprocess and Biosystems Engineering*, 43 (12): 2231–2242.

Benelmekki, M. 2015. Chapter 1: An introduction to nanoparticles and nanotechnology. In Benelmekki, M. (Ed.), *Designing Hybrid Nanoparticles*, Morgan & Claypool Publishers, San Rafael, CA, 1–14.

Bernhoft, A. 2010. A brief review on bioactive compounds in plants. *Bioactive Compounds in Plants-Benefits and Risks for Man and Animals*, 50: 11–17.

Bloch, K., Pardesi, K., Satriano, C., & Ghosh, S. 2021. Bacteriogenic platinum nanoparticles for application in nanomedicine. *Frontiers in Chemistry*, 9: 624344.

Borhade, P., Deshmukh, T., Patil, V., & Khandelwal, K. 2013. Review on Citrullus colocynthis. *International Journal of Research in Pharmacy and Chemistry*, 3 (1): 46–53.

Britannica, T. 2020. Editors of Encyclopaedia. Apocynaceae. *Encyclopedia Britannica*. Accessed 10-2-2020. https://www.britannica.com/plant/Apocynaceae

Burgdorf, W. H. C., Hoenig, L. J., & Padilla, R. S. 2015. Dangerous plants of the southwestern United States. *JAMA Dermatology*, 151 (2): 203. DOI: 10.1001/jamadermatol.2014.1769

Bystrzejewska-Piotrowska, G., Golimowski, J., & Urbana, P. L. 2009. Nanoparticles: Their potential toxicity, waste and environmental management. *Waste Manage*, 29 (9): 2587–2595.

Cho, M., & Kim, S. 2019. Multiple lines of evidence for independent origin of wild and cultivated flowering cherry (Prunus yedoensis). *Frontiers in Plant Science*, 10: 1–16.

Crews, C. 2014. Natural toxicants: Alkaloids. In Motarjemi, Y. (Ed.), *Encyclopedia of Food Safety*, Vol. 2, Academic Press, Waltham, MA, 251–260.

De la Rosa, L. A., Moreno-Escamilla, J. O., Rodrigo-García, J., & Alvarez-Parrilla, E. 2019. Chapter 12—Phenolic Compounds. In Elhadi, M. (Ed.), *Postharvest Physiology and Biochemistry of Fruits and Vegetables*, Woodhead Publishing, Sawston, 253–271. https://doi.org/10.1016/C2016-0-04653-3

Dear, J. W. 2014. Chapter 40—Poisoning. In Marshall, W. J., Lapsley, M., Day, A. P., & Ayling, R. M. (Eds.), *Clinical Biochemistry: Metabolic and Clinical Aspects* (3rd ed.), Churchill Livingstone, London, 787–807. https://doi.org/10.1016/B978-0-7020-5140-1.00040-7

Del Cueto, J., Ionescu, I. A., Pičmanová, M., Gericke, O., Motawia, M. S., Olsen, C. E., Campoy, J. A., Dicenta, F., Møller, B. L., & Sánchez-Pérez, R. 2017. Cyanogenic glucosides and derivatives in almond and sweet cherry flower buds from dormancy to flowering. *Frontiers in Plant Science*, 8: 800.

Dey, A., Mukherjee, A., & Chaudhury, M. 2017. Chapter 10—Alkaloids from apocynaceae: Origin, pharmacotherapeutic properties, and structure-activity studies. In Atta-ur-Rahman (Ed.), *Studies in Natural Products Chemistry*, Elsevier, 52: 373–488.

Dey, P., Kundu, A., Kumar, A., Gupta, M., Lee, B. M., Bhakta, T., Dash, S., & Kim, H. S. 2020. Analysis of alkaloids (indole alkaloids, isoquinoline alkaloids, tropane alkaloids). *Recent Advances in Natural Products Analysis*, 505–567. https://doi.org/10.1016/B978-0-12-816455-6.00015-9

Dykman, L. A., & Khlebtsov, N. G. 2011. Gold nanoparticles in biology and medicine: Recent advances and prospects. *Acta Naturae*, 3 (2): 34–55.

Emanuel, M. A., & Benkeblia, N. 2011. 4—Ackee fruit (Blighia sapida Konig). In Elhadi, M. (Ed.), *Postharvest Biology and Technology of Tropical and Subtropical Fruits*, Woodhead Publishing, 54–64. https://doi.org/10.1533/9780857092762.54

Enfield, B., Brooks, D. E., Welch, S., Roland, M., Klemens, J., Greenlief, K., Olson, R., & Gerkin, R. D. 2018. Human plant exposures reported to a regional (southwestern) poison control center over 8 years. *Journal of Medical Toxicology: Official Journal of the American College of Medical Toxicology*, 14 (1): 74–78.

Fahmy, S. A., Preis, E., Bakowsky, U., & Azzazy, H. 2020. Platinum nanoparticles: Green synthesis and biomedical applications. *Molecules (Basel, Switzerland)*, 25 (21): 4981.

Farooq, U., Nisar, S., Merzaia, A., & Azeem, M. 2017. Isolation of bioactive components from Calotropis procera plant latex—A review. *International Journal of Chemical and Biochemical Sciences*, 11: 95–101.

Firdhouse, M. J., & Lalitha, P. 2012. Green synthesis of silver nanoparticles using the aqueous extract of Portulaca oleracea (L.). *Asian Journal of Pharmaceutical and Clinical Research*, 6 (1): 92–94.

Heydari, M., Hashempur, M. H., Daneshfard, B., & Mosavat, S. H. 2019. Chapter 4—Bioactive foods as dietary intervention for diabetes from the perspective of Persian medicine. In Watson, R. R., & Preedy, V. R. (Eds.), *Bioactive Food as Dietary Interventions for Diabetes* (2nd ed.), Academic Press, 49–68. ://doi.org/10.1016/B978-0-12-813822-9.00004-7

Hudlikar, M., Joglekar, S., Dhaygude, M., & Kodam, K. 2012. Green synthesis of TiO2 nanoparticles by using aqueous extract of Jatropha curcas L. latex. *Materials Letters*, 75: 196–199.

Husen, A. 2019. Natural product-based fabrication of zinc oxide nanoparticles and their application. In Husen, A., & Iqbal, M. (Eds.), *Nanomaterials and Plant Potential.* Springer International Publishing AG, 193–291. https://doi.org/10.1007/978-3-030-05569-1_7

Husen, A. 2022. *Engineered Nanomaterials for Sustainable Agricultural Production, Soil Improvement and Stress Management*. Elsevier Inc., Cambridge, MA.

Husen, A., & Iqbal, M. 2019a. Nanomaterials and plant potential: An overview. In Husen, A., & Iqbal, M. (Eds.), *Nanomaterials and Plant Potential.* Springer International Publishing AG, 3–29. https://doi.org/10.1007/978-3-030-05569-1_1

Husen, A., & Iqbal, M. 2019b. *Nanomaterials and Plant Potential*. Springer International Publishing AG. DOI: 10.1007/978-3-030-05569-1

Husen, A., & Jawaid, M. 2020. *Nanomaterials for Agriculture and Forestry Applications*. Elsevier Inc. DOI: 10.1016/C2018-0-02349-X

Husen, A, & Siddiqi, K. S. 2014. Phytosynthesis of nanoparticles: Concept, controversy and application. *Nanoscale Research Letters*, 9 (229):1–24. DOI: 10.1186/1556-276X-9-229

Jaji, N., Lee, H., Hussin, M., Akil, H., Zakaria, M., & Othman, M. 2020. Advanced nickel nanoparticles technology: From synthesis to applications. Nanotechnology Reviews, 9(1): 1456–1480.

Jin-Chul, K., Madhusudhan, A., & Husen, A. 2021. *Smart Nanomaterials in Biomedical Applications*. Springer Nature Switzerland AG, Cham. DOI: 10.1007/978-3-030-84262-8

Kadam, J., Madiwale, S., Bashte, B., Dindorkar, S., Dhawal, P., & More, P. 2020. Green mediated synthesis of palladium nanoparticles using aqueous leaf extract of Gymnema sylvestre for catalytic reduction of Cr (VI). *SN Applied Sciences*, 2 (11). DOI: 10.1007/s42452-020-03663-5

Kapoor, M., Kaur, N., Sharma, C., Kaur, G., Kaur, R., Batra, K., & Rani, J. 2020. Citrullus colocynthis an important plant in Indian traditional system of medicine. *Pharmacognosy Reviews*, 14 (27): 22–27.

Kaur, S., Sharma, D., Singh, A. P., & Kaur, S. 2019. Amelioration of hepatic function, oxidative stress, and histopathologic damages by Cassia fistula L. fraction in thioacetamide-induced liver toxicity. *Environmental Science and Pollution Research International*, 26: 29930–29945.

Keeler, R. F., Van Kampen, K. R., & James, L. F. 1978. *Effects of Poisonous Plants on Livestock*. Academic Press, New York, NY, 591–600.

Kesharwani, J., Yoon, K. Y., Hwang, J., & Rai, M. 2009. Phytofabrication of silver nanoparticles by leaf extract of Datura metel: Hypothetical mechanism involved in synthesis. *Journal of Bionanoscience*, 3 (1): 39–44.

Khan, I., Kant, C., Sanwaria, A., & Meena, L. 2010. Acute cardiac toxicity of nerium oleander/indicum poisoning (kaner) poisoning. *Heart Views*, 11 (3): 115–116. DOI: 10.4103/1995-705X.76803

Kifle, Z. D., Atnafie, S. A., Yimer Tadesse, T., Belachew, T. F., & Kidanu, B. B. 2021. Methanolic crude extract of Hagenia abyssinica possesses significant antidiarrheal effect: Evidence for in vivo antidiarrheal activity. In Mancianti, F. (Ed.), *Evidence-Based Complementary and Alternative Medicine*, Hindawi, 2021: 1–8. https://doi.org/10.1155/2021/9944629

Kiran, S., Rafique, M. A., Iqbal, S., Nosheen, S., Naz, S., & Rasheed, A. 2020. Synthesis of nickel nanoparticles using Citrullus colocynthis stem extract for remediation of Reactive Yellow 160 dye. *Environmental Science and Pollution Research International*, 27: 32998–33007.

Klaassen, C. D., Halstead, B. W., & Wong, K. L. 2020. Poison. *Encyclopedia Britannica*. Accessed 12-11-2020. https://www.britannica.com/science/poison-biochemistry

Knupp, S. N. R., Knupp, L. S., Riet-Correa, F., & Lucena, R. B. 2016. Plants that cause photosensitivity in ruminants in Brazil. *Semina: Ciências Agrárias,* 37(4): 2009. DOI: 10.5433/1679-0359.2016v37n4p2009

Krenzelok, E. P., & Mrvos, R. 2011. Friends and foes in the plant world: A profile of plant ingestions and fatalities. *Clinical Toxicology (Philadelphia),* 49 (3): 142–149.

Kumar, A., Choudhary, A., Kaur, H., Guha, S., Mehta, S., & Husen, A. 2022. Potential applications of engineered nanoparticles in plant disease management: A critical update. *Chemosphere,* 295: 133798. DOI: 10.1016/j.chemosphere.2022.133798

Kumar, A., Choudhary, A., Kaur, H., Mehta, S., & Husen, A. 2021. Metal-based nanoparticles, sensors and their multifaceted application in food packaging. *Journal of Nanobiotechnology* 19(256): 1–25 https://doi.org/10.1186/s12951-021-00996-0

Kurath, M., & Maasen, S. 2006. Toxicology as a nanoscience?—Disciplinary identities reconsidered. *Partcile and Fibre Toxicology*, 3: 6. DOI: 10.1186/1743-8977-3-6

Lu, H., Wang, S. S., Zhou, Q. W., Zhao, Y. N., & Zhao, B. Y. 2012. Damage and control of major poisonous plants in the western grasslands of China—A review. *The Rangeland Journal*, 34: 329–339.

Malangu, N. 2014. Contribution of plants and traditional medicines to the disparities and similarities in acute poisoning incidents in Botswana, South Africa and Uganda. *African Journal of Traditional, Complementary, and Alternative Medicines: AJTCAM*, 11(2): 425–438.

Mata, R., Bhaskaran, A., & Sadras, S. R. 2016. Green-synthesized gold nanoparticles from Plumeria alba flower extract to augment catalytic degradation of organic dyes and inhibit bacterial growth. *Particuology*, 24, 78–86. DOI: 10.1016/j.partic.2014.12.014

Mbaveng, A. T., Hamm, R., & Kuete, V. 2014. 19—Harmful and protective effects of terpenoids from African medicinal plants. In Kuete, V. (Ed.), *Toxicological Survey of African Medicinal Plants*, Elsevier, 557–576. https://doi.org/10.1016/B978-0-12-800018-2.00019-4

Murthy, H. C. A., Abebe, B., Prakash, C. H., & Shantaveerayya, K. A. 2018. A review on green synthesis of Cu and Cuo nanomaterials for multifunctional applications. *Material Science Research India*, 15 (3). DOI: 10.13005/msri/150311

Murthy, H. C. A., Desalegn, T., Kassa, M., Abebe, B., & Assefa, T. 2020. Synthesis of green copper nanoparticles using medicinal plant hagenia abyssinica (Brace) JF. Gmel.leaf extract: Antimicrobial properties. *Journal of Nanomaterials*, 11: 1–12.

Naseer, M., Aslam, U., Khalid, B., & Chen, B. 2020. Green route to synthesize zinc oxide nanoparticles using leaf extracts of Cassia fistula and Melia azadarach and their antibacterial potential. *Scientific Reports,* 10: 9055.

Nejat, N., Valdiani, A., Cahill, D., Tan, Y. H., Maziah, M., & Abiri, R. 2015. Ornamental exterior versus therapeutic interior of Madagascar periwinkle (Catharanthus roseus): The two faces of a versatile herb. *The Scientific World Journal*, 2015: 982412.

Odeh, F., Al-Jaber, H., & Khater, D. 2014. Nanoflora—How nanotechnology enhanced the use of active phytochemicals. In Sezer A. D. (Ed.), *Application of Nanotechnology in Drug Delivery*, IntechOpen, London.

Oluwaniyi, O. O., Adegoke, H. I., Adesuji, E. T., Alabi, A. B., Bodede, S. O., Labulo, A. H., & Oseghale C. O. 2016. Biosynthesis of silver nanoparticles using aqueous leaf extract of Thevetia peruviana Juss and its antimicrobial activities. *Applied Nanoscience,* 6: 903–912.

Oluwaniyi, O. O., & Samuel, A. I. 2007. African extractability of Thevetia peruviana glycosides with alcohol mixture. *Journal of Biotechnology,* 6: 2166–2170.

Oskoueian, E., Abdullah, N., Ahmad, S., Saad, W. Z., Omar, A. R., & Ho, Y. W. 2011. Bioactive compounds and biological activities of Jatropha curcas L. kernel meal extract. *International Journal of Molecular Sciences*, 12 (9): 5955–5970.

Panter, K. E., Gardner, D. R., Lee, S. T., Pfister, J. A., Ralphs, M. H., Stegelmeier, B. L., & James, L. F. 2007. Chapter 66—Important poisonous plants of the United States. In Gupta R. C. (Ed.), *Veterinary Toxicology: Basic and Clinical Principles*, Elsevier, Cambridge, MA, 825–872.

Panter, K. E., Welch, K. D., & Gardner, D. R. 2019. Chapter 37—Poisonous plants: Biomarkers for diagnosis. In Gupta, R. C. (Ed.), *Biomarkers in Toxicology* (2nd ed.), Academic Press, Cambridge, MA, 627–652.

Perveen, S. 2018. Introductory chapter: Terpenes and terpenoids. In Perveen, S., & Al-Taweel, A. (Eds.), *Terpenes and Terpenoids*, IntechOpen, London, 1–13. DOI: 10.5772/intechopen.79683

Phan, T. V. V., Huynh, T., Manivasagan, P., Mondal S., & Oh, J. 2020. An up-to-date review on biomedical applications of palladium nanoparticles. *Nanomaterials*, 10 (1): 66.

Ponarulselvam, S., Panneerselvam, C., Murugan, K., Aarthi, N., Kalimuthu, K., & Thangamani, S. 2012. Synthesis of silver nanoparticles using leaves of Catharanthus roseus Linn. G. Don and their antiplasmodial activities. *Asian Pacific Journal of Tropical Biomedicine*, 2 (7): 574–580.

Poppenga, R. H. 2010. Poisonous plants. In Luch, A. (Ed.), *Molecular, Clinical and Environmental Toxicology*, Experientia Supplementum, 100: 123–175. DOI: 10.1007/978-3-7643-8338-1_4

Qiu, S., Sun, H., Zhang, A. H., Xu, H. Y., Yan, G. L., Han, Y., Wang, X. J. 2014. Natural alkaloids: basic aspects, biological roles, and future perspectives. *Chinese Journal of Natural Medicines*, 12: 0401–0406.

Rasheed, T., Bilal, M., Li, C., Nabeel, F., Khalid, M., & Iqbal, H. M. N. 2018. Catalytic potential of biosynthesized silver nanoparticles using Convolvulus arvensis extract for the degradation of environmental pollutants. *Journal of Photochemistry and Photobiology B: Biology*, 181: 44–52.

Ray, A., Gulati, K., Rehman, S., Rai, N., & Anand, R. 2021. Chapter 16—Role of nutraceuticals as adaptogens. In Gupta, R. C., Lall, R., & Srivastava, A. (Eds.), *Nutraceuticals* (2nd ed.), Academic Press, 229–244.

Rupiasih, N. N., Aher, A., Gosavi, S., & Vidyasagar, P. B. 2013. Green synthesis of silver nanoparticles using latex extract of Thevetia peruviana: A novel approach towards poisonous plant utilization. *Journal of Physics: Conference Series*, 423: 012032.

Sabir, S., Arshad, M., & Chaudhari, S. K. 2014. Zinc oxide nanoparticles for revolutionizing agriculture: Synthesis and applications. *The Scientific World Journal*, 2014: 925494. DOI: 10.1155/2014/925494

Santos, D. S., Silva, C. C. B., Araújo, V. O., de Fátima Souza, M., Lacerda-Lucena, P. B., Simões, S. V. D., Riet-Correa, F., & Ricardo, B. 2017. Primary photosensitization caused by ingestion of Froelichia humboldtiana by dairy goats. *Toxicon*, 125: 65–69.

Schultheiss, P. C., Knight, A. P., Traub-Dargatz, J. L., Todd, F. G., & Stermitz, F. R. 1995. Toxicity of field bindweed (Convolvulus arvensis) to mice. *Veterinary and Human Toxicology*, 37 (5): 452–454.

Seydi, N., Saneei, S., Jalalvand, A. R., Zangeneh, M. M., Zangeneh, A., Tahvilian, R., & Pirabbasi, E. 2019. Synthesis of titanium nanoparticles using Allium eriophyllum Boiss aqueous extract by green synthesis method and evaluation of their remedial properties. *Applied Organometallic Chemistry*, 33 (11).

Shafey, A. 2020. Green synthesis of metal and metal oxide nanoparticles from plant leaf extracts and their applications: A review. *Green Processing and Synthesis*, 9 (1): 304–339.

Sharma, P., Pandey, V., Sharma, M. M. M., Patra, A., Singh, B., Mehta, S., & Husen, A. 2021. A review on biosensors and nanosensors application in agroecosystems. *Nanoscale Research Letters*, 16: 136. DOI: 10.1186/s11671-021-03593-0

Siddiqi, K. S., & Husen, A. 2020. Current status of plant metabolite-based fabrication of copper/copper oxide nanoparticles and their applications: A review. *Biomaterials Research*, 24 (11): 1–15.

Singh, J., Dutta, T., Kim, K. H., Rawat, M., Samddar, P., & Kumar, P. 2018. 'Green' synthesis of metals and their oxide nanoparticles: Applications for environmental remediation. *Journal of Nanobiotechnology*, 16 (1). DOI: 10.1186/s12951-018-0408-4

Singh, R. K., Singh, D., & Mahendrakar, A. G. 2010. Jatropha poisoning in children. *Medical Journal, Armed Forces India*, 66 (1): 80–81.

Sireeratawong, S., Jaijoy, K., Khonsung, P., Lertprasertsuk, N., & Ingkaninan, K. 2016. Acute and chronic toxicities of Bacopa monnieri extract in Sprague-Dawley rats. *BMC Complementary and Alternative Medicine*, 16: 249. DOI: 10.1186/s12906-016-1236-4

Song, J. Y., & Kim, B. S. 2008. Rapid biological synthesis of silver nanoparticles using plant leaf extracts. *Bioprocess and Biosystems Engineering*, 32 (1): 79–84.

Sriapha, C., Tongpoo, A., Wongvisavakorn, S., Rittilert, P., Trakulsrichai, S., Srisuma, S., & Wananukul, W. 2015. Plant poisoning in Thailand: A 10-year analysis from Ramathibodi poison center. *Southeast Asian Journal of Tropical Medicine and Public Health*, 46 (6): 1063–1076. Erratum in: *Southeast Asian Journal of Tropical Medicine and Public Health*, 47 (2): 334.

Srividhya, M., Hemachandran, H., Veerappapillai, S., & Karuppasamy, R. 2017. Bioactive Amento flavone isolated from Cassia fistula L. leaves exhibits therapeutic efficacy. 3 Biotech, 7 (1): 33. DOI: 10.1007/s13205-017-0599-7

Tahir, K., Nazir, S., Li, B., Khan, A. U., Khan, Z. U. H., Gong, P. Y., & Khan, S. U., & Ahmad, A. 2015. Nerium oleander leaves extract mediated synthesis of gold nanoparticles and its antioxidant activity. *Materials Letters*, 156: 198–201.

Tamilselvan, N., Thirumalaia, T., Shyamalaa, P., & David, E. 2014. A review on some poisonous plants and their medicinal values. *Journal of Acute Disease*, 3 (2): 85–89.

Tamuly, C., Hazarika, M., & Bordoloi, M. 2013. Biosynthesis of Au nanoparticles by Gymnocladus assamicus and its catalytic activity. *Materials Letters*, 108: 276–279. DOI: 10.1016/j.matlet.2013.07.020

Tessou, K. Z., Lawson-Evi, P., Metowogo, K., Diallo, A., Eklu-Gadegkeku, K., Aklikokou, K., & Gbeassor, M. 2013. Acute and sub-acute toxicity studies of Plumeria alba Linn. (Apocynaceae) hydroalcoholic extract in rat. *International Journal of Biomedical Science*, 9(4): 255–259.

Thas, J. J. 2008. Siddha medicine—Background and principles and the application for skin diseases. *Clinics in Dermatology*, 26 (1): 62–78.

Tiwari, P., Mishra, B. N., & Sangwan, N. S. 2014. Phytochemical and pharmacological properties of Gymnema sylvestre: An important medicinal plant. *BioMed Research International*, 2014: 1–18.

Trancă, S. D., Szabo, R., & Cociş, M. 2017. Acute poisoning due to ingestion of Datura stramonium—A case report. *Romanian Journal of Anaesthesia and Intensive Care*, 24 (1): 65–68.

Velmurugan, P., Shim, J., Kim, K., & Oh, B.-T. 2016. Prunus×yedoensis tree gum mediated synthesis of platinum nanoparticles with antifungal activity against phytopathogens. *Materials Letters*, 174: 61–65.

Volkow, N. D., Baler, R. D., Compton, W. M., & Weiss, S. R. 2014. Adverse health effects of marijuana use. *The New England Journal of Medicine*, 370 (23): 2219–2227.

Welch, K. D., Panter, K. E., Gardner, D. R., & Stegelmeier, B. L. 2012. The good and the bad of poisonous plants: An introduction to the USDA-ARS Poisonous Plant Research Laboratory. *Journal of Medical Toxicology: Official Journal of the American College of Medical Toxicology*, 8 (2): 153–159.

Willems, & Van den Wildenberg. 2005. *Roadmap Report on Nanoparticles*. W&W Espana sl, Barcelona, Spain.

Wink, M., & Van Wyk, B. E. 2008. *Mind-Altering and Poisonous Plants of the World*. Briza Publications (RSA), Timber Press, Portland, OR.

Woodfield, H. K., & Harwood, J. L. 2017. Oilseed crops: Linseed, rapeseed, soybean, and sunflower. In Thomas, B., Murray, B. G., & Murphy, D. J. (Eds.), *Encyclopedia of Applied Plant Sciences* (2nd ed.), Academic Press, 34–38.

Yeary, L. W., Ji-Won, M., Love, L. J., Thompson, J. R., Rawn, C. J., & Phelps, T. J. 2005. Magnetic properties of biosynthesized magnetite nanoparticles. *IEEE Transactions on Magnetics*, 41 (12): 4384–4389.

4 Conservation Practices of Poisonous Plants
Present Status

Rakesh Kumar and Ram Kumar

CONTENTS

4.1 INTRODUCTION

The natural vegetation of the world is disappearing or being altered at an alarming rate. Many cultures that have lived close to nature, depending on its products for their needs, are suffering rapid cultural, social, and economic change. These people had a deep understanding of the properties of their local plants – a knowledge that is itself endangered. Ethnobotany, a branch of science that deals with the study of how people in traditional societies use plants, has immense potential to

DOI: 10.1201/b23017-5

provide new and useful plant products for the benefit of the world. Many of the plant extracts used in western medicine were discovered through their uses in traditional societies, though not necessarily for the same purpose. Equally important, however, is how ethnobotany can help local communities adapt to changing circumstances. The practices of ethnobotany are themselves being modified to ensure that the rights of traditional people to their knowledge are safeguarded and that these people benefit from any commercial discoveries made from their knowledge. The profound knowledge on plants of traditional people in wilderness areas has received much public attention and drew attention to the herbal knowledge of the urban poor in cities across the developing world. These people brought from their villages to the cities much valuable knowledge on herbal remedies that is rarely studied. Indeed, modern health care can benefit not only from the remarkable knowledge of indigenous peoples but also from the traditional practices found in all cultures around the world.

Biodiversity management is extraordinarily complex subject which contains multiple stockholders so every aspect should be handled accordingly. Poisonous or toxic plants affect cattle and human beings as well in many ways. In cattle, the symptoms include muscular weakness, weight loss, gastrointestinal and neuromotor abnormalities, photosensitivity, bleeding, abortions, and birth defects. Sometimes sudden death also occurs without the presentation of clinical signs (Plumlee 2004). Clinical signs have been attributed to various toxic compounds, including nitrates, oxalates, cyanogenic glycosides, cardiac glycosides, monofluoroacetic acid, ptaquiloside, and alkaloids (Frohne and Pfander 2005; Diaz 2010). The death of livestock in farms due to toxic plants has negative impact on economic growth of any country, and preventive or therapeutic interventions are needed (Nielsen 1988; Riet-Correa and Medeiros 2001). The first approach to avoid the negative impact of toxicoses on cattle farms is the identification of plant species that can affect animals using ethnobotany as a source of information.

The diversity of poisonous species as a source of drugs and toxins is large, but most of their potential was not identified. One of the potentials that has not been explored is the source of toxins for the manufacture of natural pesticides (biopesticides). Due to the scope of biodiversity in treating various diseases, plants are getting damaged, destroyed, and overexploited for making drugs. Due to the overexploitation, destruction, and negligence for conservation of poisonous plants, these plants are on the verge of extinction. Every plant is important for ecosystem and every plant has its own niche. So, even though we do not know the importance of any plant, all plants must be conserved at priority. The wild areas should be marked and conserved first because these areas have rich biodiversity and endemism. The details of various poisonous plants regarding toxic part and propagation methods are given in Table 4.1.

TABLE 4.1
Details of Various Poisonous Plants and Their Propagation Methods

Botanical Name of Plant	Vernacular/Common Name	Family	Toxic Part	Propagation By
Abrus precatorius	Jequirity, crab's eye, Indian licorice, ratti	Fabaceae	Seeds	Seeds
Aconitum spp.	Aconite, wolfsbane, and monkshood	Ranunculaceae	All parts of the plant, especially the roots	Seeds
Actaea pachypoda	Doll's eyes or white baneberry	Ranunculaceae	Both the berries and the entire plant	Seeds
Adenium obesum	Sabi star, kudu, or desert rose	Apocynaceae	Sap in its roots and stems	Seed or stem cuttings

(Continued)

TABLE 4.1 (*Continued*)
Details of Various Poisonous Plants and Their Propagation Methods

Botanical Name of Plant	Vernacular/Common Name	Family	Toxic Part	Propagation By
Adonis vernalis	Pheasant's eye and false hellebore	Ranunculaceae	Extract of aerial parts	Mainly vegetatively by rhizome and seeds
Aesculus hippocastanum	Horse-chestnut	Sapindaceae	Seeds and leaves	Seeds
Agave spp.	Century plant and maguey	Asparagaceae	Sap	Seeds and bulbils
Ageratina altissima	White snakeroot	Asteraceae	Whole plant	Rhizome
Agrostemma githago	Corn cockle	Caryophyllaceae	All parts of the plant	Seeds
Allium spp.	Onion, garlic, leek	Liliaceae/ Amaryllidaceae	Bulbs	Seeds and bulb
Anemone nemorosa	Wood anemone, windflower	Ranunculaceae	Sap	Rhizomes
Anthurium spp.	Anthurium, tailflower	Araceae	Sap	Seed or vegetatively by cuttings
Aquilegia spp.	Columbine	Ranunculaceae	Seeds and roots	Seeds
Areca catechu	Betel nut palm and pinang	Arecaceae	Seeds	Seeds
Argemone mexicana	Mexican poppy, flowering thistle	Papaveraceae	Seeds	Seeds
Arnica montana	Mountain arnica	Asteraceae	Complete plant	Seed
Arum maculatum	Cuckoo-pint, lords and ladies bulls	Araceae	Complete plant	Seed
Asparagus officinalis	Asparagus	Asparagaceae	Berries	Seed
Atropa belladonna	Deadly nightshade, belladonna	Solanaceae	Berries	Seeds
Brugmansia spp.	Angel's trumpet	Solanaceae	Complete plant	Seeds and cutting
Caladium spp.	Angel wings, elephant ear, and heart of Jesus	Araceae	Complete plant	Tubers and seeds
Calla palustris	Marsh calla, wild calla	Araceae	Complete plant	Rhizome
Caltha palustris	Marsh marigold and kingcup	Araceae	Leaves	Rhizome
Cascabela thevetia	Yellow oleander	Apocynaceae	Complete plant	Seeds and cuttings
Cephalanthus occidentalis	Buttonbush	Rubiaceae	Leaves	Seeds
Cerbera odollam	Suicide tree	Apocynaceae	Fruits	Ripewood cuttings and seeds
Chelidonium majus	Greater celandine	Papaveraceae	Sap of plant	Seed
Cicuta spp.	Water hemlock, cowbane	Apiaceae	Complete plant	Seeds and bulbils
Citrus limon	Lemon	Rutaceae	Juice or sap	Seeds and cuttings
Cleistanthus collinus	Garari	Phyllanthaceae	Leaves	Nodal explant
Clivia miniata	Natal lily, bush lily, and Kaffir lily	Amaryllidaceae	Sap of leaves	Seeds and cuttings
Codiaeum variegatum	Garden croton or variegated croton	Euphorbiaceae	The bark, roots, latex, seeds, and leaves	Seeds and cuttings
Colchicum autumnale	Autumn crocus and meadow saffron	Colchicaceae	Complete plant	Seeds and bulbs

(*Continued*)

TABLE 4.1 (*Continued*)
Details of Various Poisonous Plants and Their Propagation Methods

Botanical Name of Plant	Vernacular/Common Name	Family	Toxic Part	Propagation By
Conium maculatum	Hemlock, poison hemlock bad-man's oatmeal	Apiaceae	Complete plant	Seeds
Consolida spp.	Larkspur	Ranunculaceae	All parts	Seed
Convallaria majalis	Lily of the valley	Asparagaceae	All parts	Seed and vegetative
Coriaria myrtifolia	Redoul	Coriariaceae	Leaves and fruits	Seeds
Cytisus scoparius	Broom or common broom	Fabaceae	Sap of plant parts	Seeds
Daphne spp.	Daphne	Thymelaeaceae	All parts	Seeds and cuttings
Datura spp.	Jimson weed, thorn apple, stinkweed	Solanaceae	All parts	Seeds
Daucus carota	Wild carrot	Apiaceae	Leaves	Seeds
Death camas	"Deathcamas", Toxicoscordion	Melanthiaceae	All parts	Seeds and bulbs
Delphinium spp.	Larkspur	Ranunculaceae	All parts	Seeds and cuttings
Dendrocnide moroides	Stinging tree and gympie-gympie	Urticaceae	Trichomes	Seeds and cuttings
Dicentra cucullaria	Bleeding heart and Dutchman's breeches	Papaveraceae	Sap of plant	Seeds and cuttings
Dichapetalum cymosum	Gifblaar	Dichapetalaceae	All parts	Seeds
Dieffenbachia spp.	Dumbcane	Araceae	All parts	Seeds and cuttings
Digitalis purpurea	Foxglove	Plantaginaceae	Leaf, flower, and seeds	Seeds
Dioscorea communis	Black bryony	Dioscoreaceae	All parts	Seeds, bulbils and corm
Duranta erecta	Golden dewdrop, pigeon berry	Verbenaceae	Leaves and fruits	Seeds and cuttings
Erysimum cheiri	Wallflower	Brassicaceae	Seeds	Seeds and cuttings
Euonymus europaeus	Spindle, European spindle	Celastraceous	Seeds	Seeds and cuttings
Euphorbia pulcherrima	Poinsettia	Euphorbiaceae	Leaves	Seeds and cuttings
Excoecaria agallocha	Milky mangrove, river poison tree	Euphorbiaceae	Latex	Seeds
Galanthus nivalis	Snowdrop	Amaryllidaceae	Bulbs	Offset bulbs and seeds
Gelsemium sempervirens	Yellow jessamine	Gelsemiaceae	All parts	Cutting, grafting and seeds
Gloriosa superba	FLame lily, climbing lily	Colchicaceae	Every part of the plant	Seed and rhizome
Grevillea spp.	Silky oak and spider flower	Proteaceae	Flowers	Seeds and cuttings
Hedera helix	Common ivy	Araliaceae	All parts	Seeds and cuttings
Heliotropium indicum	Indian heliotrope	Boraginaceae	Sap of plant	Seeds
Helleborus niger	Christmas rose	Ranunculaceae	Root stem and leaf	Seeds and cuttings

(*Continued*)

TABLE 4.1 (*Continued*)
Details of Various Poisonous Plants and Their Propagation Methods

Botanical Name of Plant	Vernacular/Common Name	Family	Toxic Part	Propagation By
Heracleum mantegazzianum	Giant hogweed	Apiaceae	Sap of plant	Seed
Heracleum sosnowskyi	Sosnowsky's Hogweed	Apiaceae	All parts	Seeds
Hippomane mancinella	Manchineel	Euphorbiaceae	All parts	Seeds and cuttings
Hyacinthus orientalis	Hyacinth	Asparagaceae	Bulbs	Seeds, bulbs and cutting
Hydrangea spp.	Hydrangea or hortensia	Hydrangeaceae	All parts	Seeds or stem cuttings
Hyoscyamus niger	Henbane	Solanaceae	All parts	Seeds
Ilex aquifolium	European holly	Aquifoliaceae	Berries	Seeds and cuttings
Iris sibirica	Siberian iris or Siberian flag	Iridaceae	Mainly rhizome and leaves	Seeds and cuttings
Jacobaea vulgaris	Ragwort	Asteraceae	Leaves	Crowns, roots and seeds
Kalanchoe delagoensis	Mother of millions	Crassulaceae	All parts	Little plantlets and cutting
Kalmia latifolia	Mountain laurel	Ericaceae	Flowers, twigs, and pollen	Seed
Laburnum spp.	Golden chain	Fabaceae	Seeds	Seed
Lamprocapnos spectabilis	Bleeding heart	Papaveraceae	Leaves	Seeds and cuttings
Lantana camara	Big-sage, wild-sage, and tickberry	Verbenaceae	Flowers, fruits, leaves	Seeds and cuttings
Ligustrum spp.	Several species, commonly known as privet	Oleaceae	Leaves and fruits	Seeds and cuttings
Lilium spp.	Lily	Liliaceae	All parts	Seed and bulbs
Lolium temulentum	Darnel or poison ryegrass	Poaceae	Seeds	Seeds
Lupinus spp.	Lupin or lupine	Fabaceae	Seeds and leaves	Seeds
Malus domestica	Apple	Rosaceae	Seeds	Seeds, grafting and budding
Malus florentina	Florentine crab apple	Rosaceae	Seeds and leaves	Seeds
Mandragora officinarum	Mandrake	Solanaceae	Root and leaves	Seeds and offset
Mangifera indica	Mango tree	Anacardiaceae	Leaves, peels, stem, and sap	Seeds
Manihot esculenta	Cassava	Euphorbiaceae	Root and leaves	Seeds and cuttings
Melia azedarach	Chinaberry tree, Cape lilac	Meliaceae	Seeds	Seeds
Melianthus major	Honeybush	Francoaceae	All parts	Seeds and cuttings
Menispermum spp.	Moonseed	Menispermaceae	All parts	Seeds or runners
Mentha pulegium	Pennyroyal or pennyrile	Lamiaceae	Leaves	Seeds and cuttings
Myristica fragrans	Nutmeg	Myristicaceae	Seeds	Seeds
Narcissus spp.	Daffodil	Amaryllidaceae	Bulb and leaves	Bulbs
Nerium oleander	Oleander	Apocynaceae	Leaves and flowers	Seeds, cuttings and layering
Nicandra physalodes	Apple-of-Peru and shoo-fly plant	Solanaceae	All parts	Seeds

(*Continued*)

TABLE 4.1 (*Continued*)
Details of Various Poisonous Plants and Their Propagation Methods

Botanical Name of Plant	Vernacular/Common Name	Family	Toxic Part	Propagation By
Nicotiana glauca	Tree tobacco	Solanaceae	All parts	Cuttings or by seed
Oenanthe crocata	Hemlock water dropwort	Apiaceae	Root	Seed
Paris quadrifolia	Herb-paris	Melanthiaceae	Berry	Seeds and rhizome
Passiflora caerulea	The blue passion flower	Passifloraceae	Leaves and roots	Seeds or stem cuttings
Peucedanum galbanum	Blister bush	Apiaceae	Fruits, leaves, and stem	Seeds
Phaseolus lunatus	Lima bean or butter bean	Fabaceae	Legume	Seeds
Phoradendron spp.	American mistletoe	Santalaceae	All parts	Berries on host plant as it is parasitic
Physostigma venenosum	Calabar beans and ordeal beans	Fabaceae	Seeds	Seeds and rhizome
Phytolacca spp.	Pokeweed	Phytolaccaceae	Leaves, stems, roots, blossoms, berries	Seeds
Pieris japonica	Japanese pieris	Ericaceae	Leaves, stems, and flowers	Seeds and cuttings
Plumeria spp.	Frangipani	Apocynaceae	All parts	Seeds and cuttings
Podophyllum peltatum	Mayapple	Berberidaceae	All parts	Roots and seeds
Prunus cerasus	Cherry	Rosaceae	Seeds	Seeds
Prunus laurocerasus	Cherry laurel, common laurel	Rosaceae	Leaves and seed	Seeds and cuttings
Prunus padus	Bird cherry, hackberry	Rosaceae	Leaves, stems, and fruits	Seeds and cuttings
Pteridium aquilinum	Eagle fern, Bracken	Dennstaedtiaceae	Rhizome and fronds	Rhizome and spores
Pulsatilla cernua	Pasque flower, wind flower	Ranunculaceae	Flowers	Seeds
Quercus spp.	Oak	Fagaceae	The leaves and acorns	Seeds and cuttings
Rhamnus cathartica	Buckthorn	Rhamnaceae	Seeds and leaves	Seeds and root collar cuttings
Rheum rhaponticum	Rhubarb	Polygonaceae	Leaves	Seeds
Rhododendron ferrugineum	Alpenrose, snow-rose	Ericaceae	Leaves	Seeds and cuttings
Rhododendron luteum	Yellow azalea or honeysuckle azalea	Ericaceae	All parts	Seeds and cuttings
Rhododendron tomentosum	Marsh Labrador tea or wild rosemary	Ericaceae	All parts	Seeds and cuttings
Rhododendron spp.	Azalea	Ericaceae	All parts	Seeds and cuttings
Rhus spp.	African sumac	Anacardiaceae	Sap and the fruit	Seeds and root cuttings
Ricinus communis	Castor oil plant, castor bean	Euphorbiaceae	Seeds	Seeds
Robinia spp.	Black locust and false acacia	Fabaceae	All parts except flowers	Seeds
Sambucus spp.	Elder or elderberry	Adoxaceae	All parts	Seeds and cuttings

(*Continued*)

TABLE 4.1 (*Continued*)
Details of Various Poisonous Plants and Their Propagation Methods

Botanical Name of Plant	Vernacular/Common Name	Family	Toxic Part	Propagation By
Sanguinaria canadensis	Bloodroot	Papaveraceae	Rhizome	Seed or root division
Scopolia carniolica	European scopolia or henbane bell	Solanaceae	Roots	Seeds
Solanum dulcamara	Bittersweet nightshade	Solanaceae	The leaves and stems	Seed and layering
Solanum nigrum	Black nightshade	Solanaceae	Unripened berries	Seeds
Solanum pseudocapsicum	Jerusalem cherry, Madeira winter Cherry, and winter cherry	Solanaceae	All parts	Seed or cuttings
Sophora secundiflora	Mescal bean and Texas mountain Laurel	Fabaceae	Seeds	Seeds
Strychnos nux-vomica	The strychnine tree	Loganiaceae	Seeds and bark	Seeds
Taxus baccata	English yew, common yew, and graveyard tree	Taxaceae	All parts	Seed or cuttings
Toxicodendron spp.	*Toxicodendron radicans* (poison ivy), *Toxicodendron diversilobum* (poison oak)	Anacardiaceae	All parts	Creeping rhizomes or by seed
Urtica ferox	Ongaonga	Urticaceae	Spines	Seeds
Veratrum spp.	False hellebore and corn lily	Melanthiaceae	All parts	Seed or cuttings
Vernicia fordii	The tung tree	Euphorbiaceae	All parts	Bud grafting and seeds
Viscum spp.	European mistletoe	Santalaceae	Fruits	Pushing the seeds into a host Plant's bark as it is a hemiparasite
Voacanga africana	—	Apocynaceae	Bark and seeds	Seed or cuttings
Wisteria sinensis	Chinese wisteria	Fabaceae	Seed pods and seeds	Seed or cuttings
Xanthium spp.	Cocklebur	Asteraceae	The seedlings and seeds	Seeds
Zantedeschia aethiopica	Calla lily or arum lily	Araceae	All parts	Seed
Zigadenus glaberrimus	Death camas	Melanthiaceae	All parts	Seed or cuttings

4.2 NEED FOR CONSERVATION OF POISONOUS PLANTS

Biodiversity is the key indicator of the health of an ecosystem. A wide variety of species will cope better with threats than a limited number of them in large populations and makes the ecosystem more stable. Even if certain species are affected by pollution, climate change, or human activities, the ecosystem may adapt and survive. But the extinction of a species may have unforeseen impacts, sometimes snowballing into the destruction of entire ecosystems. Because every plant/animal species is indirectly dependent on some other species, if such species get threatened, ultimately other species get affected.

Biodiversity loss has been on international policy agenda for several decades now, yet the accelerating decline (MA 2005) has not been stopped and we still face many challenges regarding biodiversity conservation (Pimm et al. 1995; Stokstad 2010). There are numerous strategies to reverse the biodiversity decline, ranging from economic, through ecological, to ethical (Rands et al. 2010). Both the rationale for biodiversity conservation action and its success vary, depending on the paradigms represented by various professionals in charge of conservation, as well as social, cultural, and political context (Wilshusen et al. 2002; Waylen et al. 2010). The value of biodiversity has also been framed in different terms, ranging from purely intrinsic to various kinds of assigned values, including use and non-use instrumental values (Ehrlich and Ehrlich 1992; Raffaelli et al. 2009). Likewise, the arguments for biodiversity conservation have varied depending on the underlying reasons for maintaining biodiversity in general (Blicharska & Grandin 2015).

The diversity of poisonous plants is needed to be conserved for several reasons. Plant germplasm conservation is urgently needed because some species are going to become extinct, and many others are threatened and endangered. There is always a need to conserve plant germplasm because of human dependence on them for many different uses and indeed for survival. In fact, every plant at least helps in providing life giving oxygen, fixing carbon dioxide, and checking soil erosion. So far, only 10% of plants have been evaluated for their agricultural and medicinal potential, and so there are certainly many new drugs and new crops yet to be discovered. It is needed to leave our future options open by having as many living plant species available as possible. A great diversity of plants is needed to keep various natural ecosystems functioning and stable. No organism exists alone but all depend on a multitude of interactions that relate them together. The cultivated plants are still dependent upon the broad genetic base that exists in the wild relatives. Further, it is important to observe that natural ecosystem and the diversity of plant species that they contain are the source of pleasure and inspiration to many people (Bhatt and Singh 2004; Prance 1997).

From the above statements, it is clear that each plant is important for ecosystem stability and survival of various economically important species. So, all plants must be conserved whether they are economically important for us or not. Conservation aims at supporting sustainable development by using the biological resources in ways that do not deplete the world's variety of species or destroy their ecosystems. It involves measures such as collection, propagation, evaluation, disease identification and elimination, storage, and distribution. Conservation of plants and their genetic resources can be undertaken by *in situ* and *ex situ* conservation. *Ex situ* conservation involves conservation of medium plants outside their natural habitat in order to safeguard them from destruction, replacement, or deterioration. *Ex situ* conservation includes procedure like seed storage, DNA storage, field gene banks, and botanical gardens. (Sharma 2009).

4.3 PRINCIPLES OF BIODIVERSITY CONSERVATION

These principles were adopted in CBD (convention on biological diversity) for conservation of biodiversity. The following 12 principles are complementary and interlinked.

Principle 1: The objectives of management of land, water, and living resources are a matter of societal choices. Different sectors of society view ecosystems in terms of their own economic, cultural, and society needs. Indigenous people and other local communities living on the land are important stakeholders and their rights and interests should be recognized. Both cultural and biological diversity are central components of the ecosystem approach, and management should take this into account. Societal choices should be expressed as clearly as possible. Ecosystems should be managed for their intrinsic values and for the tangible or intangible benefits for humans in a fair and equitable way.

Principle 2: Management should be decentralized to the lowest appropriate level. Decentralized systems may lead to greater efficiency, effectiveness, and equity. Management should involve all stakeholders and balance local interests with the wider public interest.

The closer the management to the ecosystem, the greater the responsibility, ownership, accountability, participation, and use of local knowledge.

Principle 3: Ecosystem managers should consider the effects (actual or potential) of their activities on adjacent and other ecosystems. Management interventions in ecosystems often have unknown or unpredictable effects on other ecosystems; therefore, impacts need careful consideration and analysis. This may require new arrangements or ways of organization for institutions involved in decision making to make, if necessary, appropriate compromises.

Principle 4: Recognizing potential gains from management, there is usually a need to understand and manage the ecosystem in an economic context. Any such ecosystem management program should reduce those market distortions that adversely affect biological diversity; align incentives to promote biodiversity conservation and sustainable use; and internalize costs and benefits in the given ecosystem to the extent feasible.

The greatest threat to biological diversity lies in its replacement by alternative systems of land use. This often arises through market distortions, which undervalue natural systems and populations and provide perverse incentives and subsidies to favor the conversion of land to less diverse systems. Often those who benefit from conservation do not pay the costs associated with conservation and, similarly, those who generate environmental costs (e.g. pollution) escape responsibility.

Principle 5: Conservation of ecosystem structure and functioning, to maintain ecosystem services, should be a priority target of the ecosystem approach. Ecosystem functioning and resilience depends on a dynamic relationship within species, among species, and between species and their abiotic environment, as well as the physical and chemical interactions within the environment. The conservation and restoration of these interactions and processes is of greater significance for the long-term maintenance of biological diversity than simply protection of species.

Principle 6: Ecosystem must be managed within the limits of their functioning. In considering the likelihood or ease of attaining the management objectives, attention should be given to the environmental conditions that limit natural productivity, ecosystem structure, functioning, and diversity. The limits to ecosystem functioning may be affected to different degrees by temporary, unpredictable of artificially maintained conditions and, accordingly, management should be appropriately cautious.

Principle 7: The ecosystem approach should be undertaken at the appropriate spatial and temporal scales. The approach should be bounded by spatial and temporal scales that are appropriate to the objectives. Boundaries for management will be defined operationally by users, managers, scientists, and indigenous and local peoples. Connectivity between areas should be promoted where necessary. The ecosystem approach is based upon the hierarchical nature of biological diversity characterized by the interaction and integration of genes, species, and ecosystems.

Principle 8: Recognizing the varying temporal scales and lag effects that characterize ecosystem processes, objectives for ecosystem management should be set for the long term. Ecosystem processes are characterized by varying temporal scales and lag effects. This inherently conflicts with the tendency of humans to favor short-term gains and immediate benefits over future ones.

Principle 9: Management must recognize the change is inevitable. Ecosystems change, including species composition and population abundance. Hence, management should adapt to the changes. Apart from their inherent dynamics of change, ecosystems are beset by a complex of uncertainties and potential "surprises" in the human, biological, and environmental realms. Traditional disturbance regimes may be important for ecosystem structure and functioning and may need to be maintained or restored. The ecosystem approach must utilize adaptive management to anticipate and cater for such changes and events and should be cautious in making any decision that may foreclose options, but, at the same time, consider mitigating actions to cope with long-term changes such as climate change.

Principle 10: The ecosystem approach should seek the appropriate balance between, and integration of, conservation and use of biological diversity. Biological diversity is critical because of its intrinsic value and the key role it plays in providing the ecosystem and other services on which we all ultimately depend. There has been a tendency in the past to manage components of biological diversity either as protected or non-protected. There is a need for a shift to more flexible situations, where conservation and use are seen in context and the full range of measures is applied in a continuum from strictly protected to human-made ecosystems.

Principle 11: The ecosystem approach should consider all forms of relevant information, including scientific and indigenous and local knowledge, innovations, and practices. Information from all sources is critical to arriving at effective ecosystem management strategies. A much better knowledge of ecosystem functions and the impact of human use are desirable. All relevant information from any concerned area should be shared with all stakeholders and actors. Assumptions behind proposed management decisions should be made explicit and checked against available knowledge and views of stakeholders.

Principle 12: The ecosystem approach should involve all relevant sectors of society and scientific disciplines. Most problems of biological diversity management are complex, with many interactions, side effects, and implications, and therefore should involve the necessary expertise and stakeholders at the local, national, regional, and international level, as appropriate.

4.4 FACTORS RELATED TO SPECIES RARITY OF MEDICINAL/POISONOUS PLANTS

Species rarity is used to assess the extinction risk of important plants and to identify those species most at risk of extinction prior to commencement of conservation efforts (Figueiredo and Grelle 2009). It is necessary to determine how rare each species is and in which ways rare species differ from one another. Not all plants are affected in the same way by harvesting pressures (Wagh and Jain 2013; Andel and Havinga 2008). Overexploitation, indiscriminate collection, uncontrolled deforestation, and habitat destruction all affect species rarity, but are insufficient to explain individual species susceptibility or resilience to harvest pressure. Multiple biological characters correlate with extinction risk, such as habitat specificity, distribution range, population size, species diversity, growth rate, and reproductive system (Chen et al. 2016). Seed viability and longevity also play an important role in rarity of any species. If seed viability is low and seed cannot be stored for long time, there are ample chances of extinction of plant species.

4.5 TRADITIONAL METHODS OF CONSERVATION

People living in rural areas depend on their surrounding forests and vegetation for fuel wood and medicine. This population cannot afford alternative fuel and expensive modern medicines. Hence, the environmental degradation and ecological loss take place for which serious measures must be undertaken. Conservation of medicinal/poisonous plants can be done by discouraging people for cutting down trees and encouraging them to plant fast-growing indigenous trees for their domestic use. But this alone cannot majorly lead to conservation. Therefore, other newer conservation strategies must be planned and brought into action. Biotechnological techniques and procedures must be used for the same. Various methods of conservation are depicted in Figure 4.1. The conservation of plant diversity can be done by the following methods.

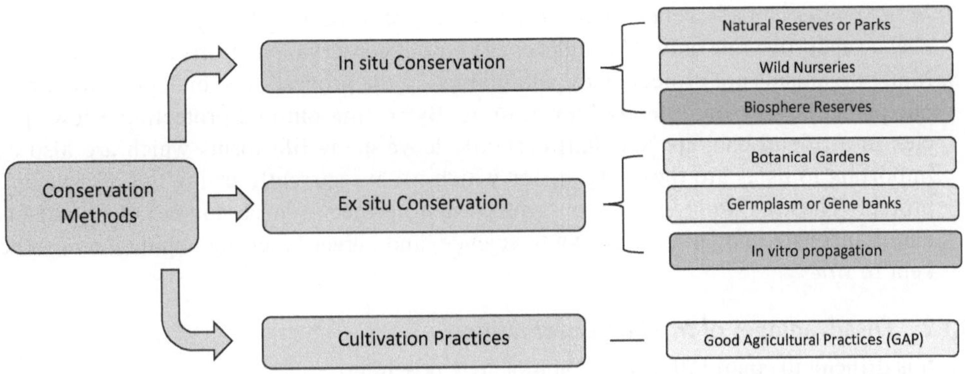

FIGURE 4.1 Schematic representation of methodological systems involved in the conservation of poisonous plants.

4.5.1 *In Situ* Conservation

There is a general perception and recommendation that biodiversity should be conserved inside their natural habitat. Species are conserved and maintained in their natural state in stable environment. *Ex situ* collections may be ineffective at preserving genetic diversity and the evolutionary potential of populations for adaptive or neutral evolution. *Ex situ* conservation led to develop some variety of non-adaptive genotypes that will depress population fitness. *In situ* conservation can be done by developing protected areas such as national park, wildlife sanctuaries, conservation reserves, and community reserves which are backbone of biodiversity conservation, while also contributing to people's livelihoods, particularly at the local level.

According to the Convention on Biological Diversity (CBD), *in situ* conservation is defined as "the conservation of ecosystems and natural habitats and the maintenance and recovery of viable populations of species in their natural surroundings and, in the case of domesticated or cultivated species, in the surroundings where they have developed their distinctive properties". *In situ* conservation aims to enable biodiversity to maintain itself within the context of the ecosystem in which it is found. *In situ* management approaches can either be species centered (targeted at populations of selected species) or ecosystem based (whole ecosystems).

4.5.1.1 Advantages of *In Situ* Conservation

1. It helps in protecting the biological diversity of the area. To ensure the survival of the species, we protect the entire natural habitat or the ecosystem.
2. Endangered species are protected against extinction.
3. It helps in restoring degraded areas by supplying biotic and abiotic components. In fact, the method involves promotion of natural system to take care of its own self – we simply provide conditions to let flourish themselves.
4. It maintains the quality of life of forest dwellers and tribals.
5. It maintains genetic diversity and provides material for genetic improvement of crops. In a natural system, organisms not only live and multiply but also evolve. A natural ecosystem allows free play of natural agencies – like drought, storms, snow, fluctuation in temperatures, excessive rains, fires, and pathogens – which provide an opportunity to the organisms to adjust to the prevailing conditions of the environment and evolve into a better adopted life form. In *ex situ* conservation, we isolate species which cause evolution process to freeze.
6. It also helps in pollution control.

7. *In situ* conservation is a cheap and convenient way of conserving biological diversity as we play a supportive role only. This reduces the cost of conservation efforts enormously.
8. Many organisms are protected and maintained in the process. The biological wealth of our planet is very imperfectly known to us. By sorting out and protecting a few species in artificial habitats, we shall certainly leave many life forms which are also as important to us as are those organisms which we are currently trying to preserve and protect. Thus, *in situ* conservation offers way to protect a large number of organisms simultaneously known or unknown to science, and bigger breeding populations can be kept *in situ*.

4.5.1.2 Disadvantages of *In Situ* Conservation

1. It is difficult to control illegal exploitation (e.g. poaching).
2. The environment may need restoring and alien species are difficult to control.
3. An important disadvantage of *in situ* conservation is that it requires large areas of earth's surface if we have to preserve the full complement of biotic diversity of a region. This involves minimizing or excluding human activity and interference from that locality which is often difficult in the face of growing demand for space.

In situ method of conservation deals with the "on-site conservation" of the wild genetic diversity in natural habitat. In India, the conservation of forest areas preserves through protected areas like national parks, wildlife sanctuaries, and biosphere reserves. *In situ* conservation of plants can be done effectively with the help of local people because they have indigenous knowledge of these plants. *In situ* conservation of plants can be done by the following means.

4.5.1.3 Natural Parks

Natural parks are protected areas of important wild resources created to preserve and restore biodiversity (Rodriguez et al. 2007; Chiarucci et al. 2001). The total number of protected area records in the April 2022 release of the World Database on Protected Areas (WDPA) is 269,667, comprising 257,388 polygons and 12,279 points and covering 245 countries and territories. Conserving medicinal/poisonous plants in their natural habitat requires evaluating the contributions and ecosystem roles of individual habitats.

4.5.1.4 Wild Nurseries

The populations of medicinal plants are under hefty burden because of overexploitation, habitat degradation, and invasive species; wild nurseries can provide an effective approach for *in situ* conservation of medicinal plants that are endemic, threatened, and in demand (Soule et al. 2005).

4.5.1.5 Biosphere Reserves

Out of 34 biodiversity rich spots in the world, three lies in India; they are the Western Ghats, the Eastern Himalayas, and Indo-Burma region. Currently, there are 18 biosphere reserves in India. These aim at securing the ecosystem by stopping irresponsible interference of humans with the ecosystem and to conserve the endemic and endangered species.

4.5.1.6 Conservation Reserves

Conservation reserves can be declared by the state governments in any area owned by the government, particularly the areas adjacent to national parks and sanctuaries and those areas which link one protected area with another. Such declaration should be made after having consultations with the local communities. Conservation reserves are declared for the purpose of protecting landscapes, seascapes, flora and fauna and their habitat. The rights of people living inside a conservation reserve are not affected.

4.5.1.7 Community Reserves

Community reserves can be declared by the State Government in any private or community land, not comprised within a national park, sanctuary, or a conservation reserve, where an individual or a community has volunteered to conserve wildlife and its habitat. Community reserves are declared for the purpose of protecting fauna, flora, and traditional or cultural conservation values and practices. As in the case of a conservation reserve, the rights of people living inside a community reserve are not affected (Kadam and Pawar 2020).

4.5.2 *Ex Situ* Conservation

Ex situ conservation deals with the off-site conservation or conservation outside of natural habitat of the wild genetic resources in natural habitat. It includes the collection, preservation, and maintenance of certain genetic resources from wild. *Ex situ* method of conservation is a complementary action to conserve the genetic diversity, thereby reducing pressure on wild environments and enhancing raw material availability. For many species of medicinal plants, since their wild population is on life-threatening level, it is not suitable for dealing *in situ* conservation action. It can serve as field gene banks and help in engaging the number of stakeholders in production and regeneration of medicinal plants. This loss of habitat makes *ex situ* conservation important method of conservation. *Ex situ* conservation is particularly important for those plants and animals which cannot be conserved by *in situ* due to distinct reasons (Kadam and Pawar 2020). There are different methods for conserving biodiversity *ex situ*.

4.5.2.1 Botanic Gardens

Botanical gardens are the most conventional methods of *ex situ* conservation, where specimens are protected for breeding and reintroduction into the wild whenever necessary and possible. Botanic gardens have had a changing role throughout history, beginning often as medicinal gardens for the study and cultivation of plants with healing properties; they have also gone through many phases including, of course, pleasure gardens, and are currently used as a site for plant conservation and people education. Botanical gardens are important places of systematic study and research on flora of the region. These are the places of great academic and economic importance. They are valuable not only to the botanists, horticulturists, and foresters but also to the students, tourists, and non-professionals. Botanical Survey of India is actively engaged in the *ex situ* conservation through its chain of botanical gardens established in different regional circles.

4.5.2.2 Gene Banks

Gene bank is a type of biorepository where genetic material is collected, stored, cataloged, and made available for redistribution. Local crop varieties, known as "landraces", are the result of generations of careful farmer selection. Due to a combination of environmental and social changes in farming communities, intensification of cultivation of a particular high yielding crop globally and inbreeding in small populations, large numbers of local crop varieties are disappearing at an alarming rate, which cause rapid decline in genetic diversity. Gene bank plays a significant role in the conservation, availability, and use of a wide range of plant genetic diversity for crop improvement for food and nutrition security. They help bridge the past and the future by ensuring the continued availability of genetic resources for research, breeding, and improved seed delivery for a sustainable and resilient agricultural system. On the basis of type of material to be conserved, gene banks are of the following types.

4.5.2.3 Seed Banks

Seed is the most convenient part of plant that helps plant to multiply. Seeds of many plant species remain viable for a long time when they are kept in reduced moisture and at low temperature.

Cryogenic preservation is suspended liveliness at sub-freezing temperature and low moisture content. A seed bank stores seeds as a source for planting in case seed reserves elsewhere are destroyed. It is a type of gene bank. The reasons for storing seeds may be varied. In the case of food crops, many useful plants that were developed over centuries are now no longer used for commercial agricultural production and are becoming rare. Storing seeds also guards against catastrophic events like natural disasters, outbreaks of disease, or war. Seed banks can be set up at the community, national, and international level. Because the crops we rely for food are grown in parts of the world distant to the centers of their domestication, the sharing of genetic material across national borders for research and plant breeding is essential. Seed banks allow quick access to plant samples for the evaluation of their properties.

4.5.2.4 Pollen Bank

This is a method in which pollen grains are stored. We can make plants which are facing extinction in the present world. Using this technique, haploid plants can be produced. Pollen can be preserved for conservation of biodiversity of important and endangered flowering plants. Pollen from flowering plants can be preserved by cryogenic technique. As these pollen grains are so tiny, millions of them can be stored in a small vial, which allows for the preservation of the full range of variations within the desired population. These stored pollens can be used in hybridization program directly whenever needed, which is not possible in the case of seeds. When seeds are subject to develop hybrids, they must be retrieved from the seed bank and then planted in the field.

4.5.2.5 DNA Bank

DNA banking is the secure, long-term storage of an individual's genetic material. In plants, DNA can be extracted from any part. Plant material is dried using silica gel and stored at −800°C to extract DNA from it. When fresh supplies are not available, DNA can be obtained from dried plant material stored in botanical research institutions that is house pressed, and the dried plant specimens are called "herbarium". DNA banks can serve many purposes in today's society. DNA banks allow for conservation of genetic material and comparative analysis of an individual's genetic information.

4.5.2.6 Field Gene Banks

Field gene banks are the repository where alive plants are maintained to fulfill the need of future generations. This method is useful for those species which have low seed viability. Germplasm conservation of vegetative propagated plants and forest species in life field gene banks require land and labor for annual or perennial replanting. On the other hand, in vitro or reduced growth storage requires little space in growth rooms which maintain thousands of genotypes together and reduce pest attack and diseases in the culture vessels. It further eliminates the need for long procedures during movement and exchange of germplasms (Kasagana and Karumuri 2011).

4.5.2.7 Cultivation Practice

This method can be very useful if plant species on high risk of extinction is encouraged to cultivate on priority basis. This can be done through either monocropping or mixed cropping systems. To meet the increasing demand, conservation and cultivation of medicinal/poisonous plants has become important. Conservation involves imposing certain regulations on their collection from the wild. If such irrational collection continues, many plant species which are widely available today will become endangered. Some may even become extinct. For their conservation, the proper cultivation methods must be encouraged. Plants collected from the wild normally vary in composition and quality due to the environmental and genetic differences. This variation and the uncertainty in their therapeutic benefit can be much reduced by their proper cultivation. The possibility of misidentification and adulteration also greatly reduces due to cultivation. Thus, medicinal/poisonous plant cultivation can help in the conservation of the wild stocks and can fulfill the global needs (Zhang 2003).

4.6 MODERN TECHNIQUES OF CONSERVATION

4.6.1 Cryobank or Cryopreservation Technique of Conservation

This is the most advanced and effective technique for storage of germplasm of various economically important plants. Most threatened plants can be conserved using this modern technique. The procedure is shown in Figure 4.2. Cryopreservation is defined as the viable freezing of biological material (seed or embryo) and their subsequent storage at ultra-low temperature (–196°C) using liquid nitrogen (Kasagana and Karumuri 2011). The conventional methods of storage fail to prevent from losses caused by attack of pathogens and pests, climatic disorders, and natural disorders. However, the conventional methods could not save the viability of short-lived seeds of economic plants and these materials can be successfully stored by cryogenic techniques due to which growth rate of cells retards; consequently, biological activities are conserved for a long time. This is helpful for the conservation of species facing extinction.

The cryopreservation practices are effective and viable for long-term benefits because it requires an initial technological investment but the maintenance costs for the application of different techniques will be lower, later. The use of cryopreservation facilitates storage and rapid multiplication of plant germplasm in a pathogen-free aseptic environment as well as optimization of physical space and labor. In a perspective of conservation strategy, the introduction of an accession into cryopreserved storage is more expensive than establishing an accession in in vitro culture or in the field, but the cryopreservation costs for the long-term vision (over 20 years) are considerably lower than those of maintenance in the field or in vitro, particularly when many accessions are preserved (Benelli 2021).

The transfer technology and validating protocols between laboratories are a complicated process (Reed et al. 2004). The critical factors involved in the cryopreservation like type of plant materials, conditions of preculture, cryopreservation technique, cooling, warming, and regrowth

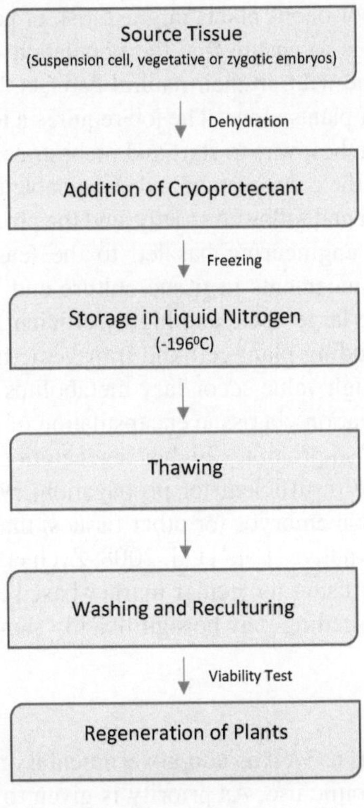

FIGURE 4.2 Steps of cryopreservation for poisonous plants.

conditions also make it difficult (Bettoni et al. 2021). The lack of reproducibility available protocols or the difficulty in adapting them can be due to numerous causes, ranging from different sources of laboratory supplies to different equipment and different levels of technical skills found in cryopreservation laboratories. The development and dissemination of increasingly simple and well-described protocols (Engelmann 2004), with adequate facilities and trained personnel, will allow new challenges in each cryopreservation laboratory or institution as well as the implementation of the cryopreservation procedures in cryobanks and for biological materials and organisms important for research and in applications (Li et al. 2021).

4.6.2 Tissue Culture Techniques Used for Conservation

Plant tissue culture is the in vitro aseptic culture of cells, tissues, organs, or whole plant under controlled nutritional and environmental conditions (Thorpe 2007). When plant species are at high risk and regeneration through conventional methods is difficult, tissue culture is applied to propagate them and prevent them from extinction. This method provides opportunity to multiply the plant in short time span and thus helpful for plants at high risk of extinction.

Plant tissue culture technique has been utilized successfully for the conservation of various toxic or poisonous plants by several workers like *Ricinus communis* (Nahar and Borna 2012), *Datura inoxia* (Waoo et al. 2013), *Abrus precatorius* (Shilwant et al. 2020), *Gloriosa superba* (Custers and Bergervoet 1994), *Aconitum balfourii* (Bist et al. 2011), *Asparagus officinalis* (Hasbulla 2019), *Malus domestica* (Grea et al. 2021), and *Nerium oleander* (Hatzilazarou et al. 2019).

4.7 PROSPECTS FOR CONSERVATION OF POISONOUS PLANTS

According to UNEP/FAO, plant diversity is being eroded at alarming rate. This is mostly due to overexploitation, habitat destruction, and unawareness among people about the importance of plant diversity. This is putting many medicinal/poisonous plants in grave risk of genetic erosion and even extinction. The best means of conservation is to ensure that the populations of species of plants and animals continue to grow and evolve in the wild, in their natural habitats. No single sector, private or public, can undertake the conservation of plants alone. The job requires a team effort, involving a wide range of disciplines and institutions. The best way to start and orchestrate such a process is for each country to prepare a national strategy for the conservation and sustainable use of its plants of various uses. A proper checklist should be formed and followed strictly, and the progress should be monitored as well.

The development of genetic engineering has led to the feasibility of large-scale biosynthesis of natural products, and advancements in tissue culture and fermentation of medicinal plants have opened new avenues for the large-scale and highly efficient production of desirable bioactive compounds. Tissue culture (including plant cell and transgenic hairy root culture) is a promising alternative to produce rare and high-value secondary metabolites of medical importance (Rao and Ravishankar 2002). Micropropagation via tissue encapsulation of propagules can not only facilitate storage and transportation, but also promotes higher regeneration rates (Baker et al. 2007). When the amounts of normal seeds are insufficient for propagation, synthetic seed technology, defined as artificially encapsulated somatic embryos (or other tissues) that could be used for cultivating in vitro or ex vitro, is a feasible alternative (Lata et al. 2008; Zych et al. 2005). Furthermore, breeding improvements can be carried out using molecular marker-based approaches applied at the genetic level, and the time required for breeding may be significantly shortened.

4.8 CONCLUSION

Various government organizations as well as non-governmental organizations are putting efforts to conserve the plants of high economic use. As priority is given to such plants, other plants remain unattended and facing high risk of extinction. Now, it is time to understand the importance of each plant and to put the efforts for its conservation. Despite the existence of various sets of recommendations for the conservation and sustainable use of plants, only a small portion of these have achieved adequate protection of plant resources through conventional conservation method. Now, latest advanced techniques like tissue culture, genetic engineering, and cryopreservation should be used on priority basis to conserve the plant species at high risk. Genetic engineering may be helpful in introducing the genes for high adaptation in local environments. Further, policies should be made and implemented effectively to conserve the plants on the verge of extinction.

NOTES

Rakesh Kumar, Department of Botany, Vijay Singh Pathik Govt. (P.G.) College, Kairana, India
Ram Kumar, Department of Botany, Vijay Singh Pathik Govt. (P.G.) College, Kairana, India

REFERENCES

Andel, T. V., & Havinga, R. 2008. Sustainability aspects of commercial medicinal plant harvesting in Suriname. *Forest Ecol Manag* 256: 1540–1545.

Baker, D. D., Chu, M., Oza, U., & Rajgarhia, V. 2007. The value of natural products to future pharmaceutical discovery. *Nat Prod Rep* 24: 1225–1244.

Benelli, C. 2021. Plant cryopreservation: A look at the present and the future. *Plants (Basel, Switzerland)* 10(12): 2744. doi: 10.3390/plants10122744.

Bettoni, J. C., Bonnart, R., & Volk, G. M. 2021. Challenges in implementing plant shoot tip cryopreservation technologies. *Plant Cell Tissue Organ Cult* 144: 21–24. doi: 10.1007/s11240-020-01846-x.

Bhatt, K. C., & Singh, R.K. 2004. On–farm in situ management of crop genetic resources. In: Dhillon, B. S., Tyagi, R. K., Arjun, L., & Saxena, S., eds. *Plant Genetic Resource Management.* Narosa Publishing House, New Delhi, India, p. 154.

Bist, R., Pathak, K., Hatwal, D., Punetha, H., & Gaur, A. K. 2011. In vitro propagation of Aconitum balfourii Stapf: A rare medicinal herb of the alpine Himalayas. *Indian Journal of Horticulture* 68(3): 394–398.

Blicharska, M., & Grandin, U. 2015. Why protect biodiversity? Perspectives of conservation professionals in Poland. *Int J Biodivers Sci Ecosyst Serv Manag* 11: 349–362. doi: 10.1080/21513732.2015.1050969.

Chen, S. L., Yu, H., Luo, H. M., Wu, Q., Li, C. F., & Steinmetz, A. 2016. Conservation and sustainable use of medicinal plants: Problems, progress, and prospects. *Chin Med* 11: 37. doi: 10.1186/s13020-016-0108-7.

Chiarucci, A., Maccherini, S., & De Dominicis, V. 2001. Evaluation and monitoring of the flora in a nature reserve by estimation methods. *Biol Conserv* 101: 305–314.

Custers, J. B. M., & Bergervoet, J. H. W. 1994. Micropropagation of Gloriosa: Towards a practical protocol. *Scientia Horticulturae* 57(4): 323–334.

Diaz, G. J., ed. 2010. *Plantas Tóxicas de Importancia en Salud y Producción Animal en Colombia.* Universidad Nacional de Colombia, Bogotá, Colombia.

Ehrlich, P. R., & Ehrlich, A. H. 1992. The value of biodiversity. *Ambio* 21: 219–226.

Engelmann, F. 2004. Plant cryopreservation: Progress and prospects. *Vitr Cell Dev Biol-Plant* 40: 427–433. doi: 10.1079/IVP2004541.

Figueiredo, M. S. L., & Grelle, C. E. V. 2009. Predicting global abundance of a threatened species from its occurrence: Implications for conservation planning. *Divers Distrib* 15: 117–121.

Frohne, D., & Pfander, H. J. 2005. *A Handbook for Pharmacists, Doctors, Toxicologists, Biologists and Veterinarians.* Timber Press, Portland, OR.

Grea, D. Z., Wuhibe, Y. A., & Zewdu, G. A. 2021. Establishment of an efficient In vitro propagation protocol for rapid regeneration of Apple (Malus domestica Borkh) root stock and scion via node culture. *Plant Cell Biotechnol Mol Biol* 22: 108–123.

Hasbullah, N. A. 2019. Micropropagation of asparagus officinalis L. (Garden Asparagus) in vitro. *Int J Life Sci Res* 7(3): 123–129.

Hatzilazarou, S., Kostas, S., & Economou, A. S. 2019. Plant regeneration of Nerium oleander L. from alginate-encapsulated shoot explants after short-term cold storage. *J Hortic Sci Biotechnol* 94(4): 441–447.

Kadam, S., & Pawar, A. 2020. Conservation of medicinal plants: A review. *Int Ayurvedic Med J* 8: 3890–3895. doi: 10.46607/iamj0807112020.

Kasagana, V. N., & Karumuri, S. S. 2011. Conservation of medicinal plants (past, present & future trends). *J Pharm Sci Res* 3(8): 1378–1386.

Lata, H., Chandra, S., Khan, I. A., & Elsohly, M. A. 2008. Propagation of Cannabis sativa L using synthetic seed technology. *Plant Med* 74: 328.

Li, H. H., Lu, J. L., Lo, H. E., Tsai, S., & Lin, C. 2021. Effect of cryopreservation on proteins from the ubiquitous marine dinoflagellate Breviolum sp. (Family Symbiodiniaceae). *Plants* 10: 1731. doi: 10.3390/plants10081731.

Millennium Ecosystem Assessment (MA). 2005. *Ecosystems and Human Well-Being: Global Assessment Reports.* Island Press, Washington, DC.

Nahar, K., & Borna, R. S. 2012. In vitro propagation from shoot tip explants of castor oil plant (*Ricinus communis* L): A bioenergy plant. *Can J Sci Ind Res* 3(5): 354–355.

Nielsen, D. 1988. Economic impact of poisonous plants on the rangeland livestock industry. *J Anim Sci* 66: 2330–2333.

Pimm, S. L., Russel, G. J., Gittleman, J. L., & Brooks, T. M. 1995. The future of biodiversity. *Science* 269: 347–350.

Plumlee, K. H. 2004. *Clinical Veterinary Toxicology.* Mosby Press, St. Louis, MO.

Prance, G. T. 1997. The conservation of botanical diversity. In: Maxted, N., Ford-Lloyd, B. V., & Hawkes, J. G., eds. *Plant Genetic Conservation.* Chapman and Hall, London, England, pp. 3–13.

Raffaelli, D., Smart, J., Austen, M., Mangi, S., Hattam, C., Termansen, M., Fraser, E., & Abson, D. 2009. *Valuation of Biodiversity? A NERC scoping study.* Natural Environment Research Council, Swindon, England.

Rands, M. R., Adams, W. M., Bennun, L., Butchart, S. H., Clements, A., Coomes, D., Entwistle, A., Hodge, I., Kapos, V., Scharlemann, J. P., & Sutherland, W. J. 2010. Biodiversity conservation: Challenges beyond 2010. *Science* 329(5997): 1298–1303.

Rao, S. R., & Ravishankar, G. A. 2002. Plant cell cultures: Chemical factories of secondary metabolites. *Biotechnol Adv* 20(2): 101–153.

Reed, B. M., Kovalchuk, I., Kushnarenko, S., Meier-Dinkel, A., Schoenweiss, K., Pluta, S., Straczynska, K., & Benson, E. E. 2004. Evaluation of critical points in technology transfer of cryopreservation protocols to international plant conservation laboratories. *Cryo-Letters* 25: 341–352.

Riet-Correa, F., & Medeiros, R. M. 2001. Intoxicações por plantas em ruminantes no Brasil e no Uruguai: importância econômica, controle e riscos para a saúde pública. *Pesquisa Veterinaria Brasilera* 21: 38–42.

Rodriguez, J. P., Brotons, L., Bustamante, J., & Seoane, J. 2007. The application of predictive modelling of species distribution to biodiversity conservation. *Divers Distrib* 13: 243–251.

Sharma, S. 2009. *Meghalaya, The land and forest: Remote sensing study.* NEHU, Shillong, Meghalaya.

Shilwant, S., Gahire, S., Bhalerao, P., & Pandhure, N. 2020. In vitro propagation of a rare medicinal plant *Abrus precatorious* E. May 2. *Biosci Disc* 11(3): 163–167.

Soule, M. E., Estes, J. A., Miller, B., & Honnold, D. L. 2005. Strongly interacting species: Conservation policy, management, and ethics. *Bioscience* 55: 168–176.

Stokstad, E. 2010. Despite progress biodiversity declines. *Science* 329: 1272–1273.

Thorpe, T. 2007. History of plant tissue culture. *J Mol Microbial Biotechnol* 37: 169–180.

Wagh, V. V., & Jain, A. K. 2013. Status of threatened medicinal plants of Jhabua district, Madhya Pradesh, India. *Ann Plant Sci* 2: 395–400.

Waoo, A. A., Khare, S. S., & Ganguli, S. 2013. In-vitro propagation of Datura inoxia from nodal and shoot tip explants. *World J Environ Eng* 1(1): 1–4.

Waylen, K. A., Fischer, A., McGowan, P. J. K., Thirgood, S. J., & Milner-Gulland, E. J. 2010. Effect of local cultural context on the success of community-based conservation interventions. *Conservation Biol* 24: 1119–1129.

Wilshusen, P. R., Brechin, S. R., Fortwangler, C. L., & West, P. C. 2002. Reinventing a square wheel: Critique of a resurgent "protection paradigm" in International Biodiversity Conservation. *Soc Nat Resour* 15: 17–40.

Zhang, X. 2003. *WHO Guidelines on Good Agricultural and Collection Practices (GACP) for Medicinal Plants.* World Health Organization, Geneva, Switzerland.

Zych, M., Furmanowa, M., Krajewska-Patan, A., Lowicka, A., Dreger, M., Mendlewska, S. 2005. Micropropagation of Rhodiola kirilowii plants using encapsulated axillary buds and callus. *Acta Biol Cracov Bot* 47: 83–87.

Section B

Poisonous Plants
(Botanical Description, Distribution, Phytochemical Constituents, Pharmacological Studies, Toxic Response, Traditional, and Other Potential Uses and Future Remarks)

5 *Abrus precatorius* (Rosary Pea)

Mansi Shah, Meenakshi K, and Indu Anna George

CONTENTS

5.1 INTRODUCTION

Abrus precatorius, a perennial climber (Family: Leguminosae) has many synonyms such as *Abrus abrus*, *Glycine abrus*, *Abrus cyaneus*, *Abrus maculates Noronha*, *Abrus minor*, and *Abrus pauciflorus* (Omoboyowa et al., 2021b), India Biodiversity Portal). A selection of its common names are listed in Table 5.1. Its distribution is cosmopolitan in the tropical regions (Cooke, 1901) and is found in several countries such as Brazil, China, India, South Africa, Sri Lanka, the West Indies (Garaniya and Bapodra, 2014), and South America (Omoboyowa et al., 2021a).

It has been reported in diverse habitats in Bangladesh such as village thickets, natural forests, and protected areas (Hassan et al., 2021). It has been seen among bushes, hedges, open lands (Gamble, 1935), and even in natural forests and sacred groves of Kerala, India (Krishna and Mg, 2019). It is mentioned in the Global Invasive Species Database. This plant could alter the soil nutrient status and it has been suspected of allelopathic effects in Florida (GISD, 2022). The main aim of this chapter is to highlight the botanical aspects, traditional uses, phytoconstituents, toxicology, and various pharmacological applications of *A. precatorius*.

5.2 BOTANICAL DESCRIPTION

A. precatorius is a climbing shrub (Gamble, 1935) that belongs to the family Fabaceae. *A. precatorius* are pinnately compound with many pairs of leaflets. Each leaflet measures about 0.75 inches in length and 0.25 inches in breadth. The rachis usually ends in a bristle. The stipules of the leaves are deciduous and the stipels are minute (Gamble, 1935). Flowers are arranged as thick racemes borne on axillary peduncles or short branches. The flowers are bracteolate and deciduous (Gamble, 1935). The calyx is campanulate with short teeth. The two upper sepals are subconnate (Cooke, 1901).

DOI: 10.1201/b23017-7

TABLE 5.1
Common Names of *Abrus precatorius*

Language/Region	Names	Reference
English	Precatory pea, rosary pea, jequirity bean, Crab's eye, John Crow beads, Indian liquorice	Balachandran and Rajendiran (2015), Garaniya and Bapodra (2014)
Gujarati	Chanothi, Gumchi	Garaniya and Bapodra (2014)
Hindi	Gundi, Rakti, Gunchi, Gunja, Gaunchi, Rati	Gamble (1935), Garaniya and Bapodra (2014)
Indonesia	Saga	Garaniya and Bapodra (2014)
Malayalam	Kunni	Garaniya and Bapodra (2014)
Nepal	Rati gedi	Garaniya and Bapodra (2014)
Sanskrit and Marathi	Gunja	Garaniya and Bapodra (2014)
Tamil	Kuntumani, Kundumani	Gamble (1935), Karthikeyan and Deepak Amalnath (2017)
Bengali	Kunch, Koonch, Chunhali	Garaniya and Bapodra (2014)
Hausa	Idon zakara	Salihu et al. (2018)
Fulfulde	Bo uhi	Salihu et al. (2018)
Nyanja	Nyumbu	Chinsembu et al. (2019)

The corolla is exserted. The standard petal is broadly ovate and narrows down to a short claw. The wing petals are narrow, oblong, and falcate. The curved keel (Gamble, 1935) is longer and broader than the wings (Cooke, 1901). The petals are usually pink, pinkish-white, or white and appear in winter, while fruits appear in summer (Garaniya and Bapodra, 2014). The flowers have nine stamens with uniform anthers and are arranged in a sheath (Gamble, 1935) with a slit (Cooke, 1901) which is slightly adherent to the standard petal. The vexillary stamen is typically absent (Garaniya and Bapodra, 2014). The ovary is subsessile with a short style and a capitate stigma. The fruits are dehiscent thick pods that are 1–1.5 inches long and 0.5 inches broad. They are thinly septate between the seeds (Gamble, 1935). The green pods ripen to a brown colour. The seeds of this plant are very attractive with a hard, shiny, and polished red seed coat with a black spot at the hilum. The details are shown in Figure 5.1. Variants of seed colour have also been reported by Balachandran and Rajendiran (2015). The two-colour variations include white and brown or white and yellow and the single colour variants are white, yellow, black, red, green, and blue (Figure 5.2). The black seeds are obtained from the cultivated variety (Cooke, 1901). The seeds do not germinate easily due to their hard seed coat. However, they germinate after either mechanical or chemical scarification. Seeds when treated with sulphuric acid for 120 minutes showed 95% germination as opposed to 0% of the control (Prakash et al., 2013).

FIGURE 5.1 The (a) leaves, (b) flowers, (c) pods, and (d) the dehisced pods of *Abrus precatorius*. (Adopted from Garaniya and Bapodra, 2014.)

FIGURE 5.2 Seed colours of *Abrus precatorius*: (a) red and black, (b) white, (c) black, and (d) brown.

5.3 TRADITIONAL USES

A. precatorius is an important medicinal plant that is used for various purposes. Its roots, leaves, and seeds are used to treat various diseases. Seeds of *A. precatorius* are reported to be poisonous and contains one of the most lethal toxins, Abrin, a toxalbumin that inhibits protein synthesis causing cell death (Gadadhar and Karande, 2013; Narayanan et al., 2005), Oral consumption of the chewed or crushed seeds will lead to immediate death. It is used for suicidal purpose in villages considering its easy availability and lack of an antidote for this poisoning (Karthikeyan and Deepak Amalnath, 2017). The toxicity of seeds is destroyed by cooking or boiling (Pandey, 1994; Rajaram and Janardhanan, 1992). Mature seeds that have a hard impermeable coat are considerably less toxic than immature seeds that have a soft and easily broken coat when ingested orally. *A. precatorius* seeds are used to treat emesis in China (Ahmed et al., 2013) and diarrhoea in Nigeria (Uzodimma Durugbo, 2014).

Seeds are famously uniform in weight (almost 1/10th of a gram) and were used as a weighing unit (Nath et al., 1992). They were once used to weigh gems and precious stones. It is also believed that *A. precatorius* was used to weigh the famous Kohinoor Diamond as well (Bhakta and Das, 2020). These seeds are used to make necklaces, rosaries, and ornaments (Cooke, 1901). They are also used in wreaths of roses (precatory beans) in Christian countries (Bhakta and Das, 2020).

Dried seed powder of *A. precatorius* is used to cure worm infestation when taken orally in a dose of one teaspoonful once a day for 2 days (Kuo et al., 1995). It is also used to treat fractures in veterinary field. Dried seeds are also administered orally as an aphrodisiac (Malhi and Trivedi, 2008). Extract of seeds in hot water is taken orally for malaria. Dried seed powder is used by various African tribal groups as oral contraceptives (Watt and Breyer Brandwiik, 1962; Arseculeratne et al., 1985; Molgaard and Nielsen, 2001). Seeds are purgative, emetic, and good source of insecticide (Khanna and Kaushik, 1989). They are also used in the treatment of tuberculosis and painful swellings. Tanzanian traditional healers claim its efficiency in the treatment of epilepsy (Moshi et al., 2005). A paste of *A. precatorius* leaves and seeds is used to treat greying hair and alopecia. Haemorrhoids are treated by taking the mixture of coconut water, seed, and root powder. The seed paste is recommended to be applied locally in sciatica, stiffness of the shoulder joint, paralysis, contusion, and inflammation (Kirtikar et al., 1918,).

The leaves of *A. precatorius* are used in the treatment of mouth ulcer, inflammation, wounds, throat infections, cough, and flu (Anant and Maitreyi, 2012; Nadkarni and Nadkarni, 1954; Bhakta and Das, 2020). The leaves and roots of this plant are used to cure tuberculosis, bronchitis, whooping cough, chest complaints, and asthma by the South African people (Kuete 2014). The leaves of *A. precatorius* are potent laxative, expectorant, and aphrodisiac agents in ayurvedic medicine (Bhakta and Das, 2020) and are used to treat urticaria, eczema, stomatitis, conjunctivitis, alopecia areata, migraine, lymphomas/leukaemia, and dysmenorrhoea (Pade, 1957). Tanzanian traditional healers are using the decoction of leaves as tea (Moshi et al., 2005) and *A. precatorius* is used to treat bronchitis, laryngitis, hepatitis, and epilepsy as folklore medicine by Chinese people (Bhakta and Das, 2020). *A. precatorius* plant is also used against schistosomiasis in Zimbabwe. The leaves are reported to have anti-suppurative properties. They are ground with lime and applied on acne sores, boils, and abscesses. The plant is also traditionally used to treat tetanus and to prevent rabies. The leaves when steeped in warm mustard oil and applied over the site of pain are beneficial against rheumatism. The juice of the fresh leaves, mixed with some bland oil, applied externally is thought to relieve local pain (Kirtikar et al., 1918,). Powdered leaves mixed with sugar were used to treat a case of leucoderma and menorrhagia (Chadha, 2004). The leaves and roots are used as diuretic and to mitigate diarrhoea, gastritis, heart diseases, kidney diseases, insomnia, cancer, and nervous system disorder (Kubiatowicz and Benson, 2003).

The roots of *A. precatorius* are used to treat jaundice and haemoglobinuric bile. They are also used to cure abdominal pains, tumours as well as abortion. A combination of *A. precatorius* root powder and pure clarified butter is recommended to cure cough if taken thrice a day for 4 days (Kirtikar et al., 1918). Root is chewed as a remedy from snake bite (Watt and Breyer-Brandwijk, 1962). Hot water extract of the fresh root can be administered orally as an antimalarial and anticonvulsant agents (Adesina, 1982). Oral administration of dried root liquid broth is found to be effective against bronchitis and hepatitis (Chukuo et al., 1995). The root is reported to have emetic, alexiteric, and expectorant activities. Root decoction with sugar is used to treat haemorrhoids. A filtrate of the root soaked in water for overnight is used to treat white vaginal discharge if taken once in the morning and once in the evening. The roots are used as an alternate for liquorices, sore throat, and rheumatism. The roots are also useful in gonorrhoea, jaundice, and other infections.

5.4 PHYTOCONSTITUENTS

Roots of *A. precatorius* contain phytoconstituents such as abrol, abrasine, precasine, precol (Khaleqe et al., 1966; Willaman and Li, 1970), abraline, abricin, abrusgenic acid, abrusgenic acid methyl ester, abruslactone, abrussic acid, anthocyanins, calcium, campesterol, cycloartenol, delphinidin, gallic acid, trigonelline, hypaphorine (Ghosal and Dutta, 1971; Ibrahim, 1980), choline, N,N-dimethyltryptophan, N,N-dimethyltryptophan metho cation methyl ester, p-coumaroylgalloyl glucodelphinidin, pectin, pentosans, phosphorus, delphinidin, gallic acid, picatorine, polygalacturonic acid, precatorine (Ghosal and Dutta, 1971), polysaccharide (Singh and Shelley, 2007), isoflavonoids, quinones – abruquinones A, B, C, D, E, F (Kuo et al., 1995), O, G, abruslactone a, abrusgenic acid–methanol solvate (Chang et al., 1983), arabinose, galactose, xylose (Ragasa et al., 2013), triterpenoids, saponins, glycyrrhizin (Akinloye and Adalumo, 1981), oleanolic acid, 7,5-Dihydroxy-6,49-Dimethoxy isoflavone and 7-O-b-D-Galactopyranoside (Saxena and Sharma, 1999). Abrusosides A, B, C, D (Choi et al., 1989) and E (Kennelly et al., 1996), galactose, arabinose, and xylose were found in the aerial parts.

The leaves are sugary when taken orally and contains glycyrrhizin, triterpene glycosides (abrusosides A, B, C, and D which are highly sweet), arabinose, galactose, xylose, Pinitol and tree glycosides based on cycloartane-type aglycone, abrutigenin (Daniel, 2006; Choi et al., 1989; Karawya et al., 1981; Ali and Malek, 1966; Akinloye and Adalumo, 1981).The Alkaloids (abrine, hepaphorine, choline, and precatorine) and flavonoids (vitexin, Liquirtiginin-7-mono- and diglycosides and Toxifolin-3-Glucosides) are also present (Ghosal and Dutta, 1971). Leaves also contain other compounds like abrine, trigonelline, abruslactone A, and hemiphloin (Ragasa et al., 2013).

Seeds are lethal and the principal compound is abrine. Seeds of this plant also contain calcium, magnesium, sodium, potassium, phosphorous, manganese, zinc, iron, copper, cellulose, and muscilase. Crystalline abrin contained 4%–9% of neutral sugar in addition to 9-3 residues of glucosamine per mole of abrin (molecular weight 65,000). Abrine (Ghosal and Dutta, 1971), abrin A, abrin B (Lin et al., 1981), abrin C (Wei et al., 1974), abrin I, abrin II, abrin III, abrus agglutinin APA-I, Abrus agglutinin APA-II (Hegde et al., 1991), abrus saponins I and II, abrisapogenol, β-Amyrin, arachidyl alcohol, brassicasterol, Decan-1-ol, Docos-13-Enoic acid, Docosan-1-ol, docosane, n-Dodecan-1-ol, dotriacontane, n-Eicos-11-Enoic acid, eicosane, n-elaidic alcohol, Heneicosan-1-ol, lignoceric acid, heneicosane, n-Heptacosan-1-ol, Heptadecan-1-ol, hexacosane, n-Hexacosan-1-ol, Hexadec-9-Enoic acid, hexadecane, n-Hexadecan-1-ol, nonacosane, n-Nonadecan-1-ol, Octacosan-1-ol, octacosane, n-Octadeca-9,12-Dienoic acid, octadecane, n-octanoic acid, Pentacosan-1-ol, pentacosane, n-pentatriacontane, n-Pentadecan-1-ol (Lefar et al., 1968; Garaniya and Bapodra, 2014), squalene, abricin, abridin (Zia et al., 1983), abrulin (Hameed et al., 1961), cycloartenol, campesterol, cholesterol, and â-Sitosterol have all been found in the seeds. Seeds contain plenty of essential amino acids like serine and other compounds such as Abrusin, Abrusin-2′-0-Apioside, hederagenin, kaikasaponin III, sophoradiol, Sophoradiol-22-0-Acetate, tryptophan (Desai et al., 1971), trimethyl (Kinjo et al., 1991), alanine (Glasby, 1991), amyrin, alpha, ursolic acid (Maiti et al., 1970), valine (Kinjo et al., 1991; Glasby, 1991), and methyl ester. It also has a glucoside abrussic acid, haemagglutinin, and urea. Alkaloids and nitrogen compounds such as methyl ester of N,N-Dimethyltryptophan metho cation (I) and precatorine (II), hypaphorine, trigonelline (Ibrahim, 1980), choline (Ghosal and Dutta, 1971), flavonoids and triterpenoids, steroids, saponins, flavones, flavonol glycosides, reducing sugars, phenolic compounds glycosides (Shatish et al., 2010; Devasagayam and Sainis, 2002; Govindarajan et al., 2005; Scartezzini and Speroni, 2000), and precatorine are present in the seeds and leaves. Lectin (Chatterjee et al., 1982; Wei et al., 1975; Roy et al., 1976), flavonoids and anthocyanins-abrectorin, Dimethoxycentaureidin-7-0-Rutinoside, precatorins I, II (Ghosal and Dutta, 1971), and III, abrectorin, centaureidin, demethoxy 7-O-beta-Drutinoside, luteolin, orientin, isoorientin (Bhardwaj et al., 1980), *A. precatorius* plant growth inhibitor (Anderson et al., 1972), and xyloglucosyl-delophinidin have been identified from the seeds. A new triterpenoid saponin 3-O-β-D-Glucopyranosyl-(1→2)-β-D-Glucopyranosyl subprogenin D together with six known terpenoids was reported from leaves and stems of *A. precatorius* (Xiao et al., 2011). C-Glucosylscutelarein 6,7-Dimethylether (abrusin), and its 2″-O-Apioside has been found as minor components in the seeds of *A. precatorius* (Markhama et al., 1989). Tetracos-15-enoic acid, Tetracosan-1-ol, tetracosane, n-Tetradecan-1-ol, tetradecanoic acid, tetratriacontane, n-Triacosan-1-ol, triacontane, n-tricosane, n-Tridecan-1-ol, Tritriacontan-1-ol, tritriacontane, n-Undecan-1-ol (Lefar et al., 1968), anthocyanins, arabinose (Karawya et al., 1981), arachidic acid, behenic acid, linolenic acid, palmitic acid: stearic acid, oleic acid (Lefar et al., 1968; Begum, 1992; Derbsey and Busson, 1968), aspartic acid, callistephin, chrysanthemin, delphin, Pelargonidin-3,5-Diglucoside (Heines, 1971), heneicosane, 7,9,15-Trimethyl, pentacosanoic acid, cholanic acid, 5-Beta (Garaniya and Bapodra, 2014; Mandava et al., 1974), cystine, galacturonic acid, glucuronic acid, leucine, tyrosine (Desai et al., 1971), delphinidin glycoside (Krishnamoorthy and Seshadri, 1962), delphinidin, (para-coumaroyl-galloyl) glucoside, Delphinidin-3-Sambubioside (Karawya et al., 1981), docosadienoic acid, docosenoic acid, eicosadienoic acid, eicosenoic acid, eicosatrienoic acid, hexadecenoic acid, lignoceric acid, octadecadienoic acid, octadecatrienoic acid, octadecenoic acid, pentadecanoic acid (Khan et al., 1970), docosatetraenoic acid, docosatrienoic acid, myristic acid (Khan et al., 1970; Derbsey and Busson, 1968), galactose, xylose (Karawya et al., 1981), gallic acid, lauric acid, linoleic acid (Desai and Sirsi, 1966), p-Sterone (Ahmad and Rahman, 1965), rhamnose, and N-N-Dimethyl metho cation (Desai et al., 1971) have been identified in the seed of this plant.

5.5 TOXICOLOGY

The seeds of *A. precatorius* contain abrin and agglutinin-I which are type II Ribosome inactivating Proteins (RIPs) (Bagaria et al., 2006a; Worbs et al., 2021). RIPs are plant-based RNA N-Glycosidases that inhibit protein synthesis in cells (Surendranath and Karande, 2008). Type II RIP consists of two chains A and B linked by disulphide bond (AB toxins) (Bagaria et al., 2006b; Cheng et al., 2010). The B-Chain has lectin-like activity that is responsible for binding to the galactose-containing receptors on eukaryotic cell surface which is essential for A – Chain internalization by endocytosis (Liang et al., 2021). The A-Chain has N-Glycosidase activity that irreversibly depurinates specific adenine from 28S ribosomal RNA leading to the inhibition of elongation factors EF-1 and EF-2 (Cheng et al., 2010). This further terminates protein synthesis leading to cell apoptosis, thus making them the most lethal toxins (Liang et al., 2021; Surendranath and Karande, 2008).

Structurally, abrin and agglutinin-I are similar to other RIPs such as ricin, botulinum, cholera, and diphtheria toxins (Pillay et al., 2005; Reedman et al., 2008). Abrin is more potent among RIP toxins and is 75 times more potent than ricin (Tiwari et al., 2017). The toxicity for humans is 1 μg/kg leading to death (Alhamdani et al., 2015). The primary structure of abrin and agglutinin-I shares a similarity of 66.9% for chain A and 80.2% for chain B. However, abrin is 200–2,000 fold more potent than agglutinin-I. The variation in secondary structure and active site of the A-Chain has significantly reduced the biological toxicity of agglutinin-I (Bagaria et al., 2006b).

The common symptoms of *A. precatorius* poisoning in humans are nausea, severe vomiting, diarrhoea, abdominal pain, and gastrointestinal bleeding (Karthikeyan and Deepak Amalnath, 2017; Wananukul et al., 2015). The onset of symptoms can occur within 24 hours of ingestion to 3–4 days after ingestion. This depends upon the type and dose of seed ingestion (Alhamdani et al., 2015; Dickers et al., 2012; Wooten et al., 2014). The oral ingestion of mature whole seeds that have a hard seed coat prevents the toxin from entering the digestive tract. The toxic effect of mature seed ingestion is limited to the gastrointestinal tract as digestion processes disturb the hard shell and expose the toxin to the GI tract (Reedman et al., 2008). In rare cases, the effects of toxins are reported to affect the central nervous system, heart, and kidney functioning. The uncommon symptoms include encephalopathy, arrhythmias, papilledema, raised intracranial pressure, and renal failure. The cause of death is majorly due to depletion of fluid volume in the body (Alhamdani et al., 2015; Karthikeyan and Deepak Amalnath, 2017; Shrestha et al., 2020; Subrahmanyan et al., 2009; Wananukul et al., 2015).

Apart from humans, *A. precatorius* poisoning is also reported in cattle, goats, and horses (Acharya and Roy, 2013; Barri et al., 1990; Simpson and Banerjee, 1932). The reported cases of poisoning in humans and animals are due to the red and black variety of seeds, rarely due to the white variety (Pillay et al., 2005). The poisoning by blue and green varieties is not reported as yet.

The reported effective treatments against *A. precatorius* poisoning are immediate correction of electrolytes imbalance, intravenous fluids, and blood transfusion if required (Patil et al., 2016), continuous renal replacement therapy (Huang et al., 2017), hemoperfusion, methylprednisolone, and other corticosteroid treatment (Ninan and James, 2019; Sahoo et al., 2010). The lack of anti-toxin is major hurdle in the treatment (Duracova et al., 2018). The development of monoclonal antibodies against abrin and agglutinin has exhibited promising results during *in vitro* and *in vivo* studies (Fabbrini et al., 2017; Li et al., 2011; Surendranath and Karande, 2008). ELISA techniques with monoclonal antibodies have been developed for the onsite detection of abrin and agglutinin in food samples, thus leading to rapid identification of toxins (Worbs et al., 2021).

5.6 PHARMACOLOGICAL ACTIVITY

Myriad of traditional uses and the presence of biological compounds in various parts of the plants have intrigued modern science to validate the pharmacological activity of *A. precatorius*. Various parts of plants have reported anti-diabetic, anti-microbial, anti-asthmatic, anti-inflammatory, anti-depressant, antioxidant, and anti-rheumatic activities.

5.7 ANTI-DIABETIC ACTIVITY

Methanolic extract of leaves has exhibited inhibition of α-Glucosidase *in vitro* with the inhibition of glucose diffusion in the gut (Altowayti et al., 2020). The ethanolic leaf extract has demonstrated pancreatic protective activity in normoglycemic and STZ/nicotinamide – induced diabetic rats by inhibiting enzymatic activities of α-amylase and α-Glucosidase (Boye et al., 2021). Further, the extract has modulated the hormonal activities by altering serum insulin, GLP-1, and Glucagon level which eventually reduced blood glucose levels and reversed the effect of diabetes (Boye et al., 2021).

5.8 ANTI-MICROBIAL AND ANTI-VIRAL ACTIVITY

Most of the part of *A. precatorius* has exhibited anti-microbial activities. The leaves, stem, seed oil, and root extracts are promising sources of anti-microbial principles. Their extracts are effective against various gram-positive bacteria, such as *Staphylococcus aureus*, *Staphylococcus epidermidis*, *Bacillus subtilis*, and gram-negative bacteria such as *Enterococcus faecalis*, *Streptococcus anginosus*, *Corynebacterium* spp., *Escherichia coli*, *Klebsiella pneumoniae*, *Proteus mirabilis*, and *Pseudomonas aeruginosa* (Bhatia and Ali Siddiqui, 2013; Elumalai et al., 2009; Singh, 2021).

Recent molecular docking studies indicated the potential of glycyrrhizin isolated from *A. precatorius* anti-viral agent against SARS-Cov-19 (Adeleye et al., 2021).

5.9 ANTI-ASTHMATIC ACTIVITY

Oral administration of ethanolic extract of leaves has exhibited anti-asthmatic activity in rat, guinea pig, and goat as evidenced by reducing the effect of carrageenan-induced paw oedema in rats and reversing the effect of histamine-induced bronchospasm in guinea pig and histamine-induced contraction of goat tracheal chain (Taur et al., 2017).

5.10 ANTI-INFLAMMATORY ACTIVITY

Oral administration of alcoholic extract of *A. precatorius* has increased the serum IL-2 level in mice (Gupta et al., 2010; Modi et al., 2021). Isoflavanquinones isolated from the roots of the plant exhibited inflammatory potential through *in vitro* phagocyte oxidative burst and pro-inflammatory cytokine TNF-α inhibition in CAL-27 (oral cancer) and Caco-2 (colon) cell lines (Okoro et al., 2021).

5.11 LARVICIDAL ACTIVITY

Methanolic extract of the seeds of *A. precatorius* exhibited potential larvicidal activity against the pod borer, *Helicoverpa armigera* (Hubner) (Ramesh Babu et al., 2018). Isoflavanquinone and abruquinone isolated from the extract of aerial parts of the *Abrus* plant showed antimalarial activity (Bhatia and Ali Siddiqui, 2013).

5.12 ANTI-PROLIFERATIVE ACTIVITY

Oral administration of ethyl acetate extracts of seeds in female Wistar rats significantly elevated the ulceration scores and upregulated the expression of COX-2, INOS, IL-1β, NF-kb, and TNF-α mRNA (Omoboyowa et al., 2021b). The fraction of ethyl acetate and methanolic extract of the leaves induced apoptosis in human acute monocytic leukaemia cell line via caspase-3/-7 and PARP cleavage (Gul et al., 2018). Stigmasterol hemihydrate and β-Monolinolein isolated from the leaves of *A. precatorius* are proved to be effective against breast cancer both *in vitro* (MDA-MB-231 breast cancer cell lines) and in vivo (7,12-dimethylbenz(a)anthracene-induced breast cancer in virgin female Sprague Dawley rats; Sofi et al., 2018). Ethyl acetate fractions of root extracts are effective against AU565 and HeLa cells respectively (Okoro et al., 2019).

5.13 CONCLUSION AND FUTURE PROSPECTS

This chapter introduces the reader to the fascinating world of *A. precatorius*. The seeds of this plant are renowned for their even weight, beauty, and toxicity. The seeds commonly have a hard-shiny seed coat that is red with a black spot at the hilum. There are variations to the seed colour ranging from monotone white, black, blue, and green to two-toned combinations of white and yellow and white and brown. They have been used to measure jewellery, make ornaments and infamously used for suicide. This climber belongs to the family Fabaceae and the plant's general description, habit, and habitat have been described in this chapter. It has several botanical synonyms and many more common and vernacular names – a testimony to its widespread distribution around the globe. It has several traditional uses ranging from treatment of diseases such as tuberculosis, malaria, bronchitis, whooping cough, chronic malaises such as asthma, eczema, rheumatism, and a host of other applications including treating greying hair and insecticide. A detailed note on its phytoconstituents and toxicology has also been included in this chapter. The lethal principle of the seed is abrin which is a Type II RIP with a potency that is 75 times more than ricin. Pharmacological investigations have revealed anti-diabetic, anti-microbial, anti-viral anti-asthmatic, anti-inflammatory, anti-proliferative, and larvicidal activities indicating the immense potential of *A. precatorius*.

ACKNOWLEDGEMENTS

The authors wish to thank Dr. Suruchi Jamkhedkar (Dept. of Life Sciences, University of Mumbai) for her efforts in sourcing the seeds of *A. precatorius*, Ms. Shraddha Dandekar and Ms. Pracheta Salunkhe (Dept. of Life Sciences, University of Mumbai) for their help with the photography. The authors would also like to acknowledge Mrs. Savita for supplying the *A. precatorius* plant.

NOTES

Mansi Shah, Department of Life Sciences, Centre for Excellence in Marine Studies, University of Mumbai, Mumbai, India
Meenakshi K, Department of Life Sciences, University of Mumbai, Mumbai, India
Indu Anna George, Department of Life Sciences, University of Mumbai, Mumbai, India

REFERENCES

Acharya, R., & Roy, S. 2013. A review on therapeutic utilities and purificatory procedure of Gunja (*Abrus precatorius* Linn.) as described in ayurveda. *Res. Rev. J. Agric. Sci. Technol.* 2:1–11.

Adeleye, O. A., Femi-Oyewo, M. N., Bamiro, O. A., Bakre, L. G., Alabi, A., Ashidi, J. S., Balogun-Agbaje, O. A., Hassan, O. M., & Fakoya, G. 2021. Ethnomedicinal herbs in African traditional medicine with potential activity for the prevention, treatment, and management of coronavirus disease 2019. *Futur. J. Pharm. Sci.* 7:72. https://doi.org/10.1186/s43094-021-00223-5

Adesina, S. K. 1982. Studies on some plants used as anticonvulsants in Amerindian and African traditional medicine. *Fitoterapia.* 53:147–162.

Ahmad, K., & Rahman, A. F. M. P. 1965. Sterone, a keto steroid from *Abrus precatorius*. *Pak. J. Biol. Agr. Sc.* 8:218.

Ahmed, S., Hasan, M. M., Ahmed, S. W., Alam Mahmood, Z., Azhar, I., & Habtemariam, S. 2013. Anti-emetic effects of bioactive natural products. *Phytopharmacology.* 4:390–433.

Akinloye, B. A., & Adalumo, L. A. 1981. *Abrus precatorius* leaves-a source of glycyrrhizin. *Niger. J. Pharm.* 12:405.

Alhamdani, M., Brown, B., & Narula, P. 2015. Abrin poisoning in an 18-month-old child. *Am. J. Case Rep.* 16:146. https://doi.org/10.12659/AJCR.892917

Ali, E., & Malek, A. 1966. Chemical investigations on *Abrus precatorius* Linn. (Beng. Kunch). Sci. Res. lll. 3:141–145.

Altowayti, W., Bohari, S., Ahmed Al-Moalemi, H., Ali Hamood Altowayti, W., & Pauliena Mohd Bohari, S. 2020. In vitro study of antidiabetic effect of *Abrus precatorius* methanol leaves extract against glucose absorption. *Int. J. Life Sci. Biotechnol.* 3:117–126. https://doi.org/10.38001/IJLSB.701093

Anant, S., & Maitreyi, Z. 2012. Pharmacognosy, phytochemistry and pharmacology of *Abrus precatorius* leaf a review. *Int. J. Pharm. Sci. Rev. Res.* 13(2):71–76.

Anderson, J. D., Mandava, N., & Gunn, C. R. 1972. Plant growth inhibitor from *Abrus precatorius* seeds. *Plant Physiol.* 49:1024–1026.

Arseculeratne, S. N., Gunatilaka, A. A. L., & Panabokke, R. G. 1985. Studies on medicinal plants of Sri Lanka Part 14, Toxicity of some traditional medicineal herbs. *J. Ethnopharmacol.* 13(3):323–335.

Bagaria, A., Surendranath, K., Ramagopal, U. A., Ramakumar, S., & Karande, A. A. 2006a. Structure-function analysis and insights into the reduced toxicity of *Abrus precatorius* Agglutinin I in relation to abrin. *J. Biol. Chem.* 281:34465–34474. https://doi.org/10.1074/JBC.M601777200

Balachandran, N., & Rajendiran, K. 2015. Multicoloured seed coat and flower in *Abrus precatorius* (Leguminosae), India. *Curr. Sci.* 109:682–684.

Barri, M. E. S., El Dirdiri, N. I., Abu Damir, H., & Idris, O. F. 1990. Toxicity of *Abrus precatorius* in Nubian goats. *Vet. Hum. Toxicol.* 32:541–545.

Begum, S. 1992. Chemical investigation of white seeded variety of *Abrus precatorius* Linn. *Pak. J. Sci. Ind. Res.* 35:270–271.

Bhakta, S., & Das, S. K. 2020. The medicinal values of *Abrus precatorius* a review study. J. Adv. Biotechnol. Exp. Ther. 3:84–91. https://doi.org/10.5455/jabet.2020.d111

Bhardwaj, D. K., Bisht, M. S., & Mehta, C. K. 1980. Flavonoids from *Abrus precatorius. Phytochemistry.* 19:2040–2041.

Bhatia, M., & Ali Siddiqui, N. 2013. *Abrus precatorius* (L.) an evaluation of traditional herb. *Indian Am. J. Pharm.* 3:3925–3315.

Boye, A., Barku, V. Y. A., Acheampong, D. O., & Ofori, E. G. 2021. *Abrus precatorius* leaf extract reverses alloxan/nicotinamide-induced diabetes mellitus in rats through hormonal (insulin, GLP-1, and glucagon) and enzymatic (α-amylase/α-glucosidase) modulation. *Biomed. Res. Int.* 2021:9920826. https://doi.org/10.1155/2021/9920826

Chadha, Y. R. 2004. *The Wealth of India. A dictionary of Indian Raw Materials and Industrial Products.* Vol 1A. National Institute of Science Communication and Information Resource, Council of Scientific and Industrial Research, New Delhi, India, 18–19.

Chang, H. M., Chiang, T. C., & Mak, T. C. 1983. New oleanene-type triterpenes from *Abrus precatorius* and x-ray crystal structire of abrusgenic acid-methanol 1:1 solvate. *Planta Med.* 49(11):165–169.

Chatterjee, B. P., Sarkar, N., & Rao, A. S. 1982. Serological and chemical investigations of the anomeric configuration of the sugar units in the D-galacto-D-mannan of fenugreek (Trigonella foenum-gracum) seed. *Carbohydr. Res.* 104(2):348–353.

Cheng, J., Lu, T. H., Liu, C. L., & Lin, J. Y. 2010. A biophysical elucidation for less toxicity of Agglutinin than Abrin-a from the seeds of *Abrus precatorius* in consequence of crystal structure. *J. Biomed. Sci.* 17:1–13. https://doi.org/10.1186/1423-0127-17-34

Chinsembu, K. C., Syakalima, M., & Semenya, S. S. 2019. Ethnomedicinal plants used by traditional healers in the management of HIV/AIDS opportunistic diseases in Lusaka, Zambia. *South African J. Bot.* 122:369–384. https://doi.org/10.1016/J.SAJB.2018.09.007

Choi, Y. H., Hussain, R. A., Pezzuto, J. M., Kinghorn, A. D., & Morton, J. F. 1989. Abrososides A-D, four novel sweet-tasting triterpene glycosides from the leaves of *Abrus precatorius. J. Nat. Prod.* 52(5):1118–1127.

Chukuo, S., Chen, S. C., Chen, L. H., Wu, J. B., Wang, J. P., & Teng, C. M. 1995. Potent anti-platelet, anti-inflammatory and anti-allergic isoflavanquinones from the roots of *Abrus precatorius. Plant Medica.* 61(4):307–312.

Cooke, T. 1901. *Flora of the Presidency of Bombay.* Vol 1. Taylor and Francis, London, England, 359.

Daniel, M. 2006. *Medicinal Plants Chemistry and Properties.* 1st ed. Oxford and IBH Publishing House Co. Pvt. Ltd, New Delhi, India, 118–119.

Derbsey, M., & Busson, F. 1968. The lipids of certain West African species. *Oleagineux.* 23:191.

Desai, V. B., & Sirsi, M. 1966. Antimicrobial activity of *Abrus precatorius. Indian J. Pharmacy.* 8:164.

Desai, V. B., Sirsi, M., Shankarappa, M., & Kasturibai, A. R. 1971. Chemical and pharmacological investigations on the seeds of *Abrus precatorius* Linn. II. Effect of seeds on mitosis and meiosis in grasshopper, Poecilocera picta and some ciliates. *Indian J. Exp. Biol.* 9(3):369–371.

Devasagayam, T. P., & Sainis, K. B. 2002. Immune system and antioxidants, especially those derived from Indian medicinal plants. *Indian J. Exp. Biol.* 40:639–655.

Dickers, K. J., Bradberry, S. M., Rice, P., Griffiths, G. D., & Vale, J. A. 2012. Abrin poisoning. *Toxicol. Rev.* 22:137–142. https://doi.org/10.2165/00139709-200322030-00002

Duracova, M., Klimentova, J., Fucikova, A., & Dresler, J. 2018. Proteomic methods of detection and quantification of protein toxins. *Toxins.* 10:99–100. https://doi.org/10.3390/TOXINS10030099

Elumalai, E. K., Sivamani, P., Thirumalai, T., Vinothkumar, P., Sivaraj, A., & David, E. 2009. In vitro antifungal activities of the aqueous and methanol extract of *Abrus precatorius* linn (fabaceae) seeds. *Pharmacology.* 2:536–543.

Fabbrini, M. S., Katayama, M., Nakase, I., & Vago, R. 2017. Plant ribosome-inactivating proteins progesses, challenges and biotechnological applications (and a few digressions). *Toxins.* 9:314. https://doi.org/10.3390/TOXINS9100314

Gadadhar, S., & Karande, A. A. 2013. Abrin immunotoxin targeted cytotoxicity and intracellular trafficking pathway. *PLoS One.* 8:e58304.

Gamble, J. S. 1935. *Flora of Madras Presidency.* Vol 1. Adlard and Son Publishing, London, England, 349.

Garaniya, N., & Bapodra, A. 2014. Ethno botanical and Phytophrmacological potential of *Abrus precatorius* L. A review. *Asian Pac. J. Trop. Biomed.* 4:S27–S34. https://doi.org/10.12980/APJTB.4.2014C1069

Ghosal, S. & Dutta, S.K. 1971. Alkaloids of *Abrus precatorius. Phytochemistry*, 10 (1): 195–198. https://doi.org/10.1016/S0031-9422(00)90270-X.

Glasby, J. H. 1991. *Dictionary of Plants Containing Secondary Metabolites.* Taylor and Francis Publishing, New York, NY, 488.

Global Invasive Species Database. 2022. Downloaded from http://www.iucngisd.org/gisd/search.php. Accessed on 24-05-2022.

Govindarajan, R., Vijayakumar, M., & Pushpangadan, P. 2005. Antioxidant approach to disease management and the role of 'Rasayana' herbs of Ayurveda. *J. Ethnopharmacol.* 99:165–178.

Gul, M. Z., Chandrasekaran, S., Manjulatha, K., Bhat, M. Y., Maurya, R., Qureshi, I. A., & Ghazi, I. A. 2018. Antiproliferative and apoptosis-inducing effects of *Abrus precatorius* against human monocytic leukaemia (THP-1) cell line. *Indian J. Pharm. Sci.* 80:307–317. https://doi.org/10.4172/PHARMACEUTICAL-SCIENCES.1000359

Gupta, V., Bansal, P., Kumar, S., Mittal, P., & Sharma, S. 2010. *Abrus precatorius*—A poison of pharmacological therapeutic potential. Research *J. Pharm. Tech 3:62–65.*

Hameed, A. K., Hasmi, M. A., & Khan, M. I. 1961. *Abrus precatorius* . I. Isolation and toxic properties of abrulin, a protein fraction from the seeds. *Pak. J. Sci. Ind. Res.* 4:53–56.

Hassan, M. A., Rahman, M. O., & Afroz, S. 2021. A new variety of *Abrus precatorius* L. (Fabaceae) from Bangladesh. *Bangladesh J. Plant Taxon.* 28:289–294. https://doi.org/10.3329/bjpt.v28i2.57127

Hegde, R., Maiti, T. K., & Podder, S. K. 1991. Purification and characterization of three toxins and two agglutinins from *Abrus precatorius* seed by using lactamyl-sepharose affinity chromatography. *Anal. Biochem.* 194(1):101–109.

Heines, V. 1971. A study of pigments in seed coat of *Abrus precatorius*, Linn. *Trans. Ky. Acad. Sci.* 32:1.

Huang, J., Zhang, W., Li, X., Feng, S., Ye, G., Wei, H., & Gong, X. 2017. Acute abrin poisoning treated with continuous renal replacement therapy and hemoperfusion successfully: A case report. *Medicine (Baltimore).* 96:e7423. https://doi.org/10.1097/MD.0000000000007423

Ibrahim, N. 1980. Phytochemical studies of *Abrus precatorius* alkaloids. *Herba Hung.* 19(3):21–26.

Karawya, M. S., El Gengaihi, S., Wassel, G., & Ibrahim, N. A. 1981. Carbohydrates of *Abrus precatorius. Fitoterapia.* 52:179–181.

Karthikeyan, A., & Amalnath, S. D. 2017. *Abrus precatorius* poisoning a retrospective study of 112 patients. *Indian J Crit Care Med.* 21(4):224–225. https://doi.org/10.4103/ijccm.IJCCM_320_16

Kennelly, E. J., Cai, L., Kim, N. C., & Kinghorn, A. D. 1996. Abrusoside e, a further sweet-tasting cycloartane glycoside from the leaves of *Abrus precatorius. Phytochemistry.* 41(5):1381–1383.

Khaleqe, A., Aminuddin, M., & Mulk, S. A. U. 1966. Investigations of *Abrus precatorius* L. constituents of dry root. Pak. C S I R Bull. Monogr. 3:203.

Khan, A. H., Khalio, Q., & Ali, S. S. 1970. Studies on the seed oil of *Abrus precatorius* L. II. composition of the lipid classes. Pak. J. Sci. Indus Res.13:391–394.

Khanna, P., & Kaushik, P. 1989. New sources of insecticides Rotenoids. *Proc. Natl. Acad. Sci.* 59(1):83–86.

Kinjo, J., Matsumoto, K., Inoue, M., Takeshita, T., & Nohara, T. 1991. A new sapogenol and other constituents in abrin semen, the seeds of *Abrus precatorius* L. 1. *Chem. Pharm. Bull.* 39(1):116–119.

Kirtikar, K. R., Basu, B. D., and I.C.S. (1918). *Indian Medicinal Plants.* Vol-I. Subindra Nath Basu, Allahabad, 430–432.

Krishna, G., & Mg, S. 2019. Indigenous knowledge of coastline sacred groves in Central Kerala, India. *Indian J. Tradit. Knowl.* 18:541–546.

Krishnamoorthy, V., & Seshadri, T. R. 1962. Survey of anthocyanins from Indian sources Part lll. *J. Sci. Ind. Res.* 21:591–593.

Kubiatowicz, R., & Benson, L. 2003. Oh no! Ethnobotany. The safe handling and storage of hazardous ethnobotanical artifacts. *Allen Press*. 18(1–2):59–73.

Kuete, V. 2014. Chapter 22: Physical, hematological, and histopathological signs of toxicity induced by African medicinal plants. *Toxicological Survey of African Medicinal Plants*. Elsevier, 635–657.

Kuo, S. C., Chen, S. C., Chen, L. H., Wu, J. B., Wang, J. P., & Teng, C. M. 1995. Potent antiplatelet, anti-inflammatory and antiallergic isoflavanquinones from the roots of *Abrus precatorius*. *Planta Med.* 61:307–312.

Lefar, M. S., Firestone, D., Coleman, E. C., Brown, N., & Shaw, D. W. 1968. Lipids from the seeds of *Abrus precatorius*. *J. Pharm. Sci.* 57:1442–1444.

Li, X. B., Yang, W., Zhang, Y., Zhang, Z. G., Kong, T., Li, D. N., Tang, J. J., Liu, L., Liu, G. W., & Wang, Z. 2011. Preparation and identification of monoclonal antibody against abrin-a. *J. Agric. Food Chem.* 59:9796–9799. https://doi.org/10.1021/JF202534Y

Liang, L. H., Yang, Y., Geng, S., Cheng, X., Yu, H. L., Liu, C. C., & Liu, S. L. 2021. Rapid differential detection of abrin isoforms by an acetonitrile- and ultrasound-assisted on-bead trypsin digestion coupled with LC-MS/MS analysis. *Toxins*. 13:358. https://doi.org/10.3390/TOXINS13050358

Lin, J. Y., Lee, T. C., Hu, S. T., & Tung, T. C. 1981. Isolation of four isotoxic proteins and one agglutinin from jequiriti bean (*Abrus precatorius*). *Toxicon*. 19:41–51.

Maiti, P. C., Mukherjea, S., & Chatterjee, A. 1970. Chemical examination of seeds of *Abrus precatorius*. *J Indian Acad. Forensic Sci.* 9:64–68.

Malhi, B. S., & Trivedi, V. P. 2008. Vegetable antifertility drugs of India. *Q. J. Crude Drug Res.* 12:1922–1928. https://doi.org/10.3109/13880207209068244

Mandava, N., Anderson, J. D., Dutky, S. R., & Thompson, M. J. 1974. Novel occurrence of 5 Beta-cholanic acid in plants isolation from jequirity bean seeds (*Abrus precatorius*). *Steroids*. 23:357–361.

Markhama, K.R., Wallace, J. W., Babut Y.N., Murty, V. K. & Rao, M.G. 1989. 8-C-glucosylscutellarein 6,7-dimethyl ether and its 2″-O-apioside from *Abrus precatorius*. *Phytochemistry* 28(1): 299–301. https://doi.org/10.1016/0031-9422(89)85069-1

Modi, C. M., Bhatt, P. R., Patel, U. D., Patel, H. B., & Pandya, K. B. 2021. Immunomodulatory effect of the hydroalcoholic extract of *Abrus precatorius* L. leaves against cyclophosphamide-induced immunosuppression in mice. *Arch. Biol. Sci.* 73:279–290. https://doi.org/10.2298/ABS210219022M

Molgaard, B., & Nielsen, S. B. 2001. Anthelmintic screening of Zimbabwean plants traditionally used against schistosomiasis. *J Ethnopharmacol.* 74(3):257–264.

Moshi, M. J., Kagashe, G. A., & Mbwambo, Z. H. 2005. Pants used to treat epilepsy by Tanzanian traditional healers. *J Ethnopharmacol.* 97(2):327–336.

Nadkarni, A. K., & Nadkarni, K. M. 1954. *Indian Materia Medica*. Vol 1. Popular Prakashan, Bombay, India, 776–784.

Narayanan, S., Surendranath, K., Bora, N., Surolia, A., & Karande, A.A. 2005. Ribosome inactivating proteins and apoptosis. *FEBS Lett.* 579:1324–1331.

Nath, D., Sethi, N., Singh, R. K., & Jain, A. K. 1992. Commonly used Indian abortifacient plants with special reference to their teratologic effects in rats. *J. Ethnopharmacol.* 36:147–154. https://doi.org/10.1016/0378-8741(92)90015-J

Ninan, E. C., & James, E. 2019. Acute disseminated encephalomyelitis due to *Abrus precatorius* poisoning—A case report. *Saudi Pharm. J.* 27:521–524. https://doi.org/10.1016/J.JSPS.2018.11.016

Okoro, E. E., Maharjan, R., Jabeen, A., Ahmad, M. S., Azhar, M., Shehla, N., Zaman, W., Shams, S., Osoniyi, O. R., Onajobi, F. D., & Choudhary, M. I. 2021. Isoflavanquinones from *Abrus precatorius* roots with their antiproliferative and anti-inflammatory effects. *Phytochemistry*. 187:112743. https://doi.org/10.1016/J.PHYTOCHEM.2021.112743

Okoro, E. E., Osoniyi, O. R., Jabeen, A., Shams, S., Choudhary, M. I., & Onajobi, F. D. 2019. Anti-proliferative and immunomodulatory activities of fractions from methanol root extract of *Abrus precatorius* L. *Clin. Phytoscience*. 51(5):1–9. https://doi.org/10.1186/S40816-019-0143-X

Omoboyowa, D. A., Balogun, T. A., Omomule, O. M., & Saibu, O.A. 2021a. Identification of terpenoids from *Abrus precatorius* against Parkinson's Disease proteins using in silico approach. *Bioinform Biol Insights*. 15:11779322211050757. https://doi.org/10.1177/11779322211050757

Omoboyowa, D. A., Omomule, O. M., Balogun, T. A., Saibu, O. A., & Metibemu, D. S. 2021b. Protective potential of ethyl acetate extract of *Abrus precatorius* (Linn) seeds against HCl/EtOH-induced gastric ulcer via pro-inflammatory regulation In vivo and in silico study. *Phytomedicine Plus*. 1:100145. https://doi.org/10.1016/J.PHYPLU.2021.100145

Pade, S. D. 1957. *Arya-Bhishekh*. Sasty Sahitya, Ahmedabad, India, 232–233.

Pandey, V. N. 1994. Leaf protein content and yield of some Indian legumes. *Plant Foods Hum. Nutr.* 46:313–322. https://doi.org/10.1007/BF01088430

Patil, M. M., Patil, S. V., Akki, A. S., Lakhkar, B., & Badiger, S. 2016. An arrow poison (*Abrus precatorius*) causing fatal poisoning in a child. *J. Clin. Diagn. Res.* 10:SD03. https://doi.org/10.7860/JCDR/2016/18234.7439

Pillay, V. V., Bhagyanathan, P. V., Krishnaprasad, R., Rajesh, R. R., & Vishnupriya, N. 2005. Poisoning due to white seed variety of *Abrus precatorius*. *J. Assoc. Physicians India.* 53:317–319.

Prakash, V., Nainwal, A., Rawat, A. S., Chauhan, J. S., & Bisht, H. 2013. Enhancement of germination in *Abrus precatorius* L. seeds by specific pre-sowing treatments. *Int. J. Conserv. Sci.* 4:237–242.

Ragasa, C. Y., Lorena, G. S., Mandia, E. H., Raga, D. D., & Shen, C. C. 2013. Chemical constituents of *Abrus precatorius*. Am. J. Essent. Oils Nat. Prod. 1(2):7–10.

Rajaram, N., & Janardhanan, K. 1992. The chemical composition and nutritional potential of the tribal pulse, *Abrus precatorius* L. *Plant Foods Hum. Nutr.* 42:285–290. https://doi.org/10.1007/BF02194088

Ramesh Babu, S., Babu, R., Meena, K., & Dudwal, R. 2018. Larvicidal activity of different solvent extracts from the seeds of *Abrus precatorius* (L.) against pod borer, Helicoverpa armigera (Hubner). *J. Entomol. Zool. Stud.* 6(1):496–499.

Reedman, L., Shih, R. D. & Hung, O. 2008. Survival after an Intentional Ingestion of Crushed Abrus Seeds. *Western Journal of Emergency Medicine.* IX (3):157–159.

Roy, J., Som, S., & Sen, A. 1976. Isolation, purification, and some properties of a lectin and abrin from *Abrus precatorius* linn. *Arch. Biochem. Biophys.* 174:359–361.

Sahoo, R., Hamide, A., Amalnath, S. D., & Narayana, B. S. 2010. Acute demyelinating encephalitis due to *Abrus precatorius* poisoning—complete recovery after steroid therapy. *Clin. Toxicol.* 46:1071–1073. https://doi.org/10.1080/15563650802334671

Salihu, T., Olukunle, J. O., Adenubi, O. T., Mbaoji, C., & Zarma, M. H. 2018. Ethnomedicinal plant species commonly used to manage arthritis in North-West Nigeria. *South African J. Bot.* 118:33–43. https://doi.org/10.1016/j.sajb.2018.06.004

Saxena, V. K., & Sharma, D. N. 1999. A new isoflavone from the roots of *Abrus precatorious*. *Fitoterapia.* 70:328–329.

Scartezzini, P., & Speroni, E. 2000. Review on some plants of Indian traditional medicine with antioxidant activity. *J Ethnopharmacol.* 71:23–43.

Shatish, M., Balaji, R., Aruna, A., Niraimathi, V., Manikadan, G., & Babu, M. B. V. 2010. Preliminary phytochemical and cytotoxic property on leaves of *Abrus precatorius*. *J. Herbal Med. Toxicol.* 4(1):21–24.

Shrestha, N., Karki, B., Regmi, S., Shrestha, P., Acharya, S., & Pathak, R. 2020. *Abrus precatorius* toxicity presenting with diarrhoea and encephalopathy a case report. *Nepal Med. Coll. J.* 22:189–192. https://doi.org/10.3126/NMCJ.V22I3.32657

Simpson, K. S., & Banerjee, P. C. 1932. Cases of poisoning in the horse with Ratti seeds (*Abrus precatorius*) by oral administration. *Indian J. Vet. Sci. Anim. Husb.* 2:59–65.

Singh, P. K. 2021. Antimicrobial activity of seeds of *Abrus precatorius* Linn. Shodhshauryam. *Int. Sci. Ref. Res. J.* 4:2581–6306.

Singh, R. B., & Shelley, D. R. 2007. Polysaccharide structure of degraded glucomannan from *Abrus precatorius* Linn. seeds. *J. Environ. Biol.* 28(2):461–464.

Sofi, M. S., Sateesh, M. K., Bashir, M., Ganie, M. A., & Nabi, S. 2018. Chemopreventive and anti-breast cancer activity of compounds isolated from leaves of *Abrus precatorius* L. *3 Biotech.* 88(8):1–14. https://doi.org/10.1007/S13205-018-1395-8

Subrahmanyan, D., Mathew, J., & Raj, M. 2009. An unusual manifestation of *Abrus precatorius* poisoning A report of two cases. *Clin. Toxicol.* 46(2):173–175. https://doi.org/10.1080/15563650601185134

Surendranath, K., & Karande, A. A. 2008. A neutralizing antibody to the A chain of abrin inhibits abrin toxicity both in vitro and in vivo. *Clin. Vaccine Immunol.* 15:737–743. https://doi.org/10.1128/CVI.00254-07

Taur, D. J., Patil, R. N., & Patil, R. Y. 2017. Antiasthmatic related properties of *Abrus precatorius* leaves on various models. *J. Tradit. Complement. Med.* 7:428–432. https://doi.org/10.1016/j.jtcme.2016.12.007

Tiwari, V., Bagaria, S., & Karande, A. A. 2017. A chimeric protein of abrin and *Abrus precatorius* agglutinin that neutralizes abrin mediated lethality in mice. *Toxicon.* 127:122–129. https://doi.org/10.1016/J.TOXICON.2017.01.008

Uzodimma Durugbo, E. 2014. South eastern Nigeria-I. *Glob. Adv. Res. J. Med. Plants.* 2:30–44.

Wananukul, W., Sriapha, C., Tongpoo, A., Wongvisavakorn, S., Rittilert, P., Trakulsrichai, S., & Srisuma, S. 2015. A 10-year analysis of plant poisoning in Thailand. *Southeast Asian J. Trop. Med. Public Heal.* 46:1063–1076.

Watt, J. M., & Breyer-Brandwijk, M.G. 1962. *The Medicinal and Poisonous Plants of Southern and Eastern Africa.* 2nd ed. E. S. Livingstone, Ltd., London, England.

Wei, C. H., Hartman, F. C., Pfuderer P, & Yang, W. K. 1974. Purification and characterization of two major toxic proteins from seeds of *Abrus precatorius*. *J. Biol. Chem.* 249:3061–3067.

Wei, C. H., Koh, C., Pfuderer, P., & Einstein, J. R. 1975. Purification, properties and crystallographic data for a principal nontoxic lectin from seeds of *Abrus precatorius*. *J. Biol. Chem.* 250:4790–4795.

Willaman, J. J., & Li, H. L. 1970. *Alkaloid-Bearing Plants and Their Contained Alkaloids*. USA Agricultural Research Service, U. S. Department of Agriculture, Washington, DC, 1–286.

Wooten, J. V., Pittman, C. T., Blake, T. A., Thomas, J. D., Devlin, J. J., Higgerson, R. A., & Johnson, R. C. 2014. A case of abrin toxin poisoning, confirmed via quantitation of l-abrine (N-Methyl-l-Tryptophan) biomarker. *J. Med. Toxicol.* 10:392–394. https://doi.org/10.1007/S13181-013-0377-9

Worbs, S., Kampa, B., Skiba, M., Hansbauer, E. M., Stern, D., Volland, H., Becher, F., Simon, S., Dorner, M. B., & Dorner, B. G. 2021. Differentiation, quantification and identification of abrin and *Abrus precatorius* agglutinin. *Toxins*. 13:284. https://doi.org/10.3390/TOXINS13040284

Xiao, Z. H., Wang, F. Z., Sun, A. J., Li, C. R., Huang, C. G., & Zhang, S. 2011. A new triterpenoid saponin from *Abrus precatorius* Linn. *Molecules*. 17:295–302.

Zia-Ul-Haque, A., Qazi, M. H., & Hamdard, M. E. 1983. Studies on the antifertility properties of active components isolated from the seeds of *Abrus precatorius* Linn I. *Pak. J. Zool.* 5(2):129–139.

6 *Aconitum napellus* (Monkshood)

Karma Yeshi and Phurpa Wangchuk

CONTENTS

6.1 INTRODUCTION

Aconitum genus belongs to the family Ranunculaceae, and members of this genus are known as aconites. This genus has three subgenera: monotypic subgenus – *Gymnaconitum, Lycoctonum*, and *Aconitum*. *Gymnaconitum* has only one annual species, *Gymnaconitum gymnandrum* Maxim, and it is reported only in China (Utelli et al., 2000). Rest two subgenera have numerous species. The distinguishing feature between *Aconitum* and *Lycoctonum* is that the former is characterized by biennial paired tuberous roots and the latter one by perennial rhizomes (Lauener & Tamura, 1978). *Aconitum* species are mostly spotted in the cooler northern hemisphere regions of Asia, North America, Great Britain, and Europe (Jamtsho et al., 2021; Stork & Marraffa, 2005). According to the Kew Royal Botanic Gardens, the *Aconitum* genus has 324 accepted species (Novikov et al., 2021). Some countries have reported several *Aconitum* species; for instance, more than 200 species are reported from China (Singhuber et al., 2009), 38 species from Nepal (Shyaula, 2011), 27 species from India (Abd, 2016), and 19 species from Bhutan (Jamtsho et al., 2021). *Aconitum* species are widely used in the scholarly traditional medicine systems in Asian countries, including Nepal (16 species; Shyaula, 2011), India (18 species; Abd, 2016), Bhutan (two species; Jamtsho et al., 2021), and China (two species; Singhuber et al., 2009). *Aconitum* is one of the most exploited medicinal plant genera in the traditional medicine system and indigenous drug discovery history (Ali et al., 2021). The medicinal

DOI: 10.1201/b23017-8

properties of *Aconitum* species are due to diverse classes of secondary metabolites, and alkaloids are one of the main compounds. Carbon-19 (C-19) and C-20 diterpene alkaloids are most common in aconites, constituting approximately 450 alkaloids (Ali et al., 2021). Many *Aconitum* species are also poisonous (e.g., *Aconitum ferox, Aconitum laciniatum, Aconitum violaceum, Aconitum chasmanthum, Aconitum luridum*, and *Aconitum napellus*) as they contain toxic diterpenoid alkaloids, including aconitine, mesaconitine, and hypaconitine which cause cardiotoxicity (Qasem et al., 2022). Oral ingestion (accidental or deliberate) is the most common route of exposure to *Aconitum* toxins, and it can also enter the body through the mucous membrane and intact skin. Whether it is oral ingestion or dermal contact, absorption of aconitine occurs rapidly within a few minutes (Stork & Marraffa, 2005). The half-maximal lethal dose (LD50) of aconitine in mice is 1.8 mg/kg via oral ingestion, while in humans, it is about 1–2 mg (Fujita et al., 2007; Sato et al., 1979). This chapter discusses *A. napellus* L. in depth, including taxonomy and distribution, ethnopharmacology, phytochemistry, reported biological properties, and associated toxicities.

6.2 BOTANICAL DESCRIPTION

A. napellus, popularly known as monkshood or wolfsbane, is a perennial herb often grown as ornamental due to its attractive blue flowers. Botanical descriptions for morphology and anatomy are referred from Akbar (2020), Duffell (2009), and Munch and Crosbie (1929). *A. napellus* belongs to the family Ranunculaceae (Grin, 2009; Heywood, 1978), and it has numerous common names. The most popularly used names are given here. **English:** Aconite, Helmet flower, Blue rocket, Devil's helmet, Monk's hood, Wolf's bane; **French:** Aconit; **German:** Blauer eisenhut, Blauer sturmhat; **Italian:** Aconitonapello, Erba luparia; **French:** Aconit napel; **Dutch:** Blauwe monnikskap; **Spanish:** Aconite, Anapelo, Stormhatt.

6.2.1 Morphology

A. napellus is a perennial herb (i.e., it has self-supporting stems) with an upright stem, round and smooth, and grows up to 2 m (Figure 6.1a).

6.2.1.1 Leaves

The leaves are simple (i.e., lobed or unlobed but not separated into leaflets). Usually, they are lobbed into three or five segments (further divided), and the edge of the leaf blade has teeth (Figure 6.1b). The surface under the leaves has only a few hairs. Leaves are petiolate and arranged alternately (i.e., one leaf per node along the stem).

6.2.1.2 Flowers

The flowers are blue to purple and usually bloom during May and July. They are bisexual and bilaterally symmetrical (i.e., only one way to divide), stalked, and racemose (Figure 6.1a). Petaloid sepals (or tepals) are five – one uppermost is helmet shaped and beaked and nearly hemispherical; the two laterals are roundish and internally hairy; and the lower two are oblong oval (Figure 6.1c).

6.2.1.3 Roots

Tuberous roots are either single or in clusters of two or more. The younger ones are smooth and yellowish white internally, while matured or older ones are deep wrinkled by side branches or branches and brown inside (Figure 6.2a, b). Root is obconical, measuring 4–10 cm long and 1–3.5 cm wide at the crown. The internal anatomy of roots (cross section) shows a distinct stellate cambial zone (Figure 6.2c).

FIGURE 6.1 (a) *Aconitum napellus* in its natural habitat, (b) palmate leaf, and (c) petaloid sepal. (Adopted and modified from Wikimedia.)

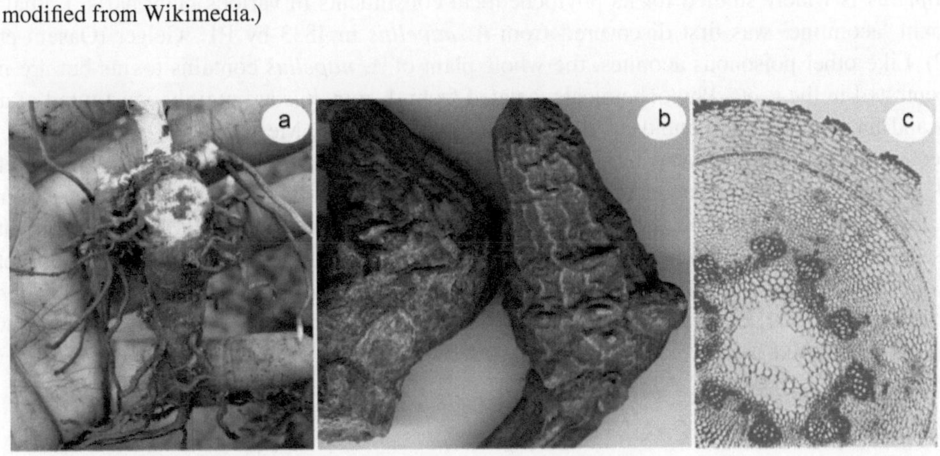

FIGURE 6.2 Roots of *Aconitum napellus*: (a) fresh young roots, (b) dried old roots, and (c) T.S. of root showing a distinct stellate cambial zone. (Adopted from Munch & Crosbie, 1929).

6.3 DISTRIBUTION

A. napellus is spotted in mountainous regions at elevations up to 3,000 m above sea level. But it usually grows in lowland areas. *A. napellus* prefers moist/wet and nutrient-rich soils, which are slightly acidic and in the shade. In terms of climatic conditions, it prefers places where the night temperature remains below 7°C and precipitation about 1 m (Johnson, 2007). *A. napellus* is native to Western and Central Europe (Duffell, 2009). It is also reported in some parts of Asia and in Oceania. Figure 6.3 shows that *A. napellus* is reported in at least 15 countries globally.

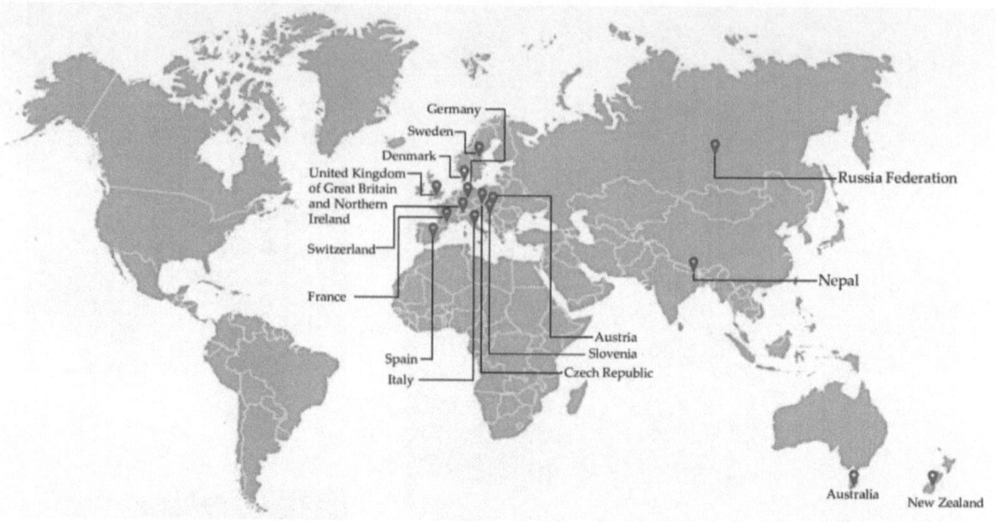

FIGURE 6.3 World map showing the countries where *Aconitum napellus* is reported. (*Country names and locations are labelled manually using information from the Australasian Virtual Herbarium homepage* (adopted from AVH, 2022).)

6.4 PHYTOCHEMICAL CONSTITUENTS

A. napellus is widely studied for its phytochemical constituents in various plant parts. Lethal cardiotoxin 'aconitine' was first discovered from *A. napellus* in 1833 by P.L. Geiger (Qasem et al., 2022). Like other poisonous aconites, the whole plant of *A. napellus* contains toxins but are more concentrated in the roots. Phytochemicals isolated from *A. napellus* are mainly constituted of alkaloids and flavonoids. Diterpenoid alkaloids, including Aconitine, Mesaconitine, Hypaconitine, and *N*-Deethylaconitine, are isolated from the root, pollen, and nectar of *A. napellus* (Arlandini et al., 1987; Dustan & Ince, 1891; Jacquemart et al., 2019). Other alkaloidal compounds such as Aconine A & B, Hypaconitine, Napelline A & B, Neoline A & B, Benzoylmesaconitine, Benzoylhypaconitine, Acosepticine A & B, 6-*O*-Acetylacosepticine A, 6-Demethyldephatine A–C, Leucostine, Acoseptine, and *N*-Acetylsepaconitine A were also isolated or identified from various parts of *A. napellus* (Jacquemart et al., 2019). Flavonoids reported from *A. napellus* are isolated exclusively from flowers (Table 6.1). Aconitine acid, which is an organic acid, was reported from the aconite juice (Figure 6.4).

6.5 PHARMACOLOGICAL STUDIES

Crude extracts and individual isolated compounds from *A. napellus* have shown a few promising biological activities, such as antidiabetic, neuroprotective, and anti-anxiety. Crude methanolic and aqueous extracts from *A. napellus* at 100, 200, and 400 mg/kg in Wistar albino rats through MTT Assay (*in vitro*) and alloxans-induced hyperglycaemic rats (*in vivo*) showed hypoglycaemic potential (Chhetree et al., 2010), thus suggesting the potential application of *A. napellus* as an alternative treatment for diabetes (Shoaib, Salem-Bekhit, et al., 2020).

In a study by Shoaib et al. (Shoaib, Siddiqui, et al., 2020) on the neuroprotective role of crude extract from *A. napellus* in streptozotocin-induced diabetic Sprague-Dawley rats, a detoxified chloroform extract showed a significant ($p < 0.05$) improvement in the myelination and degenerative changes of the nerve fibres besides behavioural improvement (locomotor activity). There were significant decrease in TBARS (thiobarbituric acid reactive substance) and increased levels of catalase,

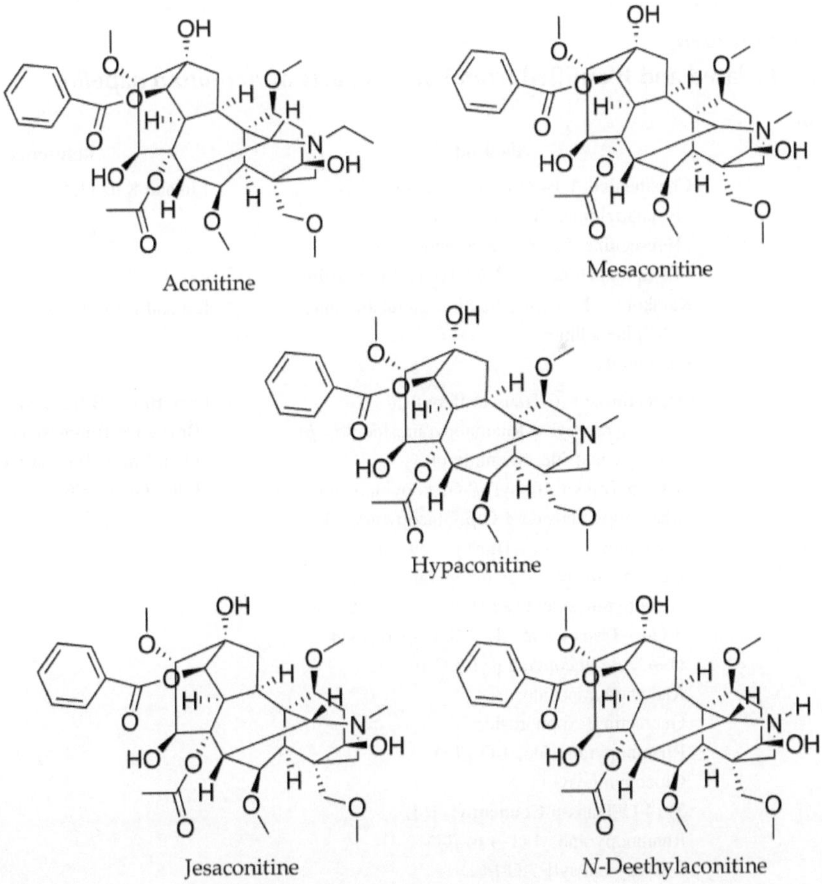

FIGURE 6.4 Chemical structures of chief alkaloidal constituents of *Aconitum napellus.*

TABLE 6.1
Compounds Isolated and Identified from Various Parts of *Aconitum napellus*

Parts Used for Isolation	Alkaloids	Reference
Pollen, and nectar	Aconitine	Dustan and Ince (1891), Jacquemart et al. (2019)
Pollen	Mesaconitine	Jacquemart et al. (2019)
Root	*N*-Deethylaconitine, Aconitine, Mesaconitine	Arlandini et al. (1987)
Pollen	Aconine A & B, Hypaconitine, Napelline B, Neoline A & B, Benzoylmesaconitine, Benzoylhypaconitine, Acosepticine A & B, 6-*O*-Acetylacosepticine A, 6-Demethyldephatine A–C, Leucostine, Acoseptine, *N*-Acetylsepaconitine A	Jacquemart et al. (2019)
Nectar	18-Demethylpubescenine	Jacquemart et al. (2019)
Pollen and nectar	6-Demethyldephatine C, Napelline A	Jacquemart et al. (2019)
Whole plant	Neoline, Napeline, Isotalatizidine, Karakoline, Senbusine A & C	Kiss et al. (2013)

(*Continued*)

TABLE 6.1 (*Continued*)
Compounds Isolated and Identified from Various Parts of *Aconitum napellus*

Parts Used for Isolation	Alkaloids	Reference
Whole plant	Chasmanine, 1,14-*O*-Diacetylneoline, Isotalatizidine, Delsoline, Delcosine, Virescenine, Songorine, Songoramine, 12-Epinapelline, 15-*O*-Acetyl-12-Epinapelline	Liu and Katz (1995)
Seeds	Karakoline, Leroyine, Neoline, Isotalatizidine, 12-Epinapelline	Liu and Katz (1995)
	Flavonoids	
Flowers	Quercetin 7-O-(6-Trans-Caffeoyl)-β-Glucopyranosyl-α-Rhamnopyranoside-3-O-β-Glucopyranoside; Kaempferol 7-O-(6-Trans-Caffeoyl)-β-Glucopyranosyl-α-Rhamnopyranoside-3-O-β-Glucopyranoside; Kaempferol-7-O-(6-Trans-*p*-Coumaroyl)-β-Glucopyranosyl-α-Rhamnopyranoside-3-O-β-Glucopyranoside; Quercetin 3-O-(6-Trans-Caffeoyl)-β-Glucopyranosyl-(1$\rightarrow$2)-β-Glucopyranosyl-7-O-α-Rhamnopyranoside; Quercetin-3-Sophoroside-7-Rhamnopyranoside; 3-O-[β-D-Glucopyranosyl-(1$\rightarrow$ 3)-(4-O-Trans-p-Coumaroyl)-α-L-Rhamnopyranosyl-(1$\rightarrow$ 6)-β-D-Glucopyranosyl]-7-O-[β-D-Glucopyranosyl-(1$\rightarrow$ 3)-α-L-Rhamnopyranosyl]kaempferol; 3-O-[β-D-Glucopyranosyl-(1$\rightarrow$ 3)-(4-O-Trans-p-Coumaroyl)-α-L-Rhamnopyranosyl-(1$\rightarrow$ 6)-β-D-Glucopyranosyl]-7-O- [β-D-Glucopyranosyl-(1$\rightarrow$ 3)-α-L-Rhamnopyranosyl]quercetin; 7-O-[β-D-Glucopyranosyl-(1$\rightarrow$ 3)-α-L-Rhamnopyranosyl]quercetin; Quercetin 3-O-(6-Trans-Caffeoyl)-β-Glucopyranosyl-(1$\rightarrow$ 2)-β-Glucopyranoside-7-O-α-Rhamnopyranoside; Kaempferol 3-O-(6-Trans-Caffeoyl)-α-Glucopyranosyl-(1$\rightarrow$ 2)-β- Glucopyranoside-7-O-α-Rhamnopyranoside; Quercetin 3-O-(6-Trans-p-Coumaroyl)-β-Glucopyranosyl-(1$\rightarrow$ 2)-β-Glucopyranoside-7-O-α-Rhamnopyranoside; Kaempferol 3-O-(6-Trans-p-Coumaroyl)-β-Glucopyranosyl-(1$\rightarrow$ 2)-β- Glucopyranoside-7-O-α-Rhamnopyranoside; Quercetin-3-Sophoroside-7-Rhamnopyranoside	Fico, Braca, Bilia, et al. (2001), Fico, Braca, De Tommasi, et al. (2001), Fico, Braca, Tommasi, et al. (2001), Luis et al. (2006)
	Organic acids	
Aconite juice	Aconitic acid	Ventre et al. (1946)

superoxide dismutase, and reduced glutathione in rats treated with *A. napellus* chloroform extract than in the diabetic control group (Shoaib, Siddiqui, et al., 2020).

Moreover, homoeopathic preparations from *A. napellus* (diluted as 12 and 30 cH in 30% cereal alcohol) also demonstrated anxiolytic effects in Wistar albino rats when treated with 0.15 mL/day for 10 consecutive days (Haine et al., 2021). Anxiolytic activity was measured by the widely used elevated plus maze (EPM) model, where EPM is a wooden device with two closed arms and two open arms perpendicular to each other. Closed arms have lateral and end walls but no walls in open arms. The greater the permanence time and entries in the open arms, the better the anxiolytic action of drugs. Diluted crude extracts of *A. napellus* (12 cH in 30% cereal alcohol) showed 39.5%±8.4% entries and 15.4%±8.6% time ($p<0.05$) in open arms, while rats treated with 30 cH dilutions showed 38%±12% entries and 11%±10% time, which was better compared to 30% alcoholic standard drug diazepam (52%±15% entries and 24.3%±6.7% time in open arms), thus exhibiting its anti-anxiety potential (Haine et al., 2021).

Compounds isolated from *A. napellus* are antioxidative, and particularly, they have free radical scavenging potential. A flavonol glycoside, quercetin 7-*O*-(6-*Trans*-Caffeoyl)-β-Glucopyranosyl-(1 $\rightarrow$ 3)-α-Rhamnopyranoside-3-*O*-β-Glucopyranoside, showed DPPH free radical scavenging antioxidant activity with an IC_{50} value of 1.94 µM (Braca et al., 2003). Flavonol glycoside, quercetin 3-*O*-(6-*Trans*-Caffeoyl)-β-Glucopyranosyl-(1 $\rightarrow$ 2)-β-Glucopyranoside-7-*O*-α-Rhamnopyranoside, showed inhibition (58.9%) of coupled oxidation of β-Carotene and linoleic acid after 1 hour (Braca et al., 2003). Two flavonol glycosides from ethanolic extract of *A. napellus* sp. *lusitanicum*, quercetin 3-O-(6-Trans-Caffeoyl)-β-Glucopyranosyl-(1$\rightarrow$2)-β-Glucopyranosyl-7-O-α-Rhamnopyranoside and Quercetin-3-Sophoroside-7-Rhaamnopyranoside, also showed DPPH free radical scavenging activity with EC_{50} values of 7.56 (better than standard compound – rutin, EC_{50}=7.67 ug/mL) and 10.56 ug/mL, respectively (Luis et al., 2006). Ultrahigh diluted homoeopathic *A. napellus* (200 c or 1,000 c, i.e., 200th or 1,000th dilution, respectively) prepared under centesimal scale (1: 99 ratio – one part of drug and 99 part of solvent) lowered baker's yeast (20%) induced fever in rabbits significantly ($p<0.05$), but the effect was lower than the standard drug paracetamol ($p<0.001$) (Ahmad et al., 2017). Flavonol glycosides from *A. napellus* sp. *lusitanicum* also showed cysticidal activity against trophozoites of *Acanthamoeba castellanii* (Martin-Escolano et al., 2021).

6.6 TOXIC RESPONSE

A. napellus was a popular poison in Rome among the elites and was a good weapon for murder by poisoning (Moog & Karenberg, 2002). The whole plant is poisonous, but the most toxic ones are roots and seeds. *A. napellus* contains three major toxic steroid alkaloids, namely aconitine, mesaconitine, and jesaconitine. Aconitine is the main alkaloid responsible for toxicity, and it is present throughout the plant but more concentrated in roots and leaves. Aconitine is a C_{19}-diester diterpenoid alkaloid present in *Aconitum* spp., which is also named aconite of *Wutou* (Vo et al., 2017; Zhang et al., 2020). Although aconitine-containing herbal medicines are well known for the treatment of rheumatoid arthritis and pains (Li et al., 2010; Yu et al., 2020), for example, *Fuzi*, which is considered 'the chief of the hundred drugs' in China, more than 5,000 cases of aconite poisoning have been reported between 2001 and 2010 (Liu, 2019), and it has caused chiefly polymorphous ventricular arrhythmias (Chan, 2009; Lin et al., 2004).

6.6.1 Toxicity in Humans

The minimum lethal dose of aconitine is 3–6 mg. One gram of fresh *A. napellus* may contain 2–20 mg of aconitine, which means a small amount of this plant can be lethal. Ingestion will instantly cause a burning sensation in the mouth (or tingling, including lips, tongue, and throat) within 10–20 minutes. Subsequently, within 2–6 hours after ingestion, it will cause nausea, salivation, weakness, violent emesis, generalized paraesthesia, and extreme pain. With more time passed by, possibly between 6 and 8 hours, a person will experience diarrhoea, cardiac rhythm disturbances, convulsions, skeletal muscle paralysis, and possibly death due to respiratory paralysis (Stork & Marraffa, 2005).

6.6.2 Mechanism of Toxicity in Human

Aconitine is a voltage-gated sodium channel activator in the myocardium. Aconitine binds to the α-subunit of voltage-gated sodium channels, causing a permanent opening of the channel (Fu et al., 2006). Aconitine also inhibits sodium and potassium ATPase to generate a cardiotonic or cardiotoxic effect (Wang et al., 2021), which results in a prolonged influx of sodium ions (Na^+) and slowing after depolarization (Friese et al., 1997) (Figure 6.5). A high concentration of intracellular Na^+ simultaneously activates the Na^+-Ca^{2+} exchanger (NCX), causing the movement of Ca^{2+} from extracellular to the cytoplasm (Fu et al., 2007). This increased influx or overload of intracellular Ca^{2+}-Na^+ ion is responsible for aconitine-induced proarrhythmic effects. Recent studies have also shown the association of the arrhythmogenic properties of aconitine with the potassium channel. Aconitine blocks the human ether-à-go-go-related gene channel (hERG), which is the primary repolarization potassium current that determines the action potential duration (APD) of cardiomyocytes (Li et al., 2010). Aconitine either blocks hERG completely during the open state or develops low affinity to the channel during the closed or resting state; either way, it prolongs the myocardial APD. Aconitine also blocks Kv1.5 channels in atrial myocytes, which causes lengthening of its action potential, thus prolonging the refractory period of atrial muscle (Li et al., 2010). All these events collectively lead to aconitine-induced cardiotoxicity in humans.

6.6.3 Clinical Management

Decontamination by administering certain syrup and charcoal is a common home remedy. Syrup of ipecac (a drug obtained from dried rhizome and roots of *Carapichea ipecacuanha*) is essentially

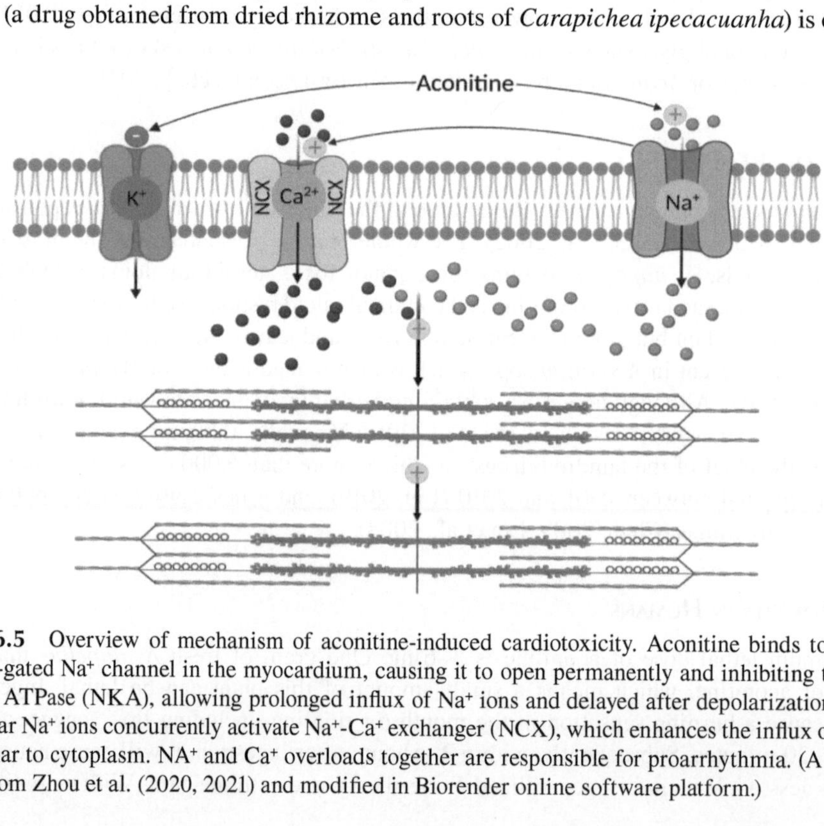

FIGURE 6.5 Overview of mechanism of aconitine-induced cardiotoxicity. Aconitine binds to α-subunit of voltage-gated Na^+ channel in the myocardium, causing it to open permanently and inhibiting the sodium potassium ATPase (NKA), allowing prolonged influx of Na^+ ions and delayed after depolarization. Elevated intracellular Na^+ ions concurrently activate Na^+-Ca^+ exchanger (NCX), which enhances the influx of Ca^+ from extracellular to cytoplasm. NA^+ and Ca^+ overloads together are responsible for proarrhythmia. (A figure was adapted from Zhou et al. (2020, 2021) and modified in Biorender online software platform.)

not advisable as it causes extensive vomiting, onsets symptoms rapidly, and has a risk of respiratory paralysis. Gastric decontamination with activated charcoal may be suitable for substantial recent ingestions. Monitoring and replacing fluid and electrolytes lost from vomiting and diarrhoea is essential during the treatment. After decontamination, a patient may or may not show symptoms but observation for another 2–4 hours is recommended, as toxicity from aconite alkaloids is unpredictable due to alkaloid variability (Stork & Marraffa, 2005). Symptomatic patients should be referred to the hospital. There is no specific antidote, and no specific laboratory tests are available (Stork & Marraffa, 2005).

6.7 TRADITIONAL AND OTHER POTENTIAL USES

A. napellus, a European native species, closely resembles *A. carmichaeli*, and it is still used in European homoeopathic preparations (Moog & Karenberg, 2002; Singhuber et al., 2009). Traditionally, *A. napellus* has been used as an arrow poison for hunting due to its toxic nature. European history recorded and believed that the wife (Julia Agrippina, 49–54 AD) of the Roman emperor Claudius poisoned him with *A. napellus* (Moog & Karenberg, 2002). This species was occasionally applied as poison to kill imprisoned criminals in European history (Shoaib, Salem-Bekhit, et al., 2020). In Chinese medicine, *A. napellus* is used to prevent cold, general weakness, and 'Yang' deficiency, and it is used as an antidote for several poisons (Singhuber et al., 2009). It is used in folklore medicine to manage facial paralysis, inflammation, musculoskeletal pains, pyrexia, gout, pericarditis (Chang & Whitaker, 2001), sciatica, and rheumatism (Venkataraghavan & Sundareesan, 1981). *A. napellus* is used against sudden development of fever, particularly when a person has been exposed to cold weather, freezing, dry wind, and fevers developed after a frightening or shocking experience (Loo, 2009). *A. napellus* is also used in Ayurvedic and Unani medicinal preparations, and other polyherbal formulations are used to treat diabetes and a nerve tonic (Shoaib et al., 2019). Aconite is considered helpful for improving subjective symptoms such as numbness, cold sensations, and extreme or chronic pains (Wang et al., 2011), which are associated with diabetic neuropathy. Aconite tubers contain toxins, including diterpenoid alkaloid aconitine (Kelly, 1990). Therefore, prior to use, boil aconite tubers for several hours or even for a couple of days based on the quantity used (Tang & Eisenbrand, 1992). In Ayurvedic medicine, shodhana method is widely applied, where *Aconitum* tuber is treated with a cow or goat milk, and the goat milk treatment gives less aconitine (Shoaib, Siddiqui, et al., 2020).

6.8 FUTURE REMARKS

A. napellus is native to Europe, and it is still used in European homoeopathic preparations. *A. napellus* is known for its ornamental (due to its attractive blue flower) and medicinal values. Traditionally, the plant was used chiefly as arrow-head poison and poisoning captured criminals in Europe, as its roots contain cardiotoxins – aconitine. Besides this, the plant is popular in indigenous medicine systems, cluing Chinese traditional medicine, Indian Ayurvedic medicine, Japanese herbal preparations, and European homoeopathic preparations for treating musculoskeletal pains. Numerous phytochemicals are isolated from different parts of *A. napellus*, and among them, diterpenoid alkaloids and flavonol glycosides are dominant constituents. Both crude extracts and pure isolated compounds (a few flavonol glycosides) have shown various pharmacological activities, including antipyretic, antidiabetic, antioxidant, anti-anxiety, and neuroprotective activities.

A. napellus (whole plant) contains toxic diterpenoid alkaloids; thus, proper detoxification is necessary prior to use as 1 g of fresh *A. napellus* may contain 2–20 mg of aconitine (lethal dose of aconitine in humans = 3–6 mg), and it can cause fatal symptoms and ultimately death. Prolonged boiling of any plant parts or treating cow or goat milk can also reduce aconitine content.

Future studies may focus on investigating the biological activities of those unstudied compounds isolated from *A. napellus* and investigating the levels of aconitine in any commercial herbal products containing *A. napellus* to ensure safety and prevent toxicities.

NOTES

Karma Yeshi, Center for Molecular Therapeutics, Australian Institute of Tropical Health and Medicine (AITHM), James Cook University, Smithfield, Australia
Phurpa Wangchuk, Center for Molecular Therapeutics, Australian Institute of Tropical Health and Medicine (AITHM), James Cook University, Smithfield, Australia

REFERENCES

Abd, S. (2016). Indian aconites: Boon or bane? *Journal of Pharmacognosy & Natural Products, 1*(1). doi:10.4172/2472-0992.1000104

Ahmad, S., Rehman, T., & Abbasi, W. M. (2017). Effects of homoeopathic ultrahigh dilutions of *Aconitum napellus* on Baker's yeast-induced fever in rabbits. *Journal of Integrative Medicine, 15*(3), 209–213. doi:10.1016/s2095–4964(17)60329-7

Akbar, S. (2020). *Aconitum napellus* L. (Ranunculaceae). In: *Handbook of 200 Medicinal Plants*. Springer, Cham, Switzerland, 81–88, doi: 10.1007/978-3-030-16807-0_9

Ali, S., Chouhan, R., Sultan, P., Hassan, Q. P., & Gandhi, S. G. (2021). A comprehensive review of phytochemistry, pharmacology and toxicology of the genus Aconitum L. *Advances in Traditional Medicine*, doi:10.1007/s13596-021-00565-8

Arlandini, E., Ballabio, M., & Gioia, B. (1987). N-Deethylaconitine from *Aconitum napellus* ssp. vulgare. *Journal of Natural Products, 50*(5), 937–939.

AVH. (2022). The Australasian Virtual Herbarium, Council of Heads of Australasian Herbaria. Accessed February 15, 2022 from https://avh.chah.org.au.

Braca, A., Fico, G., Morelli, I., De Simone, F., Tomè, F., & De Tommasi, N. (2003). Antioxidant and free radical scavenging activity of flavonol glycosides from different Aconitum species. *Journal of Ethnopharmacology, 86*(1), 63–67. doi:10.1016/s0378-8741(03)00043-6

Chan, T. Y. (2009). Aconite poisoning. *Clinical Toxicology (Philadelphia), 47*(4), 279–285.

Chang, L. K., & Whitaker, D. C. (2001). The impact of herbal medicines on dermatologic surgery. *Dermatological Surgery, 27*, 759–763.

Chhetree, R. R., Dash, G. K., Mondal, S., & Acharyya, S. (2010). Studies on the hypoglycaemic activity of *Aconitum napellus* l. roots. *Drug Invention* Today, *2*(7), 343–346.

Duffell, M. (2009). The distribution and native status of Monkshood *Aconitum napellus* sl. L. in Shropshire. A dissertation submitted to the University of Birmingham for the degree of Master of Science in Biological Recording: Collection and Management. Department of Biosciences, University of Birmingham. Accessed May 25, 2022 from http://citeseerx.ist.psu.edu/viewdoc/download?doi=10.1.1.737.7022&rep=rep1&type=pdf.

Dustan, W. R., & Ince, W. H. (1891). XXX.—Contributions to our knowledge of the aconite alkaloïds. Part I. On the crystalline alkaloïd of *Aconitum napellus*. *Journal of the Chemical Society, Transactions, 59*, 271–287.

Fico, G., Braca, A., Bilia, A. R., Tome, F., & Morelli, I. (2001). New flavonol glycosides from the flowers of *Aconitum napellus* ssp. *tauricum*. *Planta Medica, 67*(3), 287–290. doi:10.1055/s-2001-11994

Fico, G., Braca, A., De Tommasi, N., Tome, F., & Morelli, I. (2001). Flavonoids from *Aconitum napellus* subsp. *neomontanum*. *Phytochemistry, 57*(4), 543–546. doi:10.1016/S0031-9422(01)00102-9

Friese, J., Gleitz, J., Ulrike, T., Heubach, J. F., Matthiesen, M., Wilffert, B., & Selve, N. (1997). *Aconitum* sp. alkaloids: The modulation of voltage-dependent Na+ channels, toxicity and antinociceptive properties. *European Journal of Pharmacology, 377*, 165–174.

Fu, M., Wu, M., Qiao, Y., & Wang, Z. (2006). Toxicological mechanisms of *Aconitum* alkaloids. *Pharmazie, 61*(9), 735–741.

Fu, M., Wu, M., Wang, J. F., Qiao, Y. J., & Wang, Z. (2007). Disruption of the intracellular Ca2+ homeostasis in the cardiac excitation-contraction coupling is a crucial mechanism of arrhythmic toxicity in aconitine-induced cardiomyocytes. *Biochemical and Biophysical Research Communications, 354*(4), 929–936.

Fujita, Y., Terui, K., Fujita, M., Kakizaki, A., Sato, N., Oikawa, K., … Endo, S. (2007). Five cases of aconite poisoning: Toxicokinetics of aconitines. *Journal of Analytical Toxicology, 31*, 132–137.

Grin. (2009). USDA, ARS, National Genetic Resources Program. Germplasm Resources Information Network—(GRIN) [Online Database]. National Germplasm Resources Laboratory, Beltsville, Maryland. Accessed May 25, 2022 from https://npgsweb.ars-grin.gov/gringlobal/taxon/taxonomydetail?id=1347

Haine, G. B., El Ghandour, S. H., El Ghandour, S., & Fréz, A. R. (2021). Assessment of homeopathic medicine *Aconitum napellus* in the treatment of anxiety in an animal model. *International Journal of High Dilution Research, 11*(38), 33–42. doi:10.51910/ijhdr.v11i38.547

Heywood, V. H. (1978). *Flowering Plants of the World*. Batsford, London, England.

Jacquemart, A. L., Buyens, C., Herent, M. F., Quetin-Leclercq, J., Lognay, G., Hance, T., & Quinet, M. (2019). Male flowers of Aconitum compensate for toxic pollen with increased floral signals and rewards for pollinators. *Scientific Reports, 9*(1), 16498. doi:10.1038/s41598-019-53355-3

Jamtsho, T., Yeshi, K., Samten, & Wangchuk, P. (2021). Comparative analysis of two Himalayan Aconitum species for their phytopharmaceutical properties. *Journal of Herbal Medicine, 32*, 100497. doi:10.1016/j.hermed.2021.100497

Johnson, D. (2007). *Aconitum napellus* habitat. Accessed May 30, 2022 from http://bioweb.uwlax.edu/bio203/s2013/johnson_dan2/habitat.htm.

Kelly, S. (1990). Aconite poisoning. *Medical Journal of Australia, 153*, 499.

Kiss, T., Orvos, P., Bansaghi, S., Forgo, P., Jedlinszki, N., Talosi, L., … Csupor, D. (2013). Identification of diterpene alkaloids from *Aconitum napellus* subsp. firmum and GIRK channel activities of some Aconitum alkaloids. *Fitoterapia, 90*, 85–93. doi:10.1016/j.fitote.2013.07.010

Lauener, L. A., & Tamura, M. (1978). A synopsis of *Aconitum* subgenus *paraconitum*: I. *Note from the Royal Botanic Garden Eidenburgh, 37*, 113–124.

Li, Y., Tu, D., Xiao, H., Du, Y., Zou, A., Liao, Y., & Dong, S. (2010). Aconitine blocks HERG and Kv1.5 potassium channels. *Journal of Ethnopharmacology, 131*(1), 187–195.

Lin, C. C., Chan, T. Y., & Deng, J. F. (2004). Clinical features and management of herb-induced aconitine poisoning. *Annals of Emergency Medicine, 43*(5), 574–579.

Liu, H., & Katz, A. (1995). Norditerpenoid alkaloids from *Aconitum napellus* ssp. neomontanum. *Plant Medica, 62*, 191–192.

Liu, Y. (2019). *Poisonous Medicine in Ancient China. Toxicology in Antiquity.* Elsevier, New York, NY, 431–439.

Loo, M. (2009). Chapter 36—Fever. In: Loo, M., ed. *Integrative Medicine for Children*. W.B. Saunders, Philadelphia, PA, 335–341.

Luis, J. C., Valdes, F., Martin, R., Carmona, A. J., & Diaz, J. G. (2006). DPPH radical scavenging activity of two flavonol glycosides from *Aconitum napellus* sp. lusitanicum. *Fitoterapia, 77*(6), 469–471. doi:10.1016/j.fitote.2006.05.018

Martin-Escolano, R., Molero Romero, S., Diaz, J. G., Marin, C., Sanchez-Moreno, M., & Rosales, M. J. (2021). In vitro anti-Acanthamoeba activity of flavonoid glycosides isolated from Delphinium gracile, D. staphisagria, Consolida oliveriana and *Aconitum napellus*. *Parasitology, 148*(11), 1392–1400. doi:10.1017/S0031182021001025

Mitka, J., Novikov, A., & Rotten-Steiner, W. (2021). The taxonomic circumscription of Aconitum subgenus Aconitum (Ranunculaceae) in Europe Webbia. *Raccolta de Scritti Botanici, 76*, 11–45.

Moog, F. P., & Karenberg, A. (2002). Toxicology in the old testament did the High Priest Alcimus die of acute aconitine poisoning? *Adverse Drug Reactions and Toxicological Reviews, 21*, 151–156.

Munch, J. C., & Crosbie, H. H. (1929). Bioassay of Aconite and preparations. 2. The pharmacology and pharmacognosy of various species of Aconitum. *American Pharmaceutical Association, 18*(10), 886–892.

Qasem, A. M. A., Zeng, Z., Rowan, M. G., & Blagbrough, I. S. (2022). Norditerpenoid alkaloids from Aconitum and Delphinium: Structural relevance in medicine, toxicology, and metabolism. *Natural Product Reports, 39*(3), 460–473. doi:10.1039/d1np00029b

Sato, H., Yamada, C., Konno, C., Ohizumi, Y., Endo, K., & Hikkino, H. (1979). Pharmacological actions of Aconitine alkaloids. *Tohoku Journal of Experimental Medicine, 128*, 175–187.

Shoaib, A., Badruddeen, Siddiqui, H. H., Dixit, R. K., & Akhtar, J. (2019). *Aconitum Napellus*: Detoxification and acute toxicity investigation followed by sub-acute toxicity and bioavailability assessment of highest and lowest LD50 extract. *Journal of Biologically Active Products from Nature, 9*, 108–119.

Shoaib, A., Salem-Bekhit, M. M., Siddiqui, H. H., Dixit, R. K., Bayomi, M., Khalid, M., … Shakeel, F. (2020). Antidiabetic activity of standardized dried tubers extract of *Aconitum napellus* in streptozotocin-induced diabetic rats. *3 Biotech, 10*(2), 56. doi:10.1007/s13205-019-2043-7

Shoaib, A., Siddiqui, H. H., Dixit, R. K., Siddiqui, S., Deen, B., Khan, A., … Ahmad, P. (2020). Neuroprotective effects of dried tubers of *aconitum napellus*. *Plants (Basel), 9*(3). doi:10.3390/plants9030356

Shyaula, S. L. (2011). Phytochemicals, traditional uses and processing of *Aconitum* species in Nepal. *Nepal Journal of Science and Technology, 12*, 171–178.

Singhuber, J., Zhu, M., Prinz, S., & Kopp, B. (2009). Aconitum in traditional Chinese medicine: A valuable drug or an unpredictable risk? *Journal of Ethnopharmacology, 126*(1), 18–30. doi:10.1016/j.jep.2009.07.031

Stork, C., & Marraffa, J. (2005). *Aconitum* species. In: Wexler, P., ed. *Encyclopedia of Toxicology* (2nd ed.). Elsevier, Oxford, England, 39–40.

Tang, W., & Eisenbrand, G. (1992). *Chinese Drugs of Plant Origin.* Springer Verlag, Berlin-Heidelberg, 14–44.

Utelli, A. B., Roy, B. A., & Baltisberger, M. (2000). Molecular and morphological analyses of European *Aconitum* species (Ranunculace). *Plant Systematics and Evolution, 224*, 195–212.

Venkataraghavan, S., & Sundareesan, T. (1981). A short note on contraceptive in Ayurveda. *Journal of Science Research Pl Med, 2*(39).

Ventre, E. K., Ambler, J. A., Henry, H. C., Byall, S., & Paine, H. S. (1946). Extraction of aconitic acid from Sorgo. *Industrial and Engineering Chemistry, 38*(2), 201–204.

Vo, K. T., Tabas, J. A., & Smollin, C. G. (2017). Alternating ventricular complexes after overdose from an herbal medication. *JAMA Internal Medicine, 177*(8), 1199–1201.

Wang, R. H., Guo, W. S., & Liu, Y. J. (2011). The effect of modified Mahuangfuzixixin treating arthralgia on 50 cases. *China Modern Medicine, 18*, 95–96.

Wang, X. C., Jia, Q. Z., Yu, Y. L., Wang, H. D., Guo, H. C., Ma, X. D., … Wang, C. (2021). Inhibition of the INa/K and the activation of peak INa contribute to the arrhythmogenic effects of aconitine and mesaconitine in guinea pigs. *Acta Pharmacologica Sinica, 42*(2), 218–229. doi:10.1038/s41401-020-0467-6

Yu, H.-H., Li, M., Li, Y.-B., Lei, B.-B., Yuan, X., Xing, X.-K., … Cheng, B.-F. (2020). Benzoylaconitine inhibits production of IL-6 and IL-8 via MAPK, Akt, NF-κB signaling in IL-1β-induced human synovial cells. *Biological and Pharmaceutical Bulletin, 43*, 334–339.

Zhang, Y., Zong, X., Wu, J. L., Liu, Y., Liu, Z., Zhou, H., … Li, N. (2020). Pharmacokinetics and tissue distribution of eighteen major alkaloids of Aconitum carmichaelii in rats by UHPLC-QQQ-MS. *Journal of Pharmaceutical and Biomedical Analysis, 185*, 113226. doi:10.1016/j.jpba.2020.113226

Zhou, W., Liu, H., Qiu, L. Z., Yue, L. X., Zhang, G. J., Deng, H. F., … Gao, Y. (2021). Cardiac efficacy and toxicity of aconitine: A new frontier for the ancient poison. *Medicinal Research Reviews, 41*(3), 1798–1811. doi:10.1002/med.21777

7 *Acorus calamus* (Sway or Muskrat Root)

Yashashree Pradhan, Hina Alim, Nimisha Patel, Johra Khan, Belkıs Muca Yiğit, Kamal Fatima Zahra, and Ahmad Ali

CONTENTS

7.1 INTRODUCTION

There is a large number of trees, herbs and shrubs present in nature. Many of them are useful for humans in some or the other way. Some plants produce metabolites which show medicinal effect against many diseases. Due to such effects, these plants are also used in ancient world for the treatment of diseases (Ali, 2020). Due to side effects of long-term use of allopathic medicines, an increasing interest in herbal medicines contributes to factors like cultural acceptability, better

DOI: 10.1201/b23017-9

compatibility and feasibility. The traditional knowledge of many plants is more accessible, reliable and trusted by common people all over the world (Vidyarthi et al. 2013). Nowadays also 80% of world uses traditional medicine for the treatment of disease as observed in the report given by WHO. Some of these plants also show poisonous effects when used in large quantities or in inappropriate form (Balakumbahan et al. 2010). In traditional, Ayurveda and Unani medicines, many poisonous plants are used as therapeutics by reducing its poisonous effects via various procedures. One of such poisonous plants is Sway or Muskrat root, i.e., *Acorus calamus*.

A. calamus is a member of the Acoraceae family. This plant has many phytochemicals which are used as therapeutics. It is used for treatment of many diseases in ayurvedic medicine such as chronic diarrhea, cough, dysentery, tympanitis, otitis media, asthma, intermittent fevers, and glandular and abdominal tumors (Mukherjee et al. 2007). This chapter discusses about traditional knowledge, bioactive compounds, pharmacological studies and toxic response of *A. calamus*.

7.2 BOTANICAL DESCRIPTION

A. calamus is semiaquatic, aromatic and perennial herb. It has aromatic, highly branched and creeping rhizome which is cylindrical and 2.5 cm thick. Its color is light brown externally and white internally (Figure 7.1) (Mukherjee et al. 2007). The leaves are 0.7–1.7 cm wide. Its sympodial leaf is shorter than vegetative leaf. It has undulated margin (Figure 7.2). This plant rarely flowers. Its

FIGURE 7.1 Rhizomes of *Acorus calamus*. (Courtesy: ICAR–Central Coastal Agricultural Research Institute, Goa India.)

FIGURE 7.2 Leaves of *Acorus calamus*. (Courtesy: ICAR – Central Coastal Agricultural Research Institute, Goa, India.)

flowers are cylindrical, greenish brown and covered with rounded spikes and are of 3–8 cm. Its fruits are berry like, small and contain very few seeds (Balakumbahan et al. 2010).

7.3 DISTRIBUTION

A. calamus is native to eastern Europe and central Asia. It is indigenous to marshes of mountains of India. It is cultivated at higher altitudes in India about 2,200 m. It is found in the states of Maharashtra, Tamil Nadu, Manipur, Jammu Kashmir, Nagaland, Himachal Pradesh, Andhra Pradesh, Karnataka, Uttar Pradesh and Uttarakhand (Rajput et al. 2014).

7.4 PHYTOCHEMICAL CONSTITUENTS

A. calamus or muskrat root contains essential oils in its leaves and rhizomes (Tables 7.1 and 7.2). These essential oils contain various phytochemicals such as alcohols, terpenes, alkaloids, amides, terpenoids, sugars, volatile oils, lignans, esters, organic acids, steroids and glycosides (Mukherjee et al. 2007). GC-MS analysis of essential oils shows major phytochemicals such as Asarone, socalamendiol, eugenol, Camphor, Elemicin, maltose, glucose, fructose, Linolenic acid, 2-Allyl-5-Ethoxy-4-Methoxyphenol, acorenone and Nonanoic Acid. These phytochemicals have many medicinal effects (Balakumbahan et al. 2010).

TABLE 7.1
Percentage Composition of Essential Oil of *Acorus calamus* L. (Raina et al. 2003)

Sr. No.	Name of Compound	Leaf Oil (%)	Rhizome Oil (%)
1	*n*-Octane	T	t
2	*n*-Nonane	T	t
3	*α*-Pinene	T	–
4	*β*-Pinene	T	–
5	*n*-Decane	–	t
6	*β*-Phellandrene	0.1	–
7	(*Z*)-Ocimene	–	0.1
8	Linalool	4.7	0.1
9	Terpinen-4-ol	0.1	–
10	*α*-Terpineol	0.2	0.4
11	Linalyl acetate	0.2	0.2
12	Undecanol	T	–
13	Linalyl propionate	–	0.1
14	*δ*-Elemene	0.2	0.3
15	*β*-Cubebene	0.2	–
16	*β*-Caryophyllene	0.5	0.2
17	(*Z*)-Methyl isoeugenol	0.6	0.6
18	*β*-Gurjurene	–	0.4
19	*ar*-Curcumene	T	0.1
20	*α*-Muurolene	0.7	0.4
21	(*Z*)-*α*-Bisabolene	0.5	0.1
22	*β*-Bisabolene	1.9	1.2
23	Elemicin	0.4	t
24	*α*-Cadinene	–	0.1
25	*α*-Calacorene	0.1	0.1
26	GermacreneB	–	t
27	(*Z*)-Isoelimicin	1.3	1.1
28	Spathulenol	0.1	–
29	Caryophyllene oxide	0.4	0.1
30	*β*-Asarone	85.6	83.2
31	Humulene epoxide	T	–
32	T-Muurolol	0.1	–
33	*α*-Cadinol	0.3	t
34	*α*-Asarone	1.0	9.7
35	*α*-Bisabolol	–	0.6
36	*n*-Heptadecane	0.4	0.3
37	Benzyl benzoate	–	0.2
38	*n*-Octadecane	–	0.1
39	*L*-Nonadecanol	T	–

t, trace (< 0.05).

TABLE 7.2
Major Bioactive Phytochemicals Present in *Acorus calamus*

Structure	Name and Reference
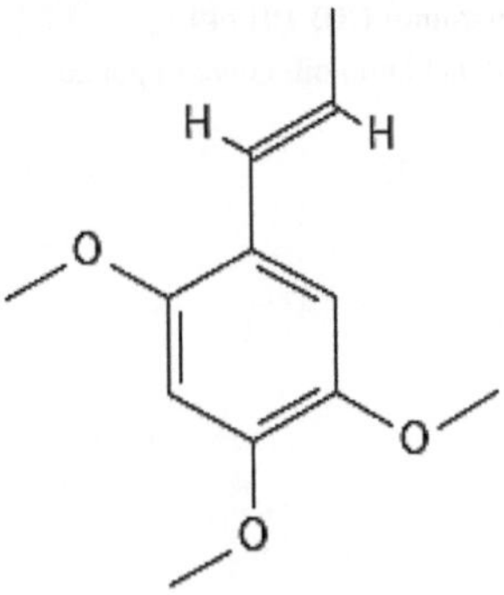	***α*-Asarone (Rhizome) CID 636822** https://pubchem.ncbi.nlm.nih.gov/compound/alpha-Asarone. Retrieved May 2, 2022
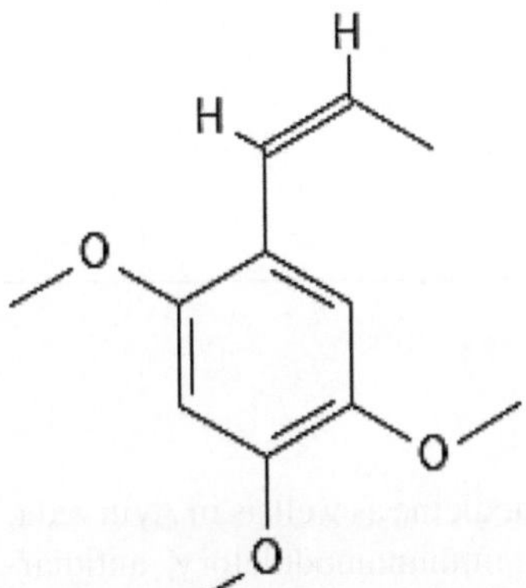	***β*-Asarone (Leaves and rhizome) CID 5281758** https://pubchem.ncbi.nlm.nih.gov/compound/beta-Asarone. Retrieved May 2, 2022
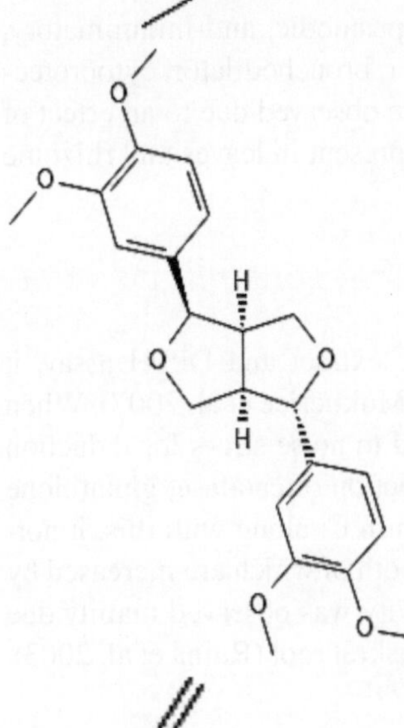	**Epieudesmin (Rhizome) CID 7000209** https://pubchem.ncbi.nlm.nih.gov/compound/Epieudesmine. Retrieved May 2, 2022
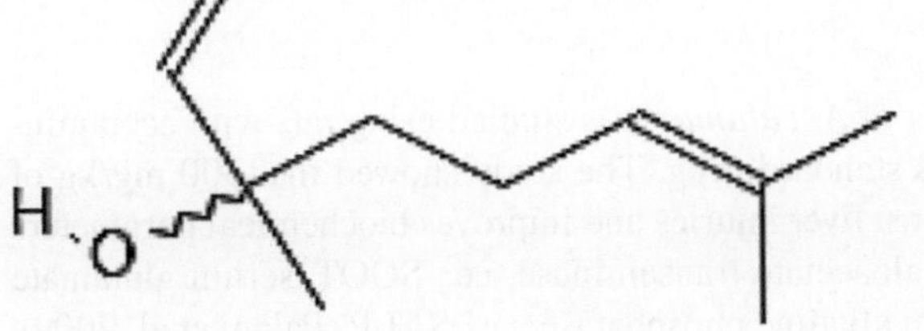	**Linalool (Leaves) CID 6549** https://pubchem.ncbi.nlm.nih.gov/compound/Linalool. Retrieved May 2, 2022

(Continued)

TABLE 7.2 (*Continued*)
Major Bioactive Phytochemicals Present in *Acorus calamus*

Structure	Name and Reference
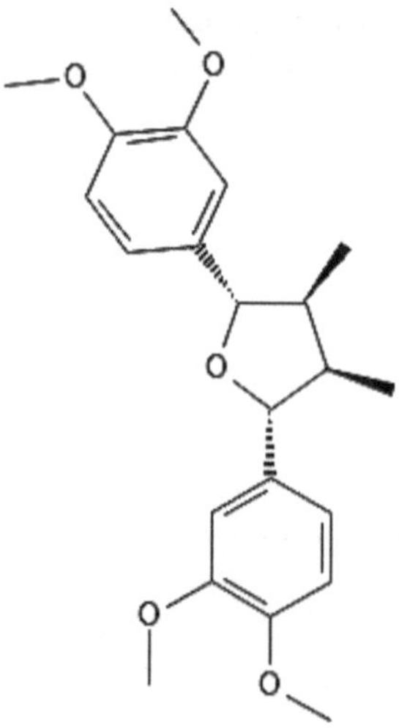	**Galgravin (Rhizome) CID 101749** https://pubchem.ncbi.nlm.nih.gov/compound/Galgravin. Retrieved May 2, 2022

7.5 PHARMACOLOGICAL STUDIES

A. calamus is one of the medicinal plants used widely in traditional medicine as well as in Ayurveda. It shows antioxidant, hepatoprotective, antibacterial, insecticidal, immunomodulatory, antidiarrheal, anticonvulsant, sedative, acetylcholinesterase inhibitory, antispasmodic, anti-inflammatory, anthelmintic, effect on cardiovascular system, hypolipidemic, antiulcer, bronchodilator, cytoprotective, anticancer, antifungal and antiviral activities. These activities are observed due to an effect of various phytochemicals present in extract as well as in essential oils present in leaves and rhizome of sway.

7.5.1 Antioxidant Activity

A. calamus when tested for antioxidant activity using ethyl acetate extract and DPPH assay, it showed efficiency in radical scavenging at concentration of 0.2 g/mL (Mukherjee et al. 2007). When methanolic and ethyl acetate extract of sway is tested on rats exposed to noise stress for induction of oxidative stress on brain, it showed that sway extracts help in reduction of catalase, glutathione peroxidase, reduced forms of glutathione (GSH), vitamin E and vitamin C; along with this, it normalizes the secretion of superoxide dismutase and lipid peroxidation both of which are increased by oxidative stress (Manikandan and Devi, 2005). This antioxidant activity was observed mainly due to the presence of α-Asarone present in large amount in rhizome of muskrat root (Raina et al. 2003).

7.5.2 Hepatoprotective Activity

Hepatoprotective activity of ethanolic extract of *A. calamus* was studied using rats with acetaminophen-induced liver injury and silymarin as standard drug. The study showed that 500 mg/kg of ethanolic extract of *A. calamus* protects against liver injuries and improves biochemical parameters of blood serum such as serum glutamate oxaloacetate transaminase, i.e., SGOT, serum glutamate pyruvate transaminase, i.e., SGPT, and serum alkaline phosphatase, i.e., SALP (Palani et al. 2009).

7.5.3 Antibacterial Activity

The antibacterial effect of ethanolic extract of *A. calamus* was studied using microbial cultures, viz., *Escherichia coli, Staphylococcus citreus, Salmonella paratyphi A* and *B, Shigella dysomei, Proteus vulgaris, Salmonella marcescens, Bacillus megaterium, Staphylococcus aureus, Corynebacterium diphtheriae, Diplococcus pneumoniae, Streptococcus viridans* and *Streptococcus pyogenes*, when tested using agar diffusion method (Mukherjee et al. 2007). α-Asarone and β-Asarone present in the ethyl acetate extract of rhizome of muskrat root show good MIC as well as inhibition zones against *S. aureus, Salmonella typhi, S. paratyphi A, Pseudomonas aeruginosa, Enterococcus faecalis, Shigella sonnei* and *E. coli* (Devi and Ganjewala, 2009). *A. calamus* volatile oils showed antibacterial activity against *Mycobacterium tuberculosis, Salmonella faecalis, Streptococcus pyogenes, Staphylococcus albus, Pseudomonas solanacearum, Bacillus pumilus* and *Corynebacterium diphtheriae* (Jain et al. 1974). Essential oils from muskrat root showed efficiency as antibacterial agent when tested against *Shigella shiga, S. paratyphi, Bacillus proteus, Staphylococcus pyogenes, Shigella boydii, E. coli* and *S. typhi* (Kapil et al. 1983).

7.5.4 Insecticidal Activity

Essential oil extracted from rhizome of *A. calamus* contains high insecticidal potential. Cis- and trans-Asarone present in this oil affects feeding ability as well as gonadal development in *Peridroma saucia* which is one of the pests on crops (Koul et al. 1990). These oils also show cidal activity against *Musca domestica, Spodoptera litura* and *Dysdercus koenigii*. It also shows complete mortality of *Culex fatigans* and *Aedes aegypti* (Mukherjee et al. 2007). Apart from essential oils, solvent extracts of rhizome of *A. calamus* show insecticidal effect against larvae of *Culex fatigans* and *Musca nebulo*. Powder form of rhizome is useful against lice, bugs and moths. Repellent activity is also observed against pulse beetle *Callosobruchus chinensis* (Mukherjee et al. 2007). Total mortality of *Atteva fabriciella* larvae is also seen with the use of ethanolic extract of rhizome of sway (Ahmad et al. 1991).

7.5.5 Immunomodulatory Activity

Ethanolic extract of *A. calamus* rhizome was tested for immunomodulatory effect using in vitro technique. Various human and rat cell lines, viz., human peripheral blood mononuclear cells (PBMCs), interleukin-2 (IL-2), intracytoplasmic interferon-γ (IFN-γ) and tumor necrosis factor-α (TNF-α), are used. The study revealed that the extract showed inhibition of antigens on PBMCs, whereas downregulation of CD25 antigen on PBMCs was observed (Mehrotra et al. 2003).

7.5.6 Antidiarrheal Activity

Castor oil induced diarrhea in Wistar albino rats is used to study the antidiarrheal effect of *A. calamus*. Methanolic extract of leaves of sway showed antidiarrheal effect with a single dose of 400 mg/kg of body weight (Kapadia and Kharat, 2012).

7.5.7 Anticonvulsant Activity

Anticonvulsant activity of *A. calamus* rhizome oil was tested using Metrazol-induced seizures in albino mice. The study suggested that dosage of volatile oil of sway before exposure to maximal electroshock to induce seizures prevents the mice from convulsion. This prevention is due to the phytochemical α-Asarone present in oil (Sharma et al. 1961). Dosage of methanolic extract of sway rhizome at concentration of 100–200 mg/kg body weight helps in preventing Pentylenetetrazol-induced seizures (Rajput et al. 2014).

7.5.8 Sedative Activity

Administration of volatile oils of *A. calamus* to the mice supplemented with pentobarbitone enhanced the sleeping time of mice. Ethanol and petroleum ether extracts of *A. calamus* also showed good sedative activity (Dandiya et al. 1959).

7.5.9 Acetylcholinesterase Inhibitory Activity

Acetylcholinesterase inhibitory activity was observed when tested using essential oil and hydro-alcohol extract of *A. calamus* and bovine erythrocytes. Methanolic extracts also showed the inhibitory activity of acetylcholinesterase at concentration of 200 μg/mL. This activity is due to the compounds α-Asarone and β-Asarone (Houghton et al. 2006).

7.5.10 Antispasmodic Activity

A. calamus rhizome essential oil when tested using dogs and rabbits for study of involuntary muscle movements of intestines showed inhibition of increased peristaltic movements. Whereas alcoholic extract showed relaxation of smooth muscles of rat intestines and also reduced contractions of heart were observed when it is administered to frog (Mukherjee et al. 2007). One of the studies also showed that n-hexane extract of *A. calamus* acts as calcium channel blockade like constituent which leads to antispasmodic effect in jejunum of guinea pigs (Gilani et al. 2006).

7.5.11 Anti-Inflammatory Activity

A. calamus rhizome extract when tested against chronic, immunologic and acute inflammation models showed significant anti-inflammatory activity (Vohra et al. 1989). When cotton pellet granuloma formation, croton oil granuloma and carrageenin-induced paw edema models of rats are administered with oral dose of *A. calamus* extract showed inhibitory response against acute and chronic inflammation (Mukherjee et al. 2007). In albino rats, 25%–75% acetone extract acted as anti-inflammatory agent. Leaf extract of *A. calamus* showed anti-inflammatory activity when tested on human keratinocyte HaCaT cells and further inhibited the production of pro-inflammatory cytokines (Rajput et al. 2014).

7.5.12 Anthelmintic Activity

Alcoholic extract of *A. calamus* showed partial paralysis in *Ascaris lumbricoides*, whereas phenolic extract showed complete paralysis (Chaudhari et al. 1981). Phenolic extract showed nematocidal activity when tested on *Meloidogyne incognita*. Clinical trial was done on children aged between 5 and 11 which showed complete cure against round worm when given 250 mg powder of *A. calamus* three times a day for 3 days. The efficiency of this trial was 83% (Mukherjee et al. 2007).

7.5.13 Effect on Cardiovascular System

Lowering of blood pressure can be observed with the use of essential oil of *A. calamus*. It also showed curing effects on atrial fibrillation, atrial flutter as well as ventricular arrhythmias when tested in dogs with quinidine as standard drug. It also causes increase in conduction time as well as refractory period in atria of rabbits (Madan et al. 1960). Alcoholic extract of *A. calamus* showed hypotensive effect on blood pressure of dogs. *A. calamus* also helps in reduction of LDL and increase in HDL. It also helps in reduction of chest pain, dyspnea on effort and body weight index (Mukherjee et al. 2007).

7.5.14 Hypolipidemic Activity

Ethanolic extract of *A. calamus* rhizome when tested on albino rats showed reduction of cholesterol biosynthesis in liver at the dose of 100–200 mg/kg. Saponins from the extract also reduce absorption and circulation of cholesterol in enterohepatic system (Parab and Mengi, 2002). α-Asarone present in the extract shows hypolipidemic mechanism by binding the active site of HMG-CoA reductase enzyme present in the liver (Rajput et al. 2014).

7.5.15 Cytoprotective and Antiulcer Activity

When the rats are exposed to cytodestructive agents such as 25% NaCl, 0.6 M HCl, 80% ethanol and 0.2 M NaOH and further administered alcoholic extract of *A. calamus*, they showed cytoprotective effect. Ethanolic extract also inhibits the gastric secretion as well as protects against gastric injuries due to reserpine, pyloric ligation, cysteamine and indomethacin leading to antiulcer effect (Mukherjee et al. 2007).

7.5.16 Bronchodilator Activity

Rhizome of *A. calamus* is given to patients of moderate to severe asthma for chewing for 2–4 weeks, and it showed positive results with bronchodilation and without any side effects. It also reduces the bronchospasms in asthmatic patients (Chandra, 1980).

7.5.17 Anticancer Activity

Methanolic extract of *A. calamus* was tested for anticancer activity using various human cancer cell lines and murine P388 lymphocytic leukemia cell line. It was observed that epieudesmin present in extract showed antineoplastic activity, whereas galgravin prevents neuronal death (Rajput et al. 2014). Lectins present in the rhizome extract of *A. calamus* show inhibition of proliferation of WEHI-279 which is a B-cell lymphoma and J774 which is a murine macrophage cancer cell line (Mukherjee et al. 2007).

7.5.18 Antifungal Activity

Antifungal activity of *A. calamus* leaves and rhizome was tested using cultures of *Penicillium crysogenum*, *Aspergillus flavus*, *Aspergillus niger*, *Candida albicans*, *Microsporum canis* and *Cryptococcus gastricus*. The organic solvent extracts showed good antifungal activity in agar diffusion method as well as in MIC technique (Devi and Ganjewala, 2009). The essential oils from *A. calamus* are tested for antifungal activity, and it showed effect against *Phomopsis destuctum*, *Diplodia natalensis*, *Candida albicans*, *Penicillium digitatum*, *Alternaria tenuis*, *Aspergillus oryzae* and *Helminthosporium oryzae* due to the presence of compounds like β-Asarone, acoradin and asaraldehyde (Mukherjee et al. 2007).

7.5.19 Antiviral Activity

A. calamus rhizome alcoholic extract showed antiviral activity against herpes simplex virus HSV-1 and HSV-2. This inhibition of virus is due to the presence of β-Asarone in the extract. It shows antiviral activity below cytotoxicity level when tested using vero cells (Mukherjee et al. 2007).

7.6 TOXIC RESPONSE

β-Asarone present in *A. calamus* has ability to act as mild co-carcinogen which interferes with normal pregnancy inter-reactions. It metabolically activates cellular damage in human lymphocytes and induces structural chromosomal aberrations. It also shows mutagenicity in Salmonella-mammalian microsome assay (Mukherjee et al. 2007). When β-Asarone was administered to rats and mice for 2 years, it was observed that it induced leiomyosarcomas of the small intestine as well as thrombosis in heart chambers. It showed LD_{50} of 1,010 mg/kg by weight in rats, whereas LD_{50} of 184 mg/kg by weight in mice when administered orally (Rajput et al. 2014).

When α-Asarone and β-Asarone are tested using Chinese hamster lung fibroblasts V79 cells and human liver hepatoma HepG2 cells, these compounds showed inhibition of cell viability which suggests cytotoxic effect of these compounds. It also induces DNA damage by breaking DNA strands (Haupenthal et al. 2017).

7.7 TRADITIONAL AND OTHER POTENTIAL USES

A. calamus or muskrat root or sway is one of the plants used in ancient medicine. It is used in many medicinal systems such as Ayurveda, western herbal medicine and Unani medicine. In these systems, rhizome of sway is mainly used in different formulations for the treatment of various ailments (Rajput et al. 2014).

In Unani medicine, *A. calamus*-containing medicine named as Ma'jun Baladur is used for the treatment of diseases like headache, coma, epilepsy, hysteria, sinusitis and drowsiness (Rajput et al. 2014). In western herbal medicine, it is used in various preparations such as powder and paste. These formulations are used for the treatment of ailments like poor digestion, bloating, gas, stomach ache and headache. A small amount of *A. calamus* mixed with honey and jaggary helps in reducing acidity. *A. calamus* also helps in the treatment of eczema, rheumatism, liver and kidney problems (Mukherjee et al. 2007). In Himalayas, tribals use decoction of rhizome of *A. calamus* as nonalcoholic beverage as well as it is also used for preventing intoxication from alcohol. Tirumala hills natives use these rhizomes for the treatment of dental problems (Rao et al. 1996).

In Ayurvedic medicine, *A. calamus* is known as Vacha in Sanskrit. It has the following properties: rasa – tikta and katu, i.e., bitter and pungent, veerya- – ushna, i.e., worm, vipaka – Katu, i.e., pungent, guna – Laghu, Tikshna and Ruksha, i.e., light, strong and dry and dosha – kaphaghna and vataghna calms kapha and vata (Deshpande et al. 2018). It acts as memory enhancer. Paste of *A. calamus* applied topically to treat arthritis and paralysis due to its pain reducing and anti-inflammatory action. Its extract is used to treat earache and tinnitus. Fumes of *A. calamus* are used to reduce pain of piles (Rajput et al. 2014).

Its katu and ushna, guna acts as appetizer but also causes emesis due to gastritis when ingested in larger amounts. Its decoction is used to quench thirst caused by diarrhea and dysentery. It helps in liquification of phlegm accumulated in laryngopharynx due to its tikshna and ushna properties. It regulates pulse rate and reduces blood pressure. Due to its uterine contracting effect, it is used in painful labor. It also increases kidney function and urine production, and hence, it is used in the treatment of renal calculi and dysuria (Deshpande et al. 2018).

Apart from the traditional use of *A. calamus*, it can also be used as biopesticide against various pests on crops. The toxic effect of chrysin shows inhibiting effect on larval stages of *Spodoptera litura* by reducing carboxylesterase activity (Wiwattanawanichakun et al. 2022). *A. calamus* extracts also show toxicities in *Callosobruchus maculatus*, *Lasioderma serricorne*, *Sitophilus oryzae* and *Tribolium confusum* larvae which are the insects causing spoilage of stored food (Su, 1991).

7.8 FUTURE REMARKS

A. calamus or muskrat root or sway is perennial shrub of Acoraceae family. It is used in many countries for treatment of many diseases. It contains many phytochemicals in its leaves and rhizomes.

α-Asarone, β-Asarone, chrysin, linalool, epieudesmin and galgravin are bioactive phytochemicals present in leaves and rhizomes of *A. calamus*. These compounds show various pharmacological activities such as antioxidant, hepatoprotective, antibacterial, insecticidal, immunomodulatory, antidiarrheal, anticonvulsant, sedative, acetylcholinesterase inhibitory, antispasmodic, anti-inflammatory, anthelmintic, effect on cardiovascular system, hypolipidemic, antiulcer, bronchodilator, cytoprotective, anticancer, antifungal and antiviral. Also, in traditional medicine, muskrat root is used for the treatment of eczema, rheumatism, stomach problems, liver and kidney problems, and mental illnesses. Apart from these activities, *A. calamus* rhizome extracts show in vitro cytotoxicity as well as genotoxicity and carcinogenicity in rats and mice. This toxicity can be used as biopesticides as it shows strong impact on various pests of crop. Due to these toxic effects, it is used in small amounts for any medicinal uses.

NOTES

Yashashree Pradhan, Department of Life Sciences, University of Mumbai, Mumbai, India
Hina Alim, Department of Life Sciences, University of Mumbai, Mumbai, India
Nimisha Patel, Department of Life Sciences, J.C. Bose University of Science & Technology, YMCA, Faridabad, India
Johra Khan, Department of Medical Laboratory Sciences, College of Applied Medical Sciences, Majmaah University, Al Majmaah, Saudi Arabia
Belkıs Muca Yiğit, Vocational School of Technical Sciences, Department of Forestry, Igdir University, Igdir, Turkey
Kamal Fatima Zahra, Faculty of Sciences and Techniques, Laboratory of Physical Chemistry of Processes and Materials/Agri-Food and Health, Hassan First University, Settat, Morocco
Ahmad Ali, Department of Life Sciences, University of Mumbai, Mumbai, India

REFERENCES

Ahmad, M., Gupta, B. K., & Bhandari, R. S. 1991. Efficacy of some plant extracts against Ailanthus web worm Atteva fabriciella. *Indian Journal of Forestry*, 14(1), 5–7.

Ali, A. 2020. Herbs that heal: The philanthropic behaviour of nature. *Annals of Phytomedicine,* 9, 7–17.

Balakumbahan, R., Rajamani, K., & Kumanan, K. 2010. Acorus calamus: An overview. *Journal of Medicinal Plants Research*, 4(25), 2740–2745.

Chandra, P. 1980. A note on the preliminary study on Acorus calamus L. in the treatment of bronchial asthma. *Journal of Research in Ayurveda and Siddha*, 1(2), 329–330.

Chaudhari, G. N., Kokate, C. K., & Nimbkar, A. Y. 1981. Search for anthelmintics of plant origin: Activities of volatile principles of Acorus calamus against Ascaris lumbricoides. *Ancient Science of Life*, 1(2), 103.

Dandiya, P. C., Baxter, R. M., Walker, G. C., & Cullumbine, H. 1959. Studies on acorus calamus, part II: Investigation of volatile oil. *Journal of Pharmacy and Pharmacology*, 11(1), 163–168.

Deshpande, A., Javalgekar, R., & Ranade, S. 2018. *Dravyaguna Vidnyan*, 4th ed. Proficient Publishing House: Pune, India.

Devi, S. A., & Ganjewala, D. 2009. Antimicrobial activity of Acorus calamus (L.) rhizome and leaf extract. *Acta Biolgica Szegediensis*, 53(1), 45–49.

Gilani, A. U. H., Shah, A. J., Ahmad, M., & Shaheen, F. 2006. Antispasmodic effect of Acorus calamus Linn. is mediated through calcium channel blockade. *Phytotherapy Research*, 20(12), 1080–1084.

Haupenthal, S., Berg, K., Gründken, M. Vallicotti, M. Hemgesberg, M. Sak, K. Schrenk, D., & Esselen, M. 2017. In vitro genotoxicity of carcinogenic asarone isomers. *Food & Function*, 8(3), 1227–1234.

Houghton, P. J., Kumar, V., Govindarajan, R., & Mukherjee, P. K. 2006. Asarones in Acorus calamus and their acetylcholinesterase inhibition. *Journal of Pharmacy and Pharmacology,* 58, 55–55.

Jain, S. R., Jain, P. R., & Jain, M. R. 1974. Antibacterial evaluation of some indigenous volatile oils. *Planta Medica*, 26(06), 196–199.

Kapadia, M. D., & Kharat, A. R. 2012. Antidiarrhoeal activity of leaves of Acorus calamus. *International Journal of Pharmaceutical Sciences and Research*, 3(10), 3847.

Kapil, V. B., Sinha, A. K., & Sinha, G. K. 1983. Antibacterial and antifungal study of some essential oils and their constituents from the plants of Kumaon and its Tarai tract. *Bulletin of Medico-Ethnobotnical Research*, 4, 124–129.

Koul, O., Smirle, M. J., & Isman, M. B. 1990. Asarones from *Acorus calamus L.* oil. *Journal of Chemical Ecology*, 16(6), 1911–1920.
Madan, B. R., Arora, R. B., & Kapila, K. 1960. Anticonvulsant, antiveratrinic and anti-arrhythmic action of Acorns calamus Linn. —An Indian indigenous drug. *Archives Internationales de Pharmacodynamie et de Therapie*, 124, 201–211.
Manikandan, S., & Devi, R. S. 2005. Antioxidant property of α-asarone against noise-stress-induced changes in different regions of rat brain. Pharmacological Research, 52(6), 467–474.
Mehrotra, S., Mishra, K. P., Maurya, R., Srimal, R. C., Yadav, V. S., Pandey, R., & Singh, V. K. 2003. Anticellular and immunosuppressive properties of ethanolic extract of Acorus calamus rhizome. *International Immunopharmacology*, 3(1), 53–61.
Mukherjee, P. K., Kumar, V., Mal, M., & Houghton, P. J. 2007. Acorus calamus.: Scientific validation of ayurvedic tradition from natural resources. *Pharmaceutical Biology*, 45(8), 651–666.
Palani, S., Raja, S., Kumar, R. P., Venkadesan, D., Devi, K., Sivaraj, A., & Kumar, B. S. 2009. Therapeutic efficacy of antihepatotoxic and antioxidant activities of Acorus calamus on acetaminophen-induced toxicity in rat. *International Journal of Integrative Biology*, 7(1), 39–44.
Parab, R. S., & Mengi, S. A. 2002. Hypolipidemic activity of Acorus calamus L. in rats. *Fitoterapia*, 73(6), 451–455.
Raina, V. K., Srivastava, S. K., & Syamasunder, K. V. 2003. Essential oil composition of Acorus calamus L. from the lower region of the Himalayas. *Flavour and Fragrance Journal*, 18(1), 18–20.
Rajput, S. B., Tonge, M. B., & Karuppayil, S. M. 2014. An overview on traditional uses and pharmacological profile of Acorus calamus Linn. (Sweet flag) and other Acorus species. *Phytomedicine*, 21(3), 268–276.
Rao, B. N. S., Rajasekhar, D., Raju, D. C., & Nagaraju. N. 1996. Ethno-medicinal notes on some plants of Tirumala hills for dental disorders. *Ethnobotany*, 8, 88–91.
Sharma, J. D., Dandiya, P. C., Baxter, R. M., & Kandel, S. I. 1961. Pharmacodynamical effects of asarone and β-asarone. *Nature*, 192(4809), 1299–1300.
Su, H. C. 1991. Laboratory evaluation of toxicity of calamus oil against four species of stored-product insects. *Journal of Entomological Science*, 26(1), 76–80.
Vidyarthi, S., Samant, S., & Sharma, P. 2013. Traditional and indigenous uses of medicinal plants by local residents in Himachal Pradesh, North Western Himalaya, India. *International Journal of Biodiversity Science and Management,* 9, 185–200.
Vohra, S. B., Shah, S. A., Sharma, K., Naqvi, S. A. H., & Dandiya, P. C. 1989. Antibacterial, antipyretic, analgesic and anti-inflammatory studies on Acorus calamus Linn. *Annals of National Academy of Medical Sciences*, 25, 13–20.
Wiwattanawanichakun, P., Saehlee, S., Yooboon, T., Kumrungsee, N., Nobsathian, S., & Bullangpoti, V. 2022. Toxicity of isolated phenolic compounds from Acorus calamus L. to control Spodoptera litura (Lepidoptera: Noctuidae) under laboratory conditions. *Chemical and Biological Technologies in Agriculture*, 9(1), 1–9.

8 *Agapanthus orientalis* (Lily of the Nile)

Godwin Anywar

CONTENTS

8.1 INTRODUCTION

Agapanthus orientalis is a synonym of *Agapanthus praecox* subsp. *orientalis* (F.M.Leight.) F.M.Leight. It belongs to the family Amaryllidaceae (Plants of the World Online, 2022) and is commonly known as lily of the Nile or African Blue lily (Barceloux, 2008). *Agapanthus* means flower of love. It is derived from the Greek *agape*—love, and *anthos*—flower (Duncan, 1985). The genus *Agapanthus* is popular for its ornamental species and the popularity continues to grow. *Agapanthus* has been known in European gardens since the seventeenth century and remains popular in cultivation throughout the world today (Duncan, 1985; Snoeijer, 2004). However, there is generally limited information on plants from this genus (Snoeijer, 2004).

8.2 BOTANICAL DESCRIPTION AND ECOLOGY

A. orientalis is an evergreen herb with a thick and short stem, flesh roots, up to 20 arching leaves per plant. It has dense umbellate inflorescence, open faced flowers, pale to medium blue or pure white, short perianth segments forming thick clumps (Quattrocchi, 2012; Barceloux, 2008; Figure 8.1).

A. orientalis is native to South Africa, from the Eastern Cape to KwaZulu-Natal (POWO, 2022; Barceloux, 2008; Quattrocchi, 2012). *Agapanthus* is easy to grow from seed but does not come true from seed. It is therefore advisable to always propagate it vegetatively (Snoeijer, 2004). It has been introduced to different parts of the world including parts of the US, Europe (Barceloux, 2008) and Australia where it is regarded as a weed (Weeds of Australia, 2022).

8.3 ETHNOBOTANY

The root infusion of *A. orientalis* is taken to induce vomiting as a cleanser among the Zulu and as a paste for the treatment of swollen legs among the Xhosa of South Africa (Pearse, 1978) and widely

DOI: 10.1201/b23017-10

FIGURE 8.1 *Agapanthus orientalis.*

used to promote child birth by herbalists in South Africa (Batten and Bokelmann, 1966). Watt and Breyer-Brandwijk (1962) reported the bulbs of *A. orientalis* to be used as an aphrodisiac with no specific records on how it is used. A close relative, *Agapanthus africanus*, is also used as traditional medicine for Zulu women in South Africa as a health tonic during pregnancy to induce or augment contractions during labour (Varga and Veale, 1997). In the Transkei region of South Africa, women use a root decoction of *A. africanus* beginning in the sixth month of pregnancy as means of ensuring an easy childbirth (Kaido et al., 1997).

8.4 TOXICITY OF *A. ORIENTALIS*

The leaves of *A. orientalis* produce sticky acrid sap that can cause severe ulceration and pain in the mouth when eaten. The sap also causes irritation of the eyes and skin (Quattrocchi, 2012; Barceloux, 2008). The sap of these plants contains a direct mucosal and skin irritant. The chemical structure of the irritant is not known (Veale et al., 1999). There are insufficient data on the toxic constituents of *Agapanthus* species to determine dose–response data (Barceloux, 2008). *A. orientalis* is also suspected of causing haemolytic poisoning in humans (Quattrocchi, 2012).

Treatment for *A. orientalis* is supportive. All plant materials should be removed from the mouth, and demulcents such as milk, antacids and cool water should be used as needed. Symptomatic irritation of the eye requires saline irrigation and examination for corneal abrasions (Barceloux, 2008).

8.5 TRADITIONAL AND MODERN MEDICINAL AND NON-MEDICINAL USES AND APPLICATIONS OF *A. ORIENTALIS*

A. orientalis has antiinflammatory, antioedema, antitussive, immunoregulatory, and uterotonic properties (van Wyk et al., 2000; Quattrocchi, 2012). The aqueous extracts of *A. africanus* showed some antagonistic effects on uterine muscarinic receptors *in vitro* and increased synthesis of prostaglandins in estrogenized rat uterus. This to some extent justifies the traditional medicinal use of *A. africanus* as a herbal oxytocic in prolonged labour (Veale et al., 1999).

8.6 PHYTOCHEMISTRY OF *A. ORIENTALIS*

The seeds of different species of *Agapanthus* contain phytoecdysteroids which have both intraspecific and interspecific variation. The commonest phytoecdysteroid identified was 20-Hydroxyecdysone, although the concentrations in *A. orientalis* were almost undetectable (Savchenko et al., 1997). The rootstocks of *Agapanthus* contain saponin (Snoeijer, 2004).

8.7 APPLICATIONS OF *A. ORIENTALIS*

The seeds of different species of *Agapanthus* contain phytoecdysteroids (Savchenko et al., 1997), which have potential applications in pest control as biopesticides. Phytoecdysteroids are plant steroids, analogous to the arthropod steroid hormones ecdysteroids. Ecdysteroids regulate all stages of insect development and reproduction. Phytoecdysteroids protect plants from insect predation by disrupting endocrine activity of dietary ecdysteroids on insects and nematodes (Bakrim et al., 2008; Dinan et al., 2001; Speranza, 2010).

8.8 PLANT BREEDING TO PRODUCE STRESS-TOLERANT VARIETIES

Two dehydrins (ApY_2SK_2 and $ApSK_3$) were isolated from *A. praecox*. The ApY_2SK_2 and $ApSK_3$ dehydrin genes from *A. praecox* can enhance stress tolerance of transgenic plant, *Arabidopsis thaliana*, by showing important protective effects under complex stresses. ApY_2SK_2 and $ApSK_3$ transformation can enhance the seedling survival ratio by up to 55%. The addition of recombinant ApY_2SK_2 and $ApSK_3$ to plant vitrification solution may increase the survival ratio of wild-type *A. thaliana* seedlings. ApY_2SK_2 and $ApSK_3$ can effectively improve cell stress tolerance (Yang et al., 2019). *ApSerpin-ZX*, a recombinant protein isolated from *A. praecox*, was characterized as a protective plant cryopreservation. It alleviated oxidative stress and cell death induced by cryopreservation (Chen et al., 2021).

8.9 CONCLUSION

Although *A. orientalis* is a popular ornamental and medicinal plant globally and locally respectively, not much research has been done on it. What is interesting to note is that their reports on its toxicity despite its popularity in traditional medicine in South Africa. A few studies have been done on its phytochemistry and pharmacology as well but the greatest or most widely used application currently is in plant breeding. Mode studies need to be conducted on the safety and toxicity as well as efficacy of *A. orientalis.*

NOTE

Godwin Anywar, Department of Plant Sciences, Microbiology & Biotechnology, Makerere University, Kampala, Uganda

REFERENCES

Bakrim, A., Maria, A., Sayah, F., Lafont, R., & Takvorian, N. (2008). Ecdysteroids in spinach (Spinacia oleracea L.): Biosynthesis, transport and regulation of levels. *Plant Physiology and Biochemistry*, *46*(10), 844–854. https://doi.org/10.1016/j.plaphy.2008.06.002

Barceloux, D. G. (2008). African blue lily (Agapanthus species). In *Medical Toxicology of Natural Substances* (pp. 743–744). https://doi.org/10.1002/9780470330319.ch118

Batten, A., & Bokelmann, A. (1966). *Wild Flowers of the Eastern Cape*. Books of Africa, Cape Town.

Chen, G., Li, R., & Shen, X. (2021). ApSerpin-ZX from Agapanthus praecox, is a potential cryoprotective agent to plant cryopreservation. *Cryobiology, 98*, 103–111. https://doi.org/10.1016/j.cryobiol.2020.11.018

Dinan, L., Savchenko, T., & Whiting, P. (2001). On the distribution of phytoecdysteroids in plants. *Cellular and Molecular Life Sciences: CMLS, 58*(8), 1121–1132. https://doi.org/10.1007/PL00000926

Duncan, G. (1985). Agapanthus species-their potential land the introduction of ten selected forms. *Veld & Flora, 71*(4), 122.

Kaido, T. L., Veale, D. J. H., Havlik, I., & Rama, D. B. K. (1997). Preliminary screening of plants used in South Africa as traditional herbal remedies during pregnancy and labour. *Journal of Ethnopharmacology, 55*(3), 185–191. https://doi.org/10.1016/S0378-8741(96)01499-7

Plants of the World Online (2022). https://powo.science.kew.org/results?q=Agapanthus%20orientalis

Pearse R. O. (1978). *Mountain splendour. Wild flowers of the Drakensberg*. Howard Timmins (Pty) Ltd, Cape Town.

Quattrocchi, U. (2012). *CRC world dictionary of medicinal and poisonous plants: Common names, scientific names, eponyms, synonyms, and etymology (5 Volume Set)*. CRC Press, Boca Raton, FL.

Savchenko, T., Whiting, P., Sarker, S. D., & Dinan, L. (1997). Phytoecdysteroids in the Genus Agapanthus (Alliaceae). *Biochemical Systematics and Ecology, 25*(7), 623–629. https://doi.org/10.1016/S0305-1978(97)00056-2

Snoeijer, W. (2004). *Agapanthus: A revision of the genus*. Timber Press, Portland, OR.

Speranza, A. (2010). Into the world of steroids: A biochemical "keep in touch" in plants and animals. *Plant Signaling & Behavior, 5*(8), 940–943. https://doi.org/10.4161/psb.5.8.12295

Van Wyk, B.-E., & Gericke, N (2000). *People's Plants. A Guide to Useful Plants of Southern Africa*. Briza Publications, Pretoria.

Varga, C. A., & Veale, D. J. H. (1997). Isihlambezo: Utilization patterns and potential health effects of pregnancy-related traditional herbal medicine. *Social Science & Medicine, 44*(7), 911–924. https://doi.org/10.1016/S0277-9536(96)00104-9

Veale, D. J. H., Havlik, I., Oliver, D. W., & Dekker, T. G. (1999). Pharmacological effects of Agapanthus africanus on the isolated rat uterus. *Journal of Ethnopharmacology, 66*(3), 257–262. https://doi.org/10.1016/S0378-8741(98)00224-4

Watt, J. M., & Breyer-Brandwijk, M. G. (1962). *The medicinal and poisonous plants of southern and eastern Africa*. S. Livingstone Ltd, London, England.

Weeds of Australia. (2022). Fact sheet index. https://keyserver.lucidcentral.org/weeds/data/media/Html/agapanthus_praecox_subsp._orientalis.htm. Accessed on 19th April 2022.

Yang, Z., Sheng, J., Lv, K., Ren, L., & Zhang, D. (2019). Y2SK2 and SK3 type dehydrins from Agapanthus praecox can improve plant stress tolerance and act as multifunctional protectants. *Plant Science, 284*, 143–160. https://doi.org/10.1016/j.plantsci.2019.03.012

9 *Antiaris toxicaria* (Upas Tree)

Suraj Dhanyakumar Umdale, Pankaj Shivnarayan Mundada, and Mahendra Laxman Ahire

CONTENTS

9.1 INTRODUCTION

Antiaris toxicaria (Pers.) Lesch. is a lofty tree with buttressed base with pale gray bark belonging to family Moraceae commonly known as the upas tree. It is remarkably distributed in tropical regions ooccurring in Australia, tropical Asia, tropical Africa, Indonesia, the Philippines and various other tropical islands. The species is of interest as a source of wood, bark cloth and pharmacological or toxic substances. *A. toxicaria* is a fairly small-scale source of timber and yields a lightweight hardwood with density of 250–540 kg/m^3 (similar to balsa). As the wood peels very easily and evenly, it is commonly used for veneer. The wood of the tree is also useful in wooden drum making in central Uganda (Omeja et al. 2005). The bark has a high concentration of tannins that are used in traditional clothes dyeing and paints. In Africa and various parts of Asia, seed, leaves and bark are used as astringents and the seeds are used to treat dysentery. In Africa and Polynesia, the bast fiber is harvested and is used in preparing strong, coarse bark cloth for clothing. The tree bark is also being used for the preparation of maravuri cloth by *Muthuvan* tribe in Kerala, India (Manithottam and Francis 2008). *A. toxicaria* is notorious as a poison for arrows, darts and blow darts. The latex and seeds of upas tree contain intensely toxic cardenolides.

9.1.1 Taxonomic and Botanical Description

A. toxicaria (from Moraceae family) is a single species of genus *Antiaris* with two varieties and four subspecies, namely *A. toxicaria* var. *africana* Scott-Elliot ex A. Chev.; *A. toxicaria* var. *usambarensis* (Engl.) C.C.Berg.; *A. toxicaria* subspp. *humbertii* (Leandri) C.C. Berg.; *A. toxicaria* subspp. *macrophylla* (R.Br.) C.C. Berg; *A. toxicaria* subsp. *madagascariensis* (H.Perrier) C.C.Berg.; and

DOI: 10.1201/b23017-11

FIGURE 9.1 *Antiaris toxicaria.* (a) Habit of plant and (b) plant twig with fruits.

A. toxicaria subspp. *welwitschii* (Engl.) C.C. Berg. (Berg and Hijman 1989; Berg et al. 2006, http://www.theplantlist.org/). It has many vernacular names, such as Ancar, Antiaris, Antjar Jafo, Antjar, Aseik, Bark Cloth Tree, Bemoe, Chaandakudaa, False Iroko, False Mvule, Hkang-awng, Hmya-seik, Jafo, Kan, Man, Mkunde, Upas Tree, Upoh Tree and Yang Yong. *A. toxicaria* is deciduous, monoecious tree that is 50 m tall with a dome-shaped canopy (Figure 9.1A); bark is 10–20 mm thick pale gray color, smooth with lenticels, the latex or exudation is milky or watery. Leaves simple, alternate, oval, 5.5–16 cm × 4–11 cm; stipule small, lateral, caducous; petiole 2–7 mm long, velvety; leaf apex acuminate or obtusely acuminate, margin entire, glabrous; lateral nerves 5–10 pairs, parallel, prominent, reticulate. Flowers unisexual, greenish-yellow; male receptacle to 2 cm across, axillary, 2–7 tepals; peduncle velvety; receptacle surrounded by imbricating bracts; stamens 2–6; filaments erect; anthers yellow; female flowers solitary, present in disk shaped heads; tepals absent; ovary superior, unilocular with single ovule; style 2, recurved. Fruit is bright red colored drupe (Figure 9.1B), 1–1.7 cm, fleshy, velvety, single deeded; seed 10–15 mm long, edible, birds, bats and humans eat and are responsible for seed dispersal. Flowering and fruiting – January to June.

9.1.2 Species Distribution

It is widely distributed in tropical areas mainly in semi-arid, rainforest and even in swamp forests and found throughout the Old-World tropics. It is native to Australia, Cameroon, China, Democratic Republic of Congo, Fiji, India and Indian subcontinents, Indonesia, Malaysia, Vietnam, Africa-Nigeria, Philippines, Sierra Leone, Sudan, Kenya, Tonga and Uganda. In India, it is sporadically distributed in Western Ghats and Eastern Ghats region (Ravikumar and Sankar 2009; Umdale et al. 2020).

9.1.3 Toxic Response/Poison

The latex of *Antiaris toxicaria* contains toxic cardiac glycoside (cardenolide) known as "Antiarin" (Kopp et al. 1992). *A. toxicaria* is considered as world's most poisonous plant (Buel 1887). The latex

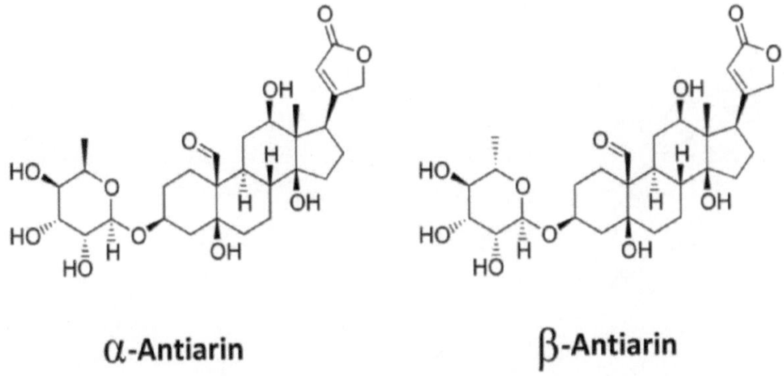

FIGURE 9.2 The structural representation of key secondary metabolites, Antiarins, in *A. toxicaria*.

is traditionally applied on darts, blow darts and arrows as a toxin during war, fish poison and for hunting from ancient times in various parts of the world (Subiono et al. 2017).

9.2 PHYTOCHEMICAL CONSTITUENTS

The leaf, stem latex, roots and seeds of *A. toxicaria* consist of several cardenolides, about 50 cardiac glycosides, antiarins (alpha-antiarin and beta-antiarin) (Figure 9.2), angiogenin, antioside, antiarosides, antiogoside, antiarojavoside, antialloside, antiosemoside, glucostrophalloside, coumarin derivatives, chalcones, cymarol, (3S,4R)-Hydroxymellein, dihydrochalcones, D-Javose, flavanones, malayoside, (R)-Peucedanol, periplogenin, peripalloside, strophalloside, prenylaurones, periplorhamnoside, strophantojavoside, toxicariosides (A, B, C, D, N, O; Figure 9.3), toxicariosides J-K-L, *β*-Sitosterol, strophanthidin, strophalloside, desglucocheirotoxin, 7-Drimen-3*β*, 11-Diol 3-O-*β*-D-Glucopyranoside; toxicariosides N-O; periplogenin, prenylflavanones, phenylpropanoid and lignin derivatives, uposide (Muehlradt et al. 1964; Hano et al. 1991; Que et al. 2009, 2010; Jiang et al. 2009; Dong et al. 2011a, b; Liu et al. 2013; Zuo et al. 2013; Shi et al. 2014; Li et al. 2015).

9.3 TRADITIONAL AND OTHER POTENTIAL USES

In Africa, the latex is applied to cuts, wounds and skin complaints, such as eczema and leprosy, and is taken internally as a purgative (Mei et al. 2007). The latex is traditionally used on darts, blow darts and arrows as a toxin during war, fish poison and for hunting. A small quantity of bark latex is used as cardiac stimulant. The seeds, leaves and bark of upas tree are used as astringent, febrifuge and antidysenteric. The ripe seeds are roasted over a fire and then eaten as a treatment for small growths on the body (Darmadi 2018). The bark of *A. toxicaria* is used as a vermifuge and in the treatment of hepatitis. The bark juice of upas tree is used in the treatment of spleen. Also, the bark is used for dyeing and making cloths and paper. The leaves and root decoctions are used to treat mental illnesses. The bark decoction is given in the treatment of cancer and leukemia (Bosu and Krampah 2005).

9.4 PHARMACOLOGICAL STUDIES

9.4.1 Anticonvulsant/Neuroprotective Activity

Epilepsy is considered as neurological brain disorder probably caused by the inappropriate functioning of the central nervous system (Mante et al. 2017). For the treatment of epilepsy, the chemical agents are frequently used and are effective. The aqueous extract of *A. toxicaria* showed

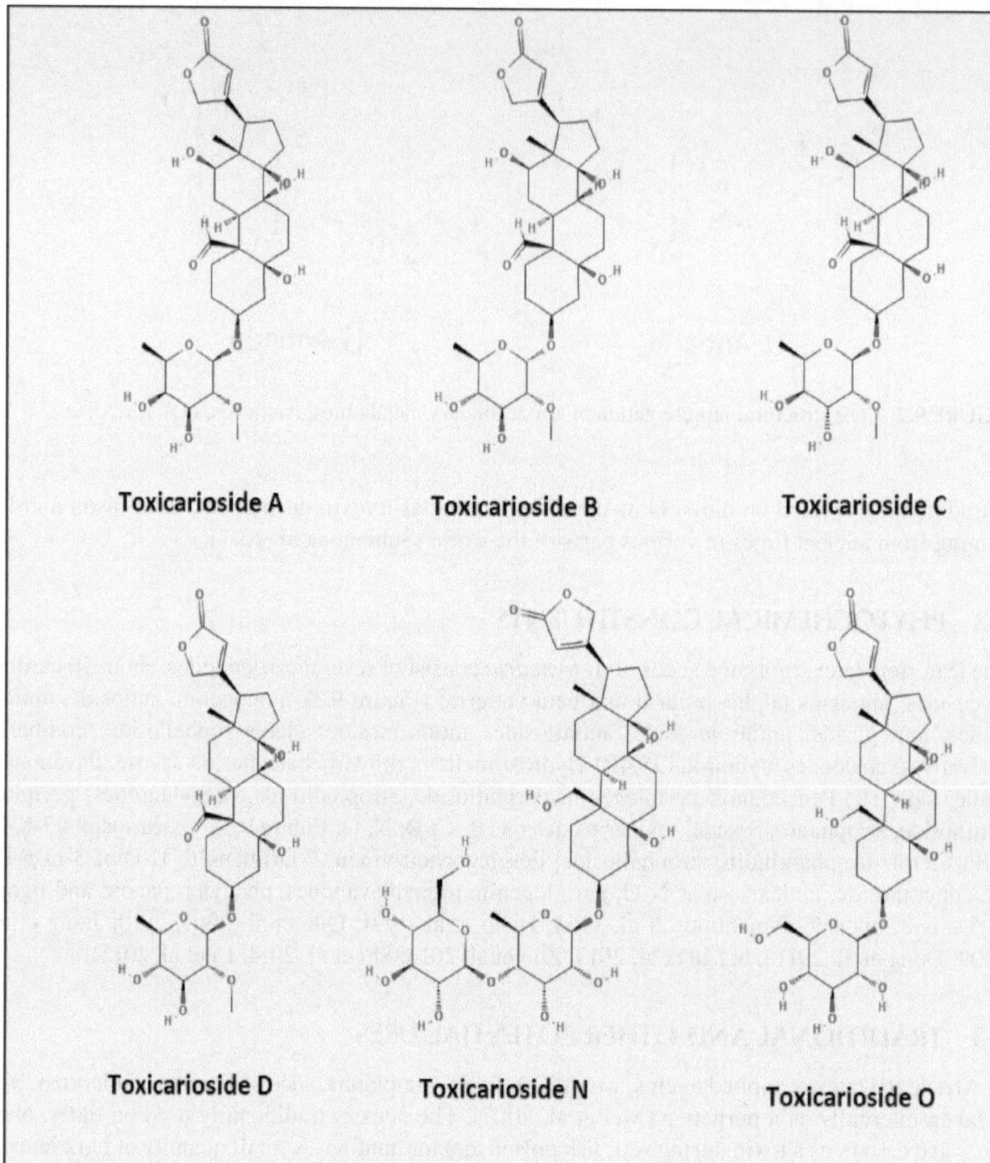

FIGURE 9.3 The structural representation of toxicariosides (A, B, C, D, N, O) in *A. toxicaria.*

anticonvulsant activity on rodents by reducing the duration of convulsions *via* inhibition of gamma amino butyric acid (GABA) and glutamate (Mante et al. 2012, 2013). The aqueous extract *A. toxicaria* used against seizures induced by pentylenetetrazole, pilocarpine and kainic acid showed the delay in duration of seizures in rodents (Mante et al. 2017). Agbaje et al. (2014) reported the anti-depressant and neuroprotective effect of *A. toxicaria* aqueous leaf extract in mice. The antidepressant activity involves the interaction of phytoconstituents of *A. toxicaria* with α_1-Adrenoceptor, D_2 dopamine receptor and nitrergic pathway. The application of decoction of Mist Antiaris in male adult Sprague-Dawley rats (SDRs) showed no any specific toxicity after biochemical, hematological and urine analysis in rats. Also, the reduction of neuropathic symptoms has been noticed after the advocating Mist Antiaris (Antwi et al. 2019; Yeboah et al. 2019). Ilesanmi et al. (2022) reported the neuroprotective activity of flavonoids isolated from leaves of *A. africana* against potassium cyanide induced oxidative damage in brain cells and further advocated in the treatment of epilepsy, seizures and other neurological disorders.

9.4.2 Antioxidant Activity

The oxidative stress is the core cause of the development of several diseases. During oxidative stress, the highly reactive chemicals formed from O_2 are known as Reactive Oxygen Species (ROS). Antioxidants are substances that prevent or delay the cell damage caused by oxidative stress. The antioxidants are responsible for elimination of ROS by scavenging of radicals during oxidation of substrates. The medicinal plants are the unique sources of natural antioxidants in the form of phenolics and flavonoids. The antioxidant activity by 2,2-Diphenyl-1-Picrylhydrazyl (DPPH) free radical scavenging method showed that the stem latex of *A. toxicaria* has potential antioxidant efficacy (Gan et al. 2008). Kuete et al. (2009) reported the antioxidant capacity of methanolic extract of stem bark of *A. africana* using DPPH radical scavenging assay. The DPPH activity showed that more than 50% inhibition at 500 µg/mL of methanolic extract further confirms the dose-dependent inhibitory potency. The methanol extract of leaves of *A. toxicaria* reported the maximum antioxidant activity tested by DPPH free radical scavenging method (Kemila and Krishnaveni 2019).

9.4.3 Antimicrobial Activity

Plants are the natural source of antimicrobial compounds that inhibit the growth or kill the microorganisms such as bacteria, viruses, fungi and algae. *A. toxicaria* contains medicinally important secondary phytoconstituents that include alkaloids, phenolics, flavonoids, tannins, terpenoids, diterpenes, steroids and other compounds. Banso and Mann (2008) reported that the flavonoids in stem bark extract showed antibacterial property and are effective against *Bacillus subtilis*, *Streptococcus pyogenes* and *Escherichia coli*. Furthermore, the flavonoid fraction from *Antiaris* could be advocated to relieve respiratory pain, stomachache and rheumatic pains. Kemila and Krishnaveni (2019) reported the antibacterial and antifungal activities of chloroform, methanol and aqueous leaf extracts of *A. toxicaria* against *E. coli*, *Staphylococcus aureus*, *Pseudomonas* sp., *Curvularia lunata* and *Aspergillus flavus*. In this study, the efficiency of methanol extract showed higher antibacterial activity. However, the aqueous extract exhibited significant antifungal activity as compared to methanol and chloroform extracts. The antimicrobial activity based on agar well diffusion essay of silver nanoparticle synthesized from the aqueous plant extract of *A. toxicaria* showed the highest antimicrobial activity against *Bacillus cereus* (Gopalakrishnan et al. 2015).

9.4.4 Anticancer/Anti-Tumor Activity

Plants medicinal properties have been utilized for treatment of various chronic diseases in modern medicine. Plants are the source of easily available, non-toxic and safe phytochemicals used in cancer treatment. The various metabolites extracted from plant parts are used to treat various types of cancer cells and tumor-related diseases by inducing the apoptosis in tumor cells. A methanol extract from the stem bark of *A. africana* Engler has various compounds that showed tumor inhibition activity against the human DU-145 and hepatocarcinoma Hep-G2 cells. Furthermore, the active phytoconstituent, ellagic acid derivative (3,3′-Dimethoxy-4′-O-β-d-Xylopyronosylellagic acid), acts as a potential antineoplastic agent (Kuete et al. 2009).

The novel nor-cardenolides (toxicariosides D, E, F, G, H) isolated from the latex of *A. toxicaria* stem showed significant growth inhibitory activity on K562 (leukemia), SGC-7901 (gastric cancer), SMMC-7721 (hepatocellular carcinoma) and HeLa (cervical cancer) cell lines (Jiang et al. 2008; Dai et al. 2009a, b; Gan et al. 2009). Dong et al. (2011) reported the isolation of a new drimane sesquiterpenoid glycoside (7-Drimen-3β, 11-Diol 3-O-β-d-Glucopyranoside) from the hydroethanolic seeds extract of *A. toxicaria*. Further, the cytotoxicity of novel compound was studied against the chronic myelogenous leukemia (K562), and human hepatoma (SMMC-7721) cell lines and reported potent cell inhibitory activity.

The two novel strophanthidol cardenolides (toxicarioside N and toxicarioside O) were isolated from the seeds of *A. toxicaria* which revealed significant cytotoxicity against Hepatocellular carcinoma (SMMC-7721) and Leukemia (K562) cells (Zuo et al. 2013). Liu et al. (2013) isolated 15 novel cardiac glycosides from latex of *A.* toxicaria and examined the expression of orphan nuclear receptor Nur77 in human lung cancer cells (NIH-H460). The treatment of isolated cardiac glycosides resulted in apoptosis of human lung cancer (NIH-H460) cells. Shi et al. (2014) isolated eight cardiac glycosides/aglycones, two coumarins and two flavanones from an ethanolic extract of fresh trunk bark of *A. toxicaria* and tested against ten human cancer cell lines, such as A549, 1A9, CAKI-1, HCT-9, KB, KB-VIN, MCF-7, PC-3, S-KMEL-2 and U-87-MG. In this study, five compounds, namely, antiaroside A, convallatoxal, convallatoxin, desglucocheirotoxin and α-Antiarin, depicted noteworthy cytotoxicity against the cancer cell lines.

Wu et al. (2014) isolated a novel periplogenin cardenolide, periplogulcoside, from the seeds of *Antiaris toxicaria* and further reported significant cytotoxicity against HeLa (cervical cancer) and HepG-2 (hepatocarcinoma) cell lines. Huang et al. (2017) reported the phytochemical toxicarioside O isolated from seeds of *A. toxicaria* which showed induction of apoptotic cell death and autophagy through the inhibition of Akt/mTOR pathway and regulation of SIRT1 in colorectal cancer cells (HCT116 and SW480). Toxicarioside N isolated from the seeds of *A. toxicaria* showed anticancer activity against the human gastrointestinal cancer cell line SGC-7901 by activating the p38MAPK pathway and further inducing apoptosis (Zhao et al. 2018). The cytotoxic activity of toxicarioside O showed the antiproliferative property against the lung cancer cells both *in vitro* and *in vivo*. Toxicarioside O inhibits cell proliferation and migration *in vitro* and also decreases the expression of trophoblast cell surface antigen 2 (Trop2), which further results in the inhibition of the PI3K/Akt pathway (Zheng et al. 2021). Hu et al. (2021) reported the cytotoxic activity of malayoside extracted from latex of *A. toxicaria*. The malayoside showed *in vitro* and *in vivo* anti-NSCLC activity indicating that it is a promising chemotherapeutic component for NSCLC.

9.4.5 Cardiovascular Activity

The cardiac insufficiency usually caused by arterial hypertension leads to the malfunction of the heart and further noticed by fatigue, shortness of breath and edema. The plant-based drugs also known as cardiotonic agents increase the capability and efficacy of contractile power of the myocardium. *A. toxicaria* is also known as "poison arrow tree" due to the presence of poisonous latex. The poisonous latex of *A. toxicaria* shows the presence of a variety of cardenolides or cardenolide glycosides. The cardenolide glycosides are also known as cardiac glycosides. The crude milky sap extracted from *A. toxicaria* using water–acetone solvent showed cardiotonic effect on anesthetized rats. Further evidence suggested that glycosides affect Na^+-, K^+-ATPase activity of heart muscle contraction (Fujimoto et al. 1983). Shi et al. (2010) reported that an ethanolic extract of *A. toxicaria* trunk bark showed potent *in vitro* cardiotonic effect on isolated guinea pig atria (heart muscle).

9.5 CONCLUSION AND FUTURE REMARKS

A. toxicaria is notorious as a poison for arrows, darts and blow darts. The latex and seeds of *A. toxicaria* contain intensely toxic cardenolides, in particular, a cardiac glycoside named antiarin. However, the phytochemical study reveals several phytoconstituents of immense pharmacological importance. *A. toxicaria* extracts were attributed to be a promising source of bioactive agents. It has been shown to possess various pharmacological properties such as antibacterial activities, anti-arrhythmic, anti-osteoporosis, epilepsy, anticonvulsant activity, antidepressant-like activity, treating mental illness and anti-tumor activity. Further research is required in the direction of pharmacological and pharmaceutical activities of various extracts. Further, researchers need to provide attention on *in vitro* multiplication via tissue culture techniques for their commercial exploitation and conservation purpose. The studies are also required for enhanced production of medicinally important metabolites from *in vitro* cultures.

ACKNOWLEDGMENTS

The authors are grateful to Yashavantrao Chavan Institute of Science, Satara for financial support under UGC-CPE, DST-FIST, RUSA (Component 8) and DBT-STAR college scheme. Financial assistance to the faculty under self-funded project as well as the provision of laboratory facilities is gratefully acknowledged.

NOTES

Suraj Dhanyakumar Umdale, Department of Botany, Jaysingpur College, Jaysingpur (Affiliated to Shivaji University), Kolhapur, India
Pankaj Shivnarayan Mundada, Department of Biotechnology, Yashavantrao Chavan Institute of Science, Satara, India
Mahendra Laxman Ahire, Department of Botany, Yashavantrao Chavan Institute of Science; and Life Science Laboratory, Rayat Institute of Research and Development, Satara, India

REFERENCES

Agbaje, E. O., Ishola, I. O., & Oniyire, J. A. 2014. Antidepressant, anxiolytic, and anticataleptic effects of aqueous leaf extract of *Antiaris toxicaria* Lesch. (Moraceae) in mice: Possible mechanisms of actions. *Journal of Basic and Clinical Physiology and Pharmacology* 25: 429–438.

Antwi, S., Asiedu-Larbi, J., & Martey, N. K. 2019. Safety and effectiveness of mist Antiaris, a herbal preparation for treatment of peripheral neuropathy. *BioMed Research International* 2019: 2607872.

Banso, A., & Mann, A. 2008. Evaluation of antibacterial properties of flavonoid fraction from Antiaris africana (Engl). *Journal of Applied Bioscience* 12: 665–670.

Berg, C. C., Corner, E. J. H., & Jarrett, F. M. 2006. *Flora Malesiana, Series I.* Vol. 17 Part 1. National Herbarium Nederland, Leiden, Netherlands, pp. 1–154.

Berg, C. C., & Hijman, M. E. E. 1989. Moraceae. In: Polhill, R. M. (Eds.), *Flora of Tropical East Africa.* CRC Press, Boca Raton, FL, pp. 1–95.

Bosu, P. P., & Krampah, E. 2005. *Antiaris toxicaria* Lesch. In: Louppe, D., Oteng-Amoako, A. A., & Brink, M. (Eds.), *PROTA (Plant Resources of Tropical Africa/Ressources végétales de l'Afrique tropicale).* Wageningen, Netherlands.

Buel, J. W. 1887. *Sea and Land: An Illustrated History of the Wonderful and Curious Things of Nature Existing Before and Since the Deluge*. Historical Publishing Company, Philadelphia, PA, pp. 470–471.

Dai, H. F., Gan, Y. J., Que, D. M., Wu, J., Wen, Z. C., & Mei, W. L. 2009a. A new cytotoxic 19-nor-cardenolide from the latex of *Antiaris toxicaria. Molecules* 14(9): 3694–3699.

Dai, H. F., Gan, Y. J., Que, D. M., Wu, J., Wen, Z. C., & Mei, W. L. 2009b. Two new cytotoxic cardenolides from the latex of *Antiaris toxicaria. Journal of Asian Natural Products Research* 11: 832–837.

Dong, W. H., Mei, W. L., Zhao, Y. X., Zeng, Y. B., Wang, H., & Dai, H. F. 2011a. A new drimane sesquiterpenoid glycoside from the seeds of *Antiaris toxicaria. Journal of Asian Natural Products Research* 13: 561–565.

Dong, W. H., Mei, W. L., Zhao, Y. X., Zeng, Y. B., Zuo, W. J., Wang, H., Li, X. N., & Dai, H. F. 2011b. Cytotoxic cardenolide glycosides from the seeds of *Antiaris toxicaria. Planta Medica* 77: 1730–1734.

Fujimoto, Y., Suzuki, Y., Kanaiwa, T., Amiya, T., Hoshi, K., & Fujino, S. 1983. Studies on the Indonesian *Antiaris toxicaria* sap. *Journal of Pharmacobio-Dynamics* 6: 128–135.

Gan, Y., Mei, W., Zeng, Y., Han, Z., & Dai, H. 2008. Liposoluble components and their antioxidant activities from *Antiaris toxicaria* latex. *Redai Yaredai Zhiwu Xuebao* 16: 144–147.

Gan, Y. J., Mei, W. L., Zhao, Y. X., & Dai, H. F. 2009. A new cytotoxic cardenolide from the latex of *Antiaris toxicaria. Chinese Chemical Letters* 20: 450–452.

Gopalakrishnan, S., Philip, K., Yamini, S., Lakshmi, S., & Fouzia, B. 2015. Antimicrobial activity of synthesized silver nanoparticles and phytochemical screening of the aqueous extract of *Antiaris toxicaria. Indian Journal of Drugs and Diseases* 4: 1–4.

Hano, Y., Mitsui, P., Nomura, T., Kawai, T., & Yoshida, Y. 1991. Two new dihydrochalcone derivatives, antiarones J and K, from the root bark of *Antiaris toxicaria. Journal of Natural Products* 54: 1049–1055.

Hu, Q. Y., Zhang, X. K., & Wang, J. N. 2021. Malayoside, a cardenolide glycoside extracted from Antiaris toxicaria Lesch, induces apoptosis in human non-small lung cancer cells via MAPK-Nur77 signaling pathway. *Biochemical Pharmacology* 190: 114622.

Huang, Y. H., Sun, Y., Huang, F. Y., Li, Y. N., Wang, C. C., Mei, W. L., Dai, H. F., Tan, G. H., & Huang, C. 2017. Toxicarioside O induces protective autophagy in a sirtuin-1-dependent manner in colorectal cancer cells. *Oncotarget* 8: 52783–52791.

Ilesanmi, O. B., Akinmoladun, A. C., Elusiyan, C. A., Ogungbe, I. V., Olugbade, T. A., & Olaleye, M. T. 2022. Neuroprotective flavonoids of the leaf of *Antiaris africana* Englea against cyanide toxicity. *Journal of Ethnopharmacology* 282: 114592.

Jiang, M. M., Dai, Y., Gao, H., Zhang, X., Wang, G. H., He, J. Y., Hu, Q. Y., Zeng, J. Z., Zhang, X. K., & Yao, X. S. 2008. Cardenolides from *Antiaris toxicaria* as potent selective Nur77 modulators. *Chemical and Pharmaceutical Bulletin* 56: 1005–1100.

Jiang, M. M., Gao, H., Dai, Y., Zhang, X., Wang, N. L., & Yao, X. S. 2009. Phenylpropanoid and lignan derivatives from *Antiaris toxicaria* and their effects on proliferation and differentiation of an osteoblast-like cell line. *Planta Medica* 75: 340–345.

Kemila, P., & Krishnaveni, C. 2019. Pharmacognostical and antioxidant properties of the leaves of *Antiaris toxicaria* Lesch. *International Journal of Pharmacy and Biological Sciences* 9: 369–373.

Kopp, B., Bauer, W. P., & Bernkop-Schnurch, A. 1992. Analysis of some Malaysian dart poisons. *Journal of Ethnopharmacology* 36: 57–62.

Kuete, V., Vouffo, B., Mbaveng, A. T., Vouffo, E. Y., Siagat, R. M., & Dongo, E. 2009. Evaluation of *Antiaris africana* methanol extract and compounds for antioxidant and antitumor activities. *Pharmaceutical Biology* 47: 1042–1049.

Li, X. S., Zhu, J. J., Zhao, H., Li, S. L., Hao, X. J., Yao, X. S., & Tang, J. S. 2015. Three new compounds from the bark of *Antiaris toxicaria. Phytochemistry Letters* 13: 182–186.

Liu, Q., Tang, J. S., Hu, M. J., Liu, J., Chen, H. F., Gao, H., Wang, G. H., Li, S. L., Hao, X. J., Zhang, X. K., & Yao, X. S. 2013. Antiproliferative cardiac glycosides from the latex of *Antiaris toxicaria. Journal of Natural Products* 76: 1771–1780.

Manithottam, J., & Francis, M. S. 2008. Preparation of Marvuri from *Antiaris toxicaria* (Pers.) Lesch. by Muthuvans of Kerala. *Indian Journal of Traditional Knowledge* 7: 74–76.

Mante, P. K., Adongo, D. W., Kukuia, K. K. E., Ameyaw, E. O., & Wood, E. 2012. Neuropharmacological assessment of an aqueous bark extract of *A ntiaris toxicaria* (Pers.) Lesch. (Moraceae) in rodents. *American Journal of Pharmacology and Toxicology* 7: 123–134.

Mante, P. K., Adongo, D. W., & Wood, E. 2017. Anticonvulsant effects of *Antiaris toxicaria* aqueous extract: Investigation using animal models of temporal lobe epilepsy. *BMC Research Notes* 10: 167.

Mante, P. K., Adongo, W. D., Wood, E., Kukuia, K. K. E., & Ameyaw, E. O. 2013. Anticonvulsant effect of *Antiaris toxicaria* (Pers.) Lesch. (Moraceae) aqueous extract in rodents. *International Scholarly Research Notices* 2013: 519208.

Mei, W. L., Gan, Y. J., & Dai, H. F. 2007. Advances in studies on chemical constituents of *Antiaris toxicaria* and their pharmacological activities. *Chinese Traditional and Herbal Drugs* 39: 151–154.

Muehlradt, P., Weiss, E. K., & Reichstein, T. 1964. Glycosides and aglycons. CCLVIII. Cardenolides of *Antiaris toxicaria* seeds. l. Separation and identification. *Helvetica Chimica Acta* 47: 2164–2186.

Omeja, P., Obua, J., & Cunningham, A. B. 2005. Demand and supply of wood for drum making in central Uganda. *The International Forestry Review* 7: 21–26.

Que, D. M., Mei, W. L., Gan, Y. J., Zeng, Y. B., & Dai, H. F. 2010. Cytotoxic cardenolides from the latex of *Antiaris toxicaria. Journal of Tropical and Subtropical Botany* 18: 440–444.

Que, D. M., Mei, W. L., Wu, J., Han, Z., & Dai, H. F. 2009. Structure elucidation of flavonoids from *Antiaris toxicaria* roots. *Chinese Journal of Organic Chemistry* 29: 1371–1375.

Ravikumar, K., & Sankar, R. V. 2009. *Antiaris toxicaria* (Moraceae)—A new distributional record to the Eastern Ghats. *Journal of Threatened Taxa* 1: 58–59.

Shi, L. S., Kuo, S. C., Sun, H. D., Morris-Natschke, S. L., Lee, K. H., & Wu, T. S. 2014. Cytotoxic cardiac glycosides and coumarins from *Antiaris toxicaria. Bioorganic and Medicinal Chemistry* 22: 1889–1898.

Shi, L. S., Liao, Y. R., Su, M. J., Lee, A. S., Kuo, P. C., Damu, A. G., Kuo, S. C., Sun, H. D., Lee, K. H., & Wu, T. S. 2010. Cardiac glycosides from *Antiaris toxicaria* with potent cardiotonic activity. *Journal of Natural Products* 73: 1214–1222.

Subiono, T., Abdadi, L., Himawan, T., & Utomo, E. P. 2017. Ligan activity of *Antiaris toxicaria* lesch. and the role of toxicarioside in crude extract: in silico, in vitro, and in vivo approaches. *Journal of Biotechnology and Biochemistry* 3: 4–10.

Umdale, S. D., Mirgal, A. B., Shinde, B. N., Sawant, R. S., Salunkhe, C. B., & Gaikwad, N. B. 2020. Evaluation of genetic diversity in *Antiaris toxicaria* Lesch. from Sacred Groves of the Western Ghats, India. *National Academy Science Letters* 43: 383–388.

Wu, X. L., Wu, Y. L., Li, H. G., Liu, H. T., Fu, X. Y., Cui, R. Q., Wang, J. H., Liu, C., & Chen, J. 2014. A new periplogenin cardenolide from the seeds of *Antiaris toxicaria. Journal of Asian Natural Products Research* 16: 418–421.

Yeboah, R., Thomford, K. P., Mensah, R., Appiah, A. A., Thomford, A. K., & Ocloo, A. 2019. A pilot study on the efficacy of an *Antiaris toxicaria* subsp. *africana* (Engl.) C.C. Berg based Ghanaian herbal product in the management of peripheral neuropathy. *Journal of Pharmacy & Pharmacognosy Research* 7: 59–66.

Zhao, H. G., Zhou, S. L., Lin, Y. Y., Dai, H. F., & Huang, F. Y. 2018. Toxicarioside N induces apoptosis in human gastric cancer SGC-7901 cell by activating the p38MAPK pathway. *Archives of Pharmacal Research* 41: 71–78.

Zheng, W. P., Huang, F. Y., Dai, S. Z., Wang, J. Y., Lin, Y. Y., Sun, Y., Tan, G. H., & Huang, Y. H. 2021. Toxicarioside O inhibits cell proliferation and Epithelial-Mesenchymal transition by downregulation of Trop2 in lung cancer cells. *Frontiers in Oncology* 10: 609275.

Zuo, W. J., Donga, W. H., Chen, J., Zhao, Y. X., Chen, H. Q., Mei, W. L., & Dai, H. F. 2013. Two new strophanthidol cardenolides from the seeds of *Antiaris toxicaria. Phytochemistry Letters* 6: 1–4.

10 *Aristolochia indica* (The Indian Birthwort)

Surjit Sen, Sudeshna Nandi, Rupa Sanyal, Abhijit Dey, and Manjula Rai

CONTENTS

10.1 INTRODUCTION

Since time immemorial, herbal remedies have been used traditionally by different communities throughout the world as healing agents for variety of ailments (Sati et al. 2011; Modak et al. 2015; Dey et al. 2017). In the twenty-first century, the growing awareness in both developed and developing countries for herbal cures indicates the global interest in traditional, complementary and alternative medicines (Borah et al. 2021). The genus *Aristolochia* belongs to the family Aristolochiaceae which is the largest genus with about 534 accepted species and *A. indica* is one of the well-documented species in traditional system of medicine to treat different ailments (Rastogi and Mehrotra 2001; Heinrich et al. 2009; Michl et al. 2013). Ethnomedicinal studies showed that the Indian subcontinent is one of the hot spots for *Aristolochia* use and *Aristolochia indica* was found to be the most frequently cited species in the literature (Heinrich et al. 2009). This endangered medicinal plant is distributed throughout the tropical to temperate regions of the world (Hwang et al. 2003). *A. indica* is used to treat ulcers, fever, cholera and bowel disorders in children during teething (Krishnaraju et al. 2005; Kanjilal et al. 2009). It is also considered as an antidote against snakebites

DOI: 10.1201/b23017-12

and poisonous insects, and it stimulates menstrual flow as well as acts as attenuant, abortifacient, anti-inflammatory, phospholipase A2 inhibitor and antimicrobial agent (Achari et al. 1981; Das et al. 2010; Dey and De 2011).

Based on the acceptance of medicinal plants in traditional healing and modern herbal healthcare, some botanical compounds have toxicological properties well documented in the literature. Despite the therapeutic effects of *A. indica*, the plants possess aristolochic acid (AA), a major active constituent which are chemically nitrophenanthrene carboxylic acids. A number of recent investigation demonstrated the cytotoxic and genotoxic activity of the compound, and they are considered as a strong carcinogen to rodents and human (Cheng et al. 2006; Slade et al. 2009; Hwang et al. 2012; Bunel et al. 2016; Youl et al. 2020). Aristolochic acid nephropathy (AAN), a renal interstitial fibrosis, which is also associated with a high incidence of upper urinary tract cancer (UUC), poses a constant health risk to the people using *A. indica* as ethnomedicine as well as for those consuming AA-rich food supplements mainly as Chinese slimming pills (Heinrich et al. 2009; Michl et al. 2013; Li et al. 2018; Han et al. 2019; Jelakovic et al. 2019; Sborchia et al. 2019; Ren et al. 2020). As a consequence of their lethal effects, the consumption of *Aristolochia* species and its derivatives were banned in many countries, including UK, Germany, USA, Canada and Australia (IARC 2002; Neinhuis et al. 2005). However, despite the well-documented warning of Food and Drug Administration regarding the safety of herbal remedies containing AA, drugs and medical preparations from species of *Aristolochia* are still available for purchase through e-commerce (Gold and Slone 2003; Schaneberg and Khan 2004; Heinrich et al. 2009). Considering the wide use of *A. indica* as medicinal plants, the aim of this study was to assess different aspects of this plant in the areas of morpho-taxonomy, phytochemistry, pharmacology, ethnomedicinal uses and specially focused on the nephrotoxicity and carcinogenic effects of the plant.

10.2 BOTANICAL DESCRIPTION

A. indica is a shrub or perennial herb with long twining stems (Kanjilal et al. 2009; Figure 10.1). Flowering and fruiting of this climbing plant is found from December to February (Neelima et al. 2011). Leaves: variable in size, glabrous, obovate-oblong to sub-panduri form, entire with somewhat undulate margins, usually obtusely acuminate, base cuneate, rounded or subcordate. Inflorescence: axillary racemes, flowers few, bracts small, ovate, acuminate, pedicles long, perianth greenish white to light purplish with globose inflated base. Stamens: 6, adnate and filaments are not distinguishable from the style. Carpel: 6 locular with two ovules. Fruits: globose, oblong, 6-valved capsule and opening from below upwards. Seeds: many, deltoid-ovate, acute, flat and winged (Kirtikar and Basu 1975).

FIGURE 10.1 *Aristolochia indica.*

10.3 DISTRIBUTION

The plant is commonly known as *Indian Birthwort* in English, *Isharmul* in Bengali and Hindi and *Sapsada* in Southern India and Sri Lanka. The species is widely distributed throughout the tropical, subtropical and Mediterranean countries manly in West Peninsula and in Ceylon. In Indian subcontinent, the species is reported from low hills and plains of India from Nepal and lower Bengal to Chittagong in Bangladesh and Coromandel Coast (Murugan et al. 2006; Kanjilal et al. 2009).

10.4 PHYTOCOMPOUNDS

To substantiate the scientific accuracy while utilising a herbal medicine, it is essential to analyse its phytochemical constituents. In the last few decades, comprehensive studies were performed on the phytochemical constituents found in various plant species and the species *A. indica* was no exception. The phytoconstituents of the genus were broadly studied and researchers have reported numerous phytocompounds of significant importance from the plants of this genus. AAs and its derivatives, such as aristolactams (ALs), benzylisoquinolines, protoberberines, isoquinolines, aporphines, diphenyl ethers, flavonoids, amides, coumarins, lignans, benzenoids, steroids, tetralones and terpenoids, were the secondary metabolites that have been characterised from the species. In this section, phytoconstituents found in *A. indica* were compiled and comprehensively presented in tabulated form (Table 10.1).

Essential oil has been extracted from the roots of the plant and preliminary studies were performed (Rao et al. 1935; Rao and Mulhara 1955). From dried ground roots, crystals of aristolochic acid-I were isolated by Kupchan and Doskotch (1962). Some novel derivatives of AA from *A. indica* were isolated and elucidated structurally by Kupchan and Merianos (1968). Novel tetracyclic sesquiterpenes namely ishwarone A (Ganguly et al. 1969), ishwarane, aristolochene (Govindachari et al. 1970), ishwarol (Govindachari and Parthasarathy 1971), ishwarane (Kelly et al. 1972) and 5βH,7β,10α-Selina-4(14),11-Diene (Govindachari et al. 1973) have been isolated and structurally characterised from the plant species. Same group of researchers have discovered two different types of novel phytocompounds from *A. indica*, a phenanthrene derivative including allatonin, aristolochine alkaline, isoaristolochic acid (Pakrashi et al. 1977) as well as a sesquiterpene(12S)-7,12-Secoishwaran-12-ol (Pakrashi et al. 1980). Lately, Cory et al. (1979) and Piers and Hall (1980) reported a stereoselective synthesis of the racemic modification of the tetracyclic sesquiterpenoid called ishwarone. Achari et al. (1981) reported a phenanthrene derivative aristolactam-N-β- D-Glucoside and two steroids 3β-Hydroxy-Stigmast-5-en-7-one and 6β-Hydroxy-Stigmast-4-en-3-one. The roots of *A. indica* contain several phytocompounds including novel phenanthroid lactone, aristololide, 5α-Stigmastane-3,6-Dione, (–)-Cubebin, (–)-Hinokinin (Achari et al. 1982), 2-Hydroxy-1-Methoxy-4H-dibenzo Quinoline-4,5-(6H)-dione, cephradione, aristolactam IIa and β-Sitosterol-β-D-Glucoside (Achari et al. 1983). AAs and a group of ALs were isolated from the roots (Mix et al. 1982). The roots of the species have afforded a novel naphthoquinone, aristolindiquinone, whose structure was deduced through NMR spectroscopy (Che et al. 1983). Che et al. (1984) isolated novel naphthoquinone, aristolindiquinone and magnoflorine, along with seven other previously reported compounds from *A. indica*. Mahesh and Bhaumik (1987) discovered methyl ester of 12-Nonacosenoic acid. Essential oil from the aerial parts was analysed by gas chromatographic-spectroscopic and olfactory methods. Sesqui- and monoterpenes were found to be dominating constituents of this essential oil, such as α-Terpinolene, aristolochene, ishwarone, ishwarol, β-Caryophyllene, ishwarane, caryophyllene oxide I, α-Humulene and linalool (Jirovetz et al. 2000). Wu et al. (2004) reported several terpenoids from *A. indica* including α-Terpinolene, β-Caryophyllene, (S)-Linalool and Caryophyllene oxide. Preliminary phytochemical analysis of the plant has unveiled the presence of steroids, cardiac glycosides, alkaloids, saponins, tannins and flavonoids (Vaghasiya and Chanda 2007). A total of 15 compounds were explored from the stem oil by gas chromatography/mass spectroscopy; the major constituents of oil were α-Pinene (16.4%), trans-pinocarveol (24.2%) and pinocarvone (14.2%)

TABLE 10.1
Active Phytoconstituents of *Aristolochia indica*

Phytoconstituents	Plant Part Used	Reference
Essential oil	Root	Rao et al. (1935)
Aristolochic acid-I	Root	Kupchan and Doskotch (1962)
Aristolochic acid-D, Aristolochic acid-n methyl ether lactam, Aristolactam β-D-Glucoside	Root	Kupchan and Merianos (1968)
Ishwarone	Root	Ganguly et al. (1969); Cory et al. (1979); Piers and Hall (1980)
Ishwarol	Root	Govindachari and Parthasarathy (1971)
Ishwarane	Root	Govindachari et al. (1970)
Aristolochene	Root	Kelly et al. (1972)
5βH,7β,10α-Selina-4(14),II-Diene		Govindachari et al. (1973)
Allatonin, Aristolochine alkaline, Isoaristolochic acid		Pakrashi et al. (1977)
(12S)-7,12-Secoishwaran-12-ol, (+)-ledol, Ishwarane	Root	Pakrashi et al. (1980)
Aristololactam N-β-D-Glucoside, 3β-Hydroxy-Stigmast-5-en-7-one, 6β-Hydroxy-Stigmast-4-en-3-one	Root	Achari et al. (1981)
Aristololide, 5α-Stigmastane-3,6-Dione, (–)-cubebin, (–)-Hinokinin	Root	Achari et al. (1982)
Aristolochic acid-I, Aristolochic acid-D, Methyl Aristolochate, Aristolactam-A II, Aristolactam I, Aristolactam, Aristolactam-C N-β-D-Glucoside, Aristolactam β-D-Glucoside	Root	Mix et al. (1982)
2-Hydroxy-1-Methoxy-4Hdibenzo quinoline-4,5-(6H)-Dione, Cephradione, aristolactam IIa, β-Sitosterol-β-D-Glucoside	Root	Achari et al. (1983)
Aristolindiquinone	Root	Che et al. (1983)
Savinin, Aristolochic acid-I, (+)-ledol, (12s)- 7,12-Secoishwaran-12-ol, Aristolactam-N-β-D-Glucoside, Aristolactam, Aristolactam-A-II, Methyl Aristolochate, Aristolochic acid-I, Aristolochic acid-D, Cepharadione A, Aristolindiquinone, Magnoflorine	Root	Che et al. (1984)
Methyl ester of 12-Nonacosenoic acid		Mahesh and Bhaumik (1987)
α-Terpinolene, Aristolochene, Ishwarone, Ishwarol, β-Caryophyllene, Ishwarane, caryophyllene oxide I, α-Humulene and Linalool	Aerial part	Jirovetz et al. (2000)
(S)-Linalool, α-Terpinolene, β-Caryophyllene, Caryophyllene oxide, Caryophyllene oxide		Wu et al. (2004)
Steroids, Cardiac glycosides, Alkaloids, Saponins, Tannins and Flavonoids		Vaghasiya and Chanda (2007)
α-Pinene, Trans-pinocarveol and Pinocarvone	Stem	Kanjilal et al. (2009)
Methyl ester of 12-Nonacosenoic acid, n-Heptadecane, n-triacontane, Palmitic acid, hexacosanoic acid, Stigmast-4-en-3-one, Friedelin, Cycloeucalenol, Rutin, Aristolactone	Root	Sati et al. (2011)
Stigmast-5-en-3β-ol (β-Sitosterol)	Aerial part	Karan et al. (2012)
Aristolochic acid-I, Aristolochic acid II, Aristolochic acid IV, Aristolochic acid-D, Aristolochic acid IIIa, Aristolochic acid Ia, Cepharadione A, Aristolactam I; N-β-D-glucopyranoside, Aristolactam AII, Aristolactam III, Aristolactam I, Aristolactam IVa, Aristolactam AIII, Aristolactam II, Aristolactam II; N-β-D-Glucopyranoside, Aristolactam IIIa; N-β-D-Glucopyranoside, Aristolactam Ia; N-β-D-Glucopyranoside, Aristoloterpenate I, Ariskanin B, 9-Methoxyaristolactam II, Norcepharadione A	Stem, Leaves	Michl et al. (2013)
Astragalin, (–) Hinokinin, Aristolochic acid-I, Aristolactam I, Aristolochic acid II	Aerial part	Desai et al. (2014)

(Kanjilal et al. 2009). Several phytoconstituents like aristolactone, *β*-Sitosterol, Stigmast-4-en-3-one, friedelin, n-Heptadecane, n-triacontane, palmitic acid, hexacosanoic acid, cycloeucalenol and rutin have been isolated from different parts of the plant (Sati et al. 2011). Stigmast-5-en-3*β*-ol (*β*-Sitosterol) was isolated from the chloroform extract of the aerial parts, and the structure and relative configuration of this compound was determined by spectroscopic methods (Karan et al. 2012). *A. indica* samples contained several aristolochic acid analogues such as aristolochic acid-I, aristolochic acid II, cepharadione A and related compounds (Michl et al. 2013). A new, efficient preparative HPLC method has been utilised to isolate astragalin along with (–) hinokinin, aristolactam I and aristolochic acids (I & II). These isolated compounds were analysed by reversed-phase HPLC method (Desai et al. 2014).

10.5 PHARMACOLOGICAL ACTIVITY OF REPORTED PHYTO-COMPOUNDS OF *ARISTOLOCHIA INDICA*

"Pharmacology" is the branch of medical science concerned with the applications, effects and mode of action of drugs. It inspects how several chemicals have an effect on biological systems (Berger and Iyengar, 2011). Over the years, several plants have been documented as well as studied for their promising medicinal essence, which unveils various pharmacological activities including anticancer, antipyretic, immunomodulation, analgesic, activation of nervous system, hepato-protection and antidiabetic (Fitzgerald et al. 2020; Singh and Sedha 2018). Scientists have commenced correlating botanical properties as well as phytochemical constituents of a plant with their pharmacological activity and this correlation will be expected to open more coordinated multidimensional research area in the upcoming days. *A. indica* is credited with numerous medicinal properties. The therapeutic value of *A. indica* has been acknowledged in several system of traditional medicine for the treatment of various ailments of human beings.

10.5.1 Antioxidant Activity

The phenolics like terpenoids were noted as vital components present in this species. Ethyl acetate, petroleum ether and chloroform were utilised as extraction solvents and 2,2-reducing power, 2,2-Diphenyl-1-Picrylhydrazyl (DPPH) radical scavenging activity, total polyphenol estimation and ammonium thiocyanate assay were conducted. Among the three extracts, the highest reducing power activity was observed in ethyl acetate (Thirugnanasampandan et al. 2008). Subramaniyan et al. (2015) studied *in vitro* antioxidant property with aqueous and chloroform extracts utilising several free radical models such as hydrogen peroxide (H_2O_2), DPPH radical scavenging and reducing power assay. The chloroform extract of the plant displayed promising antioxidative property and phytochemical screening disclosed the presence of phenol, tannins, saponins, flavonoids and glycosides in chloroform and aqueous extract which might be responsible for their mighty antioxidant activity. The ethyl alcohol extracts of leaves and flower were used to screen radical scavenging efficacy by DPPH free radical scavenging assay. Leaf extract scavenge DPPH radicals more efficiently as compared to flower extract where ascorbic acid was used as standard (Naik et al. 2015). Stigmast-5-en-3*β*-ol (*β*-Sitosterol) isolated from the chloroform extract of the aerial parts of *A. indica* displayed effective antioxidant activity in the level of DPPH and superoxide anion radicals with IC_{50} value of 7.325 and 8.498 μg/mL, respectively (Karan et al. 2012).

10.5.2 Antimicrobial Activity

Studies demonstrated that whole plant of *A. indica* with ethanol and aqueous extract exhibited a broad-spectrum antibacterial activity against multidrug resistant bacteria which include both gram positive and gram negative bacteria. Ethanol extract being the active one displayed MIC (minimum inhibitory concentration) value of 50–100 μg/mL against the bacteria tested (Venkatadri et al.

2015). Naik et al. (2015) examined antibacterial activity with leaf and flower extract of plant against both gram positive (*Staphylococcus aureus, Bacillus coagulans, Bacillus subtilis*) and gram negative (*Escherichia coli, Pseudomonas aeruginosa* and *Salmonella typhi*) bacteria. The leaf extract induced high inhibition of *B. coagulans* and *B. subtilis* and flower extract inhibited *S. typhi* and *B. coagulans* to a higher extent. Antifungal activity was also studied against *Bipolaris sorokiniana, Fusarium oxysporum* f.sp. *zingiberi Colletotrichum capsici* and *Curvularia* sp.. Both leaf and flower extracts displayed higher susceptibility towards *B. Sorokiniana, C. capsici* and *Curvularia* sp. The ethanol leaf extract showed potent antimicrobial activity of 19 mm against clinically isolated multidrug-resistance *Enterococcus faecalis* (Gopinath and Prakash 2013). Vaghasiya and Chanda (2007) screened antimicrobial activity of 14 different medicinal plants. Among the plant species screened, the best antimicrobial activity was depicted by *A. indica* against both gram positive and gram negative bacteria. The acetone extract exhibited maximum inhibitory activity against the bacterium, *S. aureus*. A report stated that essential oil containing β-Caryophyllene and α-Humulene as major constituents was observed to be moderately active against *P. aeruginosa, B. subtilis, S. aureus, E. coli, B. sphaericus* and *S. typhimurium* with concentration of 50 mg/mL (Shafi et al. 2002). From sequentially extracted water, hexane and chloroform extracts of roots, the chloroform and water extracts demonstrated potential antibacterial activity against *E. coli* and *B. pumilus*, while hexane extract was inhibitory only against *B. pumilus* (Sini and Malathy 2005). Apart from human bacterial pathogens, the butanolic extract of *A. indica* demonstrated effective inhibitory activity against the cattle pathogen *Listeria monocytogenes* (Ravikumar et al. 2005). Crude extract of root of *A. indica* possesses vast ethnomedicinal use as prospective antibacterial and antifungal compound (Kumar et al. 2006).

10.5.3 Antidiarrhoeal Activity

The ethanol and aqueous extract of *A. indica* roots has significantly reduced the severity and frequency of castor oil-induced diarrhoea via anti-secretory mechanism which was conspicuous from the decrease of total number of wet faeces in the experimental groups. Both the extracts have also retarded the gastrointestinal transit of charcoal meal in mice. Interestingly, both the extracts have successfully inhibited the castor oil-induced diarrhoea in mice model without causing any mortality of the animals or any evident clinical signs of general weakness among the experimental animals (Dharmalingam et al. 2014).

10.5.4 Antidiabetic Activity

The experimental findings of Karan et al. (2012) authenticated that aerial parts of *A. indica* possess notable antidiabetic properties. A single intravenous injection of aqueous alloxan monohydrate (150 mg/kg) induced diabetes mellitus in Swiss albino mice and immediate administration of chloroform plant extract showed maximum reduction in serum glucose level which is comparable with standard hypoglycaemic agent, glibenclamide. Similarly, the antihyperglycemic effect of methanolic and alcoholic extract of root was studied in alloxan-induced diabetic mellitus in Sprague Dawley rats and the results were compared with an oral hypoglycaemic agent glibenclamide (10 mg/kg). The methanolic extract has dose dependently reduced the blood glucose levels in normal rats showcasing the hypoglycaemic effect. Oral administration of alcohol extract at doses of 200 and 400 mg/kg reduced the blood glucose levels of the normal rat compared with disease control group (Goverdhan et al. 2008).

10.5.5 Anti-Inflammatory and Antipyretic Activity

Petroleum ether and ethanol root extracts have significantly inhibited the mast cell degranulation almost equivalent to that of standard drug ketotifen. The ethanol root extract at a dose of 150 mg/kg had lessened the rat paw oedema by 70% thereby manifesting potent anti-inflammatory activity (Mathew et al. 2011). Vedavathy and Rao (1991) reported the antipyretic activity of *A. indica* along with various other medicinal plants. For the anti-inflammatory, analgesic and antipyretic activity of antidote *A. indica*, the dried plant extract was studied against the venom of *Heteropneustes fossilis*. Inflammation was observed to be reduced when plant extract and venom Ichthyocrinotoxin were administered simultaneously on carrageenan-induced male albino rats (Das et al. 2010). Aristolactam I and (–) Hinokinin exerted efficient anti-inflammatory effects and impeded the production of cytokines TNF-α and IL-6 in LPS-stimulated THP-1 cells. The mechanism of action is considered to be NFκB dependent for (–) Hinokinin and NFκB independent for aristolactam I (Desai et al. 2014).

10.5.6 Anticancer Activity

Kupchan and Doskotch (1962) reported AA, the major active constituent found in *A. indica*, as a potent tumour inhibitor. Whole plant extract has effectively inhibited the EAC (Ehrlich Ascites Carcinoma) tumour cell growth in mice at a dose of 50 mg/kg as well as positively reversed the changes of haematological parameters post tumour inoculation (Rana and Khanam 2002). The chloroform extract from this plant displayed a pronounced anticancer activity against Ehrlich Ascites Carcinoma (MCF-7) cell line at IC_{50} 347 µg/mL compared to the standard anticancer drug Taxol (Subramaniyan et al. 2015).

10.5.7 Antivenom Activity

Various extracts of *A. indica* have been utilised as a decoction for the ailment of several diseases including snake bite treatment. Orally administered extract and partially purified fraction of *A. indica* were observed to neutralise and induce protection to mice against rattle snake venom actions potently. The purified fraction has also imparted changes in the levels of serum SOD and LPx (Samy et al. 2008). Similarly, Meenatchisundaram et al. (2009) reported that the methanolic extract of 0.14 mg was capable of completely neutralising the lethal activity of *Daboia russelli* venom by inhibiting the PLA2-dependent haemolysis of sheep RBC's induced by *D. russelli* venom. Screening of *A. indica* plant extract against scorpion (*Mesobuthus tamulus*) venom showed potent venom neutralising capacity at 4 mg by restricting the PLA2-dependent haemolysis of mice RBC's induced by red scorpion venom in a dose-dependent manner (Attarde and Apte 2013). Additionally, the pro-coagulant activity showed 1 and 1.6 mg, respectively, of plant extracts were able to entirely neutralise coagulant activity in red scorpion venom and *D. russelli* venom (Attarde and Apte 2013; Meenatchisundaram et al. 2009). The topical application of the aqueous root extract of *A. indica* has prolonged the survival of animals when administered with Russell's viper venom as the extract possesses mighty collagenase, peroxidase, gelatinolytic, l-amino acid oxidase, nuclease and protease inhibitory activities. However, the extract did not impart any acute or subchronic toxicity in animals (Bhattacharjee and Bhattacharyya 2013).

10.5.8 Anti-Feedant Activity

The methanolic extract of *A. indica* had shown remarkable larvicidal, adulticidal and repellent activity against mosquito (Kamaraj et al. 2010). Aristolochic acid-I and aristolochic acid II from the methanol extract (Pradeepa et al. 2015) and silver nanoparticles (AgNPs) biosynthesised from *A. indica* (Murugan et al. 2015) displayed potent larval toxicity against malarial vector *Anopheles stephensi*. Both the isolated compounds had severely affected the mosquito gut thereby hindering the resurrection of mosquito vectors (Pradeepa et al. 2015).

10.5.9 Anti-Pruritic Activity

To evaluate the scratching response, compound 48/80 induced scratched behaviour model was utilised. The ethanol root extract of *A. indica* at the dose of 150 mg/kg showed significant effect and decreased scratching incidences (Mathew et al. 2011).

10.5.10 Anti-Implantation and Interceptive Activity

Single dose of 100 mg kg^{-1} body weight alcoholic, crude petroleum ether and chloroform extracts demonstrated 100% interceptive activity in mature female mice with negligible cytotoxicity; the chloroform extract had the most significant effect (Pakrashi et al. 1976). Lately, a sesquiterpene was isolated from the roots which exerted 100% interceptive activity and 91.7% anti-implantation activity in mice, with a single oral dose of 100 mg kg^{-1} b.wt. (Pakrashi and Shaha 1977). A compound, p-Coumaric acid, isolated from the alcoholic extract of the roots of *A. indica* (Pakrashi and Pakrasi 1978) at the single oral dose of 50 mg kg^{-1} body weight also displayed 100% interceptive activity (Pakrashi and Pakrasi 1979). Lately, the results reported by Pakrashi and Ganguly (1982) on aristolic acid also found to be in accordance with earlier studies indicating the potent interceptive activity. Aristolic acid isolated from *A. indica* found to possess anti-implantation and anti-oestrogenic properties (Pakrashi and Chakrabarty 1978). A single oral dose of 120 and 90 mg/kg of aristolic acid resulted in a rise of acid phosphatase (AP) and fall of alkaline phosphatase (ALP) and other uterine protein in mice. Exogenous administration of progesterone along with aristolic acid could neither prevent the biochemical changes produced in the uterus nor protect the early pregnancy in mice. This alteration in uterine phosphatase levels provided some insight into the mode of action of the compound aristolic acid (Pakrashi and Ganguly 1982). The implantation impending effect of aristolic acid was further assessed by Ganguly et al. (1986) which demonstrated that compound has damaged the future implantation sites, impaired the development and reconciled with reduction in uterine weight and its total protein contents in treated animals. It was concluded that aristolic acid hinders the steroidal conditioning of the uterus and furnishes its antagonistic effect in ovum implantation.

10.5.11 Abortifacient Activity

A pure compound namely methyl ester of aristolic acid obtained from roots affected the fertility of female by inducing 100% abortions at a single oral dose of 60 mg/kg body weight administered on 6th or 7th day of pregnancy (Pakrashi and Shaha 1978). Further, Pakrashi and Shaha (1979) also studied the short-term toxicity with the same compound in mice.

10.5.12 Antifertility Activity

Aristolic acid isolated from *A. indica* found to possess antifertility effect in female albino rabbits (Pakrashi and Chakrabarty 1978). Ethanol extract of roots when administered postcoitally significantly decreased fertility in hamsters as well as rats, but chloroform, petroleum ether and aqueous extracts were found to be inactive and/or toxic. Other bioactive components were also isolated from fractions of *A. indica* like AA, aristolic acid, cis- and trans-p-coumaric acid and methyl aristolate were all inactive possessing no fertility regulation activity (Che et al. 1984). Chaudhury and Haq (1980) and Kamboj and Dhawan (1982) in their study on plants for fertility regulation in India had mentioned the name of *A. indica* by biologically evaluating the constituents extracted from its roots. The extracts were also noted to possess antispermatogenic effect on male mice (Pakrashi and Pakrasi 1977). Farnsworth and Waller (1982) reviewed plant-derived compounds that hinder the sperm production if taken orally by the male or kill sperm on contact if used vaginally by the female. *A. indica* tops their list as potent fertility inhibiting agent in male, though it will be of great importance to emerge as fertility inhibitors that are completely selective for reproductive systems and enzymes.

10.6 TOXIC RESPONSE

Since time immemorial, medicinal plants have been utilised by the traditional healers for their vast health benefits (Modak et al. 2015; Dey et al. 2017). Based on their acceptance in traditional healing as well as modern herbal healthcare, one might expect the botanicals utilised in healthcare possess low toxicity (Madic et al. 2019; Akwu et al. 2019). However, a number of latest *in vitro* and *in vivo* assays have displayed the genotoxic and cytotoxic potential of several plants used as food and medicine. *Aristolochia indica*, being a potent medicinal plant, is extensively accessible in herbal markets and most of the healers were aware about its medicinal uses. The plant has also been reported traditionally against liver ailments, dyspepsia, fever, oedema, dysmenorrhoea, rheumatic arthritis, leishmaniasis, leprosy, etc. (Michl et al. 2013). AA, a well-known major active constituent of the plant, is reported to exert antioxidant, anti-snake venom, antimicrobial, abortifacient and several other pharmacological properties (Dey and De 2011). The AAs found among the species of *Aristolochia* were highly famed for its nephrotoxicity after the tragic Belgian cohort incident. In this situation, more than 100 women who have taken the slimming pills containing a Chinese herb *Stephania tetrandra* that was accidentally replaced by AA-containing *A. fangchi*, were described to be suffering from renal interstitial fibrosis, which is often connected with urothelial malignancies (Balachandran et al. 2005; Debelle et al. 2008). Hashimoto et al. (1999) have earlier reported the association of *Aristolochia* with interstitial nephritis due to the occurrence of AAs during the treatment of analgesic, diuretic and rheumatism. Lately, *in vivo* studies reported that the high sensitivity of chronic renal failure of rats was associated with AAs (Cheng et al. 2006), and the compound was hypothesised to exert genotoxicity probably via nitric oxide (Wu et al. 2007).

AAN poses a persistent health threat to the traditional people using *A. indica* as ethnomedicine and to the urban people consuming AA-rich dietary supplements specially as Chinese slimming capsules. AAN, initially called as Chinese herb nephropathy (CHN) (Heinrich et al. 2009; Jelakovic et al. 2019; Ren et al. 2020). AAN is a type of chronic tubulointerstitial renal disease associated with upper tract urothelial carcinoma (UTUC) in almost 50% of the cases (Nortier et al. 2000). The renal failure was described by comprehensive interstitial fibrosis with loss of tubules, hyperplasia of the urothelium and atrophy (Cosyns et al. 1994; Depierreux et al. 1994). Urothelial carcinoma observed to occur in more than 40% of the patients consuming the Chinese herbs (Nortier et al. 2000; Lord et al. 2001; Nortier and Vanherweghem 2002). AAs were affirmed as the prime culprit because AA reacts with DNA to form AL-DNA adducts which were detected in the kidneys and ureteric tissues of these patients (Nortier et al. 2000; Lord et al. 2001). These lesions get accumulated in the renal cortex, where they produced a unique mutational signature in the TP53 tumour-suppressor gene (Kuo et al. 2012). Vanherweghem (1997) pointed towards the correlation between chronic interstitial nephropathies in Indians and *Aristolochia* sp. Most AAN patients exhibited an unexpected rapid progression towards end stage of renal disease. During clinical scanning severe anaemia, increased serum creatinine, mild hypertension with suppressed glomerular filtration rate, leukocyturia, proteinuria and glycosuria were observed (Meyer et al. 2000; Yang et al. 2012; Gokmen et al. 2013). During renal tract ultrasonic inspection, shrunken kidneys were visible which results in irregular and asymmetrical cortical outline (Gokmen et al. 2013). Although botanicals known or suspected to bear, AAs were no longer authorised in various countries as AAN nephropathy raised worldwide attention (Debelle et al. 2008). AAN induced by various herbal remedies and products is still stated from all over the world (Yang et al. 2006; Shaohua et al. 2010; Wu et al. 2012; Vaclavik et al. 2014; Ban et al. 2018) and long-term ingestion of herbal formula containing AAs is one of the eminent risk factors for inducing AAN (Jia et al. 2005; Vervaet et al. 2017).

In the Balkan regions also, the exposure to AAs was observed through consuming food prepared from wheat flour with the seeds of the plants of *Aristolochia* family which was primarily responsible for the so-called Balkan endemic nephropathy (BEN; Han et al. 2019; Grollman 2013;

Jelakovic et al. 2015). BEN is an endemic familial acquired chronic renal disease usually associated by urothelial carcinoma of the upper urinary tract (Stefanovic et al. 2007; Miyazaki and Nishiyama 2017). In the endemic farming villages, BEN is quite prevalent along the tributaries of the Danube river (Stiborova et al. 2016). Mostly in Asia and Balkans, the incidences of AAN are probably much higher than expectations. In Asian countries, traditional medicines were quite famous and the complexity of the pharmacopoeia depicted a high risk for AAN because of the periodic substitution of the botanical products by AA-containing herbs (Hashimoto et al. 1999; Vanherweghem 1997). In the early 1980s, products containing AA were withdrawn from the market as AA was considered to be a potent carcinogen (Pezzuto et al. 1988; Zhang et al. 2004). Despite food and drug administration's warnings concerning the safety of botanical remedies containing AA, these herbs are still sold via the Internet (Gold and Slone 2003).

Apart from nephrotoxicity, there are few reports obtainable on toxicity of *A. indica* when given orally to goats, exerting arching of the back, diarrhoea, tympany, dyspnoea and loss of hair from the back were the prominent signs. The prime lesions were haemorrhage in the mucoid abomasitis, congestion in the liver, lungs, heart, kidneys, fatty change, enteritis and straw-coloured fluid in serous cavities. An elevated concentration of urea, aspartate amino transferase activity and ammonia and a reduction in the concentrations of total protein and magnesium were detected in the serum (Sati et al. 2011). Aristolochine, the alkaloid, causes cardiac and respiratory paralysis in frogs and mice, increases respiratory rate in rabbits, and stimulates skeletal muscle in small doses but paralyses in large doses. It causes haemorrhagic nephritis in rabbits and gastrointestinal irritation in dogs (Chopra et al. 1958). AA was also reported to induce *in vitro* chromosomal aberrations and sister chromatid exchanges (SCEs) in human lymphocytes (Abel and Schimmer 1983).

10.7 TRADITIONAL USES

A. indica has been used as traditional medicine by different communities around the globe since long time. Correlating the ethnopharmacological properties with recent pharmacological and phytochemical investigation shows consistency with the latest findings (Borah et al. 2021). Many reports indicate that the plant is a potent anti-snake venomous (Samy et al. 2008). It has also been reported that the effectiveness increased many folds if *A. indica* is administered in combination with another traditional anti-snakebite plant *Rauvolfia serpentina* (Dey and De 2010a, b). Another report indicates that traditional people from Andhra Pradesh, India used this plant with pepper and garlic against snakebite (Reddy et al. 2009). The roots of *A. indica* are used to heal wounds and to enhance fertility in males by the traditional healers of Assam, India (Taid et al. 2014). In Sri Lanka, root and stem are used as febrifuge, attenuant, emmenagogue and in snake and insect bites. In the Philippines, the decoction of roots is used as a tonic, carminative and emmenagogue, and it is also an effective therapy for infantile meteorism if the crushed roots are applied to the abdomen (Casey 1960; Saha et al. 1961; Malhi and Trivedi 1972). The plant has been used in abortion by some ethnic people of Tamil Nadu, India (Balakrishnan et al. 2005). It has been reported that the roots of the plant are boiled in coconut oil and with the seed of *Centratherum anthelminticum* and are often used in various skin allergies and scabies, whereas the paste of leaf is applied to warts in Southern part of India (Harsha et al. 2003). The plant is used by the people of Assam, India for the treatment of asthma and leukoderma (Purkayastha et al. 2007; Nath et al. 2008). Some ethnic people of north-east part of India used decoction of the leaves to treat dysentery, diarrhoea and melena, and crude root extracts are used for treatment of skin disorders like leucoderma and certain asthmatic problems (Borah et al. 2006; Purkayastha et al. 2007; Nath et al. 2008). The decoction of leaf is taken orally in dermatitis and insect bite by *Kurumba* Tribals of Tamil Nadu, India (Alagesaboopathi 2011). Some ethnic people of Andhra Pradesh, India used 10–20 g of root paste externally for the treatment of scabies (Neelima et al. 2011). Several ethnomedicinal reports have been published from the studied species as shown in Table 10.2.

TABLE 10.2
Ethnomedicinal Uses of *Aristolochia indica*

Ethnomedicinal Uses	Parts Used	Formulation	Reference
Snakebite	Root	Juice obtained for root Powder made from *A. indica* (root) and *Rauvolfia serpentina* (Leaf) Pills made of root powder and fruits of *Piper nigrum*.	Samy et al. (2008), Reddy et al. (2009), Dey and De (2010a, b), Dey and De (2012), Galav et al. (2013), Michl et al. (2013), Naturu et al. (2013), Teron and Borthakur (2013), Sulochana et al. (2014)
	Root or leaves	The roots or leaves of *A. indica* are combined with the juice of *Asparagus racemosus* (root)	
	Leaves	*A. indica* leaves and 7–8 fruits of *Piper nigrum* and 4–5 *Allium sativum* bulb are made into pills	
Insect bite	Root or leaves	Roots or leaves are heated to form a paste. The paste is applied topically	Alagesaboopathi (2011)
	Leaves	Decoction of leaf taken orally	
Skin infection (allergy, scabies, wart, leucoderma)	Root	Roots of the plant are boiled in coconut oil and with the seed of *Centratherum anthelminticum* and are often used in various skin allergies and scabies	Harsha et al. (2003), Nath et al. (2008), Sivaperumal et al. (2009), Choudhary et al. (2011), Neelima et al. (2011), Bhat et al. (2014)
	Leaves	Paste of leaf is applied to warts	
Abortifacient	Root or leaves	Pills made of root or leaf powder are used	Balakrishnan et al. (2005)
Sexual problems	Root	A paste made from *A. indica* (root), *Rauvolfia serpentina* (root) and Curcuma longa (root)	Michl et al. (2013), Taid et al. (2014)
	Leaves	A paste made from fresh leaves is applied topically	
	Stem	Stem is combined with juice *Asparagus racemosus* (root)	
	Roots or leaves	Juice is made from roots and leaves	
Gastric problems/ abdominal upset/ Tympanites	Leaves	A pill made from one leaf of *A. indica* and one clove (*Syzygium aromaticum*) taken with large amounts of salt Young leaves are chewed Decoction of leaves is taken orally	Casey (1960), Saha et al. (1961), Malhi and Trivedi (1972), Borah et al. (2006), Purkayastha et al. (2007), Nath et al. (2008), Michl et al. (2013), Dharmalingam et al. (2014), Swarnalatha et al. (2017)
	Whole plant	The whole plant is chopped into pieces. A paste is made with sugar	
	Root	Crushed roots are applied to the abdomen	
Madness/ Unconsciousness	Root	A small piece of root is worn in an amulet around the neck Juice obtained from root	Michl et al. (2013)
	Leaves	Leaves are applied topically on the head Juice from leaves is dropped into the patient's eyes	
Neurotonic	Root	*A. indica* (root) and *Withania somnifera* (root) are taken with milk Powder taken with milk and honey	Michl et al. (2013)
	Leaves	Leaves are boiled with water	

10.8 CONCLUSION

The uses of *A. indica* as traditional medicine confirm that it possesses various biological activities. Moreover, scientific investigation documented it is an emmenagogue, abortifacient, antidiarrhoeal, antineoplastic, antiseptic, anti-inflammatory, antimicrobial, antioxidant and antipyretic at one side, and on the other side, it is considered as a potent nephrotoxic, antifertility and antispermatogenic agent. Despite being a potential carcinogen, the plant is still used in some herbal remedies. Most of the toxic effects associated with the plant are proved to be contributed by AA. Health experts around the world have taken action to protect the public against aristolochia and AA. Traditional use of *A. indica* as a popular abortifacient and antivenom should be restricted considering its harmful effects. In the present study, we have comprehensively reviewed the traditional knowledge along with various phytoconstituents, pharmacological attribute and toxic activity exhibited by the plant. Hopefully, these studies will explore the full potential of *A. indica* and optimise its use as a promising herbal medicine, thereby promoting global health.

NOTES

Surjit Sen, Department of Botany, Fakir Chand College, Diamond Harbour, India
Sudeshna Nandi, Molecular and Applied Mycology and Plant Pathology Laboratory, Department of Botany, University of Calcutta, Kolkata, India
Rupa Sanyal, Department of Botany, Bhairab Ganguly College, Kolkata, India
Abhijit Dey, Department of Life Sciences, Presidency University, Kolkata, India
Manjula Rai, Department of Botany, St. Joseph's College, Darjeeling, India

REFERENCES

Abel, G., & Schimmer, O. 1983. Induction of structural chromosome aberrations and sister chromatid exchanges in human lymphocytes in vitro by aristolochic acid. *Human Genetics*, 64(2): 131–133.

Achari, B., Bandyopadhyay, S., Saha, C. R., & Pakrashi, S. C. 1983. A phenanthroid lactone, steroid and lignans from Aristolochia indica. *Heterocycles*, 20: 771–774.

Achari, B., Chakrabarty, S., Bandyopadhyay, S., & Pakrashi, S. C. 1982. A new 4,5-dioxoaporphine and other constituents of *Aristolochia indica*. *Heterocycles*, 19: 1203–1206.

Achari, B., Chakrabarty, S., & Pakrashi, S. C. 1981. An N-glycoside and steroids from *Aristolochia indica*. *Phytochemistry*, 20: 1444–1445.

Akwu, N. A., Naidoo, Y., & Singh, M. 2019. Cytogenotoxic and biological evaluation of the aqueous extracts of *Grewia lasiocarpa*: An *Allium cepa* assay. *South African Journal of Botany*, 125: 371–380.

Alagesaboopathi, C. 2011. Ethnomedicinal plants used as medicine by the Kurumba tribals in Pennagaram Region, Dharmapuri District of Tamil Nadu, India. *Asian Journal of Experimental Biological Science,* 2: 140–142.

Attarde, S., & Apte, K. 2013. Studies on antivenom activity of *Aristolochia indica* plant extract against red scorpion venom by *in vivo* and *in vitro* methods. *International Journal of Pharmacognosy and Phytochemical Research*, 5(3): 168–172.

Balachandran, P., Wei, F., Lin, R. C., Khan, I. A., & Pasco, D. S. 2005. Structure activity relationships of aristolochic acid analogues: Toxicity in cultured renal epithelial cells. *Kidney International*, 67: 1797–1805.

Balakrishnan, V., Venkatesan, K., Ravindran, K. C., & Karuppusamy, S. 2005. Studies on medicinal plants used for abortion by Irulars of Coimbatore district, Tamil Nadu, India. *Bulletin of Medico-Ethno-Botanical Research,* 26: 6–9.

Ban, T. H., Min, J. W., Seo, C., Kim, D. R., Lee, Y. H., Chung, B. H., Jeong, K. H., Lee, J. W., Kim, B. S., Lee, S. H., Choi, B. S., Han, J. S., & Yang, C. W. 2018. Update of aristolochic acid nephropathy in Korea. *Korean Journal of Internal Medicine,* 33(5): 961–969.

Berger, S. I., & Iyengar, R. 2011. Role of systems pharmacology in understanding drug adverse events. *Wiley Interdisciplinary Reviews: Systems Biology and* Medicine, 3(2): 129–35. doi: 10.1002/wsbm.114.

Bhat, P., Hedge, G. R., Hedge, G, & Mulgund, G. S. 2014. Ethnomedicinal plants to cure skin diseases—An account of the traditional GR knowledge in the coastal parts of Central Western Ghats, Karnataka, India. *Journal of Ethnopharmacology,* 151: 493–502. https://doi.org/10.1016/j.jep.2013.10.062

Bhattacharjee, P., & Bhattacharyya, D. 2013. Characterization of the aqueous extract of the root of Aristolochia indica: Evaluation of its traditional use as an antidote for snake bites. *Journal of Ethnopharmacology*, 145(1): 220–226.

Borah, P. J., Borah, D., Das, U., Das, T. J., & Sarma, R. 2021. A review on ethnopharmacological utility, traditional knowledge and phytochemistry of Aristolochia species in Assam, India. *Notulae Scientia Biologicae,* 13(3): 11027. https://doi.org/10.15835/nsb13311027

Borah, P. K., Gogoi, P., Phukan, A. C., & Mahanta, J. 2006. Traditional medicine in the treatment of gastrointestinal diseases in upper Assam. *Indian Journal of Traditional Knowledge,* 5(4): 510–512.

Bunel, V., Antoine, M. H., Stévigny, C., Nortier, J, & Duez, P. 2016. New in vitro insights on a cell death pathway induced by magnolol and honokiol in aristolochic acid tubulotoxicity. *Food and Chemical Toxicology*, 87: 77–87.

Casey, R. C. 1960. 298 alleged antifertility plants of India. *Indian J Medical Sciences,* 14: 590–600.

Chaudhury, R. R., & Haq, M. 1980. Review of plants screened for antifertility activity-I. Bull Medico. *Ethnobotany Research*, 1: 408–419.

Che, C. T., Ahmed, M. S., Kang, S. S., Waller, D. P., Bingel, A. S., Martin, A., Rajamahendran, P., Bunyapraphatsara, N., Lankin, D. C., & Cordell G. A. 1984. Studies on Aristolochia III. Isolation and biological evaluation of constituents of *Aristolochia* indica roots for fertility-regulating activity. *Journal of Natural Products*, 47(2): 331–341.

Che, C. T., Cordell, G. A., Fong, H. H. S., & Evans, C. A. 1983. Aristolindiquinone—A new naphthoquinone from *Aristolochia indica* L. Aristolochiaceae. *Tetrahedron Letters*, 24: 1333–1336.

Cheng, C. L., Chen, K. J., Shih, P. H., Lu, L. Y., Hung, C. F., Lin, W. C., & Gu, J. Y. 2006. Chronic renal failure rats are highly sensitive to aristolochic acids, which are nephrotoxic and carcinogenic agents. *Cancer Letter*, 232: 236–242.

Chopra, R. N., Chopra, I. C., Handa, K. L., & Kapoor, L. D. 1958. *Indigenous drugs of India.* Dhar & Sons Pvt. Ltd., Academic Publishers, Kolkata, India.

Choudhary, M. S., Mishra, N., Upadhyay, S. T., & Upadhyay, R. 2011. Indigenous knowledge of using medicinal plants in treating skin diseases by tribal's in central Narmada valley of Madhya Pradesh (India). *Bulletin of Environment, Pharmacology and Life Sciences,* 1(1): 60–63.

Cory, R. M., Chan, D. M. T., McLaren, F. R., Rasmussen, M. H., & Renneboog, R. M. 1979. A short synthesis of Ishwarone. *Tetrahedron Letters*, 20: 4133–4136.

Cosyns, J. P., Jadoul, M., Squifflet, J. P., De Plaen, J. F., Ferluga, D., & Van Ypersele de Strihou, C. 1994. Chinese herbs nephropathy: A clue to Balkan endemic nephropathy? *Kidney International,* 45(6): 1680–1688.

Das, R., Kausik, A., & Pal, T. K. 2010. Anti-inflammatory activity study of antidote *Aristolochia indica* to the venom of *Heteropneustes fossilis* in rats. *Journal of Chemical and Pharmaceutical Research*, 2: 554–562.

Debelle, F. D., Vanherweghem, J. L., & Nortier, J. L. 2008. Aristolochic acid nephropathy: A worldwide problem. *Kidney International,* 74(2): 158–169.

Depierreux, M., Van Damme, B., Vanden Houte, K., & Vanherweghem, J. L. 1994. Pathologic aspects of a newly described nephropathy related to the prolonged use of Chinese herbs. *American Journal of Kidney Diseases,* 24(2): 172–180.

Desai, D. C., Jacob, J., Almeida, A., Kshirsagar, R., & Manju, S. L. 2014. Isolation, structural elucidation and anti-inflammatory activity of astragalin, (-) hinokinin, aristolactam I and aristolochic acids (I & II) from Aristolochia indica. *Natural Product Research,* 28(17): 1413–1417.

Dey, A., & De, J. 2011. *Aristolochia indica*: A review. *Asian Journal of Plant Sciences*, 10(2): 108–116.

Dey, A., & De, J. N. 2010a. *Rauvolfia serpentina* (L). benth. ex Kurz.—A review. *Asian Journal of Plant Sciences*, 9: 285–298.

Dey, A., & De, J. N. 2010b. Ethnoveterinary uses of medicinal plants by the aboriginals of purulia district, West Bengal, India. *International Journal of Botany*, 6: 433–440.

Dey, A., & De, J. N. 2012. Traditional use of plants against snakebite in Indian Subcontinent: A review of the recent literature. *African Journal of Traditional, Complementary and Alternative Medicines,* 9(1): 153–174. https://doi.org/10.4314/ajtcam.v9i1.20

Dey, A., Gorai, P., Mukherjee, A., Dhan, R., & Modak, B. K. 2017. Ethnobiological treatments of neurological conditions in the Chota Nagpur Plateau, India. *Journal of Ethnopharmacology*, 198: 33–44.

Dharmalingam, S. R., Madhappan, R., Ramamurthy, S., Chidambaram, K., Srikanth, M. V., Shanmugham, S., & Senthil Kumar, K. L. 2014. Investigation on antidiarrhoeal activity of *Aristolochia indica* Linn. root extracts in mice. *African Journal of Traditional Complementary and Alternative Medicines*, 11(2): 292–294. https://doi.org/10.4314/ajtcam.v11i2.11

Farnsworth, N. R., & Waller, D. P. 1982. Current status of plant products reported to inhibit sperm. *Research Frontiers in Fertility Regulation*, 2(1): 1–16.

Fitzgerald, M., Heinrich, M., & Booker, A. 2020. Medicinal plant analysis: A historical and regional discussion of emergent complex techniques. *Frontiers in Pharmacology*, 10: 1480.

Galav, P., Jain, A., & Katewa, S. S. 2013. Traditional veterinary medicine used by livestock owners of Rajasthan, India. *Indian Journal of Traditional Knowledge,* 12(1): 47–55.

Ganguly, A. K., Gopinath, K. W., Govindachari, T. R., Nagarajan, K., Pai, B. R., & Parthasarathy, P. C. 1969. Ishwarone a novel tetra cyclic sesquiterpene. *Tetrahedron Letters*, 3: 133–136.

Ganguly, T., Pakrashi, A., & Pal, A. K. 1986. Disruption of pregnancy in mouse by aristolic acid: I. Plausible explanation in relation to early pregnancy events. *Contraception*, 34: 625–637.

Gokmen, M. R., Cosyns, J. P., Arlt, V. M., Stiborova, M., Phillips, D. H., Schmeiser, H. H., Simmonds, M. S., Cook, H. T., Vanherweghem, J. L., Nortier, J. L., & Lord, G. M. 2013. The epidemiology, diagnosis, and management of aristolochic acid nephropathy: A narrative review. *Annals of Internal Medicine,* 158(6): 469–477.

Gold, L. S., & Slone, T. H. 2003. Aristolochic acid, an herbal carcinogen, sold on the Web after FDA alert. *The New England Journal of Medicine*, 349: 1576–1577.

Gopinath, R., & Prakash, M. 2013. Antibacterial activity of three medicinal plants against clinically isolated multidrug resistant *Enterococcus faecalis* (MDRE). *International Journal of Current Microbiology and Applied Sciences*, 2: 6–14.

Goverdhan, P., Sandhya Rani, M., Thirupathi, K., Rani, S., Sathesh, S., Ravi Kumar, B., & Mohan, G. K. 2008. Hypoglycemic and antihyperglycemic effect of Aristolochia indica normal and alloxan induced diabetic rats. *Pharmacologyonline*, 1: 20–29.

Govindachari, T. R., Mohamed, P. A., & Parthasarathy, P. C. 1970. Ishwarane and aristolochene, 2 new sesquiterpene hydrocarbons from *Aristolochia indica* D. *Tetrahedron*, 26: 615–619.

Govindachari, T. R., & Parthasarathy, P. C. 1971. Ishwarol, a new tetracyclic sesquiterpene alcohol from *Aristolochia indica. Indian Journal of Chemistry*, 9: 1310–1310.

Govindachari, T. R., Parthasarathy, P. C., Desai, H. K., & Mohamed, P. A. 1973. 5βH, 7β, 10α-Selina-4(14), II-diene, a new sesquiterpene hydrocarbon from *Aristolochia indica. Indian Journal of Chemistry*, 11: 971–973.

Grollman, A. P. 2013. Aristolochic acid nephropathy: Harbinger of a global iatrogenic disease. *Environmental and Molecular Mutagenesis*, 54(1): 1–7.

Han, J., Xian, Z., Zhang, Y., Liu, J., & Liang, A. 2019. Systematic overview of Aristolochic acids: Nephrotoxicity, carcinogenicity, and underlying mechanisms. *Frontiers in Pharmacology,* 10: 648. https://doi.org/10.3389/fphar.2019.00648

Harsha, V. H., Hebbar, S. S., Shripathi, V., & Hegde, G. R. 2003. Ethnomedicobotany of Uttara Kannada district in Karnataka, India: Plants in treatment of skin diseases. *Journal of Ethanopharmacology*, 84: 37–40.

Hashimoto, K., Higuchi, M., Makino, B., Sakakibara, I., Kubo, M., Komatsu, Y., Maruno, M., & Okada, M. 1999. Quantitative analysis of aristolochic acids, toxic compounds, contained in some medicinal plants. *Journal of Ethnopharmacology*, 64(2): 185–189.

Heinrich, M., Chan, J., Wanke, S., Neinhuis, C., & Simmonds, M. S. 2009. Local uses of *Aristolochia* species and content of nephrotoxic aristolochic acid 1 and 2—A global assessment based on bibliographic sources. *Journal of Ethnopharmacology,* 125(1): 108–144.

Hwang, S. M., Kelly, L. M., & Gilbert, M. G. 2003. Aristolochiaceae. In: Wu, Z., Peter, H. R., & Hong, D. (Eds.), *Flora of China.* Science Press, Beijing and Missouri Botanical Garden Press, St. Louis, MO, pp. 246–269.

Hwang, Y. H., Kim, T., Cho, W. K., Yang, H. J., Kwak, D. H., Ha, H., Song, K. H., & Ma, J. Y. 2012. *In vitro* and *in vivo* genotoxicity assessment of Aristolochia manshuriensis Kom. *Evidence-Based Complementary and Alternative Medicine,* 12: 1–9.

IARC. 2002. Working Group on the evaluation of carcinogenic risks to humans: Some traditional herbal medicines, some mycotoxins, naphthalene and styrene. *IARC Monographs on the Evaluation of Carcinogenic Risks to Humans*, 82: 1–556.

Jelakovic, B., Dika, Z., Arlt, V. M., Stiborova, M., Pavlovic, N. M., Nikolic, J., Colet, J. M., Vanherweghem, J. L., & Nortier, J. L. 2019. Balkan endemic nephropathy and the causative role of aristolochic acid. *Seminars in Nephrology,* 39(3): 284–296.

Jelakovic, B., Vukovi Lela, I., Karanovic, S., Dika, Z., Kos, J., Dickman, K., Sekoranja, M., Poljicanin, T., Misic, M., Premuzic, V., Abramovic, M., Matijevic, V., Miletic Medved, M., Cvitkovic, A., Edwards, K., Fucek, M., Leko, N., Teskera, T., Laganovic, M., Cvoriscec, D., & Grollman, A. P. 2015. Chronic dietary exposure to aristolochic acid and kidney function in native farmers from a Croatian endemic area and Bosnian immigrants. *Clinical Journal of the American Society of Nephrology,* 10(2): 215–223.

Jia, W., Zhao, A., Gao, W., Chen, M., & Xiao, P. 2005. New perspectives on the Chinese herbal nephropathy. *Phytotheraphy Research,* 19(11): 1001–1002.

Jirovetz, L., Buchbauer, G., Puschmann, C., Fleischhacker, W., Shafi, P. M., & Rosamma, M. K. 2000. Analysis of the essential oil of the aerial parts of the medicinal plant *Aristolochia indica* Linn. (Aristolochiaceae) from South India. *Scientia Pharmaceutica*, 68: 309–316.

Kamaraj, C., Rahuman, A. A., Mahapatra, A., Bagavan, A., & Elango, G. 2010. Insecticidal and larvicidal activities of medicinal plant extracts against mosquitoes. *Parasitology Research*, 107(6): 1337–1349.

Kamboj, V. P., & Dhawan, B. N. 1982. Research on plants for fertility regulation. *Indian Journal of Ethnopharmacology*, 6: 191–226.

Kanjilal, P. B., Kotoky, R., & Couladis, M. 2009. Chemical composition of the stem oil of *Aristolochia indica* L. *Journal of Essential Oil Research*, 21: 1–2.

Karan, S. K., Mishra, S. K., Pal, D., & Mondal, A. 2012. Isolation of β-sitosterol and evaluation of antidiabetic activity of *Aristolochia indica* in alloxan-induced diabetic mice with a reference to in-vitro antioxidant activity. *Journal of Medicinal Plants Research*, 6(7): 1219–1223.

Kelly, R. B., Zamecnik, J., & Beckett, B. A. 1972. Total synthesis of (f)-Ishwarane, a tetracyclic sesquiterpenoid. *Canadian Journal of Chemistry*, 50: 3455–3464.

Kirtikar, K. R., & Basu, B. D. 1975. *Indian medicinal plants*. Vol III. Periodical Experts, New Delhi, India. pp. 2120–2125.

Krishnaraju, A. V., Rao, T. V. N., Sundararaju, D., Vanisree, M., Tsay, H. S., & Subbaraju, G. V. 2005. Assessment of bioactivity of Indian medicinal plants using Brine shrimp (*Artemia salina*) lethality assay. *International Journal of Applied Science and Engineering*, 3: 125–134.

Kumar, V. P., Chauhan, N. S., Padh, H., & Rajani, M. 2006. Search for antibacterial and antifungal agents from selected Indian medicinal plants. *Journal of Ethnopharmacology*, 107(2): 182–188.

Kuo, P. C., Li, Y. C., & Wu, T. S. 2012. Chemical constituents and pharmacology of the Aristolochia (mǎdōu ling) species. *Journal of Traditional and Complementary Medicine,* 2(4): 249–266.

Kupchan, S. M., & Doskotch, R. W. 1962. Tumor inhibitors. I. Aristolochic acid, the active principle of *Aristolochia indica. Journal of Medicinal Chemistry*, 5: 657–659.

Kupchan, S. M., & Merianos, J. J. 1968. Tumor inhibitors. XXXII. Isolation and structural elucidation of novel derivatives of aristolochic acid from *Aristolochia indica. The Journal of Organic Chemistry*, 33: 3735–3738.

Li, W., Chan, C. K., Wong, Y. L., Chan, K. J., Chan, H. W., & Chan W. 2018. Cooking methods employing natural anti-oxidant food additives effectively reduced concentration of nephrotoxic and carcinogenic aristolochic acids in contaminated food grains. *Food Chemistry*, 264, 270–276.

Lord, G. M., Cook, T., Arlt, V. M., Schmeiser, H. H., Williams, G., & Pusey, C. D. 2001. Urothelial malignant disease and Chinese herbal nephropathy. *Lancet,* 358(9292): 1515–1516.

Madic, V., Stojanović-Radic, Z., Juskovic, M., Jugovic, D., Zabar Popovic, A., & Vasiljevic, P. 2019. Genotoxic and antigenotoxic potential of herbal mixture and five medicinal plants used in ethnopharmacology. *South African Journal of Botany*, 125: 290–297.

Mahesh, V. K., & Bhaumik, H. L. 1987. Isolation of methyl ester of 12-nonacosenoic acid from *Aristolochia indica. Indian Journal of Chemistry*, 26: 86–86.

Malhi, B. S., & Trivedi, V. P. 1972. Vegetable antifertility drugs of India. *Quarterly Journal of Crude Drug Research,* 12: 1922.

Mathew, J. E., Kaitheri, S. K., Dinakaranvachala, S., & Jose, M. 2011. Anti-inflammatory, antipruritic and mast cell stabilizing activity of *Aristolochia indica. Iranian Journal of Basic Medical Sciences*, 14: 422–427.

Meenatchisundaram, S., Parameswari, G., & Michael, A. 2009. Studies on antivenom activity of *Andrographis paniculata* and *Aristolochia indica* plant extracts against Daboia russelli venom by *in vivo* and *in vitro* methods. *Indian Journal of Science and Technology*, 2: 76–79.

Meyer, M. M., Chen, T. P., & Bennett, W. M. 2000. Chinese herb nephropathy. *Proceedings (Baylor University. Medical Center*, 13 (4): 334–337.

Michl, J., Jennings, H. M., Kite, G. C., Ingrouille, M. J., Simmonds, M. S., & Heinrich, M. 2013. Is aristolochic acid nephropathy a widespread problem in developing countries? A case study of *Aristolochia indica* L. in Bangladesh using an ethnobotanical-phytochemical approach. *Journal of Ethnopharmacology*, 149(1): 235–244.

Mix, D. B., Guinaudeau, H., & Shamma, M. 1982. The aristolochic acids and aristolactams. *Journal of Natural Products*, 45: 657–666.

Miyazaki, J., & Nishiyama, H. 2017. Epidemiology of urothelial carcinoma. *International Journal of Urology*, 24(10): 730–734.

Modak, B. K., Gorai, P., Dhan, R., Mukherjee, A., & Dey, A. 2015. Tradition in treating taboo: Folkloric medicinal wisdom of the aboriginals of Purulia district, West Bengal, India against sexual, gynaecological and related disorders. *Journal of Ethnopharmacology*, 169: 370–386.

Murugan, K., Labeeba, M. A., Panneerselvam, C., Dinesh, D., Suresh, U., Subramaniam, J., Madhiyazhagan, P., Hwang, J. S., Wang, L., Nicoletti, M., & Benelli, G. 2015. *Aristolochia indica* green-synthesized silver nanoparticles: A sustainable control tool against the malaria vector Anopheles stephensi? *Research in Veterinary Science*, 102: 127–135.

Murugan, R., Shivanna, K. R., & Rao, R. R. 2006. Pollination biology of *Aristolochia tagala*, a rare species of medicinal importance. *Current Science,* 91: 795–798.

Naik, A. S., Siddiqua, S., Shrunga, M. N., Sunitha, K. L., Prashith, K. T. R., & Raghavendra, H. L. 2015. Antimicrobial and radical scavenging efficacy of leaf and flower of *Aristolochia indica* Linn. *Science Technology and Arts Research Journal,* 4(1): 103–108.

Nath, K. K., Deka, P., Borkataki, S., & Borthakur, S. K. 2008. Traditional remedies of respiratory disorders from Assam, India. *Pleione*, 2(2): 211–216.

Naturu, S., Pulicherla, Y., Mattigunta, L., & Devi, V. R. C. 2013. Traditional phytotherapy treatment for snake bite and scorpion sting by ethnic groups of Kadapa district, Andhra Pradesh, India. *International Journal of Pharmaceutical Sciences Review and Research,* 20(1): 64–70.

Neelima, M., Prasad, G. P., Sudarsanam, G., Pratap, G. P., & Jyothi, B. 2011. Ethnobotanical studies in Rapur forest division of Nellore district in Andhra Pradesh. *Life Science Leaflets*, 11: 333–345.

Neinhuis, C., Wanke, S., Hilu, K.W., Müller, K., & Borsch, T. 2005. Phylogeny of Aristolochiaceae based on parsimony, likelihood and Bayesian analyses of trnL-trnF sequences. *Plant Systematics and Evolution*, 250: 7–26.

Nortier, J. L., Martinez, M. C., Schmeiser, H. H., Arlt, V. M., Bieler, C. A., Petein, M., Depierreux, M. F., De Pauw, L., Abramowicz, D., Vereerstraeten, P.. & Vanherweghem, J. L. 2000. Urothelial carcinoma associated with the use of a Chinese herb (*Aristolochia fangchi*). *The New England Journal of Medicine,* 342(23): 1686–1692.

Nortier, J. L., & Vanherweghem, J. L. 2002. Renal interstitial fibrosis and urothelial carcinoma associated with the use of a Chinese herb (Aristolochia fangchi). *Toxicology*, 181–182: 577–580.

Pakrashi, A., & Chakrabarty, B. 1978. Antifertility effect of aristolic acid from *Aristolochia indica* (Linn) in female albino rabbits. *Experientia*, 34(10): 1377.

Pakrashi, A., & Ganguly, T. 1982. Changes in uterine phosphatase levels in mice treated with aristolic acid during early pregnancy. *Contraception*, 26: 635–643.

Pakrashi, A., & Pakrasi, P. 1978. Biological profile of p-coumaric acid isolated from *Aristolochia indica* Linn. *Indian Journal of Experimental Biology,* 16: 1285–1287.

Pakrashi, A., & Pakrasi, P. 1979. Antifertility efficacy of the plant *Aristolochia indica* Linn. on mouse. *Contraception*, 20: 49–54.

Pakrashi, A., & Pakrasi, P. L. 1977. Antispermatogenic effect of the extract of *Aristolochia indica* Linn on male mice. *Indian Journal of Experimental Biology*, 15(4): 256–259.

Pakrashi, A., & Shaha, C. 1977. Effect of a sesquiterpene from *Aristolochia indica* Linn. on fertility in female mice. *Experientia,* 33: 1498–1499.

Pakrashi, A., & Shaha, C. 1979. Short term toxicity study with methyl ester of aristolic acid from *Aristolochia indica* Linn. in mice. *Indian Journal of Experimental Biology*, 17(4): 437–439.

Pakrashi, A., Chakrabarty, B., & Dasgupta, A. 1976. Effect of the extracts from *Aristolochia indica* Linn. on interception in female mice. *Experientia,* 32: 394–395.

Pakrashi, S. C., Dastidar, P. G., Basu, P., & Achari, B. 1977. New phenanthrene derivatives from *Aristolochia indica. Phytochemistry*, 16: 1103–1104.

Pakrashi, S. C., Dastidar, P. G., Chakrabarty, S., & Achari, B. 1980. (12S)-7,12-Secoishwaran-12-01, a new type of sesquiterpene from *Aristolochia indica* Linn. *Journal of Organic Chemistry*, 45: 4765–4767.

Pakrashi, A., & Shaha, C. 1978. Effect of methyl ester of aristolic acid from Aristolochia indica Linn. on fertility of female mice. *Experientia*, 34: 1192–1193.

Pezzuto, J. M., Swanson, S. M., Mar, W., Che, C. T., Cordell, G. A., & Fong, H. H. 1988. Evaluation of the mutagenic and cytostatic potential of aristolochic acid (3,4- methylenedioxy-8-methoxy-10-nitrophenathrene-1-carboxylic acid) and several of its derivatives. *Mutation Research*, 206: 447–454.

Piers, E., & Hall, T. W. 1980. Total synthesis of tetra cyclic sesquiterpenoids: (±)-Ishwarone. *Canadian Journal of Chemistry*, 58: 2613–2623.

Pradeepa, V., Sathish-Narayanan, S., Kirubakaran, S. A., Thanigaivel, A., & Senthil-Nathan, S. 2015. Toxicity of aristolochic acids isolated from *Aristolochia indica* Linn (Aristolochiaceae) against the malarial vector Anopheles stephensi Liston (Diptera: Culicidae). *Experimental Parasitology*, 153: 8–16.

Purkayastha, J., Dutta, M., & Nath, S. C. 2007. Ethnomedicinal plants from Dibru-Saikhowa biosphere reserve, Assam. *Indian Journal of Traditional Knowledge,* 6(3): 477–480.

Rana, A. Y. K. M. M., & Khanam, J. A. 2002. *Aristolochia indica* whole plant extract as an antineoplastic agent. *Journal of Medical Sciences*, 2: 202–205.

Rao, A. S., & Mulhara, M. S. 1955. Preliminary studies on the essential oil from *Aristolochia indica. Journal of the Indian Chemical Society*, 37: 266–271.

Rao, U. S. K., Manjurath, B. L., & Menon, K. N. 1935. Essential oil from the roots of *Aristolochia indica. Journal of the Indian Chemical Society*, 12: 494–498.

Rastogi, R. P., & Mehrotra, B. N. P. 2001. *Compendium of Indian medicinal plants.* Vol II. CDRI and National Institute of Sciences Communication, New Delhi India, p. 660.

Ravikumar, S., Nazar, S., Nuralshiefa, A., & Abideen, S. 2005. Antibacterial activity of traditional therapeutic coastal medicinal plants against some pathogens. *Journal of Environmental Biology*, 26(2): 383–386.

Reddy, C. S., Reddy, K. N., Murthy, E. N., & Raju, V. S. 2009. Traditional medicinal plants in Seshachalam hills, Andhra Pradesh, India. *Journal of Medicinal Plants Research*, 3: 408–412.

Ren, J., Rudemiller, N. P., Wen, Y., Lu, X., Privratsky, J. R., & Crowley, S. D. 2020. The transcription factor Twist1 in the distal nephron but not in macrophages propagates aristolochic acid nephropathy. *Kidney International*, 97(1): 119–129.

Saha, J. C., Savini, E. C., & Kasinathan, S. 1961. Ecbolic properties of Indian medicinal plants, I. *Indian J Medical Sciences*, 49: 130.

Samy, R. P., Thwin, M. M., Gopalakrishnakone, P., & Ignacimuthu, S. 2008. Ethnobotanical survey of folk plants for the treatment of snakebites in Southern part of Tamilnadu, India. *Journal of Ethnopharmacology*, 115(2): 302–312.

Sati, H., Sati, B., Saklani, S., Bhatt, P. C., & Mishra, A. P. 2011. Phytochemical and pharmacological potential of Aristolochia indica: A review. *Research Journal Pharmaceutical, Biological and Chemical Sciences,* 2(4): 647–654.

Sborchia, M., De Prez, E. G., Antoine, M. H., Bienfait, L., Indra, R., Valbuena, G., & Arlt, V. M. 2019. The impact of p53 on aristolochic acid I-induced nephrotoxicity and DNA damage *in vivo* and *in vitro. Archives of Toxicology*, 93(11): 3345–3366.

Schaneberg, B. T., & Khan, I. A. 2004. Analysis of products suspected of containing *Aristolochia or Asarum* species. *Journal of Ethnopharmacology*, 94 (2–3): 245–249. https://doi.org/10.1016/j.jep.2004.06.010

Shafi, P. M., Rosamma, M. K., Jamil, K., & Reddy P. S. 2002. Antibacterial activity of the essential oil from *Aristolochia indica. Fitoterapia*, 73: 439–441.

Shaohua, Z., Ananda, S., Ruxia, Y., Liang, R., Xiaorui, C., & Liang, L. 2010. Fatal renal failure due to the Chinese herb "GuanMuTong" (*Aristolochia manshuriensis*): Autopsy findings and review of literature. *Forensic Science International,* 199(1–3): e5–e7.

Singh, S., & Sedha, S. 2018. Medicinal plants and their pharmacological aspects. *FPI*, 1: 156–170.

Sini, S., & Malathy, N. S. 2005. Antimicrobial properties of roots of medicinal plants. *Ancient Science of Life*, 25(2): 62–65.

Sivaperumal, R., Ramya, S., Ravi, A. V., Rajasekaran, C., & Jayakumararaj, R. 2009. Herbal remedies practiced by Malayali's to treat skin diseases. *Environment and We—An International Journal of Science and Technology,* 4: 35–44.

Slade, N., Moll, U. M., Brdar, B., Zoric, A., & Jelakovic, B. 2009. P53 mutations as fingerprints for aristolochic acid: An environmental carcinogen in endemic (Balkan) nephropathy. *Mutation Research*, 663(1–2): 1–6. https://doi.org/10.1016/j.mrfmmm.2009.01.005

Stefanovic, V., Jelakovic, B., Cukuranovic, R., Bukvic, D., Nikolic, J., Lukic, L., Gluhovschi, G., Toncheva, D., Polenakovic, M., & Cosyns, J. P. 2007. Diagnostic criteria for Balkan endemic nephropathy: Proposal by an international panel. *Renal Failure*, 29 (7): 867–880.

Stiborova, M., Arlt, V. M., & Schmeiser, H. H. 2016. Balkan endemic nephropathy: An update on its aetiology. *Archives of Toxicology,* 90 (11): 2595–2615.

Subramaniyan, V., Saravanan, R., Baskaran, D., & Ramalalingam, S. 2015. In vitro free radical scavenging and anticancer potential of *Aristolochia indica* against MCF-7 cell line. *International Journal of Pharmacy and Pharmaceutical Sciences*, 7(6): 392–396.

Sulochana, A. K., Raveendran, D., Krishnamma, A. P., & Oommen, V. 2014. Ethnomedicinal plants used for snake envenomation by folk traditional practitioners from Kallar forest region of South Western Ghats, Kerala, India. *Journal of Intercultural Ethnopharmacology,* 4(1): 47–51. https://doi.org/10.5455/jice.20141010122750

Swarnalatha, A. M., Buraka, K. S., & Kalavatamma, P. C. 2017. Indigenous healthcare practices of rural women for digestive disorders in Andhra Pradesh. *Journal of Pharmacognosy and Phytochemistry*, SP1: 882–886.

Taid, T. C., Rajkhowa, R. C., & Kalita, J. C. 2014. A study on the medicinal plants used by the local traditional healers of Dhemaji district, Assam, India for curing reproductive health related disorders. *Advances in Applied Science Research*, 5(1): 296–301.

Teron, R., & Borthakur, S. K. 2013. Folklore claims of some medicinal plants as antidote against poisons among the Karbis of Assam, India. *Pleione*, 7(2): 346–356.

Thirugnanasampandan, R., Mahendran, G., & Bai, V. N. 2008. Antioxidant properties of some medicinal *Aristolochiaceae species*. *African Journal of Biotechnology*, 7: 357–361.

Vaclavik, L., Krynitsky, A. J., & Rader, J. I. 2014. Quantification of aristolochic acids I and II in herbal dietary supplements by ultra-high-performance liquid chromatography-multistage fragmentation mass spectrometry. *Food Additives & Contaminants: Part A: Chemistry, Analysis, Control, Exposure & Risk Assessment*, 31(5): 784–791.

Vaghasiya, Y., & Chanda, S. V. 2007. Screening of methanol and acetone extracts of fourteen Indian medicinal plants for antimicrobial activity. *Turkish Journal of Biology*, 31: 243–248.

Vanherweghem, J. M. 1997. *Aristolochia* sp. and chronic intersetial nephropathies in Indians. *Lancet*, 349: 1399–1399.

Vedavathy, S., & Rao, K. N. 1991. Antipyretic activity of six indigenous medicinal plants of Tirumala hills. *Journal of Ethanopharmacology*, 33: 193–196.

Venkatadri, B., Arunagirinathan, N., Rameshkumar, M. R., Ramesh, L., Dhanasezhian, A., & Agastian, P. 2015. *In vitro* antibacterial activity of aqueous and ethanol extracts of *Aristolochia indica* and *Toddalia asiatica* against multidrug-resistant bacteria. *Indian Journal of Pharmaceutical Sciences*, 77(6): 788–791.

Vervaet, B. A., D'Haese, P. C., & Verhulst, A. 2017. Environmental toxin-induced acute kidney injury. *Clinical Kidney Journal*, 10(6): 747–758.

Wu, F. L., Chen, Y. M., Lai, T. S., Shen, L. J., Ho, Y. F., Lee, Y. T., Wu, M. S., Lin, S. L., & Wu, K. D. 2012. Does Chinese herb nephropathy account for the high incidence of end-stage renal disease in Taiwan? *Nephron Clinical Practice,* 120 (4): c215–c222.

Wu, K., Jiang, L., Cao, J., Lei, J. C., & Zou, G. L. 2007. Genotoxic effect and nitrative DNA damage in HepG2 cells exposed aristolochic acid. *Mutation Research*, 630: 97–102.

Wu, T. S., Damu, A. G., Su, C. R., & Kuo, P. C. 2004. Terpenoids of *Aristolochia* and their biological activities. *Natural Product Reports*, 21(5): 594–624.

Yang, H. Y., Lin, J. L., Chen, K. H., Yu, C. C., Hsu, P. Y., & Lin, C. L. 2006. Aristolochic acid-related nephropathy associated with the popular Chinese herb Xixin. *Journal of Nephrology*, 19(1): 111–114.

Yang, L., Su, T., Li, X. M., Wang, X., Cai, S. Q., Meng, L. Q., Zou, W. Z., & Wang, H. Y. 2012. Aristolochic acid nephropathy: Variation in presentation and prognosis. *Nephrology Dialysis Transplantation,* 27(1): 292–298.

Youl, E. N., Husson, C., El Khattabi, C., El Mere, S., Declèves, A. E., Pochet, S., & Antoine, M. H. 2020. Characterization of cytotoxic effects of aristolochic acids on the vascular endothelium. *Toxicology in Vitro*, 65: 104811.

Zhang, H., Cifone, M. A., Murli, H., Erexson, G. L., Mecchi, M. S., & Lawlor, T. E. 2004. Application of simplified *in vitro* screening tests to detect genotoxicity of aristolochic acid. *Food Chemical Toxicology*, 42: 2021–2028.

11 *Atractylis gummifera* (Stemless Atractylis)

Noureddine Chaachouay and Lahcen Zidane

CONTENTS

11.1 INTRODUCTION

Remedy plants, also known as phytotherapy, can help prevent and even treat some types of diseases in the body without drugs (Pal and Shukla 2003). This therapy, therefore, should be used with care since misusing a medicinal plant may not result in illness improvement. Many renal and hepatic damage and death incidents have been linked to herbal treatments (Benbella and Abdellaoui 2017; Stickel et al. 2005). The safety of some commercially available plants has lately been called into question after reports of unfavorable responses to prescription medications and excessive dosages far above the fatal dose. As a result, medicinal plants' effective and safe use has been highlighted as a crucial investigation focus (Burns et al. 2010).

Plant poisoning is uncommon in various parts of the globe, yet it causes significant clinical mortality and morbidity. *Atractylis gummifera* is one of the medicinal herbs periodically implicated in human and animal poisonings. In the traditional system of medicine, stemless atractylis has been used to manage sleepiness, bronchitis, snakebite toxicity, ulcers, and intestinal parasites. Additionally, it was well known for its emetic, purgative, diuretic, and antipyretic effects (Bruneton 1996; Larrey and Pageaux 1995).

Stemless atractylis intoxications have been documented in Mediterranean regions since the mid-nineteenth century (Lefranc 1866). This plant is poisonous due to the presence of two diterpenoid glucosides (carboxyatractyloside and atractyloside), which inhibit mitochondrial oxidative phosphorylation. According to the Poison Control Center of Morocco (PCCM), atractylis is the plant that causes the most poisoning. There were 487 cases of intoxication between 1980 and 2013 (PCCM 2009). This chapter presents an overview of the results found about the botany, traditional uses, phytochemical, and pharmacological activities of *A. gummifera* and discusses possible pharmacological therapy for intoxication.

11.2 BOTANICAL DESCRIPTION

Carl Linnaeus identified *A. gummifera* Salzm. ex L. and classified it among the Asteraceae family, the genus *Atractylis*, and the *gummifera* species. *A. gummifera* var. *gummifera, A. gummifera* var.

DOI: 10.1201/b23017-13

macrocephala (Desf.) Pott.-Alap., *Carlina gummifera* (L.) Less., and *Acarna gummifera* Willd. are also synonyms of this species. *A. gummifera* Salzm. ex L. was known as Addad or Chouk El alk (Arabic), Masticogna (Italian), Stemless Atractylis (English), and Chardon à glu (French). Other vernacular names are "lxia," "blue thistle," "birdlime," and "chameleon leukos" due to its color shift.

A. gummifera is an acaulescent, perennial, and thorny plant. The herb has a long rhizome that may grow 30–40 cm long and 7–8 cm wide. The parenchymal cells of the rhizome are densely filled with crystalline masses. In contrast, the cortical parenchyma is densely filled with secretory lacunae. Its stem is reduced or unapparent from which tears flow, which constitutes a kind of glue, from where the name pine thistle came. The leaves are sharply split into thorny lobes and clustered into rosettes. The flowers are pink, purple, or pinkish and huddled in a capitulum surrounded by spike-covered bracts (Figures 11.1 and 11.2). A yellowish-white latex is exuded from the bracts' bases (Bellakhdar 2006; Bruneton 1999; Vallejo et al. 2009).

11.3 DISTRIBUTION

A. gummifera is a thistle that may be found all over the globe, although it is particularly prevalent in the Mediterranean area (Morocco, Algeria, Tunisia, and Libya) and Southern Europe (Malta, Italy, Greece, France, Portugal, and Spain) (Daniele et al. 2005, Chaachouay et al. 2020). The pine thistle grows naturally in wet soils of cultivated lands and humid regions, and the sap of the rhizome is toxic (Vallejo et al. 2009).

11.4 PHYTOCHEMICAL CONSTITUENTS

The underground parts of *A. gummifera* include two poisonous diterpenoid glucosides: atractyloside and carboxyatractyloside. Lefranc was the first to isolate atractyloside (ATR) from the roots of stemless atractylis (Lefranc 1866). Carboxyatractyloside (CATR) was first isolated in 1964 as gummiferin

FIGURE 11.1 Inflorescence of pine thistle.

FIGURE 11.2 Achenes and flowers of pine thistle.

and was later identified as 4-Carboxyatractyloside (Danieli et al. 1972). Carboxyatractyloside is distinguished from atractyloside by a second carboxylic group in the diterpene ring at position C-4′. Carboxyatractyloside is found in fresh plants but not in dry plants due to decarboxylation to atractyloside during aging or dehydration. Additionally, carboxyatractyloside is more poisonous than atractyloside (Daniele et al. 2005; Luciani et al. 1978). Other chemical substances were discovered in pine thistles, such as inulin, sugars, cellulose, mineral matter, water, amino acids, organic acids, hydrochloric acid, sulfuric acid, carbonic acid, flavones, flavonols, anthocyanins, tannins, and essential oils, as shown in Table 11.1.

11.5 PHARMACOLOGICAL STUDIES

A. gummifera is a well-known herbal remedy that has been utilized for centuries. Many pharmacological investigations into its bioactivities have shown that it might be the basis for novel medications and therapy implements. Numerous pharmacological effects, including antioxidants, antipyretic, antifungal, anti-diabetic, anti-leishmanial, burn healing, diuretic, emetic, insecticidal, and purgative, have been documented for the extracts of pine thistle (Table 11.2).

11.6 TOXIC RESPONSE

A. gummifera has traditionally been used for medicinal purposes. In the first century, the Greek pharmacologist Dioscorides described the plant and its benefits. Theophrastus (300 B.C.) also recognized the plant's toxicological properties on animals; he distinguished two varieties of "chameleon," "white" and "dark," based on their root color and characteristics. Theophrastus reported that the white root was also employed in ancient times to remedy worms and as a poison for pigs and dogs (Daniele et al. 2005; Matthioli 1957).

TABLE 11.1
Constituents of *Atractylis gummifera* by Chemical Class

Metabolites	Chemical Class	References
Organic matter	Inulin, sugars, cellulose, mineral matter, water, amino acids (aspartic and glutamic acid, proline, leucine, valine, and tryptophan), organic acids (acetic, isovaleric, oxalic, and malic acid), hydrochloric acid, sulfuric acid, carbonic acid, silica	Lefranc (1866)
Diterpene glycosides	Atractyloside, carboxyatractyloside	Danieli et al. (1972), Lefranc (1866)
Flavonoids	Flavonols, flavones (orientin, homoorientin, corymboside, neocorymboside, isoschaftoside), anthocyanins	Benbouziane and Beneddra (2016), Khadhri et al. (2013), Khadija (2016)
Tannins	Condensed tannins	Benbouziane and Beneddra (2016), Khadhri et al. (2013), Khadija (2016)
Essential oils	Acetylene compounds (carlina oxide and 13-Methoxy carlina oxide), non-terpenic aliphatic compounds, non-terpene compounds.	Mejdoub et al. (2020)

TABLE 11.2
Pharmacological Features of *Atractylis gummifera*

Pharmacological Properties	Extract	References
Antioxidant	Essential oils	Mejdoub et al. (2020)
	Aqueous and organic extracts	Bouabid et al. (2020)
	Traditional and methanolic extracts	Khadhri et al. (2013)
Antipyretic	Traditional extracts	Daniele et al. (2005), Georgiou et al. (1988), Larrey and Pageaux (1995)
Antifungal	Essential oils	Mejdoub et al. (2020)
Anti-diabetic	Aqueous and methanolic macerated extracts	Bouabid et al. (2019)
Anti-leishmanial	Chloroform extracts	Deiva et al. (2020)
Burn healing	Aqueous extract	Sifour et al. (2012)
Diuretic	Traditional extracts	Daniele et al. (2005), Georgiou et al. (1988), Larrey and Pageaux (1995)
Emetic	Traditional extracts	Daniele et al. (2005), Georgiou et al. (1988), Larrey and Pageaux (1995)
Insecticidal	Essential oils	Mejdoub et al. (2020)
Purgative	Traditional extracts	Daniele et al. (2005), Georgiou et al. (1988), Larrey and Pageaux (1995)

Though the poison of *A. gummifera* is widely recognized, it is nevertheless a prevalent source of toxicity. The poisoning of pine thistle is caused by the presence of two diterpenoid glucosides, carboxyatractyloside and atractyloside, which may block mitochondrial oxidative phosphorylation by permeabilizing them. The glucosides of *A. gummifera* induce severe hepatitis, with catastrophic liver problems occurring often. Even at concentrations that are unlikely to cause cytolysis, the plant's deadly glucosides interact with detoxication or transformation systems in the liver, limiting ADP–ATP conversion via blockage of P450 cytochrome (Daniele et al. 2005; Haouzi et al. 2002).

Anxiety, convulsions, diarrhea, headache, epigastric, nausea, stomach pain, and vomiting are common poisoning symptoms, often followed by a coma. Clinical symptoms are associated with hypoglycemia caused by neurovegetative diseases and renal failure. Georgiou and colleagues described the poisoning caused by *A. gummifera* in a 7-year-old kid who drank a medicinal herb extract prepared from the plant's root. The documented physical symptoms included overall uneasiness, epigastric discomfort, vomiting, and unconsciousness (Daniele et al. 2005; Georgiou et al. 1988).

No particular pharmaceutical therapy is available for *A. gummifera* poisoning. All therapeutic interventions, including electrolyte and fluid replacement, seizure management, cardiopulmonary and respiratory support, and standard procedures for severe hepatic and renal failure, are symptomatic (Stewart and Steenkamp 2000). Certain professionals indicate that typical treatment procedures include inducing intestinal evacuation, gastric lavage, the injection of activated charcoal, and vomiting (Ben Salah et al. 2001, Chaachouay et al. 2021, Orch et al. 2021). Various substances were tested to create better pharmacological therapies for pine thistle intoxication. Furthermore, it has been revealed in vitro that certain chemical, such as dithiothreitol, hydrocortisone, dinitrophenol, or verapamil, may prevent the dangerous consequences of atractyloside. These substances demonstrated protective benefits when administered 30 or 60 minutes before ATR but were ineffective after ATR exposure (Catanzano et al. 1969; Georgiou et al. 1988).

11.7 TRADITIONAL AND OTHER POTENTIAL USES

A. gummifera was very well known throughout the Renaissance, particularly as the "chameleon." Matthioli describes the chameleon bianco as a fragrant, sweet shrub with a powerful odor in his book. He proposed utilizing its root in various ways, including as a vermifuge when cooked in vinegar, wild marjoram, and wine; as a drink when combined with wine to combat snake venom (Matthioli 1957). Another Renaissance work precisely characterized the "chameleon nero" and the "chameleon bianco." He prescribed "chameleon nero" root decoction for toothache relief, breath refreshment, skin discoloration, such as freckles, and "chameleon bianco" decoction root for urinary retention and tiredness. *A. gummifera* has been used in traditional therapy to cure various diseases, including hydropsy, sleepiness, ulcers, intestinal parasites, and snakebites poisoning. It was employed in traditional Arabic treatment to cauterize abscesses. The dried rhizome is also often employed as an incense in Arabic societies to fend off ill fortune (Daniele et al. 2005; Hamouda 2004). Furthermore, the herb was valued for its diuretic, antipyretic, emetic, and purgative effects (Bruneton 1996; Larrey and Pageaux 1995, Chaachouay et al. 2021). In Northern African folk medicine, it is occasionally employed to heal syphilitic sores, induce miscarriage, and bleach teeth (Chaachouay et al. 2021; Chaachouay et al. 2022; Daniele et al. 2005; Georgiou et al. 1988). It is also used in folk veterinary treatment to treat parasites (Chaachouay et al. 2022; Viegi et al. 2003).

11.8 CONCLUSION AND FUTURE REMARKS

A. gummifera poisoning is a recurring concern in regions where the plant grows naturally. The poisoning is difficult to avoid since it is often inadvertent and includes youngsters drawn to the delicious juice contained in the roots of *A. gummifera.* Due to the absence of specific antidotes, investigations have been conducted to evaluate different substances capable of lowering atractyloside toxicity. A few, notably verapamil and dithiothreitol, proved helpful, but only when administered before atractyloside exposure. To design more effective pharmacological therapies, we need to learn more about the pharmacokinetics, long-term toxicity, reproductive toxicity, and mutagenicity of atractyloside. Immunotherapy studies may provide a novel method. Antitoxin antibodies have been created and are now utilized in clinical practice to treat intoxication caused by colchicine and cardiac glycosides containing *A. gummifera* and oleander toxicity. Another practical strategy would be to disseminate information on pine thistle toxicity and educate youngsters to identify harmful plants they should avoid.

NOTES

Noureddine Chaachouay, Agri-Food and Health Laboratory (AFHL), Faculty of Sciences and Techniques of Settat, Hassan First University, Settat, Morocco
Lahcen Zidane, Plant, Animal Productions and Agro-Industry Laboratory, Department of Biology, Faculty of Sciences, Ibn Tofail University, Kenitra, Morocco

REFERENCES

Bellakhdar, J. 2006. *Plantes médicinales au Maghreb et soins de base. Précis de phytothérapie moderne*. Editions le Fennec, Casablanca, Morocco, p. 386.

Ben Salah, N., Zaghdoudi, I., Zhioua, M., Hamouda, C., Amamou, M., & Thabet, H. 2001. Quelques spécialités de chez nous : Intoxications par les plantes, le chloralose et le méthanol. Article available online at: http://www.darhached.org/articles/JAMU/jamu2001/Bensalah11.pdf

Benbella, M., & Abdellaoui, A. 2017. Attenuation of Toxic Power of Atractylis Gummifera, 8th International Conference on Chemical, Agricultural, Biological and Health Sciences—Paris (France). Article available online at: https://doi.org/10.17758/EIRAI.F1017216.

Benbouziane, F., & Beneddra, M. 2016. *Contribution à une étude botanique et chimique du chardon à glu, Atractylis gummifera L., famille des Asteraceae. Mémoire de fin d'étude pour l'obtention du diplôme de docteur en pharmacie*. Université Abou Bekr Belkaîd Faculté de medicine, Tlemcen, Algérie.

Bouabid, K., Lamchouri, F., Toufik, H., Boulfia, M., Senhaji, S., & Faouzi, M. E. A. 2019. In vivo anti-diabetic effect of aqueous and methanolic macerated extracts of Atractylis gummifera. *Bangladesh Journal of Pharmacology*, 14(2), 67–73.

Bouabid, K., Lamchouri, F., Toufik, H., & Faouzi, M. E. A. 2020. Phytochemical investigation, in vitro and in vivo antioxidant properties of aqueous and organic extracts of toxic plant: Atractylis gummifera L. *Journal of Ethnopharmacology*, 253, 112640.

Bruneton, J. 1996. *Plantes toxiques : Vegetaux dangereux pour l'homme et les animaux/Jean Bruneton*. Tec Doc, Paris, France.

Bruneton, J. 1999. *Toxic plants dangerous to humans and animals*. Intercept Limited, Andover, England.

Burns, J. J., Zhao, L., Taylor, E. W., & Spelman, K. 2010. The influence of traditional herbal formulas on cytokine activity. *Toxicology*, 278(1), 140–159.

Catanzano, G., Delons, S., & Benyahia, T. D. 1969. 2 cases of poisoning due to "gum thistle" (Atractylis gummifera L.). Clinical development and anatomo-pathologic lesions. *Maroc Medical*, 49(529), 651–655.

Chaachouay, N., Azeroual, A., Bencharki, B., Douira, A., & Zidane, L. 2022. Ethnoveterinary medicines plants for animal therapy in the Rif, North of Morocco. *South African Journal of Botany,* 147, 176–191.

Chaachouay, N., Douira, A., & Zidane, L. 2021. Herbal medicine used in the treatment of human diseases in the Rif, Northern Morocco. *Arabian Journal for Science and Engineering*, 47, 1–23.

Chaachouay, N., Benkhnigue, O., Douira, A., & Zidane, L. 2021. Poisonous medicinal plants used in the popular pharmacopoeia of the Rif, northern Morocco. Toxicon, 189, 24-32.

Chaachouay, N., Benkhnigue, O., & Zidane, L. 2022. Ethnomedicinal study of medicinal and aromatic plants used against dermatological diseases by the People of Rif, Morocco. *Journal of Herbal Medicine,* 5, 100542.

Chaachouay, N., Douira, A., Hassikou, R., Brhadda, N., Dahmani, J., Belahbib, N., Ziri, R., & Zidane, L. 2020. Etude floristique et ethnomédicinale des plantes aromatiques et médicinales dans le Rif (Nord du Maroc)" [Phdthesis, Département de Biologie - Université Ibn Tofail - Kénitra]. https://tel.archives-ouvertes.fr/tel-03376377

Daniele, C., Dahamna, S., Firuzi, O., Sekfali, N., Saso, L., & Mazzanti, G. 2005. Atractylis gummifera L. poisoning: An ethnopharmacological review. *Journal of Ethnopharmacology*, 97(2), 175–181.

Danieli, B., Bombardelli, E., Bonati, A., & Gabetta, B. 1972. Structure of the diterpenoid car yatractyloside. *Phytochemistry*, 11(12), 3501–3504.

Deiva, S., Ferguson, L., Rateb, M. E., Williams, R., & Brucoli, F. 2020. 2-furyl (phenyl) methanol isolated from Atractilis gummifera rhizome exhibits anti-leishmanial activity. *Fitoterapia*, 140, 104420.

Georgiou, M., Sianidou, L., Hatzis, T., Papadatos, J., & Koutselinis, A. 1988. Hepatotoxicity due to Atractylis gummifera-L. *Journal of Toxicology: Clinical Toxicology*, 26(7), 487–493.

Hamouda, C. 2004. A review of acute poisoning from Atractylis gummifera L. *The Journal of Emergency Actors*, 3, 23–26.

Haouzi, D., Cohen, I., Vieira, H. L. A., Poncet, D., Boya, P., Castedo, M., Vadrot, N., Belzacq, A.-S., Fau, D., & Brenner, C. 2002. Mitochondrial permeability transition as a novel principle of hepatorenal toxicity in vivo. *Apoptosis*, 7(5), 395–405.

Khadhri, A., El Mokni, R., & Smiti, S. 2013. Composés phénoliques et activités antioxydantes de deux extraits de chardon à glu : Atractylis gummifera. *Revue Soc Sci Nat de Tunisie*, 39, 44–52.

Khadija, B. 2016. Etude phytochimique et pharmacologique d'Atractylis gummifera.

Larrey, D., & Pageaux, G. P. 1995. Hepatotoxicity of herbal remedies and mushrooms. *Seminars in Liver Disease*, 15(03), 183–188.

Lefranc, M. E. 1866. Étude Botanique, Chimique Et Toxicologique Sur L'atractylis Gummifera. *Bulletin de la Société Botanique de France*, 13(3), 146–157.

Luciani, S., Carpenedo, F., & Tarjan, E. M. 1978. *Effects of atractyloside and carboxyatractyloside in the whole animal. Atractyloside: Chemistry, biochemistry and toxicology*. Piccin Medical Books, Padova, Italy, pp. 109–124.

Matthioli, P. A. 1957. Nei sei libri della materia medicinale di Pedacio Dioscoride Anazarbeo. Venezia, p. 333.

Mejdoub, K., Mami, I. R., Belabbes, R., Dib, M. E. A., DJabou, N., Tabti, B., Benyelles, N. G., Costa, J., & Muselli, A. 2020. Chemical variability of Atractylis gummifera essential oils at three developmental stages and investigation of their antioxidant, antifungal and insecticidal activities. *Current Bioactive Compounds,* 16(4), 489–497.

Orch, H., Chaachouay, N., Douiri, E. M., Faiz, N., Zidane, L., & Douira, A. 2021. Use of medicinal plants in dermato-cosmetology : An ethnobotanical study among the population of Izarène. *Jordan Journal of Pharmaceutical Sciences*, Volume 14, No. 3, 323–340.

Pal, S. K., & Shukla, Y. (2003). Herbal medicine: Current status and the future. *Asian Pacific Journal of Cancer Prevention*, 4(4), 281–288.

PCCM. 2009. Intoxication aux plantes. Centre Antipoison du Maroc, Rapport annuel. Toxicologie Maroc, p. 16.

Sifour, M., Ouled-Hadddar, H., Ouitas, L., & Kerfa, A. (2012). Burn healing activity of aqueous extract of Atractylis gummifera (L). Séminaire international: Cancer, stress cellulaire et substances bioactives. *Séminaire internationale*, 23-24 Sep, 71_74

Stewart, M. J., & Steenkamp, V. 2000. The biochemistry and toxicity of atractyloside : A review. *Therapeutic Drug Monitoring*, 22(6), 641–649.

Stickel, F., Patsenker, E., & Schuppan, D. 2005. Herbal hepatotoxicity. *Journal of Hepatology*, 43(5), 901–910.

Vallejo, J. R., Peral, D., Gemio, P., Carrasco, M. C., Heinrich, M., & Pardo-de-Santayana, M. 2009. Atractylis gummifera and Centaurea ornata in the province of Badajoz (Extremadura, Spain)—Ethnopharmacological importance and toxicological risk. *Journal of Ethnopharmacology*, 126(2), 366–370.

Viegi, L., Pieroni, A., Guarrera, P. M., & Vangelisti, R. 2003. A review of plants used in folk veterinary medicine in Italy as basis for a databank. *Journal of Ethnopharmacology*, 89(2–3), 221–244.

12 *Atropa belladonna* (Deadly Nightshade)

Arundhati Singh, Vedanshi Pal, Shreyshi Aggarwal, and Manu Pant

CONTENTS

12.1 INTRODUCTION

Atropa belladonna (L.) is a bushy herb that belongs to the family Solanaceae. The name has an interesting meaning wherein 'Atropa' refers to 'Atropos' – the Greek goddess meaning 'to cut the thread of life' and belladonna means 'the beautiful lady' in the Italian language. This can also be linked to the fact that in ancient days the plant extract was used by women to dilate the pupils to look attractive (Jordan 1976). The plant bears greenish-purple flowers and the berries are purplish black with a sweet taste (Joshi et al. 2003; Cikla et al. 2011). The berries are frequently confused with the blackcurrant and blueberries (Laffargue et al. 2011). *A. belladonna* is widely known for its highly toxic nature earning them the name 'deadly nightshade berries' whose ingestion leads to anticholinergic toxic syndrome (ATS) that affects central and the peripheral nervous systems. Symptoms like short-term memory loss, respiratory or cardiovascular failure, hallucination, ataxia, disorientation, seizures, or coma are associated with the central nervous system (CNS) being affected by tropane alkaloids and are usually age and dose dependent (Cikla et al. 2011; Tulin and Ismet 2011). The effects on the peripheral nervous system are seen in the form of drying of the mucous membrane, hypo- and hypertension, tachycardia, flushed skin, urine retention, ileus, hyperreflexia, etc. (Tulin and Ismet 2011). In the case of children, symptoms are manifested as meaningless speech, lethargy, flushing, and coma, as per the level of poisoning (Çaksen et al. 2003).

The toxicity of the plant is attributed to the tropane alkaloids which include atropine, scopolamine, and hyoscyamine (Ulbricht et al. 2004; Fatur and Kreft 2020). Although the plant is

DOI: 10.1201/b23017-14

categorized as a poisonous plant, its clinical utility has also been realized. The low dosage of plant alkaloids has proven to be extremely efficient in medical treatments. For instance, atropine is used as a muscle relaxant, heartbeat regulator, pupil dilation for ophthalmic treatment, and for nullifying the effect of chloroform and opium poisoning; scopolamine is used in the treatment of stomach ailments, and when combined with other medicines, the plant's extracts are used for treating medical conditions like stomach ulcers, irritable bowel syndrome, excessive night-time urination, spastic colon, motion sickness, diverticulitis, pink eye, and Parkinson's disease.

This chapter discusses the morphology, distribution, propagation strategies, phytoconstituents, pharmacological behavior, toxicity, and potential medicinal usage of the plant *A. belladonna.*

12.2 BOTANICAL DESCRIPTION

12.2.1 Habit and Habitat

A. belladonna is a perennial herbaceous plant with an erect, branched stem bearing purplish-black berries, kidney-shaped seeds, and a thick, well-developed root system that is over 15 cm long. Plants can grow up to 2 m in height (Lee 2007). Leaves are oval shaped and have smooth edges with pointed ends. Leaves grow in stalks to about 18 cm in length, in an alternate pattern, and are poisonous. Leaves are harvested in late spring and dried for later use. The flowers are 2.5–3 cm long, purplish or purplish yellow in color (Figure 12.1a). The corolla is campanulate shaped with five petals hanging downward. The flowers start blooming in June, mature in early September, and produce fruits in the second year of growth. The berries are shiny in appearance, have five sepals, and are seen in the part where the fruit attaches to the plant (Figure 12.1b). The plants are often seen as bushes on wooded hills or limestones or in the shade of trees. Calcareous soil with sufficient shade, cold zones, saline soil, and good drainage is reportedly the best combination for *A. belladonna* growth (Dutta and Rita 2011). Plants growing in light calcareous soil rich in the nitrogenous matter are found to be most poisonous due to high alkaloid content (Rowson 1950). The preferable temperature range is 20°C–25°C, with a moist and cool environment. Escalation in temperature conditions retards plant growth and development, and extreme temperatures lead to plant death. Although *A. belladonna* can grow over a range of pH conditions, acidic soil with pH 6 is most favorable.

All the parts of the plant contain tropane alkaloids (Chevallier 1996; Dutta and Rita 2011), the content varying with the developmental stage of the plant. The alkaloid content is reportedly very high in green and immature berries, while roots and leaves contain about 0.6% and 0.4% of the active alkaloid, respectively (Chopra et al. 1986).

12.2.2 Propagation Strategies

Conventional propagation of *A. belladonna* is majorly done by seeds, and in some cases, by cuttings. Besides, plant tissue culture technology has also been used for mass propagation of the species.

12.2.2.1 Seed-Based Propagation

Seeds are extracted from the berries. Berries are crushed and dried in sunlight usually from September to November for seed extraction. The seeds are sown in raised flat beds made on well-plowed land surrounded by irrigation channels. Since the seeds have a hard testa, pretreatment of the seeds has been reported to enhance the germination rate. Soaking the seeds in boiling water for around 30 minutes, dilute hydrogen peroxide solution (around 60% v/v) for 18–24 hours (Sievers 1914), or gibberellic acid solution and exposure to gamma rays are commonly used methods for seed pretreatment (Sievers 1914; Nikolaeva 1982; Abdel-Hady et al. 2008; Rani and Prasad 2014). In many cases, fungicide treatment to the seeds is also given to avoid fungal contamination. General requirements for rapid seed germination are chalky, fine-limed, permeable soil, temperature range of 15°C (about 18 hours exposure) to 30°C (about 6 hours exposure), shade conditions, and high humidity

FIGURE 12.1 (a) *Atropa belladonna* plant displayed flower and (b) berries.

(Rowson 1950; Genova et al. 1997). Decomposed farmyard manure or chemical fertilizers are added to ensure optimal nitrate supply to the soil which enhances plant growth and yield. The seedbeds are covered with farmyard manure and with straw followed by immediate watering after sowing. However, excess irrigation must be avoided in all cases. Germination takes an average of 4–6 weeks (Dutta and Rita 2011), and seedlings usually attain a height of 15–20 cm within 8–12 weeks after which they are transplanted to the field conditions at a spacing of 60×60 or 60×45 cm and watered till the plantlets grow. An ideal time of transplantation in the field is March to April or October to November (https://agriinfo.in/package-of-practices-for-cultivation-of-belladonna-1396/).

Harvesting is done as soon as the flowering starts as alkaloid content is higher during this time. Leaves are harvested in 3 months of plantation and dried under shade. Root harvesting is done at intervals of over 3 years, followed by washing, splitting, and shade/sun drying.

12.2.2.2 Cutting-Based Propagation

A. belladonna can also be propagated by cuttings of the green branch (Dutta and Rita 2011). A soft woodcutting can be taken in spring to early summer and dipped in auxin solution (rooting hormone). The cutting is then planted in soil rich in compost. The cuttings are covered with polybags which are periodically removed to provide optimal aeration. Cuttings are kept moist until they root well. Once roots develop and new shoots emerge, cuttings are shifted to individual pots for better growth and plant development. However, as compared to seeds, cuttings are less responsive in terms of the rate of propagation.

12.2.2.3 *In Vitro* Propagation

In vitro plant propagation refers to growing plants under laboratory conditions using plant tissue culture technology. The technique of micropropagation involves growing plant cells, tissue, organs, protoplast, and seeds (known as explants) *in vitro* on an artificial nutrient medium under aseptic and controlled conditions of temperature, light, and humidity. The essential nutrients and plant growth regulators in the medium facilitate organogenic differentiation of culture plant tissue, resulting in mass propagation of the plant from a very small piece of tissue. Table 12.1 highlights different micropropagation strategies utilized for the propagation of *A. belladonna* from different explants.

12.3 DISTRIBUTION

The Atropa genus is well distributed from western Europe to the Himalayan region (Dutta and Rita 2011). *A. belladonna* is native to Europe, the Middle East (in Iran and Turkey), and North Africa (in Algeria, Morocco) (Rowson 1950; Davis et al. 1965; Tutin et al. 1972; Zhang et al. 2005; Majid et al. 2012). The species is usually found at an altitude of 50′ and 55′ N and an altitudinal range of 300–600 ft. Its distribution in Britain was documented by Butcher (1947) who reported the occurrence of the species in areas of chalk and limestone formation like South Downs, North Downs, from Yorkshire to Wiltshire. *A. belladonna* was subsequently introduced in several other countries and is now cultivated in the regions of Europe, Northern and Southern America, and many corridors of South Eastern Asia (Lee 2007). In the latest study, the occurrence of *A. belladonna* has been recorded in Lebanon in various areas of Wadi Jhannam (Zein and Dagher-Kharrat 2021). In India, *A. belladonna* is found at higher altitudes of the Himalayan belt. Jammu and Kashmir, Himachal Pradesh, and Sikkim are the major *A. belladonna* harboring regions. However, the species is also being cultivated in parts of Uttarakhand, West Bengal, Bihar, and Punjab (Anonymous 1948).

12.4 PHYTOCHEMICAL CONSTITUENTS

A. belladonna has been widely studied for its phytochemical constituents which account for the bioactive properties of the plant. These bioactive alkaloids have been regarded as potential acetylcholine antagonists (Guggisberg and Hesse 1983). The major constituents are reported to be alkaloids like atropine, norhyoscyamine, apoatropine, hyoscyamine, 6β-Hydroxyhyoscyamine, and scopolamine which are extracted from the plant's leaf (Baralle and Gross 1969). Hartmann et al. (1986) identified 13 alkaloids from the roots and 7 alkaloids from the stem of *A. belladonna* through gas chromatography and mass spectrometry. Among these, scopolamine was isolated from roots, norhyoscyamine from shoots, and hyoscyamine from all plant parts. Liquid chromatography and mass spectrometric analysis have also reported glycoside flavonoids from the plant extract, wherein kaempferol glycosides are the most prominent constituents (Millet et al. 2010).

Atropine is the primary alkaloid and is present in high content in all parts of the plant including mature fruits (Zarate et al. 2006). It is a racemate made of equal molar concentrations of (S)- and (R)-atropine with the chemical formula $C_{17}H_{23}NO_3$, and it has a molecular weight of 289.4 (Figure 12.2). Under aqueous fermentation, the atropine degrades into known products like tropic

TABLE 12.1
***In Vitro* Propagation of *Atropa belladonna* from Different Explants**

Explant	Response	Induction Medium and Plant Growth Regulators Used	Reference
Root callus	Formation of embryo-like structures	Synthetic Culture Medium (SSM)+NAA	Thomas and Street (1970)
Dicotyledonous embryo from ovules	Callus-derived culture	LS Medium+Kinetin	Zenkteler (1971)
Anther (Microspores)	Haploid embryoids	Medium containing FeEDDHA+Kin+Aux +Casamino acids	Rashid and Street (1973)
Cell suspension derived protoplast	Embryo development	MS Medium+Naphthalene Acetic Acid (NAA)+Kinetic+Sorbitol	Gosch et al. (1975)
Leaves of anther-derived haploid plant	Callus-derived plants	Wood and Braun's basal medium+Lin and Staba vitamins+NAA+Kinetin	Eapen et al. (1978)
Shoot tip and axillary meristem	Shoot organogenesis	MS Medium+BA	Benjamin et al. (1987)
Cell suspension	Leaf, stem, and root cultures	MS Medium+NAA+BA	Taha (2003)
Leaves and axillary buds	Adventitious bud initiation	MS medium+6-BA+NAA	Xirong et al. (2011)
Hairy roots	Adventitious bud induction	MS medium+B5 Medium+Kinetin	Mera and Takayama (2012)
Leaf, lamellae (5 mm), midrib (5–7 mm) and petiole (5 mm) in length	Direct root organogenesis	MS Medium+Naphthalene Acetic Acid (NAA)+IBA+Kinetic	Rani and Prasad (2013)
Leaves	Callus induction	MS medium+B5-Vitamins+2,4-D+Kinetin	Khater et al. (2013)
Nodes	Shoot-derived culture	MS medium+BA	Al-Akaidi and Zeki Ameen Kassab Bashi (2017)
Seeds	Root induction	MS+ IBA+IAA	Stateva and Desheva (2017)
Leaf, stem and root	Leaf callus initiated	MS medium+2,4- D	Elshorbagy et al. (2018)
Leaf	Callus and plant differentiation induction	Calli induction: MS Medium+Kinetin+BA (Culture I) and MS medium+BA+NAA (Culture II) Differentiation: MS Medium+BA +IAA	Al-Ashaal et al. (2018)
Seeds	Shoot-derived culture	MS medium+BA	Ahmed (2020)
Shoot tip and nodes	Shoot organogenesis	MS medium	Ahmed et al. (2021)

acid, apoatropine, noratropine, and anisodamine (Eger et al. 1999; Kirchhoff et al. 2004). Atropine is known to play a wide variety of roles, working as an anesthetic adjuvant, mydriatic, parasympatholytic, and bronchodilator (National Library of Medicine n.d.-a). The drugs derived from atropine are classified as anticholinergic drugs and include antimuscarinic agents. A high dose of atropine leads to parasympathetic organs paralysis, drying of the mucous membrane, fever, etc. Severe toxification leads to respiratory depression and can lead to coma, circulatory collapse, and in extreme cases, even death (Drugbank n.d.).

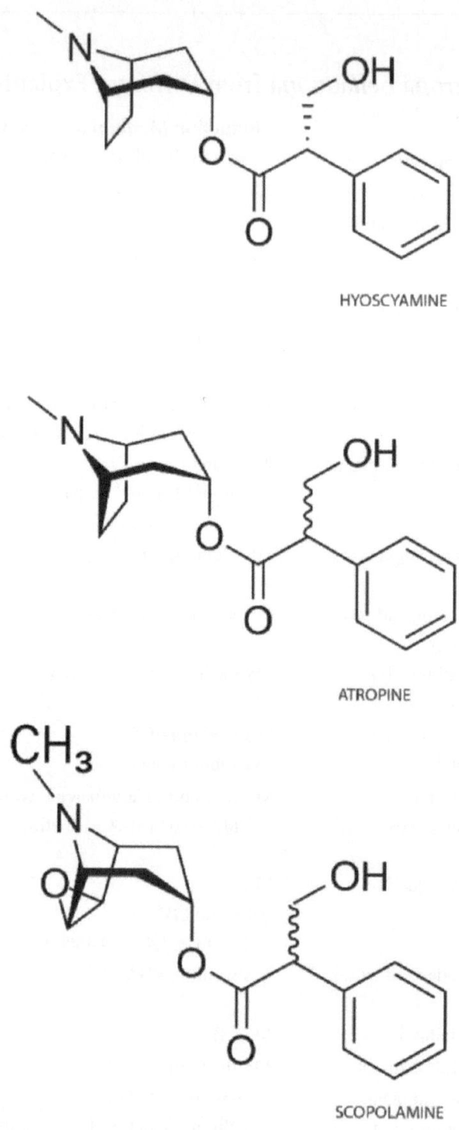

FIGURE 12.2 Chemical structures of prominent phytoconstituents of *Atropa belladonna*, namely hyoscyamine, atropine, and scopolamine.

Hyoscyamine is S-Atropine and has anticholinergic properties. It is generally used to treat nausea, motion sickness, allergies related to rhinitis, and hyperactive bladder by inhibiting the action on the smooth muscles and is used intravenously. It has also been employed in imaging the gastrointestinal tract for developing better insight into the treatment (National Library of Medicine n.d.-c).

Scopolamine is of particular interest as compared to hyoscine due to its high pharmacological application and fewer side effects (Hashimoto and Yamada 1994). Scopolamine alkaloid is known for its anticholinergic, antiemetic, and anti-vertigo properties. The molecular formula of Scopolamine is $C_{17}H_{21}NO_4$, with a molecular weight of 303.35. It is utilized to control discharge of saliva, gastric convulsions, and persistent vomiting (National Library of Medicine n.d.-b). Large-scale production of scopolamine by hairy root culture has also been investigated (Palazon et al. 1995). Table 12.2 displays the main phytoconstituents of *A. belladonna* and their therapeutic uses.

TABLE 12.2
Main Phytoconstituents of *Atropa belladonna* and Their Therapeutic Uses

Phytoconstituent	General Therapeutic Use	Reference
Atropine	Treatment of near-sightedness in children, bradycardia, respiratory and heart related ailments, antidote for overdose of cholinergic drugs, mushroom/opium/chloroform poisoning	Yen et al. (1989), Chou et al. (1997)
Scopolamine	Treatment of a number of health conditions including irritable bowel syndrome, spastic colon, stomach ulcers, Parkinson's disease, diverticulitis, motion sickness, excessive night-time urination, pink eye Utilized in anticholinergic medications for the treatment of central nervous system (CNS) disorders and insomnia	Heinrich and Gibbon (2001), Habibi et al. (2015), Morris (2017)
Hyoscyamine	Treatment of moderate nausea, motion sickness, hyperactive bladder, allergic rhinitis, arrhythmias and organophosphate harming, infective shock and gastrointestinal colic	Hartmann et al. (1986), Yun et al. (1992), Sweta and Lakshmi (2015), Morris (2017)
Hyoscine	Utilized as cervical antispasmodic specialist during labor and in the treatment of patients with nausea and vomiting	Sirohiwal et al. (2005)

12.5 PHARMACOLOGICAL STUDIES

The natural compounds of *A. belladonna* have been reported to have pharmacological properties indicating their potential use as a drug. Investigations have reported critical neuro-pharmacological, pain relieving and mitigating properties of the plant (Owais et al. 2014). As a result, the plant has been in use in the treatment of disorders like gout, chickenpox, asthma, hypertension, internal contamination (Duke et al. 1987; Kumar et al. 1997; Bettermann et al. 2001), muscle and joint pain, scarlet fever, pancreatitis, intense irritation, peritonitis, neuro-inflammatory problems and Parkinson's disease (King 1966; Kahn et al. 1991), stomach ailments, hyperkinesis, hyperhidrosis, hemorrhoids (Ramoutsaki et al. 2002; Joshi et al. 2003), aspiratory breakdown during pneumonia and typhoid fever (Tombs and Silverman 2004), treatment of wounds, toxin neutralizer (Toporcer et al. 2006), encephalitis fever, conjunctivitis, snake and scorpion bite (Ceha et al. 1997; Bousta et al. 2001; Duncan and Collison 2003; Samy et al. 2008; Mishra and Tiwari 2011), neutralizing effects of opium, chloroform, and insecticide poisoning (Murray 2017).

Atropine is used by ophthalmologists to dilate the pupils during retina assessment. This is possible because atropine acts on muscarinic receptors in the eye, briefly incapacitating the eye muscles that loosens up the focal point which in turn helps the eyes to fix gaze at one point. Atropine and scopolamine are used as a sedative and act as a drug against vomiting and nausea as they modulate involuntary movements in the body by controlling muscarinic acetylcholine receptors (Ulbricht et al. 2004; Owais et al. 2014). They are also used in the treatment of conditions of mental imbalance, anxiety, perspiration, saliva release, respiratory problems, neurological disorders controlling heartbeats by decreasing the sympathetic effects on salivary glands, eyes, heart, gastrointestinal tract, and bronchial glands (Çaksen et al. 2003; Joshi et al. 2003; Kwakye et al. 2018). Atropine also has potent antiviral properties and helps in curtailing the spread of infection in the body (Duke et al. 1987) and is also used to treat organophosphate poisoning (Eddleston et al. 2008). Scopolamine has been utilized as a tranquilizer or as an anti-insomnia, as it pushes down the parasympathetic sensory system. It is used to assuage stomach cramps and convulsions, pulse rate, muscle contractions, prevention of vertigo, and recovery from anesthesia (Dobrescu 1971; Beyer et al. 2009; Kersten and Wyller 2013).

12.6 TOXIC RESPONSE

The tropane alkaloids present in *A. belladonna* are responsible for its toxicity. The leaves and berries carry a high number of alkaloids which affect the central and parasympathetic nervous systems. Poisoning can occur in infants, adolescents, and adults. The danger of intoxication in children is prevalent because of confusion with other edible berries. The toxic responses are manifested in a broad spectrum from causing ATS, psychosis, fatigue, hyperglycemia, and even unconsciousness. The alkaloids are known to plug the muscarinic receptors in the CNS and postganglionic parasympathetic muscarinic receptors which results in hallucinations (Trabattoni et al. 1984; Ulbricht et al. 2004). ATS characterized by hyperactive delirium involving confusion, visual and hearing hallucinations, tiredness, persistent vomiting, seeing and picking up imaginary objects is another prominent effect of *A. belladonna* poisoning (Demirhan et al. 2013). The syndrome is caused by *A. belladonna* alkaloids causing acetylcholine to become ineffective at the central and peripheral muscarinic receptors thereby influencing neurotransmission in the body.

Manifestation of deadly nightshade poisoning in different patients has been recorded as incoherent speech, inability to recognize members of the family, persistent vomiting, visual and hearing hallucinations, unconsciousness, fatigue, mydriatic and isochoric pupils, irregular breathing, dryness of the oral mucosa, flushed skin, redness of the neck, leukocytosis, metabolic acidosis, pyuria hyperglycemia, hypertension, hypotension hyperreflexia, ileus or lowered bowel sounds, urinary retention, and tachycardia (Çaksen et al. 2003; Berdai et al. 2012; Demirhan et al. 2013; Glatstein et al. 2014).

The toxicity of *A. belladonna* is extended to domestic animals or grazing animals and causes paralysis and necrosis. In case of pet animals like cats and dogs, poisoning causes excessive salivation, vomiting, and diarrhea (Hardin et al. 2018).

The treatment for this toxicity involves maintenance of respiratory, circulatory support and airway patency. ATS is generally treated by the use of benzodiazepine. Physostigmine extracted from Calabar Bean (*Physostigma venenosum*) is another antidote for belladonna poisoning (Berdai et al. 2012). Physostigmine inhibits acetylcholinesterase (causing breakdown of acetylcholine) in both central and peripheral nervous systems resulting in accumulation of acetylcholine within the synapse, thereby activating muscarinic and nicotinic receptors. Consequently, anticholinergic effects are suppressed and nervous system gets stimulated for optimal functioning (Figure 12.3).

12.7 TRADITIONAL AND OTHER POTENTIAL USES

A. belladonna phytochemicals with bioactive properties have been extracted from different parts of the plants and have been widely utilized in traditional medications (Bansode and Salalkar 2015). In Roman times, diluted solutions of *A. belladonna* were frequently used by ladies to widen the pupil to look more attractive, additionally giving them reddish face blush. However, these practices were declared unhealthy as it led to increased cases of cardiac problems and blindness.

From the middle of the nineteenth century to the late 1950s, the therapeutic properties of *A. belladonna* were increasingly realized. The medication assortments, mortars, etc. prepared from the plant were highly popular among the masses and could be easily bought over the counter, although from only pharmaceutical scientific experts and pharmacists. *A. belladonna* therapeutic concoctions reportedly showed better relief than other traditional herbal treatments being tried in those times (Lee 1983). The leaves were generally used for treatment of worms in children, while cigarettes prepared from the leaves were given to the patients to get relief from convulsive asthma (Shultes 1962). The roots have been in use as a soothing agent (Rhodes et al. 1978), for treating sore throat, ulcerative colitis (Shanafelt et al. 2002), and cardiac ailments (Walach et al. 2001; Nisar et al. 2013).

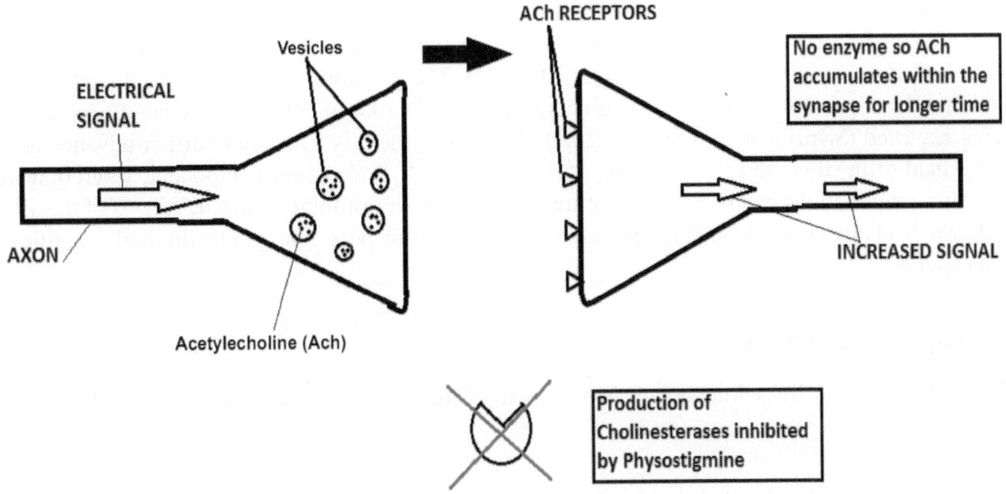

FIGURE 12.3 Mechanism of action of physostigmine against *Atropa belladonna* induced toxicity.

The traditional systems of medicines across the world have been using *A. belladonna* preparations for the treatment of a wide array of ailments like cardiac and neural disorders, spams, pain, muscle contractions, inflammation, headache, chest pain, intercostal neuralgia, muscular spasms, joint and muscle pain, pain due to gall bladder stone, digestive health, diabetes, abdominal pain, asthma, persistent cough anal fissure, and hemorrhoids.

For the preparation of *A. belladonna*-based drugs, plant alkaloids are combined with different compounds (synthetic/natural) to treat specific ailments (Fletcher 2017). In the homeopathy system of medicine, *A. belladonna* is prominently used as a painkiller and to cure nerve pain, migraine, nasal congestion, etc. In herbal medicines (based on traditional systems like Ayurveda and Unani), the dosage varies depending on the plant and plant part from which extraction has been done. These herbal preparations are available either as juices or powder (Singh 2015). The allopathic supplements are available as creams, ointments, and solid pills with different concentrations of *A. belladonna* extracts (Taneja et al. 2019). The plants' extracts are also used in cosmetic preparations like ointments and lotions. Belladonna Naturkosmetik Company, Berlin, manufactures different types of face, body creams, and cosmetics using *A. belladonna.* In all cases, frequent usage of *A. belladonna* therapeutic and cosmetic products is suggested to be avoided for their possible side effects.

12.8 FUTURE REMARK

It is evident that *A. belladonna* is a phytochemically active plant that has been utilized for a vast range of medicinal activities. However, there is more in the plant that needs to be discovered and harnessed. An important aspect is large-scale propagation of the plant to facilitate the availability of raw material for medicinal preparations. Since conventional system of propagation is time and labor consuming, advanced techniques of plant tissue culture need to be utilized for rapid mass propagation of disease-free plant material in a limited space and time. Recent interest has been toward improving constituent concentration in *A. belladonna* (Zeng et al. 2021). This calls for use of biotechnological methods to enhance secondary metabolite production through plant transformation strategies like hairy root culture and CRISPR/Cas9. Besides, there is scope for identification and extraction of lesser known plant metabolites. For instance, glucoside synthesis from hairy roots of *A. belladonna* has been tried (Singh et al. 2021). The possibility of producing *A. belladonna* medicinally important alkaloids in engineered organisms like *Saccharomyces cerevisiae* is also being explored (Srinivasan and Smolke 2020). Moreover, there is a need for further advancements in the field of in silico and *in vitro* drug designing from *A. belladonna* component, wherein novel effective combinations can be prepared to treat diseases with minimal or no side effects of the alkaloids. Such advances would lead to optimal utilization for this poisonous yet medicinally significant plant species.

ACKNOWLEDGMENT

The authors gratefully acknowledge Mr. Steward Cain, Scotland for providing photographs of *Atropa belladonna* plant.

NOTES

Arundhati Singh, School of Agriculture and Environment, The University of Western Australia, Perth, Australia

Vedanshi Pal, Department of Life Sciences, Graphic Era Deemed to be University, Dehradun, India

Shreyshi Aggarwal, Department of Life Sciences, Graphic Era Deemed to be University, Dehradun, India

Manu Pant, Department of Life Sciences, Graphic Era Deemed to be University, Dehradun, India

REFERENCES

Abdel-Hady, M. S., Okasha, E. M., Soliman, S. S. A., & Talaat, M. 2008. Effect of gamma radiation and gibberellic acid on germination and alkaloid production in *Atropa belladonna* L. *Australian Journal of Basic and Applied Sciences 2*(3): 401–405.

Ahmed, H. N. 2020. Effect of BA on initiation and multiplication of *Atropa Belladonna* L. shoot tips produced *in vitro* and content protein. *International Journal of Agricultural and Statistical Science 16*(1): 1137–1141.

Ahmed, Z. S., Salim, A. M., & Alsoufi, A. S. 2021. *In vitro* micropropagation of *Atropa Belladonna* L. using plant growth regulators. *Plant Cell Biotechnology and Molecular Biology 22*(3–4): 169–174.

Al-Akaidi, H. N. A., & Zeki Ameen Kassab Bashi, B. 2017. *Propagation of Belladonna Plants Atropa Belladonna L. by Tissue Culture and Determination of Some Alkaloids in Callus*. College of Agriculture and Forestry, University of Mosul, Mosul, Iraq.

Al-Ashaal, H. A. A. H., Abuelnaga, A., & El-Beih, A. A. 2018. Antifood poisoning pathogenic bacteria in correlation to coumarins and flavonoids of *Atropa belladonna* field plant and *in vitro* cultures. *Journal of Pharmaceutical Sciences and Research 10*(2): 235–239.

Anonymous. 1948. *The Wealth of India: A Dictionary of Indian Raw Materials and Industrial Products (Raw Materials–Volume 1)*. Council of Scientific and Industrial Research, New Delhi, India, pp. 135–137.

Bansode, T. S., & Salalkar, B. K. 2015. Phytochemical analysis of some selected Indian medicinal plants. *International Journal of Pharma and Bio Sciences 6*(1): 550–556.

Baralle, F. E., & Gross, E. G. 1969. Biosynthesis of cuscohygrine in *Atropa belladonna* from sodium acetate-1–14C. *Phytochemistry 8*(5): 849–851.

Benjamin, B. D., Roja, P. C., Heble, M. R., & Chadha, M. S. 1987. Multiple shoot cultures of *Atropa belladonna*: Effect of physico-chemical factors on growth and alkaloid formation. *Journal of Plant Physiology 129*(1–2): 129–135.

Berdai, M. A., Labib, S., Chetouani, K., & Harandou, M. 2012. Case report-*Atropa belladonna* intoxication: A case report. *Pan African Medical Journal 11*(1): 72.

Bettermann, H., Cysarz, D., Portsteffen, A., & Kümmell, H. C. 2001. Bimodal dose-dependent effect on autonomic, cardiac control after oral administration of *Atropa belladonna*. *Autonomic Neuroscience 90*(1–2): 132–137.

Beyer, J., Drummer, O. H., & Maurer, H. H. 2009. Analysis of toxic alkaloids in body samples. *Forensic Science International 185*(1–3): 1–9.

Bousta, D., Soulimani, R., Jarmouni, I., Belon, P., Falla, J., Froment, N., & Younos, C. 2001. Neurotropic, immunological and gastric effects of low doses of *Atropa belladonna* L., *Gelsemium sempervirens* L. and poumon histamine in stressed mice. *Journal of Ethnopharmacology 74*(3): 205–215.

Butcher, R. W. 1947. *Atropa Belladonna* L. *Journal of Ecology 34*(2): 345–353.

Çaksen, H., Odabaş, D., Akbayram, S., Cesur, Y., Arslan, Ş., Üner, A., & Öner, A. F. 2003. Deadly nightshade (*Atropa belladonna*) intoxication: An analysis of 49 children. *Human & Experimental Toxicology 22*(12): 665–668.

Ceha, L. J., Presperin, C., Young, E., Allswede, M., & Erickson, T. 1997. Anticholinergic toxicity from nightshade berry poisoning responsive to physostigmine. *The Journal of Emergency Medicine 15*(1): 65–69.

Chevallier, A. 1996. *The Encyclopedia of Medicinal Plants: An Excellent Guide to over 500 of the More Well-Known Medicinal Herbs from around the World.* Dorling Kindersley, London.

Chopra, R. N., Nayar, S. L., & Chopra, I. C. 1986. Glossary of Indian medicinal plants. *Quarterly Review of Biology 33*(2): 156.

Chou, A. C., Shih, Y. F., Ho, T. C., & Lin, L. L. K. 1997. The effectiveness of 0.5% atropine in controlling high myopia in children. *Journal of Ocular Pharmacology and Therapeutics 13*(1): 61–67.

Eddleston, M., N.A. Buckley, P. Eyer & A. H. Dawson. 2008. Management of acute organophosphorus pesticide poisoning. *Lancet. 371*(9612): 597–607.

Cıkla, U., Turkmen, S., Karaca, Y., Ayaz, A. F., Turedi, S., & Gunduz, A. 2011. An *Atropa belladonna* L. poisoning with acute subdural hematoma. *Human & Experimental Toxicology 30*(12): 1998–2001.

Davis, P. H., Miller, R. R., & Tan, K. 1965. *Flora of Turkey and the Aegean Islands*. Vol. 1–9. University Press, Edinburgh, Scotland.

Demirhan, A., Tekelioğlu, Ü. Y., Yıldız, İ., Korkmaz, T., Bilgi, M., Akkaya, A., & Koçoğlu, H. 2013. Anticholinergic toxic syndrome caused by *Atropa belladonna* fruit (Deadly nightshade): A case report. *Turkish Journal of Anaesthesiology and Reanimation 41*(6): 226.

Dobrescu, D. I. 1971. Propranolol in the treatment of disturbances of the autonomic nervous system. *Current Therapeutic Research, Clinical and Experimental 13*(1): 69–73.

Drugbank. n.d. Atropine. https://go.drugbank.com/drugs/DB00572.

Duke, J. A., Bogenschutz-Godwin, M. J., DuCellier, J., & Duke, P. K. 1987. *Handbook of Medicinal Herbs*. CRC, Press Inc., Boca, Raton, FL.

Duncan, G., & Collison, D. J. 2003. Role of the non-neuronal cholinergic system in the eye: A review. *Life Sciences 72*(18–19): 2013–2019.

Dutta, A. K., & Rita, P. 2011. An updated overview on *Atropa belladonna* L. *International Research Journal of Pharmacy 2*: 11–17.

Eapen, S., Rangan, T. S., Chadha, M. S., & Heble, M. R. 1978. Biosynthetic and cytological studies in tissue cultures and regenerated plants of haploid *Atropa belladonna. Canadian Journal of Botany 56*(21): 2781–2784.

Eger, K., Troschütz, R., & Roth, H. J. 1999 (Eds). Ester aliphatischer Carbonsäuren. In: *Arzneistoffanalyse*, 4th ed. Deutscher Apotheker Verlag, Stuttgart, Germany, pp. 138–144.

Elshorbagy, M. I., Ibrahim, S. M., Abo El-Seoud, K. A., & Abd El-Maksoud, A. I. 2018. Tropane alkaloids production from callus culture of *Atropa belladonna* L. as affected by elicitors and precursor feeding. *International Journal of Pharmaceutical Sciences Review and Research 9*: 116.

Fatur, K., & Kreft, S. 2020. Common anticholinergic solanaceaous plants of temperate Europe—A review of intoxications from the literature (1966–2018). *Toxicon 177*: 52–88.

Fletcher, J. 2017. Uses and risks of belladonna. MedicalNews Today. https://www.medicalnewstoday.com/articles/318180.

Genova, E., Komitska, G., & Beeva, Y. 1997. Study on the germination of *Atropa belladonna* L. seeds. *Bulgarian Journal of Plant Physiology 23*(1–2): 61–66.

Glatstein, M., Danino, D., Wolyniez, I., & Scolnik, D. 2014. Seizures caused by ingestion of *Atropa belladonna* in a homeopathic medicine in a previously well infant case report and review of the literature. *American Journal of Therapeutics 21*(6): e196–e198.

Gosch, G., Bajaj, Y. P. S., & Reinert, J. 1975. Isolation, culture, and induction of embryogenesis in protoplasts from cell-suspensions of *Atropa belladonna. Protoplasma 86*: 405–410.

Guggisberg, A., & Hesse, M. 1983. Putrescine, spermidine, spermine, and related polyamine alkaloids. In: Brossi, A. (Ed.), *The Alkaloids.* Vol. 22. Academic Press, New York, NY, pp. 85–188.

Habibi, P., Piri, K., Deljo, A., Moghadam, Y. A., & Ghiasvand, T. 2015. Increasing scopolamine content in hairy roots of *Atropa belladonna* using bioreactor. *Brazilian Archives of Biology and Technology 58*(2): 166–174.

Hardin, J. W., Brownie, C. F., & Krings, A. 2018. *Plants Poisonous to Livestock and Pets in North Carolina, 2.0*. North Carolina State University, Raleigh, NC.

Hartmann, T., Witte, L., Oprach, F., & Toppel, G. 1986. Reinvestigation of alkaloid composition of *Atropa belladonna* plant, root cultures and cell suspension cultures. *Planta Medica 52*(5): 390–395.

Hashimoto, T., & Yamada, Y. 1994. Alkaloid biogenesis: Molecular aspects. *Annual Review of Plant Biology 45*: 257–285.

Heinrich, M., & Gibbon, S. 2001. Ethnopharmacology in drug discovery: An analysis of its role and potential contribution. *Journal of Pharmacy and Pharmacology 53*(4): 425–432.

https://agriinfo.in/package-of-practices-for-cultivation-of-belladonna-1396/.

Jordan, M. 1976. *A Guide to Wild Plants: The Edible and Poisonous Species of the Northern Hemisphere*. Millington Books Ltd., London, England.

Joshi, P., Wicks, A. C., & Munshi, S. K. 2003. Recurrent autumnal psychosis. *Postgraduate Medical Journal 79*(930): 239–240.

Kahn, A., Rebuffat, E., Sottiaux, M., Muller, M. F., Bochner, A., & Grosswasser, J. 1991. Prevention of airway obstructions during sleep in infants with breath-holding spells by means of oral belladonna: A prospective double-blind crossover evaluation. *Sleep 14*(5): 432–438.

Kersten, H., & Wyller, T. B. 2013. Anticholinergic drug burden in older people's brain—How well is it measured? *Basic and Clinical Pharmacology & Toxicology 114*(2): 151–159.

Khater, M. A., Soliman, S. S. A., Abdel-Hady, M. S., & Fayed, A. H. 2013. Tropane alkaloid production via new promising *Atropa belladonna* L. lines by *in vivo* and *in vitro*. *Nature and Science 11*(3): 32–40.

King, J. C. 1966. Anisotropine methylbromide for relief of gastrointestinal spasm: Double-blind crossover comparison study with belladonna alkaloids and phenobarbital. *Current Therapeutic Research, Clinical and Experimental 8*(11): 535–541.

Kirchhoff, C., Bitar, Y., Ebel, S., & Holzgrabe, U. 2004. Analysis of atropine, its degradation products and related substances of natural origin by means of reversed-phase high-performance liquid chromatography. *Journal of Chromatography 1046*: 115–120.

Kumar, S., Shukla, Y. N., Lavania, U. C., Sharma, A., & Singh, A. K. 1997. Medicinal and aromatic plants: Prospects for India. *Journal of Applied Research on Medicinal and Aromatic Plants 19*(2): 361–365.

Kwakye, K. F., Jimenez, J., Jimenez, J. A., & Aschner, M. 2018. *Atropa belladonna* neurotoxicity: Implications to neurological disorders. *Food and Chemical Toxicology 116*(Pt B): 346–353.

Laffargue, F., Oudot, C., Constanty, A., Bedu, A., & Ketterer-Martinon, S. 2011. Deadly nightshade (*Atropa belladonna*) intoxication in a 2-year-old child. *Archives de Pédiatrie 18*(2): 186–188.

Lee, M. R. 1983. Atropine. In: Polson, C. J., & Green, M. A. (Eds.), *Clinical Toxicology*, 3rd ed. Pitman Books Ltd, London, England, pp. 355–368.

Lee, M. R. 2007. Solanaceae IV: *Atropa belladonna*, deadly nightshade. *Journal of the Royal College of Physicians of Edinburgh 37*(1): 77–84.

Majid, R., Zargar, M. A., & Ahmad, L. 2012. Evaluation of anti-inflammatory potential of *Atropa acuminata* in carrageenan induced inflammation in rats. *Journal of Medicinal Plant Research 6*(43): 5586–5592.

Mera, N., & Takayama, S. 2012. Effects of medium components and shear conditions on the formation and growth of adventitious bud derived from hairy roots of *Atropa belladonna* L. *Environmental Control in Biology 50*(4): 393–406.

Millet, A., Stintzing, F., & Merfort, I. 2010. Flavonol quantification and stability of phenolics in fermented extracts from fresh *Betula pendula* leaves. *Journal of Pharmaceutical and Biomedical Sciences 53*: 137–144.

Mishra, B. B., & Tiwari, V. K. 2011. Natural products: An evolving role in future drug discovery. *European Journal of Medicinal Chemistry 46*: 4769–4807.

Morris, S. Y. 2017. Belladonna: Remedy with a dark past. https://www.healthline.com/health/belladonna-dark-past.

Murray, J. J. 2017. Belladonna—It's more than a pretty woman. https://jamesjmurray.com/2017/04/05/belladonna-its-more-than-a-pretty-woman/.

National Library of Medicine. n.d.-a. Atropine. https://pubchem.ncbi.nlm.nih.gov/compound/Atropine.

National Library of Medicine. n.d.-b. Hyoscyamine. https://pubchem.ncbi.nlm.nih.gov/compound/Hyoscine.

National Library of Medicine. n.d.-c. Scopolamine. https://pubchem.ncbi.nlm.nih.gov/compound/%28S%29-atropine#section=MeSH-Pharmacological-Classification.

Nikolaeva, M. G. 1982. Seed dormancy and factors of its control. In: Kan, N.A. (Ed.), *Physiological and Biochemistry of Seed Dormancy and Germination*. Colos Press, Moscow, Russia, pp. 72–96.

Nisar, A., Malik, A. H., & Zargar, M. A. 2013. *Atropa acuminata* blunts production of pro-inflammatory mediator's eicosanoids, leukotrienes, cytokines *in vitro* and *in vivo* models of acute inflammatory responses. *Journal of Ethnopharmacology 147*(3): 584–594.

Owais, F., Anwar, S., Muhammad, S., Ishtiaque, S., & Mohiuddin, O. 2014. Analgesic, anti-inflammatory and neuropharmacological effects of *Atropa belladonna*. *Pakistan Journal of Pharmaceutical Sciences 27*(6): 2183–2187.

Palazon, J., Altabella, T., Cusido, R. M., Ribo, M., & Pinol, M. T. 1995. Growth and tropane alkaloid production in *Agrobacterium* transformed roots and derived callus of *Datura*. *Biologiaplantarum 37*: 161–168.

Ramoutsaki, I. A., Askitopoulou, H., & Konsolaki, E. 2002. Pain relief and sedation in roman byzantine texts: *Mandragoras officinarum*, *Hyoscyamos niger* and *Atropa belladonna*. *International Congress 1242*: 43–50.

Rani, N. S. A., & Prasad, M. P. 2013. Studies on the organogenesis of *Atropa belladonna* in *in-vitro* conditions. *International Journal of Biotechnology and Bioengineering Research 4*(5): 457–464.

Rani, N.S. A., & Prasad, M. P. 2014. *In-vitro* studies on the germination of *Atropa belladonna* seeds under different conditions. *International Journal of Scientific and Research Publication 3*(10): 552–555.

Rashid, A., & Street, H. E. 1973. The development of haploid embryoids from anther cultures of *Atropa belladonna* L. *Planta 113*: 263–270.

Rhodes, J. B., Abrams, J. H., & Manning, R. T. 1978. Controlled clinical trial of sedative anticholinergic drugs in patients with the irritable bowel syndrome. *Journal of Clinical Pharmacology 18*(7): 340–345.

Rowson, J. M. 1950. The pharmacognosy of *Atropa belladonna* Linn. *Journal of Pharmacy and Pharmacology 2*(4): 201–216.

Samy, R. P., Thwin, M. M., Gopalakrishnakone, P., & Ignacimuthu, S. 2008. Ethnobotanical survey of folk plants for the treatment of snakebites in southern art of Tamil Nadu, India. *Journal of Ethnopharmacology 115*: 302–312.

Shanafelt, T. D., Barton, D. L., Adjei, A. A., & Loprinzi, C. L. 2002. Pathophysiology and treatment of hot flashes. *Mayo Clinic Proceedings 77*(11): 1207–1218.

Shultes, R. E. 1962. The role of the ethnobotanist in search medicinal plants. *Lloydia 25*(4): 257–266.

Sievers, A. F. 1914. The possibility of increasing the alkaloidal content of belladonna plants through selection. *American Journal of Pharmaceutical Education 86*: 483–505.

Singh, J. 2015. *Atropa belladonna* (deadly nightshade). https://www.ayurtimes.com/atropa-belladonna-deadly-nightshade/.

Singh, S., Pandey, P., Akhtar, M. Q., Negi, A. S., & Banerjee, S. 2021. A new synthetic biology approach for the production of curcumin and its glucoside in *Atropa belladonna* hairy roots. *Journal of Biotechnology 328*: 23–33.

Sirohiwal, D., Dahiya, K., & De, M. 2005. Efficacy of hyoscine-N butyl bromide (Buscopan) suppositories as a cervical spasmolytic agent in labour. *Australian and New Zealand Journal of Obstetrics and Gynaecology 45*(2): 128–129.

Srinivasan, P., & Smolke, C. D. 2020. Biosynthesis of medicinal tropane alkaloids in yeast. *Nature 585*(7826): 614–619.

Stateva, S., & Desheva, G. 2017. *Adaptation of the Species Atropa Belladonna L. Grown In Vitro to the Environment*. Jubilee International Scientific Conference, Sustainable Regional Development Perspectives, Plovdiv, Bulgaria, pp. 504–510.

Sweta, V. R., & Lakshmi, T. 2015. Pharmacological profile of tropane alkaloids. *Journal of Chemical and Pharmaceutical Research 7*(5): 117–119.

Taha, H. S. 2003. Effect of biotic stress (*Aspergillus niger*) on the production and accumulation of total alkaloids in *Atropa belladonna* via tissue culture. *International Conference on Medicinal and Aromatic Plants (Part II) 1*(597): 257–264.

Taneja, D., Khurana, A., Vichitra, A., Sarkar, S., Gupta, A. K., Mittal, R., Bawaskar, R., Sahoo, A. R., Prusty, U., Singh, S., Sharma, M., Pant, R., Singh, U., Upadhyay, A. K., Sehegal, S., Patnaik, S., Nath, T., & Manchanda, R. K. 2019. An assessment of a public health initiative of homeopathy for primary teething. *Homeopathy 108*(1): 2–11.

Thomas, E., & Street, H. E. 1970. Organogenesis in cell suspension cultures of *Atropa belladonna* L. and *Atropa belladonna* cultivar *lutea* döll. *Annals of Botany 34*: 657–669.

Tombs, S., & Silverman, I. 2004. Pupillometry: A sexual selection approach. *Evolution and Human Behavior 25*(4): 211–228.

Toporcer, T., Grendel, T., Vidinský, B., Gal, P., Sabo, J., & Hudák, R. 2006. Mechanical properties of skin wounds after *Atropa belladonna* application in rats. *Journal of Metals, Materials and Minerals 16*(1): 25–29.

Trabattoni, G., Visitini, D., Terzano, G. M., & Lechi, A. 1984. Accidental poisoning with deadly nightshade berries: A case report. *Human and Experimental Toxicology 3*(6): 513–516.

Tulin, F., & Ismet, K. 2011. Psychiatric aspects of a case with deadly nightshade intoxication. *Academic Emergency Medicine Journal 10*(2): 86–88.

Tutin, T. G., Heywood, V. H., Burges, N. A., Moore, D. M., Valentine, D. H., Walters, S. M., & Webb, D. A. 1972. *Flora Europaea. Vol. 2, Rosaceae to Umbelliferae*. Cambridge University Press, Cambridge, England, p. 399.

Ulbricht, C., Basch, E., Hammerness, P., Vora, M., Wylie, J., & Woods, J. 2004. An evidence-based systematic review of belladonna by the natural standard research collaboration. *Journal of Herbal Medicine 4*(4): 61–90.

Walach, H., Koster, H., Hennig, T., & Haag, G. 2001. The effects of homeopathic belladonna 30CH in healthy volunteers—A randomized, double-blind experiment. *Journal of Psychosomatic Research 50*(3): 155–160.

Xirong, W., Wei, Q., Yan, L., Di, L. J., & Hua, L. Z. 2011. Establishment of high frequency regeneration system of *Atropa belladonna* L. & screening of kanamycin resistance. *Agricultural Science and Technology 12*(3): 372–374.

Yen, M. Y., Liu, J. H., Kao, S. C., & Shiao, C. H. 1989. Comparison of the effect of atropine and cyclopentolate on myopia. *Annals of Ophthalmology 21*(5): 180–182, 187.

Yun, D. J., Hashimoto, T., & Yamada, Y. 1992. Metabolic engineering of medicinal plants: Transgenic *Atropa belladonna* with an improved alkaloid composition. *Proceedings of the National Academy of Sciences 89*(24): 11799–11803.

Zarate, R., Jaber-Vazdekis, N. J., Medina, B., & Ravelo, A. G. 2006. Tailoring tropane alkaloid accumulation in transgenic hairy roots of *Atropa baetica* by over-expressing the gene encoding hyoscyamine 6 beta-hydroxylase. *Biotechnology Letters 28*(16): 1271–1277.

Zein, H. E., & Dagher-Kharrat, M. B. 2021. New records of vascular plants for the flora of Lebanon: A rare species rediscovered after seventy years, *Daphne pontica* L. (Thymelaeaceae), and three new occurrences, *Atropa belladonna* L. (Solanaceae), *Circaealutetiana* L. (Onagraceae), and *Euonymus latifolius* (L.) Mill. (Celastraceae). *Check List 17*(2): 655–667.

Zeng, L., Zhang, Q., Jiang, C., Zheng, Y., Zuo, Y., Qin, J., Liao, Z., & Deng, H. 2021. Development of *Atropa belladonna* L. plants with high-yield hyoscyamine and without its derivatives using the CRISPR/Cas9 system. *International Journal of Molecular Sciences 22*(4): 1731.

Zenkteler, M. A. 1971. Development of new plants from leaves and roots of *Atropa belladonna* L. in the *in vitro* culture. *Acta Societatis Botanicorum Poloniae 40*(2): 305–313.

Zhang, L., Kai, G. Y., Lu, B. B., Zhang, H. M., Tang, K. X., Jiang, J. H., & Chen, W. S. 2005. Metabolic engineering of tropane alkaloid biosynthesis in plants. *Journal of Integrative Plant Biology 47*(2): 136–143.

13 *Calotropis gigantea* (Crown Flower or Giant Milkweed)

Prabhjot Kaur, Amarjit Singh, and Jitender Sharma

CONTENTS

13.1 INTRODUCTION

Calotropis gigantea is commonly known as giant milkweed due to the presence of milky white latex in different parts of the plant and its larger size than congeneric plants. Genus epithet "Calotropis" is a compound Greek term derived from "kalos" referring to "beautiful" and "tropis" meaning "ship keel". The name might have been given referring to the rigid coronal scales on the flowers. Species epithet "gigantea" indicates "large" probably referring to the size of the plant (NParks 2022). Some other common names include yercum fiber, crown flower, ivory plant, and swallow-wort. It belongs to the family Apocynaceae and sub-family Asclepiadoideae which comprises nearly 2,000 species spanning in more than 280 genera (Ali-Seyed and Ayesha 2020; Kaur et al. 2021; Negi and Bisht 2021; Mushir et al. 2016; Wadhwani et al. 2021).

Apocynaceae covers a large number of tropical trees, shrubs, and vines with its sub-family Asclepiadoideae plants producing milky latex and hence called milkweeds (Al Sulaibi et al. 2020; GRIN-Global 2020). These species present rich source of secondary metabolites including cardenolides, triterpenoids, alkaloids, and iridoids. A wide range of pharmacological activities has been related to them, some of which include anticancer, cardioprotective, neuroprotective,

DOI: 10.1201/b23017-15

hepatoprotective, anti-inflammatory, and antimalarial properties. *Calotropis* along with eight other genera has been identified with antiproliferative properties, while antiplasmodial properties have been observed in only seven genera including *Calotropis* (Chan et al. 2016). Two most studied and commonly found sister species of this genus are *Calotropis gigantea* (Sweta arka) and *Calotropis procera* (Rakta arka), while other two less commonly found species are *Calotropis sussuela* and *Calotropis acia* (Wadhwani et al. 2021). Both *C. gigantea* and *C. procera* have been demonstrated to possess vital therapeutic properties as well as toxic characteristics.

Though *C. gigantea* possesses ornamental and esthetic values, it does not require special cultivation techniques. The plant can withstand harsh environmental conditions including abiotic stress of water scarcity and salinity. Such characteristics make the plant grow at less habitable conditions like overgrazed pastures, desert area, roadside embankment, and sandy soils. Ease of growing in unfavorable conditions and pollination via animals and wind make this plant a self-cultivating weed. However, people cultivate it also due to medicinal properties of its constituents and religious significance especially in India. Although different constituents of this plant as well as latex have been shown to have multiple therapeutic uses, these are poisonous too. Latex contains calcium oxalate which upon contact with the eyes or skin can cause conjunctivitis, irritation, and corneal opacity (Kharat and Kharat 2019; Khosravi and Kumar 2021). Alongside, cardiac glycosides or cardenolides present in the latex and other parts are found to be cytotoxic.

Yet, *C. gigantea* harbors and provides nourishment to many insects and birds including butterflies, bees, and sunbird. Monarch caterpillars feed exclusively on milkweeds and store the ingested cardenolides to deter their predators (National Wildlife Federation 2022; USDA 2022). The toxin is retained in the adult butterfly as well which is reflected by bright orange color of the wings as a warning (USDA 2022). Besides, grasshopper *Poecilocerus pictus* also feeds on this plant. Aphids that fed on the leaves were having longer lifespan and higher fecundity (Al Sulaibi et al. 2020).

For such reasons, this plant holds ethnobotanical significance since ancient times and continues to offer the scope for further exploration. Thus, in the present review, a detailed insight of its bioactive constituents and therapeutic uses has been provided. Non-traditional and novel fields of applications have also been discussed.

13.2 BOTANICAL DESCRIPTION

C. gigantea is an evergreen, perennial, branched shrub most often confused with its congeneric plant *C. procera* (NParks 2022; Nguyen et al. 2017; Kaur et al. 2021). But the distinction can be made even with a keen observation of macro-morphological features. *C. gigantea*, being taller than *C. procera*, can grow up to 4.5 m and above while leaves can be smaller and stems thicker. Besides, the branches of the former are thicker and can be up to 6 cm in diameter as compared to 1 cm thickness of that in the latter. Major distinction between these two can also be made by observing the Corolla which are star shaped in the former and can lean backward, while in the case of *C. procera* they are cup shaped. Figure 13.1 shows different stages of plant life.

Roots are branched, tortuous, and cylindrical, and they can form suckers for propagation. Root bark is mucilaginous, short and curved. Stem is aerial, solid, erect, cylindrical, and branched with the diameter reaching 20–25 cm in thickness. All parts of the plant contain latex. Leaves are simple, subsessile, cauline and ramal, phyllotaxy opposite and decussate, exstipulate, ovate, entire, having reticulate unicostate venation and wax coating. Inflorescence shows flowers arranged in axillary and pedunculate, umbelliform cymes. Flowers are mostly white or sometimes lavender in color, bracteate, ebracteolate, complete, hermaphrodite, actinomorphic, pentamerous, with hypogynous condition. Sepals are five in number and polysepalous, while Petals are also five in number, gamopetalous, and star shaped. Androecium has five stamens and gynandrous condition. Five coronary growth radiate from the stamina tube making corona. Gynoecium has bicarpellary ovary, apocarpous, styles united at the apex, stigma is peltate, and anthers fuse to form gynostegium. Fruits are

FIGURE 13.1 *Calotropis gigantea*: (a) phyllotaxy, (b) flowering, (c) fruits, (d) dehiscing fruit, (e) seeds with silky hair, and (f) seeds.

etaerio of follicles, oval in shape, about 5–7 cm in diameter, and 10–14 cm in length. Seeds are many in number, flat, obovate, compressed having tuft of silky white hair (coma).

13.3 DISTRIBUTION

The plant is widely distributed in arid, semi-arid, and waste regions owing to its high adaptability and tolerance toward various abiotic stresses such as drought and salinity. It is native to Indian subcontinent (India, Nepal, Bhutan, Sri Lanka, and Pakistan), Indo-China (Myanmar, Laos, Vietnam, and Thailand), Malesia (Malaysia and Indonesia) in Asia-tropical and China, Western Asia (Iran) in Asia-temperate (GRIN-Global 2020). The plant grows well in overgrazed pastures, exhausted or sandy soils, coastal dunes, abandoned areas and waste lands, road sides, river flats, railway embankments, etc. Sympatric occurrence of *C. gigantea* and *C. procera* can be observed as both are congeneric and share close homology (Kaur et al. 2021; CABI 2021).

13.4 PHYTOCHEMICAL CONSTITUENTS

Phytoconstituents are generally bioactive molecules found in plants with specific biological activity (Alafnan et al. 2021). *C. gigantea* and its sister species are rich in bioactive phytoconstituents including different hydrolytic enzymes and cardiac glycosides. Hydrolytic enzymes mainly comprise cysteine proteases, while cardiac glycosides are the steroids having specific impact on cardiac activity. Major cardiac glycosides identified are uscharin, calotropin, voruscharin, uschridin, calotoxin, uzarigenin, proceroside, syriogenin, etc. (Khosravi and Kumar 2021). Latex of these plants contains significant amount of hydrolytic enzymes especially proteases (Boller 1986; Rajesh et al. 2005). Sengupta et al. (1984) isolated two different types of calotropins I and II from the latex of the plant which had protease activity. Both cysteine proteases shared close homology but their amino acid sequence, auto-digestion behavior, and peptide mapping were different (Bhattacharya et al. 1985). Latex is known to contain a total of four cysteine proteases namely Calotropin FI, FII, DI, and DII (Juncker et al. 2009). These have been used by plants to serve different purposes such as for development, fruit ripening, enabling protection, and degrading seed storage proteins. (Juncker et al. 2009). Moreover, they showed papain and ficin like specificities toward oxidized Insulin B chain (Sengupta et al. 1984). GC-MS analysis of ethanolic leaf extract of *C. gigantea* revealed 22 constituents with different functional groups. These groups included methyl, carbonyl, hydrocarbon, and hydroxyl (Dhivya and Manimegalai 2016; Kiruthiga and Kumar 2019).

Recently, isolation and extraction of three new phytochemicals were carried out from dried leaves of *C. gigantea* along with previously reported nine compounds (Nguyen et al. 2017). Soluble extract of phytochemicals was obtained in methyl alcohol after reflux and partitioned in different organic solvents. Newly found phytochemicals were separated and purified from chloroform and ethyl acetate fractions. These included a new lignan 9′-methoxypinoresinol and two glycosylated 5-hydroxymethylfurfurals namely calofurfuralside A and calofurfuralside B. Besides, nine previously known compounds from the leaves of *C. gigantea* were isolated alongside. These included calotropone, 5-hydroxymethylfurfural, pregna-4,6-diene-3,20-dione, 2-methylpyridin-3-ol, isoliquiritigenin, (+)-syringaresinol, (+)-pinoresinol, (+)-pinoresinol 4-O-[6″-O-vanilloyl]-β-D-glucopyranoside, and 14α-hydroxypregna-4,6-diene-3,20-dione. Earlier, Misra et al. (1993) identified several phytochemicals from leaf extract. These included holarrhetine, cyanidin-3-rhamnoglucoside, taraxasterol isovalerate, mudarine, uscharin, calotropin, calotoxin, and phenol (Mushir et al. 2016). HPLC analysis of methanolic extract of *C. gigantea* leaves has shown abundance of alkaloids, flavonoids, and terpenoids (Damodaran et al. 2019). Different instruments and techniques have been used to identify the constituents present in different parts of the plant. Major techniques employed include HPLC, NMR and ECD spectroscopy. Some of the phytoconstituents from different parts of *C. gigantea* identified using different techniques are listed in Table 13.1.

TABLE 13.1
Some of the Phytoconstituents Identified from Different Parts of *Calotropis gigantea*

Part of the Plant	Solvent for Extraction	Identification Technique	Constituents	Reference
Root	Chloroform	NMR and ECD spectroscopy	Caloside G	Nguyen et al. (2021)
Root bark	Ethyl alcohol-water (7:3)	NMR and ESIMS analysis	Six new cardenolides	You et al. (2013)
Bark	Ethyl alcohol	NMR, LC-MS	Three new cardenolides	Van Khang et al. (2014)
Leaf	Dichloromethane	FTIR, NMR	19-Nor- and 18,20-Epoxy-cardenolides	Lhinhatrakool and Sutthivaiyakit (2006)
	Methanol	HPLC	Gallic acid, Metamitron, Piperidine, Bumetanide, Octanoic acid, Calactin, Ornithine, Catechin, Benzoic acid, Tannic acid, Beta-sitosterol, Ethylene, Holarrhetine, Calotropogenin, Sapogenins, Cyanidin, Quercetin-3-rutinoside, Cinnamic acid, Calotropogenin, Quercetin, Coumarin, Taraxasterol	Damodaran et al. (2019)

13.5 PHARMACOLOGICAL STUDIES

There are different therapeutic uses from traditional as well as modern medical sciences assigned to the phytochemical constituents of the genus *Calotropis* including *C. gigantea.* But the plants belonging to genus *Calotropis* are not only medicinal but also poisonous. Both *C. gigantea* and *C. procera* are rich in phytochemical constituents with most of them possessing similar properties, but the latter has been reported to be more poisonous. Some of the reported pharmacological studies on the former for its different therapeutic applications are as follows:

13.5.1 Potent Inhibitor of Corona Virus (Covid 19)

Calotropin from the latex of *C. gigantea* has been studied *in silico* for molecular docking with M^{pro} COVID-19 protein using PatchDock tool (Sharma and Kaur 2020). It was found that hydrophobic interactions can form M^{pro} and calotropin complex with docking score of –253.66. In this study, 3D (dimensional) structure of calotropin was docked against receptor structure of main protease M^{pro} or 3C like protease ($3CL^{pro}$). Calotropin was first screened as a potential ligand using Lipinski's rule of five. In docking studies, binding pocket S1 of M^{pro} was found to be the best docking site which involved two hydrophobic interactions and hydrogen bonding in complex formation. Though the study was completely based on computational techniques and no wet-lab data for validation were available, it presented the potential of calotropin as a potent drug candidate (Sharma and Kaur 2020). In another study, Dutta et al. (2021) analyzed 30 out of 74 phytoconstituents extracted in methanol from *C. gigantea* for the inhibition of main protease by targeting $M^{pro}/3CL^{pro}$. Among the studied compounds, juniper camphor showed the most significant interaction. Furthermore, it exhibited the likelihood of both interactions, i.e., hydrogen bonding and hydrophobic interactions.

13.5.2 Anticancer and Cytotoxic Properties

Several phytochemicals from *C. gigantea* have been studied for their cytotoxic effect on human pancreatic cancer cell line PANC-1 (Nguyen et al. 2017). In normoglycemic condition, growth of PANC-1 cancer cell line was significantly inhibited by a new lignan named 9′-methoxypinoresinol with IC_{50} value 3.7 µM. Winitchaikul et al. (2021) investigated the extract from stem bark of *C. gigantea* against HCT116 and HT-29 cell lines so as to explore anti-colorectal cancer activity. The highest potency was exhibited by dichloromethane extract of stem bark as compared to ethyl acetate and water fractions. Moreover, when used along with 5-fluorouracil, this crude extract expressed enhanced apoptosis of cancer cells through mitochondria-dependent pathway while simultaneously having minimal toxic effects on healthy cells (Winitchaikul et al. 2021).

Cardenolides or cardiac glycosides from this plant have shown to exert anticancer effect by inhibiting transcriptional activity of hypoxia inducible factor-1 (HIF-1) (Zheng et al. 2021). Based on their chemical structure, two types of cardiac glycosides are found in *C. gigantea* which include single linkage or classical (e.g. digoxin) and double linkage or non-classical (e.g. 19-dihydrocalotoxin) cardiac glycosides. The main difference lies in the chemical linkage between the aglycone and glycosyl moieties (Zheng et al. 2021). However, both types of cardenolides have been reported to exhibit the inhibitory effect on HIF-1 transcriptional activity with a few showing cytotoxicity to multiple cancer cell lines. MCF-7 cell lines causing breast cancer in humans underwent apoptosis when treated with methanolic extract of stems and leaves of *C. gigantea.* A dose of 40 caused apoptosis in 56.9% of cell population which was confirmed by analyzing DNA fragmentation and expression of pro-apoptotic genes by PCR (Kharat and Kharat 2019). Some of the cancer cell lines which have shown growth inhibition and suppression by *C. gigantea* are shown in Table 13.2.

TABLE 13.2
Anticancer Properties of Different Parts of *Calotropis gigantea*

Part of the Plant	Extract	Compound	Cell Line	Reference
Root	Ethyl alcohol	Calotropone	Human gastric cancer SGC-7901 cell lines and Chronic myelogenous leukemia K562	Wang et al. (2008)
	Ethyl acetate	Calotroposid A	WiDr colon cancer cells	Mutiah et al. (2017)
	Chloroform	Caloside G	PANC-1 human pancreatic and HeLa human cervical carcinoma cell lines	Nguyen et al. (2021)
Root bark	Methanol and its chloroform-soluble fraction	Extract	Ehrlich ascites carcinoma	Habib (2011)
	Ethyl alcohol and water	Cardenolides	Human A549 and Hela cell lines	You et al. (2013)
Leaves and stems	Methanol	Cardenolides	Human breast carcinoma cell line (MCF-7 cells)	Kharat and Kharat (2019)
Leaves	Hexane, Dichloromethane, and Methanol	Cardenolides	MCF-7 cancer cells, NCL-H18 lung cancer cell, and KB skin cancer cell	Seeka and Sutthivaiyakit (2010)
	Dichloromethane		HeLa cell, MCF-7 cancer cell, HT-29 cancer cell, MDA-MB-231 cancer cell, SKOV-3 cancer cell, and Hep-G2 cancer cells	Wong et al. (2011)
	Ethyl acetate	Lignan (9′-methoxypinoresinol)	Pancreatic cancer cell line PANC-1	Nguyen et al. (2017)
	Methanol	Alkaloids, terpenoids, and flavonoids	HeLa, MCF7, and A549 cell lines	Damodaran et al. (2019)
Flower	Ethyl acetate	Anhydrosophoradiol-3-acetate (A3A)	Ehrlich ascites carcinoma	Habib and Karim (2013)
—	—	Calotropin	Nonsmall cell lung cancer	Tian et al. (2018)

13.5.3 Antimicrobial Activity

Different parts of *C. gigantea* have been explored for activity against several genera of bacteria and fungi. Strong antimicrobial activity has been exhibited by methanolic extract of leaves containing amines, ketones, alcohols, and aldehydes (Sachin et al. 2018; Singh and Javed 2015). Antibacterial activity was exhibited by extract from leaves of this plant (Kumar et al. 2010; Suchita et al. 2014) along with antifungal activity for different clinical strains of *Candida* sp., namely *Candida albicans*, *Candida krusei*, *Candida tropicalis*, and *Candida parapsilosis* (Senthil Kumar et al. 2012).

Alam et al. (2008) reported antibacterial activity of root bark extract of *C. gigantea* against *Shigella sonnei*, *Bacillus subtilis*, *Bacillus megaterium*, *Sarcina lutea*, *Pseudomonas aeruginosa*, and *Escherichia coli*. Kumar et al. (2010) explored aqueous extract of leaves for the inhibitory effect on growth of six clinical bacterial strains against positive control of 10 μg/disc three antibiotics. Zone of inhibition by aqueous leaf extract was higher than positive control of respective antibiotic for *E. coli*, *P. aeruginosa*, and *Bacillus cereus*. Though lesser than positive control, *Micrococcus luteus*, *Klebsiella pneumoniae*, and *Staphylococcus aureus* also showed inhibition by aqueous leaf extract of the plant.

Antifungal activity of concoctions constituting aerial parts of *C. gigantea* has been demonstrated in a study by Usha et al. (2009). Concoctions contained two more plants namely *Datura stramonium* and *Azadirachta indica*. The study was conducted to observe the antifungal activity against *Fusarium mangiferae* which causes floral malfunctioning in mango (*Mangifera indica*). It was observed that the malfunctioning was controlled due to suppression of fungus and not by its complete elimination (Usha et al. 2009). Other fungi shown to be susceptible against *C. gigantea* include *Phyllactinia corylea, Peridiopsora mori, Pseudocercospora mori, Myrothecium roridum, Colletotrichum graminicola, Drechslera sorokiniana, Fusarium solani, Macrophomina phaseolina*, and *Phomopsis sojae* (Al Sulaibi et al. 2020). The studies related to antimicrobial potential of the plant confirm the bactericidal and fungicidal activities for a vast range of pathogens.

13.5.4 Wound Healing and Procoagulant Properties

Wound healing is a multi-step process involving several cells, proteins, matrix, etc. to interact in a defined manner. Tissue regeneration occurs at the final stage after hemostasis, inflammation, and proliferation. Even these processes also include several steps which require different signal molecules, growth factors, and cell interactions. Ali-Seyed and Ayesha (2020) reviewed wound healing potential of *C. gigantea* and *C. procera*. Different model studies using phytoconstituents from *Calotropis* established their role in healing process of different kinds of wounds including chronic ones (Ali-Seyed and Ayesha 2020). Latex of *C. gigantea* has been a source of proteolytic enzymes with several potential applications.

Some of the studies related to wound healing (Ashwani 1999) and procoagulant properties (Rajesh et al. 2005) of crude enzymes extract from latex have been carried out. The enzymes hydrolyzed casein, crude fibrin clot, and fibrinogen but were inhibited by indole acetic acid (IAA) indicating that they were cysteine proteases (Ashwani 1999; Rajesh et al. 2005). Furthermore, higher concentration of crude latex extract containing cysteine proteases induced hemorrhage at the site of administration. Clot dissolution, a pre-requisite in early coagulation mechanism of wound healing, was the highest in case of treatment with crude latex extract as compared to papain and trypsin (Rajesh et al. 2005). Excision model study of ethanol extract of root bark showed enhanced wound contraction and reduced fibrosis (Deshmukh et al. 2009). Topical use of extract in an ointment base to albino rat was found to exhibit superior healing efficiency (Samy and Chow 2012). Several studies related to wound healing have been extensively reviewed and cited (Ali-Seyed and Ayesha 2020).

13.5.5 Anthelmintic and Nematocidal Activities

Ethnomedical use of phytoconstituents of *C. gigantea* has also been explored for deworming. Anthelmintic activity of alcoholic and aqueous extracts of roots was investigated against earthworm, i.e., *Pheretima posthuma*, and compared with albendazole (Argal and Sachan 2009). It was found that 100 mg/mL dose of root extract produced almost the same effect as of 80 mg/mL albendazole that caused paralysis and death. De et al. (2019) compared different doses of methanolic and aqueous extracts of leaves of *C. gigantea* with albendazole and carboxyl methyl cellulose as control. It was found that both aqueous and methanolic leaf extract at 100 mg/mL dose caused early paralysis and death in *P. posthuma* as compared to 20 mg/mL dose of albendazole. Latex of this shrub has also shown nematocidal activity against *Heterodera cajani* (cowpea cyst nematode) and *Meloidogyne incognita* (cotton root-knot nematode; Devi and Gupta 2000).

13.5.6 Anti-Tussive and Anti-Asthmatic Properties

Different parts of *C. gigantea* have been used to extract phytoconstituents of potential therapeutic use. Jaliwala et al. (2011) studied anti-tussive, expectorant, and anti-asthmatic effects of aqueous and ethanol extract of its flowers. Anti-tussive properties of aqueous extract at doses of 200 and

500 mg/kg were investigated for induced coughing in guinea pigs and mice. Aqueous extract at dose of 500 mg/kg body weight exhibited significant anti-tussive property against coughing induced by ammonia and sulfur dioxide in mice and citric acid in guinea pig. For detection of anti-asthmatic activity of the extract, 2 mL of histamine (0.1%) and acetylcholine chloride (2%) in 50% ratio was sprayed for inhalation by guinea pigs. Frequency of delitescence of symptoms like convulsions and tumbling was significantly impacted by administration of aqueous extract. Results showed around 20% increase in delitescence of symptoms indicating anti-asthmatic effect of the extract (Jaliwala et al. 2011). Bairagi et al. (2021) have also mentioned anti-tussive properties of the leaf extract due to the presence of bioactive compounds.

13.5.7 Sedative, Anti-Depressant, and Anxiolytic Activities

With a large variety of phytoconstituents, *C. gigantea* has many therapeutic uses to offer. Aqueous leaf extract of the plant when administered to rats at dose of 20 mL/kg body weight showed anti-depressant activity (Rani et al. 2020). In another study on Swiss mice, ethanolic extract of *C. gigantea* induced sleep. Sedative and anxiolytic properties were confirmed by Hole cross test, Open field test, and Elevated plus maze (EPM) test (Khan et al. 2014).

In addition to this, *C. gigantea* has been found to be useful in causing vasodilation and hair growth (Mushir et al. 2016). Several therapeutic applications and studies related to them have been reviewed and cited by Bairagi et al. (2018). Some other pharmacological studies and therapeutic uses of *C. gigantea* are listed in Table 13.3.

13.6 TOXIC RESPONSE

There have been numerous studies on the therapeutic uses of *C. gigantea*, but the dose higher than the recommended dose or inappropriate administration can lead to severe toxicity. Nausea, vomiting, and diarrhea are some common symptoms of low dose toxicity, while higher dose can cause burning micturition, headache, and intestine weakness (Mushir et al. 2016). Pregnant women may suffer with certain issues leading to abortion, lungs and liver ailments (Mushir et al. 2016). Some of the constituents present in different parts of *Calotropis* are reported to be toxic. These toxic compounds are calotoxin, uscharin, calotropin, calotropage, etc. which make all parts of the plant toxic (Iyadurai et al. 2020). Even latex has also been found to contain such compounds which can induce toxic response. A case study reported cardiac toxicity in a patient due to oral ingestion of latex of *C. gigantea* (Iyadurai et al. 2020). It was administered initially to the subject by traditional medical practitioner for the treatment of envenomation by snake bite. It might be believed that latex of *Calotropis* can treat snake envenomation since it possesses a snake-repellent odor and used by snake charmers to repel snakes.

C. gigantea though has been used as a medicinal plant, the toxic aspect of its phytoconstituents cannot be ignored. Ingestion of latex caused severe burning and irritation in throat of the subject instantly. With the time, the subject developed cardiovascular failure and treated for severe *C. gigantea* poisoning (Iyadurai et al. 2020). Mechanism of *Calotropis*-induced cardiac toxicity involved Na+/K+ ATPase pump which was inhibited by cardenolides such as calotropin. Insects like monarch caterpillars which exclusively feed on its leaves have developed resistance toward cardenolides by possessing antisensitive Na+/K+ ATPase pump (Iyadurai et al. 2020).

Cytotoxicity study of *C. procera* on animals has revealed symptoms like arrhythmias and tachycardia in sheep within 4 hours of leaves ingestion (de Lima et al. 2011). Similar effect was found with *C. gigantea* for developing cardiogenic shock in the subject (Iyadurai et al. 2020). There is scanty of literature found on toxicity study of *C. gigantea* in contrast to its closest congeneric plant *C. procera* since the latter being more toxic. Nevertheless, the effects caused by *C. procera* can be related to *C. gigantea* for sharing a large profile of their constituents. Reddy (2019) observed that the most common symptom of oral ingestion of latex of *C. procera* at lower dose was abdominal pain. Higher dose of leaf extract and latex caused throat burning, stomach irritation, vertigo, diarrhea, and convulsions.

TABLE 13.3
Therapeutic Uses of Different Parts of *Calotropis gigantea*

Part of the Plant	Solvent	Therapeutic Use	Test Method	Reference
Root	Methanol	Antioxidant activity	2, 2-Diphenyl-1-picrylhydrazyl and florescence recovery after photobleaching method	Elakkiya and Prasanna (2012)
Root bark	Aqueous	Antidiarrheal activity	Charcoal-induced gut motility	Ali et al. (2010)
Aerial parts	Water:ethanol (50:50)	Antidiarrheal activity	Castor oil-induced diarrhea by enteropooling method	Chitme et al. (2004)
Latex and fruits	Ethanol	Antiviral	Cytopathic effect (CPE) inhibition assay	Parhira et al. (2014)
Fresh leaves		Molluscicidal activity		Lobo and Llagas (1991)
Leaves	Successive extraction: Petroleum ether, chloroform, ethyl acetate, n-butanol, ethanol, and distilled water	Anti-inflammatory activity	Albumin denaturation technique	Vijay et al. (2010)
	Methanol and chloroform	Hepatoprotective activity	Acetaminophen-induced hepatotoxicity models	Usmani and Kushwaha (2010)
Leaf, Flower	Chloroform	Hypoglycemic activity	Streptozotocin-induced diabetic rats	Rathod et al. (2011)
	Aqueous	Ovicidal activity	Inhibition of egg hatchability of *Helicoverpa armigera*	Prabhu et al. (2012)
Flower	Alcohol	Analgesic activity	Writhing test, Hot plate method	Pathak and Argal (2007)
Whole plant	Methanol	Antivenom and antinecrotic activity	*In vivo* neutralization of *Vipera russelli* venom in mice, *in vitro* studies	Chacko et al. (2012)

13.7 TRADITIONAL AND OTHER POTENTIAL USES

Potential uses of *C. gigantea* have been explored since ancient times. Traditionally, different parts of the plant were used in folk medicines. In India and some other parts of the world, this plant has been used as traditional medicine for the treatment of various ailments including fever, muscular spasm, rheumatism, dysentery, and asthma and as an expectorant (Chaudhuri and Malodia 2017). In combination with other plants, *C. gigantea* has been traditionally used for wound healing (Ali-Seyed and Ayesha 2020).

In Ayurveda, a formulation Loha rasayanam contains *C. gigantea* root bark as one of the ingredients and is used as antioxidant and anti-aging agent (Shree Nagindas Chaganlal Shah Rasavaidya 1921; Kadiyala et al. 2013). Unani traditional medicine Jauhar-e-kafoor contain latex of the plant and is used to treat piles, hemorrhoids and anorectal mass (Kadiyala et al. 2013). The plant has been mentioned in several books in Unani system which includes *Kitabul-Hashaish* and *Makhzan-al-Advia* (Dioscorides 78 AD; Najmul Ghani 1902 AD; Mushir et al. 2016). Ethnomedical uses of root of *C. gigantea* in Unani system have been found as astringent, tonic, expectorant, diaphoretic, anthelmintic, anti-inflammatory, sedative, wound healer, antidote, and digestive. In addition to other constituents, root bark has been found effective in ailments like diarrhea, epilepsy, amenorrhea, cholera, dysentery, toothache, elephantiasis, and fever (Najmul Ghani 1902 AD; Mushir et al. 2016).

In Siddha system, many formulations contain *C. gigantea* and parts thereof to treat different ailments. These medicines include Moolathirkku pugai, Muzhangal Vathathirku Poochu, Sannikku Pottanam, Thiriloga Centuram, Abraha Parpam, and Chukku Thailam (Kadiyala et al. 2013; TKDL 2022). Diseases such as vata, ozaena, sinusitis, headache, cough, bronchitis, leprosy, colic disease, gonorrhea, dysuria, sannipataja, ama, hernia, tuberculosis, phthisis, diarrhea, osteoarthritis of knee joint are known to be treated with Siddha medicines. Furthermore, it has been found that administration of same medicine with different routes produces different pharmacological effects (Kadiyala et al. 2013). Homeopathic treatment also utilizes this plant for curing different ailments (Bhardwaj and Misra 2018). In *Materia Medica*, its uses have been mentioned to treat elephantiasis, toothache, purging, and vomiting (Bhardwaj and Misra 2018; Gautam and Bhadauria 2009).

Besides, *C. gigantea* is commercially called as bowstring of India as it provides a durable fiber used for making carpets, ropes, sewing thread, and fishing nets (Oudhia 2001). Seeds provide floss which is utilized as stuffing material. Extracts obtained from different parts of plant can alter the productivity of several agricultural crops by impacting their process of germination and seedling vigor. Moisture binding capacity and soil nutrients also improve if the plant is used as green manure (Chaudhuri and Malodia 2017). Some of the other potential applications are as follows.

13.7.1 Nanoparticle Synthesis

Nanoparticles have been widely used nowadays for various high end applications (Jin-Chul et al. 2021; Siddiqi and Husen 2021; Husen 2022; Taghiyari et al., 2022; Husen and Siddiqi 2022). These are nano-range particles with at least one dimension in 1–100 nm. These particles exhibit characteristics different than their bulk material which can be exploited in different industries and sectors. Recently, Chaudhuri and Malodia (2017) synthesized zinc oxide (ZnO) nanoparticles from leaf extract of *C. gigantea*. Earlier, Vidya et al. (2013) used $ZnNO_3$ (zinc nitrate) for production of ZnO nanoparticles (30–35 nm) from leaf extract of *C. gigantea*. Green-synthesized nanoparticles have found applications such as nanofertilizers, nanonutrients, and nanopesticides to increase agricultural crop productivity (Chaudhuri and Malodia 2017). To cite a few, nanoparticles have been used in drug delivery and in manufacturing materials for different industries such as pharmaceuticals, micro-electronics, aerospace, and energy generation (Khan et al. 2019).

13.7.2 Biofuel Generation

Owing to relatively high oil content present in the seeds, *C. gigantea* is also studied as a potential source for biodiesel production (Phoo et al. 2014). Since the seeds also contain cardenolides which are toxic for many vertebrates, seed oil is considered to be inedible. This brings to *C. gigantea* the benefit of being a non-competitor to food supply and hence a sustainable source. Studies by Sundar Rao et al. (1983) showed that seeds of this plant contain 30.8% oil content by weight. Phoo et al. (2014) also confirmed the high oil content in the seeds which was found to be 33.3%. The major constituents that its seeds contain by weight were 36.3% linoleic acid followed by 30.3% oleic acid, 15.5% palmitic acid, and 10.5% stearic acid. Minor constituents were found to be 0.8% asclepic acid, 0.8% linolenic acid, 0.6% arachidic acid, 0.4% lignoceric acid, 0.3% palmitoleic acid, and 0.1% behenic acid (Phoo et al. 2014).

For biogas production, co-digestion of mixture of fresh and chopped leaves of *C. gigantea* with buffalo dung was studied (Shilpkar et al. 2007). A significant increase of 109.8% in biogas production was observed along with higher CO_2 content and fertilizer value (Shilpkar et al. 2007). Phoo et al. (2014) extracted the oil from *C. gigantea* seeds and produced biodiesel by alkali-catalyzed transesterification of free fatty acids (27.5%). All fuel properties except oxidation stability were found to conform four main standards, i.e., European Standard EN 14214, Philippine National Standard PNS2020:2003, American Society for Testing Materials ASTM D6751, and Japanese Automotive

Standards Organization JASO M360 (Al Sulaibi et al. 2020; Phoo et al. 2014). This indicated that biodiesel generated from seeds of *C. gigantea* is suitable for tropical countries so as to avoid poor cold flow properties (Phoo et al. 2014).

13.7.3 Fiber Source

Calotropis has been stated as a sustainable source of fiber for use as an adsorbent and building material. Hydrophobic and water-resistant properties of its fibers make them a potential adsorbent for oil and other hydrocarbons. Besides, the wood can be used for making roof-tops and in paper-making (Al Sulaibi et al. 2020). Since the cultivation of this plant does not require fertile land or special care, it can provide a sustainable alternate to fiber source.

13.7.4 Reclamation of Barren Land

This plant can grow in xeric conditions and abandoned wastelands where nothing else would grow without adequate culture and irrigation. This makes it a suitable choice for reclaiming drifting sands and bringing abandoned lands under tillage. Even the leaves can be used to increase the organic matter of poor and sandy soils by mulching and green manuring. It can also be used as soil binder or a hedge to protect the soil from eroding (PFAF 2022).

13.7.5 Esthetic and Ornamental Value

Flowers of *C. gigantea* are crown shaped, white or lavender, and beautiful which is also reflected in the name of the plant, i.e., *Calotropis*. The shrub has a religious significance in India and the flowers are generally offered in temples. Moreover, the flowers have long shelf life for which they have gained ornamental value and used in floral decoration (Khosravi and Kumar 2021). In India, *C. gigantea* is commonly grown outside the Shiv temples for its esthetic value.

13.8 CONCLUSION AND FUTURE REMARKS

The genus *Calotropis* is mainly recognized by the presence of its milky latex which contains a wide variety of bioactive compounds. Being used traditionally by Ayurveda, Siddha, and Unani systems since ancient times, *C. gigantea* is gaining interest in Allopathy as well. The therapeutic uses of this plant and its various parts have been explored in the treatment of several types of cancers and other ailments. Recently, molecular docking studies have found the constituents of this plant to be potential drug against COVID-19. But the therapeutic use of this medicinal plant has to be cautiously undertaken as the cardenolides present in it have shown to cause toxicity in many living forms including vertebrates. Non-therapeutic applications, however, have also been studied which indicate that it can be potentially used for nanoparticle synthesis, biofuel generation and alternate source of fiber for construction, papermaking, etc. Further exploration in therapeutics and non-therapeutics is expected to add on to the potential of this underutilized weed.

NOTES

Prabhjot Kaur, CSIR-Research Associate, Department of Biotechnology, Kurukshetra University, Kurukshetra, India
Amarjit Singh, Department of Botany, Guru Nanak Khalsa College, Yamuna Nagar, India
Jitender Sharma, Department of Biotechnology, Kurukshetra University, Kurukshetra, India

REFERENCES

Al Sulaibi, M. A., Thiemann, C., & Thiemann, T. 2020. Chemical constituents and uses of *Calotropis procera* and *Calotropis gigantean*—A review (Part I—The plants as material and energy resources). *Open Chemistry Journal* 7: 1–15.

Alafnan, A., Sridharagatta, S., Saleem, H., Khurshid, U., Alamri, A., Ansari, S. Y., Abidin, S. A. Z., Ansari, S. A., Alamri, A. S., Ahemad, N., & Anwar, S. 2021. Evaluation of the phytochemical, antioxidant, enzyme inhibition, and wound healing potential of *Calotropis gigantea* (L.) dryand: A source of a bioactive medicinal product. *Frontiers in Pharmacology* 12: 701369. doi: 10.3389/fphar.2021.701369

Alam, M. A., Habib, M. R., Nikkon, R., Rahman, M., & Karim, M. R. 2008. Antimicrobial activity of Akanda (*Calotropis gigantea* L.) on some pathogenic bacteria. *Bangladesh Journal of Scientific and Industrial Research* 43(3): 397–404.

Ali, R., Amin, K. M. Y., Ahmad, G., Wadud, A., & Jahan, N. 2010. Efficacy of post Beekhe Madar (*Calotropis gigantea*) root bark in experimentally induced diarrhoea. *Hippocratic Journal of Unani Medicine* 5(3): 1–8.

Ali, R., Jahan, N., & Amin, K. M. Y. 2012. Anti diarrhoeal activity of *Calotropis gigantea* root bark in experimental animals. *Journal of Research in Unani Medicine* 1(1): 1–5.

Ali-Seyed, M., & Ayesha, S. 2020. *Calotropis*—A multi-potential plant to humankind: Special focus on its wound healing efficacy. *Biocatalysis and Agricultural Biotechnology* 28: 101725.

Argal, A. & Sachan, R. 2009. Anthelmintic activity of *Calotropis gigantea* roots. *Journal of Pharmacy Research* 2(6):1085–6.

Ashwani, K. 1999. *Ayurvedic Medicines: Some Potential Plants for Medicine from India. A Meeting of the International Forum on Traditional Medicines*. Toyama Medical and Pharmaceutical University, Toyama, Japan.

Bairagi, S., Takarkhede, S., Giri, V., Gupta, C., Ghadge, A., Dwivedi, A., & Ghadi, P. 2021. Potential medicinal properties of *Calotropis gigantea* linn: A review. *World Journal of Advance Healthcare Research* 5(2): 71–6.

Bairagi, S. M., Ghule, P., & Gilhotra, R. 2018. Pharmacology of natural products: An recent approach on *Calotropis gigantea* and Calotropis procera. *Ars Pharmaceutica* 59(1): 37–44.

Bhardwaj, A., & Misra, K. 2018. Homeopathic remedies. In: Misra, K., Sharma, P., & Bhardwaj, A. (Eds.), *Management of High Altitude Pathophysiology*, Academic Press, London, England, pp. 217–29.

Bhattacharya, D., Sengupta, A., & Sinha, N. K. 1985. Phenolic hydroxyl ionization in calotropins from *Calotropis gigantea* latex. *Phytochemistry* 24(11): 2537–9.

Boller, T. 1986. Roles of proteolytic in interactions of plant with other organisms. In: Dalling, M. J. (Ed.), *Plant Proteolytic Enzymes*. CRC Press, Boca Raton, FL, pp. 67–96.

CABI. 2021. Invasive Species Compendium. Available online at: https://www.cabi.org/isc/datasheet/16848 (Accessed on 13 March 2022).

Chacko, N., Ibrahim, M., Shetty, P., & Shastry, C. S. 2012. Evaluation of antivenom activity of *Calotropis gigantea* plant extract against *Vipera russelli* snake venom. *International Journal of Pharmaceutical Sciences and Research* 3(7): 2272–9.

Chan, E.W.C., Wong, S.K. & Chan, H.T. 2016. Apocynaceae species with antiproliferative and/or antiplasmodial properties: A review of ten genera. *Journal of Integrative Medicine* 14(4): 269–84.

Chaudhuri, S. K., & Malodia, L. 2017. Biosynthesis of zinc oxide nanoparticles using leaf extract of *Calotropis gigantea*: Characterization and its evaluation on tree seedling growth in nursery stage. *Applied Nanoscience* 7(8): 501–12.

Chitme, H. R., Chandra, R., & Kaushik, S. 2004. Studies on anti-diarrhoeal activity of *Calotropis gigantea* R. Br. in experimental animals. *Journal of Pharmacy and Pharmaceutical Sciences* 7(1): 70–5.

Damodaran, B., Nagaraja, P., Jain, V., Wimalasiri, M. M. V., Sankolli, G., Kumar, G., & Prabhu, V. 2019. Phytochemical screening and evaluation of cytotoxic activity of *Calotropis gigantea* leaf extract on MCF7, HeLa, and A549 cancer cell lines. *Journal of Natural Science, Biology and Medicine* 10(2): 131–8.

de Lima, J. M., de Freitas, F. J. C., Amorim, R. N. L., Câmara, A. C. L., Batista, J. S., & Soto-Blanco, B. 2011. Clinical and pathological effects of Calotropis procera exposure in sheep and rats. *Toxicon* 57(1): 183–5.

De, S., Dutta, S., Das, D. C., & Das, M. 2019. In-vitro anthelmintic activity of Akanda (*Calotropis gigantea* L.) whole plant methanolic extract in Indian adult earthworm. *Journal of Phytopharmacology* 8(4): 152–4.

Deshmukh, P. T., Fernandes, J., Atul, A., & Toppo, E. 2009. Wound healing activity of *Calotropis gigantea* root bark in rats. *Journal of Ethnopharmacology* 125(1): 178–81.

Devi, L. S., & Gupta, P. 2000. Evaluation of some plant latices against *Heterodera cajani* on cowpea (*Vigna sinensis*). *National Academy Science Letters-India* 23: 65–7.

Dhivya, R., & Manimegalai, K. 2016. Phytochemical screening and analysis of active secondary metabolites present in the ethanolic extract of *Calotropis gigantea* leaves using GCMS technique. *World Journal of Pharmacy and Pharmaceutical Sciences* 5(10): 1510–23.

Dioscorides 78 A.D. Kitabul Hashaish. (Manuscripts). Patna Khuda Baksh library. V-II. 305, 1080.

Dutta, M., Nezam, M., Chowdhury, S., Rakib, A., Paul, A., Sami, S. A., Uddin, M., Rana, M., Hossain, S., Effendi, Y., & Idroes, R. 2021. Appraisals of the Bangladeshi medicinal plant *Calotropis gigantea* used by folk medicine practitioners in the management of COVID-19: A biochemical and computational approach. *Frontiers in Molecular Biosciences* 8: 625391.doi: 10.3389/fmolb.2021.625391

Elakkiya, P., & Prasanna, G. 2012. A study on phytochemical screening and in vitro antioxidant activity of *Calotropis gigantea* L. *International Journal of Pharm Tech Research* 4(4): 1428–31.

Gautam, A. K., & Bhadauria, R. 2009. Homeopathic flora of Bilaspur district of Himachal Pradesh, India: A preliminary survey. *Ethnobotanical Leaflets* 2009(1): 14.

GRIN-Global. 2020. National Genebank of Tunisia—Plant Germplasm: Taxonomy. Available online at: https://npgsweb.ars-grin.gov/gringlobal/taxon/taxonomydetail?id=8652 (Accessed on 3 November 2022).

Habib, M. R. 2011. Evaluation of antitumour activity of *Calotropis gigantea* L. root bark against Ehrlich ascites carcinoma in Swiss albino mice. *Asian Pacific Journal of Tropical Medicine* 4(10): 786–90.

Habib, M. R., & Karim, M. R. 2013. Effect of anhydrosophoradiol-3-acetate of *Calotropis gigantea* (Linn.) flower as antitumoric agent against Ehrlich's ascites carcinoma in mice. *Pharmacological Reports* 65(3): 761–7.

Husen, A. 2022. *Engineered Nanomaterials for Sustainable Agricultural Production, Soil Improvement and Stress Management*. Elsevier Inc., Cambridge, MA.

Husen, A., & Siddiqi, K. S. 2022. *Advances in Smart Nanomaterials and their Applications*. Elsevier Inc., Cambridge, MA.

Juncker, T., Schumacher, M., Dicato, M., & Diederich, M. 2009. UNBS1450 from *Calotropis procera* as a regulator of signaling pathways involved in proliferation and cell death. *Biochemical Pharmacology* 78(1):1–10.

Iyadurai, R., Gunasekaran, K., Jose, A. & Pitchaimuthu, K. 2020. Calotropis poisoning with severe cardiac toxicity A case report. *Journal of Family Medicine and Primary Care* 9(8): 4444.

Jaliwala, Y.A., Neha, C., Bhatt, N.K., Panda, P.K. & Mohanti, P.K. 2011. Pharmacological evaluation of anti-tussive, anti-asthmatic and expectorant activities of *Calotropis gigantea* R. Br. in experimental animals. *Journal of Pharmacy Research* 4(10): 3383–5.

Jin-Chul, K., Madhusudhan, A., & Husen, A. 2021. *Smart Nanomaterials in Biomedical Applications*. Springer Nature Switzerland AG, Cham.

Kadiyala, M., Ponnusankar, S. & Elango, K. 2013. *Calotropis gigantea* (L.) R. Br (Apocynaceae): A phytochemical and pharmacological review. *Journal of Ethnopharmacology* 150(1): 32–50.

Kaur, A., Batish, D. R., Kaur, S., & Chauhan, B. S. 2021. An overview of the characteristics and potential of Calotropis procera from botanical, ecological, and economic perspectives. *Frontiers in Plant Science* 12: 1188.

Khan, I., Saeed, K., & Khan, I. 2019. Nanoparticles: Properties, applications and toxicities. *Arabian Journal of Chemistry* 12(7): 908–31.

Khan, I. N., Sarker, M. M. I., & Ajrin, M. 2014. Sedative and anxiolytic effects of ethanolic extract of *Calotropis gigantea* (Asclepiadaceae) leaves. *Asian Pacific Journal of Tropical Biomedicine* 4: S400–S404.

Kharat, K. R., & Kharat, A. S. 2019. The *Calotropis gigantea* methanolic extract induces apoptosis in human breast carcinoma cells. *Iranian Journal of Medical Sciences* 44(6): 483–92.

Khosravi, Z., & Kumar, A. H. S. 2021. Pharmacognosy and pharmacology of *Calotropis gigantea* for discovery of anticancer therapeutics. *Pharmacognosy Magazine* 17(6): 123–8.

Kiruthiga, B., & Kumar, P. S. 2019. Disparities and similarities in the phytochemical Profiling of white *Calotropis gigantea* leaves amassed from six coastal locales across coleroon delta. *Plant Archives* 19(2): 3393–403.

Kumar, G., Karthik, L., & Rao, K. V. B. 2010. Antibacterial activity of aqueous extract of *Calotropis gigantea* leaves—An in vitro study. *International Journal of Pharmaceutical Sciences Review and Research* 4(2): 141–4.

Lobo, P. P. G., & Llagas, F. D. 1991. Evaluation of starflower (*Calotropis gigantea*) against golden apple snail (*Pomacea canaliculata*) in lowland transplanted rice. *Philippine Journal of Crop Science (Philippines)* 16: 103–7.

Misra, M. K., Mohanty, M. K. & Das, P. K. 1993. Studies on the method–ethnobotany of *Calotropis gigantea* and *C. procera*. *Ancient Science of Life* 13(1–2): 40.

Mushir, A., Jahan, N., & Ahmed, A. 2016. A review on phytochemical and biological properties of *Calotropis gigantea* (Linn.) R. Br. *Discovery Phytomedicine* 3(3): 15–21.

Najmul, G. 1902A.D. *Khazain ul Advia*. Idara Kitabul Shifa; (YNM), New Delhi, India, pp. 175–178.
National Wildlife Federation. 2022. Monarch Butterfly. Available online at: https://www.nwf.org/Educational-Resources/Wildlife-Guide/Invertebrates/Monarch-Butterfly#:~:text=As%20adults%2C%20monarchs%20feed%20on, their%20only%20caterpillar%20host%20plant (Accessed on 11 March 2022).
Negi, D., & Bisht, A. S. 2021. A review on brief study of *Calotropis gigantea* Linn. *Journal of Drug Delivery and Therapeutics* 11(5): 224–8.
Nguyen, K. D., Dang, P. H., Nguyen, H. X., Nguyen, M. T., Awale, S., & Nguyen, N. T. 2017. Phytochemical and cytotoxic studies on the leaves of *Calotropis gigantea. Bioorganic & Medicinal Chemistry Letters* 27(13): 2902–6.
Nguyen, M. T., Nguyen, K. D., Dang, P. H., Nguyen, H. X., Awale, S., & Nguyen, N. T. 2021. A new cytotoxic cardenolide from the roots of *Calotropis gigantea. Natural Product Research* 35(23): 5096–101.
NParks. 2022. *Calotropis Gigantea*. Available online at: https://www.nparks.gov.sg/florafaunaweb/flora/1/7/1774 (Accessed on 11 March 2022).
Oudhia, P. 2001. *Calotropis gigantea*: Useful Weed. Available online at: https://hort.purdue.edu/newcrop/CropFactSheets/Calotropis.html (Accessed on 11 March 2022).
Parhira, S., Yang, Z. F., Zhu, G. Y., Chen, Q. L., Zhou, B. X., Wang, Y. T., Liu, L., Bai, L. P., & Jiang, Z. H. 2014. In vitro anti-influenza virus activities of a new lignan glycoside from the latex of *Calotropis gigantea. PloS One* 9(8): 1–13.
Pathak, A. K., & Argal, A. 2007. Analgesic activity of *Calotropis gigantea* flower. *Fitoterapia* 78(1): 40–2.
PFAF. 2022. *Calotropis gigantean*—(L.) Dryand. ex W.T.Aiton. Available online at: https://pfaf.org/user/Plant.aspx?LatinName=Calotropis+gigantea (Accessed on 11 May 2022).
Phoo, Z. W. M. M., Razon, L. F., Knothe, G., Ilham, Z., Goembira, F., Madrazo, C. F., Roces, S. A, & Saka, S. 2014. Evaluation of Indian milkweed (*Calotropis gigantea*) seed oil as alternative feedstock for biodiesel. *Industrial Crops and Products* 54: 226–32.
Prabhu, S., Priyadharshini, P., & Veeravel, R. 2012. Effect of aqueous extracts of different plant parts of milkweed plant (*Calotropis gigantea* R. Br.) against ovicidal activity on Helicoverpa armigera (Hubner). *International Journal of Advanced Life Sciences* 2: 39–44.
Rajesh, R., Gowda, C. R., Nataraju, A., Dhananjaya, B. L., Kemparaju, K., & Vishwanath, B. S. 2005. Procoagulant activity of *Calotropis gigantea* latex associated with fibrin (ogen) olytic activity. *Toxicon* 46(1): 84–92.
Rani, K. N., Kalpana, D., Syamala, N., Divya, T., Vasu, G., & Rao, P. V. 2020. Evaluation of anti depressant activity of *Calotropis gigantea* by using experimental rats. *The Pharma Innovation Journal* 9(7): 251–4.
Rathod, N. R., Chitme, H. R., Irchhaiya, R., & Chandra, R. 2011. Hypoglycemic effect of *Calotropis gigantea* Linn. leaves and flowers in streptozotocin-induced diabetic rats. *Oman Medical Journal* 26(2): 104–8.
Reddy, C. Y. 2019. Clinical manifestations in Calotropis poisoning: A prospective study in Government General Hospital Nalgonda, India. *International Journal of Advances in Medicine* 6(4): 1314–6.
Sachin, S., Rani, A., Amresh, N., Rajadurai, M., & Sathyamurthy, B. 2018. Phytochemical studies on the methanolic extract of *Calotropis gigantea* leaves. *Indo American Journal of Pharmaceutical Sciences* 5(7): 6248–60.
Samy, P.R, & Chow, V.T.K. 2012. Pilot Study with regard to the Wound Healing Activity of Protein from Calotropis procera (Ait.) R. Br., *Evidence-Based Complementary and Alternative Medicine*, 2012: 294528, 11 pages. https://doi.org/10.1155/2012/294528
Seeka, C., & Sutthivaiyakit, S. 2010. Cytotoxic cardenolides from the leaves of *Calotropis gigantea. Chemical and Pharmaceutical Bulletin* 58(5): 725–8.
Sengupta, A., Bhattacharya, D., Pal, G., & Sinha, N. K. 1984. Comparative studies on calotropins DI and DII from the latex of *Calotropis gigantea. Archives of Biochemistry and Biophysics* 232(1): 17–25.
Senthil Kumar, S., Sivamani, P., Baskaran, C., & Jamal Mohamed, M. 2012. Evaluation of antimicrobial activity and phytochemical analysis of organic solvent extracts of *Calotropis gigantea. IOSR Journal of Pharmacy* 2(3): 389–94.
Sharma, A. D., & Kaur, I. 2020. Calotropin from milk of *Calotropis gigantea* a potent inhibitor of COVID 19 corona virus infection by Molecular docking studies. *Research & Reviews in Biotechnology & Biosciences* 7(2): 52–7.
Shilpkar, P., Shah, M., & Chaudhary, D. R. 2007. An alternate use of *Calotropis gigantea*: Biomethanation. *Current Science* 92(4): 435–7.
Shree Nagindas Chaganlal Shah Rasavaidya. 1921. *Bharat Bhaishajya Ratnakar* (vol 4). B. Jain Publishers, New Delhi, India.
Siddiqi, K. S., & Husen, A. 2021. Plant response to silver nanoparticles: A critical review. *Critical Reviews in Biotechnology* 42(7): 773–990. https://doi.org/10.1080/07388551.2021.1975091

Singh, M., & Javed, K. 2015. Comparative study of chemical composition of *Calotropis gigantea* flower, leaf and fruit essential oil. *European Chemical Bulletin* 4(10): 477–80.

Suchita, S., Sanchita, S., Mishra, R. M., & Mahesh Pal, S. 2014. Preliminary phytochemical screening of *Calotropis gigantea* leaf. *International Journal of Scientific and Research Publications* 4(2): 2–4.

Sundar Rao, K., Pantulu, A. J., & Lakshminarayana, G. 1983. Analysis of *Calotropis gigantea, Acacia caesia* and *Abelmoschus ficulneus* seeds. *Journal of the American Oil Chemists' Society* 60(7): 1259–61.

Taghiyari, H. R., Morrell, J. J., & Husen, A. 2022. *Emerging Nanomaterials: Opportunities and Challenges in Forestry Sectors*. Springer Nature Switzerland AG, Cham.

TKDL. 2022. Title Demo. Available online at: http://www.tkdl.res.in/tkdl/LangGerman/siddha/Utility/TitleDemo/S-Title.htm (Accessed on 11 February 2022).

USDA (United States Department of Agriculture). 2022. Monarch Butterfly. Available online at: https://www.fs.usda.gov/wildflowers/pollinators/pollinator-of-the-month/monarch_butterfly.shtml (Accessed on 3 November 2022).

Usha, K., Singh, B., Praseetha, P., Deepa, N., Agarwal, D. K., Agarwal, R., & Nagaraja, A. 2009. Antifungal activity of *Datura stramonium, Calotropis gigantea* and *Azadirachta indica* against *Fusarium mangiferae* and floral malformation in mango. *European Journal of Plant Pathology* 124(4): 637–57.

Usmani, S., & Kushwaha, P. 2010. Hepatoprotective activity of extracts of leaves of *Calotropis gigantea. Asian Journal of Pharmaceutical and Clinical Research* 3(3): 195–6.

Van Khang, P., Zhang, Z. G., Meng, Y. H., Guo, D. A., Liu, X., Hu, L. H., & Ma, L. 2014. Cardenolides from the bark of *Calotropis gigantea. Natural Product Research* 28(15): 1191–6.

Vidya, C., Hiremath, S., Chandraprabha, M.N., Antonyraj, M.L., Gopal, I.V., Jain, A. & Bansal, K., 2013. Green synthesis of ZnO nanoparticles by *Calotropis gigantea. International Journal of Current Engineering and Technology* 1(1): 118–20.

Vijay, J., Usman, M. R. M., Salunkhe, P. S., & Gagrani, M. B. 2010. Anti-inflammatory activity of *Calotropis gigantea* Linn. leaves extract on in-vitro models. *International Journal of Clinical Practice—Review and Research* 1(2): 1–5.

Wadhwani, B. D., Mali, D., Vyas, P., Nair, R., & Khandelwal, P. 2021. A review on phytochemical constituents and pharmacological potential of *Calotropis procera. RSC Advances* 11(57): 35854–78.

Winitchaikul, T., Sawong, S., Surangkul, D., Srikummool, M., Somran, J., Pekthong, D., Kamonlakorn, K., Nangngam, P., Parhira, S., & Srisawang, P. 2021. *Calotropis gigantea* stem bark extract induced apoptosis related to ROS and ATP production in colon cancer cells. *Plos One* 16(8): e0254392. doi: 10.1371/journal. pone.0254392

Wong, S. K., Lim, Y. Y., Abdullah, N. R., & Nordin, F. J. 2011. Assessment of antiproliferative and antiplasmodial activities of five selected Apocynaceae species. *BMC Complementary and Alternative Medicine* 11(1): 1–8.

You, H., Lei, M., Song, W., Chen, H., Meng, Y., Guo, D., Liu, X., & Hu, L. 2013. Cytotoxic cardenolides from the root bark of *Calotropis gigantea. Steroids* 78(10): 1029–34.

Zheng, Z., Zhou, Z., Zhang, Q., Zhou, X., Yang, J., Yang, M. R., Zhu, G. Y., Jiang, Z. H., Li, T., Lin, Q., & Bai, L. P. 2021. Non-classical cardenolides from *Calotropis gigantea* exhibit anticancer effect as HIF-1 inhibitors. *Bioorganic Chemistry* 109: 104740.

14 *Colchicum autumnale* (Naked Ladies)

Neha Sharma and Riddhima Singh

CONTENTS

14.1 INTRODUCTION

Colchicum autumnale is a perennial hysteranthous geophyte commonly called as the naked lady, autumn crocus, and white saffron belonging to the family Colchicaceae that includes the genus *Colchicum* L., which was previously placed in the Liliaceae s.l. Colchicaceae family members. It originally comes from Colchis or Colchis, a historic Georgian state and kingdom on the Black Sea's eastern shore, where flora were abundant (Imazio et al. 2009). The genus Colchicum (which includes Merendera Ramond and Bulbocodium) has 99 species, according to a recent overview (Persson 2007). The majority of these are restricted to tiny locations, mostly in the north. Western Asia connects Africa, southern Europe, and the Middle East to the Central Asian boundaries. *Colchium autumnale* is one of the few remaining species that can be found in the northern hemisphere, for example, in Europe and the United Kingdom (Persson 1993). While Stefanoff (1926) classified *Colchium alpinum* DC as a separate subgenus (Archicolchicum), Bowles (1952) identified it as one of *C. autumnale's* closest relatives, alongside *Colchium neapolitanum* Ten. and *Colchium corsicum* Baker. *C. autumnale* is closely related to the Iberian *Colchicum multiflorum* Brot. In the Indian market, there are two types: one that is sweet and the other that is bitter. *Colchicum luteum*, also known as Suranjan Talkh, is a bitter type of Colchicum and Shirin Suranja. Its corms have a reticulated look, have a bitter taste, and are smaller in size. Many attempts have been made to introduce *C. autumnale* into India, but with little success. Though *C. autumnale* does not grow in India, it can be substituted with Baker's *C. luteum* that is available (Chopra et al. 1958).

C. autumnale contains the alkaloid colchicine, which inhibits mitosis and so prevents cell division. Gout, rheumatism, arthritis, dropsy, gonorrhoea, and an enlarged prostate were all treated with corm and seeds by the ancient Greeks and Romans. Corms are toxic and dangerous, according to Dioscorides and Galen, because they taste good and make you want to eat more; the leaves, corm, and seeds are all poisonous. It was used in Babylonia to treat swelling poison in the limbs, scorpion stings, head and eye illnesses, and breast pain. Throughout history, the plant has been referred to as ephemera, Hermes' finger, pater noster, and tue-chiens. Modern phytonyms refer to the fabled land of Colchis, which is located near Armenia. According to Byzantine historians, Jacob Psychristus,

DOI: 10.1201/b23017-16

a fifth-century physician, was the first to use *C. autumnale* in the treatment of gout. The disease (podagra) was widespread at the period, with the major causes being excessive consumption of alcoholic beverages and food. It was dubbed as the "disease of kings" by Hippocrates in the fifth century BC, who named it "the unwalkable disease."

14.2 BOTANICAL DESCRIPTION

C. autumnale (Figure 14.1) is a geophyte with a 3.5–7.0 cm diameter corm that is irregularly shaped on one side and convex on the other (Heimann-Winawer 1919). At the bottom, the flattened beak-shaped feature on the corm side is more or less noticeable. Jaehn et al. (1985) named it protuberance. The prophyll develops as a leaf primordium and is only detectable at an early stage of development. Other authors, such as Irmisch (1850) and Kirchner et al. (1934), missed the prophyll and mistook the protuberance of the daughter corm for an extension of the mother corm. The internode between the prophyll and the first sheath leaf is compacted and acts as a place for the roots to be inserted. Two separate root types distinguish *C. autumnale's* unbranched root system. Plants of all ages can have either type of root. Spring root formation, on the other hand, is dependent on corm depth in the soil and does not occur in corms that reach a depth of 15–20 cm (Rimbach 1897; Franková et al. 2004). Spring roots have a tetrarch or pentarch structure, with a vascular bundle that is more than twice the diameter of September roots (Rimbach 1897). Although no contraction can be seen, slight undulation of the endodermis in the enlarged basal region indicates a contractile activity (Rimbach 1897). Two different sheath leaves formed along the basal section of the innovation branch of *C. autumnale* were described for the first time by Irmisch (1850). The initial leaf, also known as the ephemeral sheath leaf, is 1 cm long and fleshy. It resembles an extended hood and is made up of a thin white membrane with no vascular bundles (Irmisch 1850; Heimann-Winawer 1919; Wehsarg 1929). The first leaf splits open at the abaxial side of the top as the shoot elongates, indicating an addorsed position like a prophyll (Irmisch 1850; Jaehn et al. 1985). The whitish second sheath leaf has a green upper brim, is about ten times longer than the first, and ends in a little tip (Irmisch 1850). The leaves that form a lamina after the second sheath leaf are the green, photosynthetically active leaves. The second foliar leaf's insertion line is bent down in the median position. As a result, the reserve bud is somewhat below the general insertion line of the adjacent second foliar leaf. While the sheath leaves and the first foliar leaf are distichous, the second and subsequent foliar leaves adopt an alternative phyllotaxis with a divergence angle of 140°–150° (Irmisch 1850; Wehsarg 1929).

FIGURE 14.1 *Colchicum autumnale.*

Flowers are placed throughout the upper portion of the annual stem in an umbel-like (Schumann 1895; Wehsarg 1929) thick raceme (Persson 1993). Additional flowers emerge from the axils of the following plants. Scales, rather than foliar leaves, are the most common type of leaf (Wehsarg 1929). Six connate petaloid tepals form a narrow 20–35 cm long tube, and six thin tepal lobes with rounded tips make up the perianth. Tepal colour changes from white to light-purple or purple when floral tissue is exposed to sunshine (Wehsarg 1929). Tepal lobes can grow to be 40–50 mm long (Heimann-Winawer 1919). Flowers are hypogynous, with ovaries located just above the new corm, underground. Three connate carpels create one ovary, each with its own loculus and axillary placentation. The ovary matures into a 2–7.5 cm long and 1–2.5 cm wide beaked oval capsule (Muntean et al. 1979), subdivided into three septa. Seeds are roundish, brown, and have a huge white proliferation between micropyle and chalaza on one side of the seed. The endotesta is made up of two layers of cutinized cells that contain brown pigment, oil, and aleurone grains (Heimann-Winawer 1919). The exotesta is made up of two to three layers of thin-walled parenchymatic cells with starch granules, as well as one layer of epidermal cells (Heimann-Winawer 1919).

14.3 DISTRIBUTION

C. autumnale is essentially a Middle European species. Its native distribution stretches from Ireland and Northern England in the Northwest region over North of France, South of Belgium, and Central Germany to Southern Poland in the Northeast. Thereafter, it extends southwards over westernmost Ukraine and the Romanian Carpathians to Southeast Bulgaria. The southern distribution limit of the plant expands from Northern Greece and Southern Albania over the Northern Apennines in Italy to the Eastern Pyrenees, the Asturias, and the Sierra de Gredos in Spain, where the species apparently has its southwestern most occurrences (Akhtar and Siddiqui 2018).

C. autumnale's distribution range corresponds to ranges of the submediterranean/montane-middle European distribution range type 8 (Meusel et al. 1965). This particular range type consists mainly species of the temperate deciduous forest flora. The range of *C. autumnale* can be regarded as an anthropogenic extended representative, within the *Hippocrepis-comosa-subtype 8.9*.

However, there are certain limits to the distribution of *C. autumnale*. Within the continuous continental European distribution range, *C. autumnale* is missing to a great degree in a large gap in the Great Hungarian Plain, where precipitation is low and soils are partly salinized. Another distribution gap is located in the Bavarian/Bohemian Forest region of the Czech Republic, Austria and Germany, and in the Eastern Central Alps between the Ötztaler Alps and the Lower Tauern (Hendrych and Danilevsk 1985; Chytrý and Rafajová 2003; Niklfeld and Schratt-Ehrendorfer 1967; BIB 2010). A smaller gap is located in Central France in the Sologne (Dept. Loir-et-Cher, Arrond. Romorentin-Lantenay) due to the dominant occurrence of infertile, acidic soils in these regions, as *C. autumnale* prefers base rich, fertile soils (Butcher 1954; Oberdorfer 1994).

The species is popular in gardens and becomes naturalized when planted or discarded into suitable habitats, hence the naturalized occurrences are to be found along the northern distribution limit reaching from Inverness in Scotland and a single naturalized occurrence from the Shetlands (BSBI 2010), to coastal Norway (max. 62°N), the Uppsala region in Sweden (60°N), and Southwest Finland (Åland/Ahvenanmaa). From what has been noted, it is inferred that the frequency and abundance of *C. autumnale* in the northern lowland regions and at the southern distribution limit is lower than in central parts of its distribution area (D'Amato 1955; Pignatti 1982; Oberdorfer 1994).

It is also found growing on grassy slopes in the temperate regions like Himalayas, Afghanistan, and Turkestan.

It grows in wet meadows, woodland, clearings, and shady rocky habitats on non-calcareous substrates, and it may be found up to an altitude of 2,000 m (Thakur et al. 2021). Also, the local abundance of *C. autumnale* is determined by several elements such as type of grassland management (Winter and Kriechbaum 2009, stand height (Smith 2004), and location within the distribution area (Hengeveld and Haeck 1982). Its average cover percentages generally range from 1% to 5%, but it can also reach 6%–25% (Bassler et al. 1998). In moist meadows, the species can cover up to 50% of the total vegetation (Stebler and Schröter 1892; Braungart 1899). And in grasslands in England, *C. autumnale* is usually abundant or sub-dominant but never forms closed stands (Butcher 1954).

14.4 PHYTOCHEMICAL CONSTITUENTS

The major chemical constituents present in the *C. autumnale* plant are colchicine, colchicoside, and 3-Demethylcolchicine (Poutaraud and Girardin 2002; Ellington et al. 2003). Colchicine, the primary alkaloid of the plant, is an amorphous, yellow white alkaloid that dissolves rapidly in water, alcohol, or chloroform. It has been utilized in the treatment of gout (Rueffer and Zenk 1998) and has been investigated in many other conditions, including familial Mediterranean fever (FMF), cirrhosis and Sweet's syndrome, asthma, liver fibrosis, pericarditis with infusion, and Behcet's disease. Recently, allocolchicines (i.e., derivatives of colchicine) and other analogues have reported some intriguing effects in cancer cells. This is largely due to allocolchicine's ability to halt mitosis by inhibiting tubulin polymerization into microtubules hindering the progress of cells through the cell cycle and leading to the induction of apoptosis. This suppression of microtubule formation is especially useful in cancer cells that proliferate rapidly and uncontrollably (Larocque et al. 2014; Gründemann et al. 2015). The other alkaloids present in *C. autumnale* are cornigerine, 2- and 3-Demethyl-N-Deacetyl-N-Formylcolchicine, 2- and 3-Demethyldemecolcine, and 2-Demethylcolchifoline. Without the tropolone ring, no alkaloids are discovered in the plant (Suhail et al. 2017).

Phytochemical analysis of hydroalcoholic extract of *C. autumnale* flower and root showed the presence of flavonoids, tannins, terpenoids, and polyphenols, while carbohydrate and protein were absent (Suica-Bunghez et al. 2017).

A study reported the presence of six chemical constituents in the flowers of *C. autumnale* which was introduced from Europe to China, i.e., colchicine, 2-Demethyl colchicine, 2-Demethyl colchicine, 2-Demethyl β-Lumicolchicine, 2-Demethyl demecolcine, and β-Lumicolchicine. It is also reported in the study that the content of colchicine was low, while the content of 2-Demethyl colchicine was high in the flowers of *C. autumnale* (Hongping et al. 1999).

14.5 PHARMACOLOGICAL AND THERAPEUTIC ACTIVITIES OF *COLCHICUM AUTUMNALE*

Colchicum species, which have been widely used as a remedy for many years, are still useful in practice. According to numerous studies, this plant is also cultivated for use in the pharmaceutical industry. *C. autumnale* has been reported to be beneficial in all three types of arthritis: rheumatoid, osteoarthritis, and gouty arthritis. It has anti-inflammatory and anti-arthritic activity in all three forms of arthritis, which is comparable to the potent conventional anti-inflammatory drug Diclofenac sodium (Viquar et al. 2012). Colchicum species contains tropolone alkaloid which is used to treat FMF, gout, amyloidosis, cirrhosis, Behcet's illness, psoriasis, Hodgkin lymphoma, myeloid leukaemia, and skin malignancies. A group of pharma professors from Istanbul University, Turkey, is still working on species studies using Colchicum. The primary goal of colchicine synthesis and isolation research is to discover compounds that have similar properties but are less harmful. There is a lot of literature about synthesis methods, but none about colchicine. Because of their complicated method, they are cost effective.

The antioxidant activity of ethanolic extract and chloroform extract of *C. autumnale* corms was reported by Ahmad et al. (2010). Colchicum showed promising antioxidant properties phytochemically. The maximum activity, 91%, was discovered in the chloroform fraction, while the total range was 56%–91%. The enzymes acetylcholinesterase (AChE), butyrylcholinesterase (BChE), lipoxygenase, and urease were evaluated against the crude methanolic extract and several fractions of *Colchicum* species, including chloroform, ethyl acetate, n-Butanol, and aqueous. The crude methanolic extract of the species had considerable enzyme inhibitory activity against lipoxygenase, Colchicum BChE, and AChE, but no activity was observed against urease (Lin et al. 2020). Lin et al. (2020) investigated on colchicine's anticancer efficacy in gastric carcinoma (GC) cells. Colchicine reduces the growth of gastric cell lines (AGS and NCI-N87 cells) in a dose-dependent manner. Furthermore, the in vivo study revealed that colchicine treatment inhibits tumour growth in nude mice.

Using ELISA, Sevim et al. (2010) found that methanol extract of seeds and corms of species had anti-activity Alzheimer's against Colchicum AChE and BChE. Colchicine was found to have anti-Alzheimer activity in 20 patients by Aisen et al (2001). For 11 weeks, these patients were given hydroxychloroquine 200 mg twice daily, or hydroxychloroquine 200 mg twice daily plus colchicine 0.6 mg twice daily. These regimens show anti-Alzheimer action with no side effects, according to the findings. Furthermore, the antioxidant activity of extracts was tested for their scavenging activity using a 2,000 g/mL-1 concentration of 2,2-Diphenyl-1-Picrylhydrazyl (DPPH) by Sevim et al. (2010). The methanol extract has significantly higher antioxidant activity than the other extracts. Colchicine was approved for gout treatment in 1987 after a double-blind placebo-controlled research. Patients with chronic arthritis gout pain were studied. The patients were divided into two groups, one of which was received 0.6 mg of colchicine twice a day and another group was administered placebo for 3 months in a randomized double-blind research. In the first group, there was a considerable reduction in acute gout crisis. While 59 attacks were observed in 9 patientson placebo treatment, only 2 patients with 5 attacks observed in colchicine-treated patients. The constant treatment with colchicine was found to be more effective in preventing arthritis attacks (Goldstein and Schwabe 1974). Colchicine was used to treat and prevent serositis in a patient with FMF (Zemer et al. 1991). Three hundred and fifty people were studied for FMF. A daily dose of 1–2 mg/day colchicine was administered to youngsters between the ages of 6 and 13. As a result, none of the children developed amyloidosis.

Colchicine and silymarin were found to have hepatoprotective effects in rats by Favari and Pérez-Alvarez (1997). Colchicine and silymarin dosages were given to rats who had liver damage. Both compounds show hepatoprotective properties against chronic liver injury, according to the findings. Colchicine is a common treatment for Behcet's syndrome. Colchicine's efficacy was evaluated in a 2-year randomized, double-blind, placebo-controlled research involving a wider number of patients of both sexes. Colchicine was found to be significantly effective in both groups of patients for Behcet's syndrome.

14.6 TOXIC RESPONSE

C. autumnale, also known as autumn crocus, wild saffron, and naked lady, contains the antimitotic alkaloid colchicine, which prevents DNA synthesis and tubulin polymerization during mitosis. Following a latent period of 4–12 hours, clinical signs of colchicine poisoning appear in three stages. Peripheral leukocytosis, gastrointestinal symptoms with fluid loss, and hypovolemic shock characterize the first phase. Heart failure, arrhythmias, renal failure, liver injury, respiratory distress, coagulopathies, bone marrow depression, and neuromuscular involvement all occur during the second stage of intoxication, which lasts for 24–72 hours. The second phase might last for 5–7 days before the third phase, which is marked by leukocytosis and baldness. Colchicine is rapidly absorbed from the gastrointestinal system and processed largely by the liver in a first-order process after ingestion. Biliary excretion and enterohepatic recirculation are both substantial. Renal excretion accounts for only around 20% of unaltered colchicine elimination, while this percentage may be higher in patients with liver impairment. *C. autumnale* is highly toxic to Solenopsis Invasive Invicta Buren, often known as red imported fire ants. In 3 days, ants fed sugar water containing 5,000 mg/L bulb powder died 54.67% of the time, compared to 45.33% of the time when fed sugar water containing 50 mg/L colchicine. Due to the sub-lethal concentration, the total colony weight loss reached 44.63% and 58.73% after 15 days of feeding. *C. autumnale* could be used as a botanical pesticide to suppress RIFAs. Colchicine has been related to numerous deaths and intoxications. Colchicine is used to treat acute gouty arthritis, and the most prevalent cause of colchicine poisoning is a suicide colchicine tablet overdose.

14.7 TRADITIONAL AND OTHER POTENTIAL USES

Unani medicine is one of the oldest medical systems in use now, owing to its effective medications generated from animal, plant, and mineral resources. *C. autumnale* is an ancient herb that has been used in the Unani school of medicine for generations and has a good reputation for treating a variety of ailments (Kabeeruddin 1957; Khan 1957). This plant's chemical formulation is used as a digestive tonic, purgative, anti-inflammatory, appetizer, stomachic, deobstruent, anti-arthritic, and nervine tonic. It's also used to cure phlegm (Baitar 1985).

C. autumnale is primarily used to reduce pain and inflammation in acute gout and certain gouty illnesses, as well as to minimize their duration (Wallis 1985). It is also used to treat myeloid leukaemia (Ali 2012). It enhances skin, liver, and kidney secretions, as well as bile flow. It is a very effective therapy for ascites caused by liver illness. It works well as a purgative in cases of cerebral and hepatic congestion. It's also effective against genital diseases like gonorrhoea (Khory and Katarak 1993). Colchicine is extracted and used orally in tablet form to treat arthritis and FMF, while the corm and seeds are used to treat enlarged prostate, dropsy and gout, rheumatism, and arthritis (Bhattacharjee 2010). It's also used to relieve sciatica and boost aphrodisiac activity (Ghani 2010; Kabeeruddin 2007).

14.8 CONCLUSION

The health benefits of *C. autumnale* are numerous. The corm or rhizome, root, and seeds of the plant are utilized for therapeutic purposes. Sedative and resolvent qualities are also documented for this plant. According to published research, each of its chemical constituents (like the alkaloids such as colchicine, colchicoresin, and demecolcine) is associated with various pharmacological properties. It has been traditionally used for headache, arthralgia, gout, rheumatism, worm infestation, piles, chronic ulcers, constipation, diseases of the liver and spleen etc. Many reported studies have proved its efficacy in various ailments. Several preliminary studies have reported its effectiveness in different forms of arthritis. Therefore, *C. autumnale* should be studied further for high-throughput screening, scientific validation, and comparative effectiveness in various forms of arthritis such as osteoarthritis, rheumatoid arthritis, and gouty arthritis. However, more scientific studies and clinical trials are required on this plant to ensure its scientific validation for clinical use in patients. Since various studies have already reported its poisoning in humans as well as in cattle, now its optimal, effective and safe dose must be established. Prior to evaluating its pharmacological activity, it should also be standardized to assure uniformity in therapeutic efficacy.

NOTES

Neha Sharma, Amity Institute of Biotechnology, Amity University, Noida, India
Riddhima Singh, Amity Institute of Biotechnology, Amity University, Noida, India

REFERENCES

Ahmad, B. (2010). Antioxidant activity and phenolic compounds from *Colchicum luteum* Baker (Liliaceae). *African Journal of Biotechnology*, 9(35), 5762–5766.

Aisen, P. S., Marin, D. B., Brickman, A. M., Santoro, J., & Fusco, M. (2001). Pilot tolerability studies of hydroxychloroquine and colchicine in Alzheimer disease. *Alzheimer Disease & Associated Disorders*, 15(2), 96–101.Akhtar, S., & Siddiqui, M. Z. (2018). Suranjan shirin (*Colchicum autumnale*): A review of an anti-arthritic Unani drug. *The Pharma Innovation*, 7(12), 09–12.

Ali, M. (2012). Medicinal Plants Used in the Folk Medicines of Kammarpally Forest Range of Nizamabad Forest Division, Telangana State. In: Nandini Kumar (Ed.), *Textbook of Pharmacognosy*. Shri Balaji Printers, Delhi, India, pp. 340–342.

Baitar, I. (1985). *Kitab-ul-Jamia Al Mufridaat al AdviawalAghzia*. CCRUM, New Delhi, India, pp. 127–130.

Bassler, G., Karrer, G., & Lichtenecker, A. (1998). *Endbericht zum MAB-Pilotprojekt "Das Grünland im Berggebiet Österreichs. Teilprojekt 2: Grünlandtypen im Transekt von Oppenberg bis Tauplitz"*. Institut für Botanik, Universität für Bodenkultur, Wien, Austria.

Bhattacharjee, S. K. (2010). The Potential Role of Quercus Infectoria Gall Extract on Osteoblast Function and Bone Metabolism. In: M. Z. Siddiqui (Ed.), *Handbook of Medicinal Plants*. 4th ed., Pointer Publisher, Jaipur, India, pp. 108–109.

BIB. (2010). Botanischer Informationsknoten Bayern (last access 07.08.2010) http://www.bayernflora.de/.

Bowles, E. A. (1952). *Handbook of Crocus and Colchicum for Gardeners*. 2nd ed. The Bodley Head, London, UK.

Braungart, R. (1899). *Handbuch der rationellen Wiesen- und Weiden-Kultur und Futterverwendung, entwickelt und ausgestaltet auf den Grundlagen der modernen Fütterungslehre*. Theodor Ackermann, München, Germany.

BSBI. (2010). Botanical Society of the British Isles (last access 02.12.2010) http://www.bsbi.org.uk/.

Butcher, R. W. (1954). Biological flora of the british isles: *Colchicum autumnale* L. *Journal of Ecology*, 42, 249–257.

Chopra, R. N., Chopra, K. C., & Kapoor, L. D. (1958). Colchicum luteum. In: L. Handa (Ed.), *Indigenous Drugs of India*. 2nd ed., U.N. Dhur& Sons Pvt. Ltd, Calcutta, India, pp. 131–133.

Chytrý, M., & Rafajová, M. (2003). Czech National Phytosociological Database: Basic statistics of the available vegetation-plot data. *Preslia*, 75, 1–15.

D'Amato, F. (1955). Revisione Citosistematica del Genere Colchicum. I: C. Autumnale L., C. Lusitanum Brot. E C. Neapolitanum Ten. (con Tavole XVII-XXII, 11 Tabelle e 3 Figure nel testo). *Caryologia*, 7(2), 292–349.

Ellington, E., Bastida, J., Viladomat, F., & Codina, C. (2003). Supercritical carbon dioxide extraction of colchicine and related alkaloids from seeds of Colchicum autumnale L. *Phytochemical Analysis: An International Journal of Plant Chemical and Biochemical Techniques*, 14(3), 164–169.

Favari, L., & Pérez-Alvarez, V. (1997). Comparative effects of colchicine and silymarin on CCl4-chronic liver damage in rats. *Archives of Medical Research*, 28(1), 11–17.

Franková, L., Cibírová, K., Bóka, K., Gašparíková, O., & Pšenák, M. (2004). The role of the roots in the life strategy of Colchicum autumnale. *Biologia, Bratislava*, 59(13), 87–93.

Ghani, N. (2010). Organoleptic and physicochemical characterization of Süranjän Shérén (Colchicum autumnale), an anti-arthritic unani drug. In: Asim Ali Khan (Ed.), *Khzain-ul-Advia*. CCRUM, New Delhi, India, pp. 477–478.

Goldstein, R. C., & Schwabe, A. D. (1974). Prophylactic colchicine therapy in familial Mediterranean fever: A controlled, double-blind study. *Annals of Internal Medicine*, 81(6), 792–794.

Gründemann, C., Diegel, C., Sauer, B., Garcia-Käufer, M., & Huber, R. (2015). Immunomodulatory effects of preparations from Anthroposophical Medicine for parenteral use. *BMC Complementary and Alternative Medicine*, 15(1), 1–8.

Heimann-Winawer, P. (1919). *BeiträgezurEmbryologie von Colchicum autumnale L.* Vol. 21, Speyer & Kaerner, Speyer, Germany.

Hendrych, R., & Danilevsky, M. L. (1985). *Karpatische Migrationen und Florenbeziehungen in den tschechischen Ländern der Tschechoslowakei*. Univerzita Karlova, Czechia.

Hengeveld, R., & Haeck, J. (1982). The distribution of abundance. I. Measurements. *Journal of Biogeography*, 9(4), 303–316.

Hongping, H., Fuchu, L., Lin, H., & Hongyou, Z. (1999). Alkaloids from the flowers of Colchicum autumnale. *Acta Botanica Yunnanica*, 21(3), 364–368.

Imazio, M., Brucato, A., Trinchero, R., Spodick, D., & Adler, Y. (2009). Colchicine for pericarditis: Hype or hope. *European Heart Journal*, 30(5), 532–539.

Irmisch, T. (1850). Zur Morphologie der monokotylen Knollen-und Zwiebelgewächse. *Botanische Zeitung*, 8(12), 286.

Jaehn, F., Pfirsch, E., & Roux, J. (1985). Zur Architektur des Jahressprosses der Herbstzeitlose (Colchicum autumnale L.). *Beiträge zur Biologie der Pflanzen*, 60, 303–311.

Kabeeruddin, H. M. (1957). Majoon Suranjan: A potent unani formulation for arthritis. In: Jamal Akhtar (Ed.), *Al Qarabadeen*. Malik Son's Publishers, Faisalbad, Pakistan, p. 413.

Kabeeruddin, H. M. (2007). Majoon Halaila. In: Khavasul Adviya (Ed.), *Makhzan-ul-Mufridat*. Idara Kitab-ul-Shifa, New Delhi, India, p. 270.

Khan, M. A. (1957). Toxicity effects of autumn crocus (*Colchicum autumnale*). In: Qarabadeen-e-Azam (Ed.), *Qarabadeen-e-Azam*. Aijaz Publishing House, New Delhi, India, p. 601.

Khory, R. N., & Katarak, N. N. (1993). *Materia Medica of India and Therapeutics*. 3rd Reprint ed. Neeraj Publishing House, Delhi.

Kirchner, O., Loew, E., & Schrter, C. (1934). *Lebensformen der BltenpflanzenMitteleuropas, Bd. 1, Abt. 3*. Eugen Ulmer, Stuttgart, Germany.

Larocque, K., Ovadje, P., Djurdjevic, S., Mehdi, M., Green, J., & Pandey, S. (2014). Novel analogue of colchicine induces selective pro-death autophagy and necrosis in human cancer cells. *PLoS One*, 9(1), e87064.

Lin, S., Qin, D., Zhang, Y., Zheng, Q., Yang, L., Cheng, D., & Zhang, Z. (2020). Toxicity and sublethal effects of autumn crocus (Colchicum autumnale) bulb powder on red imported fire ants (Solenopsisinvicta). *Toxins*, 12(11), 731.

Meusel, H., Jäger, E. J., & Weinert, E. (1965). *Vergleichende chorologie der zentraleuropaischen flora.* G. Fischer.

Muntean, L. S., Salontai, A., Botez, C., & Tamas, M. (1979). Contribution to the biological study of Colchicum autumnale L. *NotulaeBotanicaeHortiAgrobotanici Cluj-Napoca*, 10, 81–88.

Niklfeld, H., & Schratt-Ehrendorfer, L. (Eds.) (1967). *Unpublizierte Daten des Projektes "Floristische Kartierung Österreich".*

Oberdorfer, E. (1994). *Pflanzensoziologische Exkursionsflora. 7., überarb. und erg. Aufl., Stuttgart (Ulmer).* Eugen Ulmer, Stuttgart.

Persson, K. (1993). Reproductive Strategies and Evolution in Colchicum. In Proc. 5th OPTIMA Meeting, pp. 394–414.

Persson, K. (2007). Nomenclatural synopsis of the genus Colchicum (Colchicaceae), with some new species and combinations. *Botanische Jahrbücherfür Systematik, Pflanzengeschichte und Pflanzengeographie*, 127, 165–242.

Pignatti, S. (1982). Flora d Italia vol. primo. Bologna. *Edagricole*, 1, 3.

Poutaraud, A., & Girardin, P. (2002). Alkaloids in meadow saffron, *Colchicum autumnale* L. *Journal of Herbs, Spices & Medicinal Plants*, 9(1), 63–79.

Rimbach, A. (1897). BiologischeBeobachtungen an Colchicum autumnale. *Berichte der Deutschen Botanischen Gesellschaft,* 15, 298–303.

Rueffer, M., & Zenk, M. H. (1998). Microsome-mediated transformation of O-methylandrocymbine to demecolcine and colchicine. *FEBS Letters*, 438(1–2), 111–113.

Schumann, K. (1895). In E. P. Nat.Pfl.sam. IV/2.

Sevim, D., Senol, F. S., Budakoglu, E., Orhan, I. E., Sener, B., & Kaya, E. (2010). Studies on anticholinesterase and antioxidant effects of samples from Colchicum L. genus of Turkish origin. *FABAD Journal of Pharmaceutical Sciences*, 35, 195–201.

Smith, R. J. (2004). Conservation Biology of Colchicum Autumnale L. and Campanula Trachelium L. in the Nore Valley, Southeast Ireland (Doctoral Dissertation, Trinity College Dublin).

Stebler, F. G., & Schröter, C. (1892). *Beiträge zur Kenntnis der Matten und Weiden der Schweiz.* Eidgenössisches Volkswirtschaftsdepartement. Wyss, Bern.

Stefanoff, B. (1926). Monographiïanaroda Colchicum L. *Sb BlgarAkadNaukKniga*, 22, 3–102.

Suhail, S., Shakir Jamil, S., & Jilani, S. (2017). Phytochemical and pharmacological review of Suranjan Shireen (Colchicum autumnale). *Indo American Journal of Pharmaceutical Research*, 7(04), 8492–8496.

Suica-Bunghez, I. R., Ion, R. M., Teodorescu, S., Sorescu, A. A., Stirbescu, R. M., & Stirbescu, N. M. (2017). Fitochemical and antioxidant characterization of Autumn Crocus (Colchicum autumnale) flowers and roots plant extracts. *Journal of Science and Arts*, 17(3), 539–546.

Thakur, S., Chaudhary, G., & Kaurav, H. (2021). Colchicum autumnale (Suranjan): A cytotoxic plant with anti-arthritis properties. *Advance Pharmaceutical Journal*, 6(3), 80–86.

Viquar, S., Amin, K. M. Y., & Yunus, M. N. (2012). Comparative AntiArthritic Profile of Some Unani Drugs, M.D.U. Thesis, Dept. of IlmulAdvia, A.M.U, Aligarh.

Wallis, T. E. (1985). Antirheumatics. In: Hashim Amir Ali (Ed.), *Textbook of Pharmacognosy*. 5th ed., CBS Publishers and Distributors, Shahdara, Delhi, pp. 398–400.

Wehsarg, O. (1929). *Die Verbreitung und Bekämpfung der Ackerunkräuter in Deutschland. Die Bekämpfung des Unkrautes Siebzehntes Stück, Band II: Einzelunkräuter, ihr Vorkommen und ihre Bekämpfung, Lieferung III: Herbstzeitlose und Weißer Germer.* Berlin, Deutsche Landwirtschaftsgesellschaft.

Winter, S., & Kriechbaum, M. (2009). Demographische Untersuchungen an Colchicum autumnale im Rahmen eines angewandten Naturschutzprojektes. *Sauteria* 18, 277–298.

Zemer, D., Livneh, A., Danon, Y. L., Pras, M., & Sohar, E. (1991). Long-term colchicine treatment in children with familial mediterranean fever. *Arthritis & Rheumatism: Official Journal of the American College of Rheumatology*, 34(8), 973–977.

15 *Conium maculatum* (Hemlock or Poison Hemlock)

Upagya Gyaneshwari, Akhilesh Kumar Singh, Priti Pal, and Brijesh Pandey

CONTENTS

15.1 INTRODUCTION

Conium maculatum L. (*C. maculatum*; Figure 15.1), sometimes known as poison hemlock ($2n=22$), is a biennial herbaceous plant. It can also be cultivated as a short-lived perennial or a wintry annual. The sturdy, straight, and branching (mostly upward) stem can reach a height of 1–2 m. The most essential key features for species identification are smooth texture and pale green (with purple dots) stems. The terete or angled basal section is 4–8 mm thick. The stem's cross-section is vacuous or seldom solid, other than at the nodes, where it exhibits thin, shallow vertical ridges. The inflorescences are enormous, open compound umbels measuring 4–6 cm in diameter. The topmost terminal inflorescence is the first to blossom. The five petals of the poison hemlock are white and incurved. The fruit is an oblong to spherical schizocarp with two grey or brown mericarps that create a complex (seeds). The seeds are narrow circular, measuring 1.2–2 mm in width and 2–3 mm in length. These weigh roughly 0.5 mg and have a somewhat extended apical and a highly convex dorsal side with five visible wave ridges from topmost to lowermost. It has a tall, pale, unbranched taproot that is 5–11 mm thick. From the crown, the leaves form a rosette and alternate on the stem. The petiole base is smaller on upper leaves and extends to sheath the stem (Holm et al. 1997; Fröberg 2010). The little white blooms are borne in umbrella-like flat clusters ("umbelliferae"). Root is tuberous and white. This plant blooms in the spring and bears ripe fruit in the summer (Vetter 2004; De Landoni 1990). *C. maculatum* requires wet and nitrogen-rich soils. A single species may generate between 35,000 and 40,000 seeds, which fall nearer the original plant but can also be distributed by water and animals (Mitich 1998).

C. maculatum is a popular nitrophile weed species that belongs to the Apiaceae (previously Umbelliferae) family. Amongst higher plants, it is the most hazardous. Coniine, N-Methyl-coniine, conhydrine, pseudoconhydrine, and γ-Coniceine, which are piperidine alkaloids, are produced by cyclizing an eight-carbon chain derived from four acetate molecules. It's worth mentioning that γ-Coniceine is the precursor of the other hemlock alkaloids. All vegetative tissues, flowers, and

DOI: 10.1201/b23017-17

FIGURE 15.1 *Conium maculatum* plant.

fruits contain these alkaloids. The concentrations of various alkaloids (both absolute and relative) are biologically, genetically, and spatiotemporally controlled (Vetter 2004). *C. maculatum* is found in continental Europe, southwest Asia, and Africa and the Middle East. It has expanded to nearby places such as various regions of America, China, Southern Africa, Australia, and New Zealand from its native habitat. *C. maculatum* is an invasive and pioneer plant that competes with grazing and crops while also expanding on natural species. It is also a major health risk to almost all cattle and people (Hotti and Rischer 2017).

15.2 DISTRIBUTION

C. maculatum comes from a family of very widespread, extensively distributed weed species, thus its development (specifically, the biological conditions) is a key research topic. *C. maculatum* seeds are spread in north-central Kentucky (USA) from the middle of September until the middle of February. As per the report of Baskin and Baskin (1990), 40%–85% of newly produced seeds were morphologically dormant. A wet substrate, a photoperiod of 14 hours of light, and alternate temperature phases of 30°C/15°C are all that is required for proper germination. *C. maculatum* is a resistant species that grows in humid weather. It behaves as pioneer species, colonizing the damaged areas. *C. maculatum* is said to be the European species that has adapted to America and Asia. It now has a global distribution, primarily in temperate climates below 5,000 feet in elevation (De Landoni 1990). It has been found in a variety of places, including wet ground, hedgerows, stream and river banks, roadsides, woods, pastures, meadows, waste ground, coarse areas, and everywhere with enough moisture (Frank et al. 1995; Di Napoli et al. 2019). Despite the fact that the plant is reported to be a prevalent weed throughout Europe, America, Africa, and Australia (Mekkonen 1994; Al-Snafi 2016), it was originally imported to the United States from Europe as a garden/ornamental plant. The establishment of the plant in other nations, such as Norway, is a result of grain transportation (Kielland and Anders 1998). The plant is one of Hungary's most prevalent weeds and dangerous plants, with poisonings mostly occurring in the spring and summer. This herb is widespread and thrives in damp waste areas that aren't maintained through cultivation or pruning. As a result, dense populations can be found along ditch banks, fence rows, low-lying rocky outcrops, and wooded regions (De Landoni 1990). Grassland, forest margins, riparian habitats, freshwater wetlands, waste ground, disturbed areas, field margins, and fallows are just a few of the habitats where *C. maculatum* thrives. Infestations are very short lived, lasting only 1–2 years. This plant prefers moist, fertile soils and avoids acidic soils, and direct sunlight. *C. maculatum* is expected to be more resistant of heavy metal-contaminated soils (arsenic, cadmium, and lead) than some native species in the United States, which may help it compete in such a circumstance (Centre for Agriculture and Bioscience International 2022; Pitcher 1989; Lopez et al. 1999).

15.3 PHYTOCHEMICAL CONSTITUENTS

The presence of various piperidine alkaloids (Figure 15.2) such as coniine, N-Methyl-coniine, conhydrine, pseudoconhydrine, and γ-Coniceine, as well as some additional hemlock alkaloids precursors, makes *C. maculatum* one of the most dangerous and lethal plants (Vetter 2004; Boskabadi et al. 2021). Table 15.1 presents the physicochemical properties, toxic and therapeutic effects of different types of alkaloids present in *C. maculatum.* These alkaloids, notably coniine, are a discriminatory agonist for nicotinic-type acetylcholine receptors (nAChRs) that bind to the receptors and disturb the activities of the central nervous system. nAChRs may be found throughout the body, including the central and sympathetic nervous systems, neuromuscular junctions, and the adrenal medulla (Mondal et al. 2014; Boskabadi et al. 2021). Coniine, γ-Coniceine, and N-Methyl coniine show teratogenic effect in humans and animals due to their molecular structure. These alkaloids have a side chain containing propyl or bulkier group. Since coniceine is more toxic than coniine and 2-Propylpyridine is non-teratogenic, partial unsaturation appears to enhance toxicity. Aromatization of the ring, on the other hand, appears to lower toxicity (Lopez et al. 1999).

Piperidine alkaloids have a piperidine nucleus, which is a six-membered saturated heterocyclic ring (Hotti and Rischer 2017). *Nicotiana, Sedum, Lobelia, Conium, Pinus, Punica, Duboisia, Hydrangea, Withania, Carica, Prosopis, Dichroa, Cassia, Ammodendron, Genista, Lupinus, Liparia,* and *Collidium* are only a few of the genera that contain piperidine alkaloids. Outside of the plant kingdom, fire ants (*Solenopsis* sp.) also carry simple piperidine alkaloids. The presence of

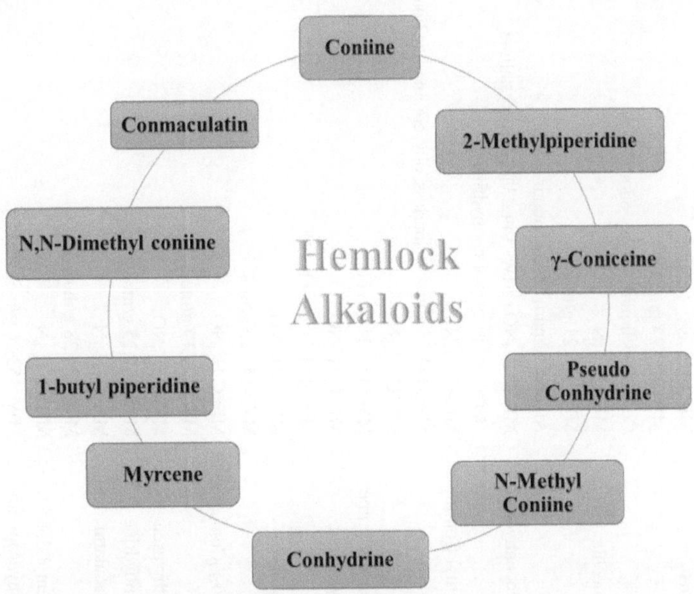

FIGURE 15.2 Different alkaloids of *Conium maculatum.*

TABLE 15.1
Properties and Effects of Different Alkaloids of *Conium maculatum*

Alkaloid	Physiochemical Property	Toxic Effect	Therapeutic Activities	References
2-Methylpiperidine	MF: $C_6H_{13}N$ MW: 99.17 g/mol Colourless to yellow liquid; Pepper-like aroma	Toxic effect on rabbit	Used as a food additive, as a flavouring agent	Hotti and Rischer (2017), Dionisio et al. (2018), pubchem.ncbi.in.
Coniine	MF: $C_8H_{17}N$ MW: 127.23 g/mol Colourless liquid; Mousy odour	Fatal for human, teratogenic	Active ingredient in products for pain and fatigue relief	The Merck Index (1976); Panter et al. (2017), Hotti and Rischer (2017), pubchem.ncbi.in.
γ-Coniceine	MF: $C_8H_{15}N$ MW: 125.21 g/mol Alkaline liquid; Mousy odour	Teratogenic	Not reported	Lopez et al. (1999), Lee et al. (2013), Hotti and Rischer (2017), pubchem.ncbi.in.
Pseudoconhydrine	MF: $C_8H_{18}NO+$ MW: 144.23 g/mol; crystalline alkaloid; isomer of conhydrine	Poisonous	Not reported	Hotti and Rischer (2017), Merriam-Webster (2022), pubchem.ncbi.in.
N-Methyl Coniine	MF: $C_9H_{19}N$ MW: 141.25 g/mol; colourless, oily, coniine-like liquid	Teratogenic, Neurotoxin	Not reported	Lopez et al. (1999), Schep et al. (2009), Hotti and Rischer (2017), pubchem.ncbi.in.
N,N-Dimethylconiine	MF: $C_{10}H_{22}N$ MW: 156.24 g/mol	Not reported	Not reported	Hotti and Rischer (2017), pubchem.ncbi.in.
Conhydrinone	MF: $C_8H_{15}NO$ MW: 141.21 g/mol	Not reported	Not reported	Hotti and Rischer (2017)
Conhydrine	MF: $C_8H_{17}NO$ MW: 143.23 g/mol	Acute toxic; irritant	Not reported	Hotti and Rischer (2017), pubchem.ncbi.in.
1′-Oxo-γ-Coniceine	MF: $C_8H_{13}NO$ MW: 139.19 g/mol	Not reported	Not reported	Hotti and Rischer (2017), pubchem.ncbi.in.
N-Methylpseudoconhydrine	MF: $C_9H_{19}NO$ MW: 157.25 g/mol	Not reported	Not reported	Hotti and Rischer (2017), pubchem.ncbi.in.
Conmaculatin	MF: $C_{10}H_{21}N$ MW: 155.28 g/mol	Lethal at higher dose	Antinociceptive	Radulovic et al. (2012), Hotti and Rischer (2017), pubchem.ncbi.in.
2-n-Pentyl-3,4,5,6-Tetrahydropyridine	MF: $C_{10}H_{19}N$ MW: 153.26 g/mol	Not reported	Not reported	Hotti and Rischer (2017), pubchem.ncbi.in.
5-Hydroxy-2-n-Pentylpiperidine	MF: $C_{10}H_{21}NO$ MW: 171.28 g/mol	Not reported	Not reported	Hotti and Rischer (2017), pubchem.ncbi.in.
1-Butyl piperidine	MF: $C_9H_{19}N$ MW: 141.25 g/mol	Not reported	Antimicrobial agent	Di Napoli et al. (2019), pubchem.ncbi.in.
Myrcene	MF: $C_{10}H_{16}$ MW: 136.23 g/mol; Yellow oily liquid with a pleasant odour	Mutagenicity, Genotoxicity, Reproductive toxicology	Antimicrobial agent, antinociceptive, anti-inflammatory, analgesic	NTP (2010), Di Napoli et al. (2019), Surendran et al. (2021), pubchem.ncbi.in.

hemlock alkaloids in plants is limited to some genera that are not related (Jones et al. 1982; Panter et al. 1988). Piperidine alkaloids are biosynthetically produced from lysine, acetate, or mevalonate as a precursor. The piperidine nucleus formed after nitrogen is added at some point in the route. Acetate-derived piperidine alkaloids are discovered in *C. maculatum* which are generated through the cyclization of an eight-carbon chain derived from four acetate units (Vetter 2004).

The quantities and relative proportions of distinct Conium alkaloids appear to be influenced by a variety of circumstances (heat, humidity, duration, and age of the plant). During the rainy season, c-Coniceine seems to be the primary alkaloid component, while coniine is prominent during the dry season. During the ripening period of the fruit, the c-Coniceine content exhibited down fall, while the coniine content exhibited upward trend dramatically in an old study (Fairbairn and Challen 1959). The proportional concentration of alkaloids varies depending on the stage of plant development and genotype. Cromwell (1956) reported little amount of total alkaloid, mostly c-Coniceine in the seedlings. C-Coniceine was the most abundant alkaloid in the leaves of new growth (i.e. during the active growth stage), with conhydrine levels being low. In the active stage, the roots contained minimal levels of alkaloids, while in the dormant stage, a higher proportion of c-Coniceine and coniine was found than in the roots of young plants with substantial metabolic activity. The distribution of alkaloids changes dramatically during flowering, pollination, and fertilization stages. The quantity of c-Coniceine in the body is inversely related to the amount of coniine in the body. When the fruits are mature, the primary component is N-Methylconiine. Coniine and c-Coniceine concentrations exhibit diurnal variations, with inverse relation (Fairbairn and Suwal 1961). In the seeds and flowers of *C. maculatum*, coniine is found in the maximum concentration as nicotinic alkaloids. The blood–brain barrier, the placenta, and breast milk all carry coniine and other alkaloids throughout the body. These alkaloids are acetylcholine receptor agonists of the nicotinic type (nAChRs). nAChRs can be found in the central and autonomic nervous systems, neuromuscular junctions, and the adrenal medulla (Lopez et al. 1999; Vetter 2004; Schep et al. 2009). The hydro-distilled essential oils of *C. maculatum* leaves and inflorescences had diverse chemical profiles, with 1-Butyl piperidine and myrcene in the inflorescence and predominantly (E)-Caryophyllene in the leaves (Di Napoli et al. 2019). GC-MS was used in the final phase of development to identify 24 distinct steroids as the primary chemical elements of root extracts of this plant species: Ergosta-5,7,9(11), 22-Tetraen-3-ol, Ergosta-5,8,22-Trien-3-ol, Ergosta-5,7,22-Trien-3-ol, Ergosta-5,7,22-Trien-3-ol, 5-Ergosta-7,22-Dien-3-ol, Ergost-5-En-3-ol, Stigmasta-5,22-Dien-3 5α-Stigmasta-22-Ene-3,6-Dione, 3-Hydroxystigmast-5-En-7-One, 5-Stigmastane-3,6-Dione Stigmast-4-En-3-One, Stigmasta-4,6-Dien-3-One, Stigmasta-4,22-Diene-3,6-Dione, Stigmast-4-Ene-3,6-Dione, 5α-Stigmasta-22-Ene-3,6-Dione, 3Hydroxystigmast-5-En-7-One, and 5-Stigmastane-3,6-Dione (Al-Snafi 2016). The acyclic monoterpenes (Z)-Ocimene and (E)-Ocimene were found in seed oil, whereas (Z)-Ocimene and myrcene were revealed as the second and third most abundant components in the inflorescence oil, respectively (Al-Snafi 2016). The roots of *C. maculatum* included five furocoumarins, two prenylated coumarins, two aliphatic C17-Polyacetylenes, and the phenylpropanoid elemicin. Falcarindiol, xanthotoxin, and isopimpinellin had the greatest amounts, while elemicin had the lowest (Chizzola and Lohwasser 2020). Aliphatic C17-Polyacetylenes of the falcarinol type have antibacterial, antimycobacterial, and antifungal effects, as well as anti-inflammatory and anti-platelet aggregatory capabilities. Falcarinol has also been shown to have neuritogenic and neuroprotective properties (Christensen 2011). Falcarindiol has a strong GABA-receptor modulating effect (Wyrembek et al. 2012). More importantly, falcarindiol and falcarinol have been shown to have anticancer properties. These compounds were able to reduce the number of neoplastic lesions as well as the pace of polyp development in the rat stomach. It suggests that these chemicals may help to prevent colorectal cancer from developing (Kobaek-Larsen et al. 2017). Because the roots of *C. maculatum* do not produce alkaloids, there is no or very little toxicity associated with these plant components. Piperidine alkaloids found in the aerial sections of *C. maculatum* make it a particularly hazardous plant. Alkaloids could not be discovered in diverse tap root samples in a first try (Chizzola and Lohwasser 2020).

15.4 PHARMACOLOGICAL STUDIES

The plant was utilized by Greek and Arabic physicians to treat indolent tumours, swellings, and joint problems. Toxic hemlock juice was blended with seeds of betony (*Stachys officinalis*) and fennel to treat a mad dog bite (rabies). Religious cults/sects employed roasted roots to treat gout pains in the 1400s and 1500s. Strychnine and other deadly toxins have been treated using the plant as a last resort antidote (Daugherty 1995). From the sixth decade of the eighteenth century onwards, the herb was used to treat malignant ulcers. Hemlock tinctures and extracts have been used as a sedative and analgesic. Asthma, epilepsy, tetanus, whooping cough, angina, chore, and stomach aches were all treated with its antispasmodic properties. Poison hemlock has been used as a powder, plaster, and poultice in Finnish folk medicine to treat stiffened glands, cramps, and malignant lesions (Madaan and Kumar 2012). Likewise, it has been used to cure herpes, erysipelas (also known as Ignis sacer, holy fire, and St. Anthony's fire; a bacterial skin illness caused by *Streptococcus pyogenes Rosenbach*), and breast cancer (Arihan et al. 2009). Succus conii, a narcotic, sedative, analgesic, spasmolytic, anti-aphrodisiac, and anticancer agent, is incorporated in numerous herbals despite its lethal nature (Reynolds 2005). In ethnomedicine, this species has been discovered to be effective as an analgesic and anti-inflammatory (Al-Snafi 2016) in the treatment of diabetes in Turkey and typhoid fever and sterility in Morocco (Kharchoufa et al. 2018). Its use to treat herpes and inflamed joints by external application is also in practice. Its effect in individuals with epilepsy, asthma, angina, rheumatism, and tetanus has also been studied and explored (Hotti and Rischer 2017). The two toxic alkaloids, coniine and coniceine, have been detected largely in seeds, according to research on the chemical components of poison hemlock. Sedative and anti-inflammatory properties of the species are also due to these substances (Vetter 2004; Panter et al. 2011; Al-Snafi 2016; Kharchoufa et al. 2018). Apart from alkaloids, *C. maculatum* has a lot of flavonoids (antioxidants), coumarins (antimicrobial, anti-inflammatory), polyacetylenes, vitamins, and oils (Al-Snafi 2016). In *Romanian ethnomedicine*, *C. maculatum* was used as a sedative and to treat neuralgia (Bojor 2018; Grosu and Ichim 2020).

Conium's anticancer properties and its application in traditional medical systems have been validated by modern research (Mondal et al. 2014). At 48 hours, conium administration reduced cell viability and colony formation, slowed cell proliferation, and stopped the cell cycle at the sub-G stage. Conium treatment in HeLa cells induces an increase in reactive oxygen species (ROS), MMP depolarization, morphological defects, and DNA damage after 24 hours, as well as phosphatidylserine externalization after 48 hours. In HeLa cells, downregulation of Akt and NFkB inhibited cell growth, but cytochrome-c release and Caspase-3 activation caused apoptosis, suggesting that the signalling route is conducted through the mitochondria-mediated Caspase-3-Dependent pathway. Conium interacts with DNA molecules, according to CD spectroscopy. Furthermore, the analgesic and anti-inflammatory properties of the alkaloidal fraction of *C. maculatum* aerial parts were investigated. At a dose of 200 mg/kg, the alkaloidal portion of the plant displayed substantial analgesic efficacy. At a dose of 200 mg/kg, it suppressed paw oedema in rats by 71% and decreased paw volume by one-fourth when compared to the regulate group during the first hour of the trial (Madaan and Kumar 2012). Conmaculatin (2-Pentyl piperidine), a new volatile alkaloid related to coniine discovered in *C. maculatum*, had substantial peripheral and central antinociceptive action in rats at doses of 10–20 mg/kg. But it was discovered that doses greater than 20 mg/kg were deadly (Radulovic et al. 2012; Al-Snafi 2016). Also, the highly diluted medication using *C. maculatum* reduces tissue parasitism and inflammation in mice infected by *Trypanosoma cruzi* (Lopes et al. 2016). The immune response and apoptosis are modulated by extremely diluted medicines, modifying the morbidity of animals infected with a highly virulent strain of *T. cruzi*, allowing for a shorter infection course and more treatment choices for Chagas disease (Falkowski-Temporini et al. 2017).

The essential oils are hydro-distilled from the leaves and inflorescences of *C. maculatum* growing in (Sicily, Italy) have been shown antimicrobial action. The composition of essential oils was evaluated using gas chromatography-mass spectrometry (GC-MS), and the inhibitory effects on the growth of two Gram negative pathogens, *Escherichia coli* and *Pseudomonas aeruginosa*, were assessed using two different techniques. The essential oils have different chemical profiles (1-Butyl piperidine and myrcene in the inflorescence, mostly (E)-Caryophyllene in the leaf). The latter oil was particularly efficient at inhibiting the growth of *P. aeruginosa*. These findings suggest that hemlock essential oils could be used as antibacterial agents (Di Napoli et al. 2019). Piperidine derivatives are used as anticancer, antiviral, antimalarial, antibacterial, antifungal, antihypertensive, analgesic, anti-inflammatory, anti-Alzheimer, antipsychotic, and/or anticoagulant medicines in a variety of applications (Abdelshaheed et al. 2021).

15.5 MODE OF ACTION

Major bioactive components of *C. maculatum* are various piperidine alkaloids, including coniine, which have qualities comparable to nicotine. It binds to nicotinic acetylcholine receptors and impairs central nervous system activities. Coniine has a nicotine-like effect in activating and then depressing autonomic ganglia, as well as a curare-like impact in paralysing skeletal muscle motor nerve terminals (Clarke et al. 1981; Mondal et al. 2014). Nicotine, anabasine, lobeline, coniine, and other pyridine and piperidine alkaloids are all hazardous in the same way. The stimulation and then inhibition of nicotinic acetylcholine receptors is their major activity. The conditions such as coma and seizures are caused by the activation of nicotinic receptors in the cortex, thalamus, interpeduncular nucleus, and other parts of the CNS. Nicotine has been demonstrated to stimulate presynaptic cholinergic receptors, which increases rapid excitatory neuronal transmission in the CNS. It promotes both cholinergic and glutamatergic transmission by increasing presynaptic calcium (Furbee 2009).

The nicotinic acetylcholine receptor channel complex is considered to be blocked by piperidine alkaloids. They also block the binding sites linked with sodium, potassium, and calcium ion channels in brain membranes (Ojima and Iula 1999). Another study by Mishra et al. (2015) supported this fact as a piperidine derivative, piperine, showed an inhibitory effect on Na^+ channels. In a comparative study, coniine was observed to show its effect by depolarizing cell membrane potential of SH-SY5Y (predominately express autonomic nAChRs) and TE-671 (express the human foetal muscle-type nAChR) cells. Coniine's ability to stimulate and then desensitize nicotinic acetylcholine receptors may be responsible for its teratogenic effects (nAChRs; Green et al. 2010). It blocks the action of acetylcholine of the nicotinic receptor channel complex in a non-competitive manner. Piperidine alkaloids with a carbon side chain of at least three carbons or greater linked to the carbon alpha to the piperidine nitrogen are teratogenic in nature, according to several studies. The presence of a double bond close to the nitrogen, as in coniceine, boosts the alkaloid's teratogenic potential. They are teratogens because they impede foetal mobility by desensitizing the foetal muscle-type nAChR (Lee et al. 2013). In rat anococcygeus muscle, Erkent and colleagues (2016) discovered that coniine suppresses the nicotinic receptor-mediated nitrergic and noradrenergic transmitter response. Nicotine causes this by blocking presynaptic nicotinic acetylcholine receptors and increasing their concentration (nAChR). It has been discovered that coniine inhibits the noradrenergic response in guinea pig atria via blocking the same receptors. Coniine inhibited the release of noradrenergic and nitrergic transmitters mediated by the nAChR. Noradrenaline release inhibition was more significant than NO release inhibition. In pharmacological investigations, coniine can be a valuable, particular resource as a presynaptic nicotinic receptor antagonist (Erkent et al. 2016). However, piperine and piperidine act as an anticancer agent by inhibiting the Bcl-2 expression (anti-apoptotic) and increasing the Bax expression. A high Bax:Bcl-2 ratio leads to an increase in the amount of cytochrome-c released from mitochondria into the cytosol consequent on the cellular apoptosis (Figure 15.3) (Han et al. 2017; Mitra et al. 2021).

15.6 TOXIC RESPONSE

The whole plant is toxic, but more toxic parts of plants include roots, stems, young shoots, leaves, flowers, mature fruit, and seeds (Al-Snafi 2016; De Landoni 1990). Vomiting, shaking, mobility

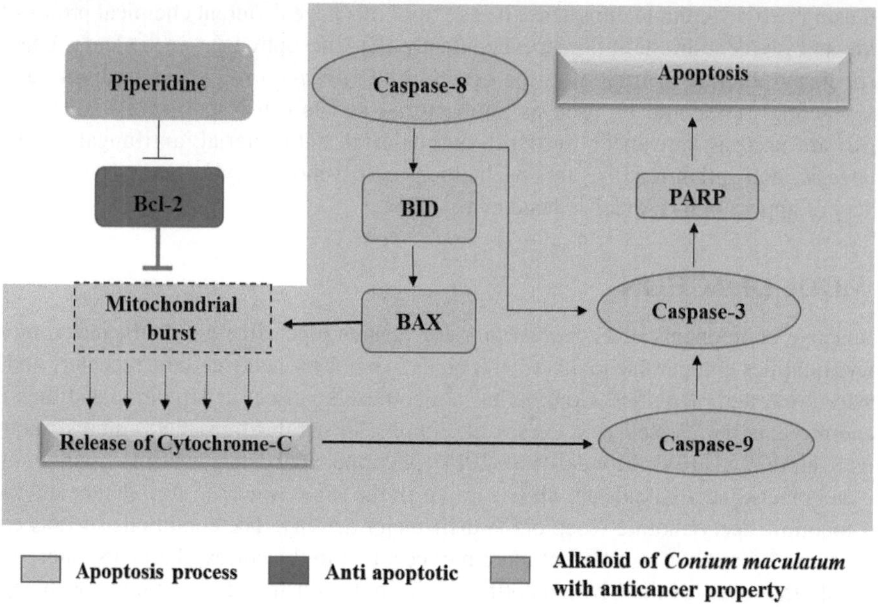

FIGURE 15.3 Anticancer property/activity of piperidine, an alkaloid of *Conium maculatum*. Piperidine inhibits Bcl-2, which is anti-apoptotic in nature, i.e., it plugs the pore of mitochondria, thereby no release of Cyt-C and no apoptosis in cell. It may lead to cancer if there is no apoptosis inside the cell if in need to be repaired. PARP – Poly(ADP-ribose) polymerase – cleaves PARP-1 into fragments and works as a hallmark of cell death.

issues, slow and weak subsequently quick pulse, rapid breathing, salivation, excretion, nausea, convulsions, coma, and death are the most common symptoms of hemlock poisoning (Figure 15.4) (Vetter 2004). Convulsions, coma, extreme delirium, salivation, and involuntary bladder and bowel discharges are all possible signs. Irritation of the oral mucosa, nausea, diarrhoea (rare), bradycardia, hypotension, mydriasis, seizures after progressive muscular paralysis, and respiratory failure are the most prevalent clinical consequences in people today (De Landoni 1990).

As per a reported case study of such poisoning, the green tips of "wild carrots" were consumed by a healthy 4-year-old boy, where he became sleepy and took a long nap. After being taken to hospital and on physical examination, his pulse rate and respiratory rate were found to be 100 beats/minute and 26/minute, correspondingly. His pupils were tiny and sensitive, and his vision was asymmetrical. He had a flexible neck. There was no visible evidence of injury. The patient's stomach was lavaged, and activated charcoal was given to him. During the first 90 minutes, he grew increasingly less responsive, although his gag reflex remained intact. The presence of piperidine alkaloids, which are characteristic of poison hemlock, was confirmed by gas chromatography and mass spectrometry. 850 mg of y-Coniceine per gramme of fresh plant was found in the leaves (Frank et al. 1995). A 49-year-old woman with dry mouth, faintness, and weariness was taken to the hospital. She began to exhibit these symptoms within 5 minutes after ingesting the raw plant. Plant fragments were discovered after a gastric lavage. When admitted, she was unconscious with dilated and unresponsive pupils. There was no attempt to breathe on its own. Pupils became responsive after 10 hours in critical care. She started working on her breathing. She died 9 days after consuming hemlock (Ferah et al. 2006). A postmortem finding suggests *C. maculatum* as a profound cause to be associated with accidental and deliberate deaths in humans (Diaz 2015). Another example described the likely toxicity of an uncertain sudden death (USD) of a lady based on medicolegal features. The causes for this sudden death were declared after the remnants of a vegetal peeling were found at that place. After the autopsy, multiorgan failures were reported

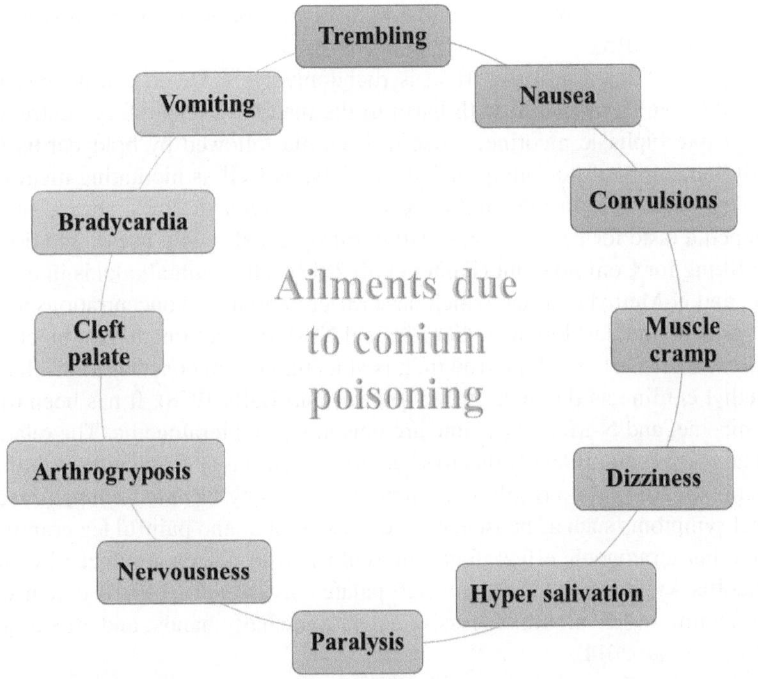

FIGURE 15.4 Poisoning effects of alkaloids of *Conium maculatum*.

on the basis of macroscopic findings. The danger of cyanide in postmortem blood was initially explored; nevertheless, cyanide levels were determined to be below normal limits. Because the plant residues found lacked morphological traits, a genetic nrDNA ITS2 sequencing research was conducted. The evidence might be identified as water dropwort (*Oenanthe* spp.), which contains oenanthotoxin, a strong lethal poison, based on the DNA sequence. It might be related to the hemlock Conium or cowbane Cicuta species, which are more regularly encountered. Later, the vegetal extract and stomach content were analysed using liquid chromatography-tandem high resolution mass spectrometry (LC-QTOF MS), which convincingly proved the toxin's presence (Martínez et al. 2021).

Ingestion of poison hemlock (*C. maculatum*) by pregnant cows during the gestation interval of days 50–75 induces multiple congenital contractures (MCCs) in calves. Crooked calf disease is the common name for this condition. During days 50–75 of pregnancy, pregnant cows that consumed the fresh green herb developed calves with arthrogryposis and spine curvature. These defects are similar to those induced by administration of the isolated teratogenic compound coniine (Casteel 2007). At least five piperidine alkaloids are found in poison hemlock, all of which are associated with changes to the toxicity. The teratogenic impacts are aided by coniine and coniceine. The teratogenic consequences are similar to those caused by lupines in cattle, which result in cleft palates and various congenital skeletal contractures (MCC) (Panter et al. 2017). Nervousness, frequent urination and faeces, shaking, staggering, ataxia, hyperpnea, and tachycardia are all prominent clinical characteristics of cow and horse poisoning. This is in addition to anxiety and incontinence. Possible effects include coma and death from respiratory failure (Anadón et al. 2012; Panter et al. 2012). Furthermore, as indicated by certain cow cases, poison hemlock is teratogenic. Cleft palate and different congenital skeletal contractures are examples of birth defects (Panter et al. 1999, 2013). Poison hemlock intoxication at large scale in calves has been recorded in the Bulgarian province of Kotel. Anorexia, tachycardia, hypersalivation, muscle cramps, hyperthermia, staggering, ataxia, and hyperreflexia were amongst the clinical indications of intoxication seen in 38 calves.

Convulsions were found in three of the animals involved, followed by death (Binev et al. 2007; Cortinovis and Caloni 2015).

Coniine's most significant adverse effect is rhabdomyolysis. Death can happen from coniine-induced respiratory paralysis paired with harm to the medulla's respiratory centre. Conium alkaloids, which act like biphasic nicotine, cause tachycardia followed by bradycardia (Boskabadi et al. 2021). Early indicators of poisoning include anxiety, as well as nictitating membrane blockage of the eyes (most noticeable in pigs and once in a while seen in cows, sheep, and goats). Once they've developed a taste for it, pigs, goats, cattle, European elk, wild geese, and domestic turkeys have shown a liking for Conium plant (Panter et al. 2017). The main alkaloids in conium are coniine, coniceine, and N-Methyl coniine, which have different relative concentrations according to the stage of plant growth and development. Coniine and N-Methyl coniine have a biochemical precursor called γ-Coniceine. Early in plant growth, it is at its maximum concentration, changing to coniine and N-Methyl coniine as the plant grows (Keeler and Balls 1978). It has been discovered that coniine, γ-Coniceine, and N-Methyl coniine are poisonous and teratogenic. The relative toxicity of these compounds differs significantly due to structural variations (γ-Coniceine > coniine > N-Methyl coniine; Panter et al. 2017). As a result of unknowingly utilising *C. maculatum*, a case report documented clinical symptoms such as nausea, rash, disorientation, and painful leg cramps (Colombo et al. 2009). The other teratogenic effect, foetal musculoskeletal defects produced by coniine include scoliosis, torticollis, kyphosis, lordosis, and cleft palate (Green et al. 2010). A 6-year-old girl patient had a burning feeling in her mouth, excessive salivation, shaky hands, and ataxia, according to a case report (Konca et al. 2014).

Poisoning effects have been documented in different mammals after consuming poison hemlock: pigs (Markham 1985), tule elks, sheep, rabbits, goats (Vetter 2004), horses (Nice et al. 2005), cattle and deer (Swerczek and Swerczek 2012). Cows are more toxic than other animals when exposed to hemlock. Cows were shown to be deadly to fresh plant harvested at the same location at a dose of 5.3 g plant/kg body mass (Keeler and Balls 1978). Conium alkaloids were found in the hay, as well as the plants in the hayfield and the urine of the animals afflicted. Hemlock intake has been linked to congenital skeletal abnormalities in calves (Keeler and Balls 1978). Several species, such as birds, hens, and turkeys, are poisoned by coniine (Frank and Reed 1986). Some common signs of hemlock toxicity in mammals include muscle weakness, trembling, nervousness, ataxic gait, cyanotic surfaces, dilating pupils, early central nervous system stimulation, excessive salivation, bloating, intestinal discomfort, quick and weak pulse, lack of appetite, cyanotic membranes, and dilating pupils. This results in depression and finally death from respiratory paralysis (Keeler et al. 1980; Nice et al. 2005). Some of these symptoms arise immediately after intake, and 2–3 hours later, respiratory paralysis occurs, while others appear later (3–4 days). Coniine affects different species in different ways. Coniine, for example, is toxic to cows at a day-to-day dose of 3.3–6.6 mg/kg, and around 1 kg of poison hemlock is fatal (Keeler et al. 1980; Nice et al. 2005). Pigs are fatal at 1 g/kg of seeds and 8 g/kg of plant. Non-toxic substances include conyrine, 3-Methylpiperidine, and N-Methylpiperidine (Keeler and Balls 1978). Inhaling coniine or smashed green organic material does not cause toxicity, but eating coniine and γ-Coniceine is most dangerous to cattle (Hotti and Rischer 2017).

Intoxication with hemlock can cause a headache, pharyngeal discomfort, blurred vision, vomiting, lethargy, steady paralysis of the extremities, and death. Convulsions, intense delirium, salivation, coma, and involuntary bladder and bowel discharges are all possible signs. Oral mucosa soreness, salivation, nausea, slight stomach discomfort, diarrhoea (rare), bradycardia, and other symptoms like hypotension, mydriasis, seizures after progressive muscular paralysis, and respiratory failure are the most prevalent clinical concerns in people. Blood gases, electrolytes, and a plant sample are used in the diagnosis. Humans are poisonous to coniine, and 3 mg causes symptoms. Coniine doses of up to 150–300 mg, or 6–8 leaves, can be tolerated (6 g) (Biberci et al. 2002). Mistakenly made herbal medicine or the use of an inappropriate plant due to misidentification can potentially result in accidental consumption of harmful plants (e.g., poison hemlock) (Beyer et al. 2009; Hotti and Rischer 2017).

15.7 ECOLOGICAL ROLE

Hemlock alkaloids have an important role in maintaining ecological balance. It can act as a pioneer species and rapidly colonizes disturbed sites (National Park Service 2022). One of the flowery smell components in poison hemlock is coniine (Roberts 1998). It is a food plant for the larvae of various lepidoptera, such as silver-ground carpet moths and the poison hemlock moth (*Agonopterix alstroemeriana*). The hemlock moth has been involved in managing the plant biologically (Castells and Berenbaum 2006). Coniine may entice flies (Diptera), which massively fertilize poison hemlock blossoms (Nitao 1987). Volatile alkaloids are recognized by the hemlock moth as part of its host plant identification. Coniine is also required for the correct development of hemlock moth larvae (Castells and Berenbaum 2006). Coniine may, on the one hand, act as an insect-paralysing agent as it was capable of paralysing fire ants (Mody et al. 1976); on the other hand, it also attracts insects into the pitcher (Harborne 1982).

15.8 TRADITIONAL AND OTHER POTENTIAL USE

Criminals were given the juice or essence of the *C. maculatum* plant. It was the death poison that Socrates, the Greek philosopher, was forced to consume (399 BC). Herpes, erysipelas (a kind of superficial cellulitis), and breast cancers were all treated with *C. maculatum*. Conium seeds that haven't ripened have been dried and stored for use as an analgesic, antispasmodic, or sedative. From 1864 through 1898, the plant's preserved leaf and juice were mentioned in pharmacopoeia in London and Edinburgh, and The British Pharmaceutical Codex (1934) was the final recognized medicinal acknowledgement (Bowman and Sanghvi 1963). To reduce surgical pain, the ancients utilized four plants: datura (*Datura stramonium*), withania (*Withania somniferum*), hemlock (*Conium maculatum*), and opium poppy (*Papaver somniferum*). These plants were used as narcotics and anodynes, with decoctions, poultices, suppositories, lozenges, tablets, and topical treatments being used. The extracts from hemlock (with other herbs) were used to relieve the pain of surgery in ancient times (Kennedy and Longnecker 1990). *C. maculatum*, sometimes known as "hemlock tea", has a long and illustrious history. In ancient Greece, it was used to murder people, and the decoction was used to execute the philosopher Socrates (Daugherty 1995). The piperidine alkaloids coniine and g-Coniceine poisoned people. It was utilized by Greek and Arabian physicians to cure indolent tumours, swellings, and joint problems. The bitter juice of the plant, combined with betony (*Stachys officinalis*) and fennel (*Foeniculum vulgare*) seeds, was exploited to heal a mad dog bit. This juice was used as an antidote for strychnine and other very dangerous chemicals in the past (Boericke 1982). Religious groups in the 15th and 16th centuries utilized roasted roots to relieve the aches of gout. In the instances of asthma, epilepsy, whooping cough, angina, chorea, and stomach aches, plant tinctures and extracts were utilized due to their calming, anodyne, and antispasmodic qualities. Hemlock is a well-known homoeopathic remedy with a variety of applications (Forrington 1997; Vetter 2004). *C. maculatum* has been used to treat spasmodic diseases, mental excitation, rheumatic pains in the elderly and infirm, stomach ache, pain from a gastric ulcer, agitation, and restlessness. The analgesic and anti-inflammatory properties of the alkaloidal fraction of *C. maculatum* upper portions were investigated. The published literature shows that *C. maculatum*'s analgesic and anti-inflammatory activities are due to its alkaloids (Madaan and Kumar 2012). *C. maculatum* extract is used as a treatment for cervical cancer in traditional homoeopathy (Mondal et al. 2014).

15.9 CONCLUSION AND FUTURE REMARKS

Because *C. maculatum* contains many piperidine alkaloids, including coniine, a piperidine alkaloid, it has the potential to treat a range of diseases. Piperidine derivatives are a huge and main class of nitrogen-bearing heterocyclic chemicals. Piperidines are naturally occurring chemicals found in *C. maculatum* leaves that have been demonstrated to have anticancer, anti-inflammatory, and antihypertensive activities, as well as functioning as antioxidants. Synthetic piperidine derivatives, on the other hand, have been shown in numerous studies to have anticancer, antiviral, antibacterial, antimalarial, and antifungal properties. Piperidine alkaloids from *C. maculatum* are being studied further as possible sources of pharmaceutically active chemicals. Piperidine and its derivatives can be engineered and produced to have additional biological effects such as neuroprotective, antihypertensive, anti-Alzheimer's, analgesic, anti-inflammatory, antioxidant, antipsychotic, and anticoagulant properties. Because of the poisonous effects of coniine, coniceine, and N-Methyl coniine, *C. maculatum* is not recommended for human or animal ingestion. Because they have a side chain containing propyl or a larger group, these alkaloids from *C. maculatum* are teratogenic. Further, as there is no specific remedy, the best way to deal with the plant's output reductions is to prevent them. As a remedy, herbicides and grazing with less sensitive animals (such as sheep) have been suggested. *C. maculatum* alkaloids can be passed through milk and fowl muscle tissue, allowing the latter to reach the human food chain. Because it's difficult to link abnormalities in kids to much earlier maternal poisoning, the harms caused by *C. maculatum*, in most cases, long-term toxicity, are likely to be grossly underestimated. To prevent these hazardous alkaloids from entering our food chain, certain measures must be taken.

NOTES

Upagya Gyaneshwari, Departments of Biotechnology, School of Life Sciences, Mahatma Gandhi Central University, Motihari, India

Akhilesh Kumar Singh, Departments of Biotechnology, School of Life Sciences, Mahatma Gandhi Central University, Motihari, India

Priti Pal, Department of Civil Engineering, Shri Ramswaroop Memorial College of Engineering & Management, Lucknow, India

Brijesh Pandey, Departments of Biotechnology, School of Life Sciences, Mahatma Gandhi Central University, Motihari, India

REFERENCES

Abdelshaheed, M. M., Fawzy, I. M., El-Subbagh, H. I., & Youssef, K. M. 2021. Piperidine nucleus in the field of drug discovery. *Future Journal of Pharmaceutical Sciences* 7(1): 1–11. https://doi.org/10.1186/s43094-021-00335-y.

Al-Snafi, A. E. 2016. Pharmacology and toxicology of *Conium maculatum*—A review. *The Pharmaceutical and Chemical Journal* 3(2): 136–142.

Anadón, A., Martínez-Larrañaga, M. R., & Castellano, V. 2012. Poisonous plants of Europe. In: Gupta, R. C., ed. *Veterinary Toxicology: Basic and Clinical Principles*. 2nd ed. Elsevier Inc., San Diego, CA, pp. 1080–1094.

Arihan, O., Boz, M., Iskit, A. B., & Ilhan, M. 2009. Antinociceptive activity of coniine in mice. *Journal of Ethnopharmacology* 125: 274–278.

Baskin, M., & Baskin, C. 1990. Seed germination ecology of poison hemlock, Conium maculatum. *Canadian Journal of Botany* 68: 2018–2024.

Beyer, J., Drummer, O. H., & Maurer, H. H. 2009. Analysis of toxic alkaloids in body samples. *Forensic Science International* 185: 1–9.

Biberci, E., Altuntas, Y., Cobanoglu, A., & Alpinar, A. 2002. Acute respiratory arrest following hemlock (Conium maculatum). *Journal of Toxicology: Clinical Toxicology* 40: 517–518.

Binev, R., Mitev, J., & Miteva, T. 2007. Intoxication with Poison Hemlock (Conium maculatum L.) in calves. *Trakia Journal of Sciences* 5: 40–50.

Boericke, W. 1982. *Homeopathic Materia Medica*. 9th ed. B Jain Publishers Pvt Ltd, Uttar Pradesh, India.

Bojor, O. 2018. *Plantele medicinale si aromatice de la A la Z*. [*The Medicinal and Aromatic Plants from A to Z*]. Dharana Publishing House, Bucharest, Romania.

Boskabadi, J., Askari, Z., Zakariaei, Z., Fakhar, M., & Tabaripour, R. 2021. Mild-to-severe poisoning due to *Conium Maculatum* as toxic herb: A case series. *Clinical Case Reports* 9(7): e04509. https://doi.org/10.1002/ccr3.4509.

Bowman, C., & Sanghvi, S. 1963. Pharmacological actions of hemlock (*Conium maculatum*) alkaloids. *Journal Pharmaceutical Pharmacology* 15: 1–25.

Casteel, S. W. 2007. Reproductive toxicants. In: Youngquist R. S., & Threlfall W. R. (eds.), *Current Therapy in Large Animal Theriogenology.* WB Saunders, Philadelphia, PA, pp. 420–427.

Castells, E., & Berenbaum, M. R. 2006. Laboratory rearing of Agonopterix alstroemeriana, the defoliating poison hemlock (*Conium maculatum* L.) moth, and effects of piperidine alkaloids on preference and performance. *Ecological Entomology* 35: 607–615.

Centre for Agriculture and Bioscience International (CABI). 2022. *Conium maculatum*. In: *Invasive Species Compendium.* CAB International, Wallingford, England. Available at: www.cabi.org/isc (Accessed on February 25, 2022).

Chizzola, R., & Lohwasser, U. 2020. Diversity of secondary metabolites in roots from *Conium maculatum* L. *Plants (Basel, Switzerland)* 9(8): 939. https://doi.org/10.3390/plants9080939.

Christensen, L. P. 2011. Aliphatic C17-polyacetylenes of the falcarinol type as potential health promoting compounds in food plants of the Apiaceae family. *Recent Patents on Food, Nutrition & Agriculture* 3: 64–77.

Clarke, M. L., Harvey, D. G., & Humphreys, D. J. 1981. *Veterinary Toxicology*. 2nd ed. Bailliere Tindall, London, England, p. 258.

Colombo, M. L., Marangon, K., Locatelli, C., Giacchè, M., Zulli, R., & Restani, P. 2009. Hemlock poisoning due to plant misidentification. *Journal of Pharmaceutical Sciences and Research* 1(4): 43–47.

Cortinovis, C., & Caloni, F. 2015. Alkaloid-containing plants poisonous to cattle and horses in Europe. *Toxins* 7(12): 5301–5307. https://doi.org/10.3390/toxins7124884.

Cromwell, J. 1956. The separation, microestimation and distribution of alkaloid of poison hemlock (*Conium maculatum*). *Biochemical Journal* 64: 259–266.

Daugherty, C. G. 1995. The death of Socrates and the toxicology of hemlock. *Journal of Medical Biography* 3(3): 178–182.

De Landoni, J. H. 1990. Conium Maculatum, L. (PIM 144). Available at: http://www.inchem.org/documents/pims/plant/conium.htm (Accessed on March 5, 2020).

Di Napoli, M., Varcamonti, M., Basile, A., Bruno, M., Maggi, F., & Zanfardino, A. 2019. Anti-Pseudomonas aeruginosa activity of hemlock (*Conium maculatum*, Apiaceae) essential oil. *Natural Product Research* 33(23): 3436–3440. https://doi.org/10.1080/14786419.2018.1477151.

Diaz, G. J. 2015. Toxicosis by plant alkaloids in humans and animals in Colombia. *Toxins* 7(12): 5408–5416. https://doi.org/10.3390/toxins7124892.

Dionisio, K. L., Phillips, K., Price, P. S., Grulke, C. M., Williams, A., Biryol, D., & Isaacs, K. K. 2018. The chemical and products database, a resource for exposure-relevant data on chemicals in consumer products. *Scientific Data* 5(1): 1–9. https://doi.org/10.1038/sdata.2018.125.

Erkent, U., Iskit, A. B., Onur, R., & Ilhan, M. 2016. The effect of coniine on presynaptic nicotinic receptors. *Zeitschrift fur Naturforschung. C, Journal of Biosciences* 71(5–6): 115–120. https://doi.org/10.1515/znc-2015-0194.

Fairbairn, W., & Challen, B. 1959. The alkaloids of hemlock (*Conium maculatum* L.) distribution in relation to the development of the fruit. *Biochemical Journal* 72: 556–561.

Fairbairn, W., & Suwal, N. 1961. The alkaloids of hemlock (*Conium maculatum* L.) II. Evidence for a rapid turnover of the major alkaloids. *Phytochemistry* 1: 38–46.

Falkowski-Temporini, G. J., Lopes, C. R., Massini, P. F., Brustolin, C. F., Ferraz, F. N., Sandri, P. F., Hernandes, L., Aleixo, D. L., Barion, T. F., Esper, L. G., & de Araújo, S. M. 2017. Increase of the hepatocytes and splenocytes apoptosis accompanies clinical improvement and higher survival in mice infected with *Trypanosoma cruzi* and treated with highly diluted *Lycopodium clavatum*. *Microbial Pathogenesis* 110: 107–116. https://doi.org/10.1016/j.micpath.2017.06.027.

Ferah, A. U., Dilek, M. O., Ercument, Y., & Oktay, D. 2006. Hemlock poisoning (case report): Die like Socrates. *Resuscitation* 70(2): 337. https://doi.org/10.1016/j.resuscitation.2006.06.125.

Forrington, E. A. 1997. *Clinical Materia Medica.* 4th ed. B Jain Publishers Pvt Ltd., Uttar Pradesh, India, pp. 232–233.

Frank, A. A., & Reed, W. M. 1986. Conium maculatum (poison hemlock) toxicosis in a flock of range turkeys. *Avian Diseases* 31: 386–388.

Frank, B. S., Michelson, W. B., Panter, K. E., & Gardner, D. R. 1995. Ingestion of poison hemlock (*Conium maculatum*). *The Western Journal of Medicine* 163(6): 573–574.

Fröberg, L. 2010. *Conium maculatum*. In: Jonsell, B., & Karlsson, T., eds. *The Swedish Museum of Natural History*. Flora Nordica, Stockholm, Sweden, pp. 210–211.

Furbee, B. 2009. Neurotoxic plants. In: Dobbs M.R. (ed.), *Clinical Neurotoxicology.* WB Saunders, Philadelphia, PA, pp. 523–542.

Green, B. T., Lee, S. T., Panter, K. E., Welch, K. D., Cook, D., Pfister, J. A., & Kem, W. R. 2010. Actions of piperidine alkaloid teratogens at fetal nicotinic acetylcholine receptors. *Neurotoxicology and Teratology* 32(3): 383–390. https://doi.org/10.1016/j.ntt.2010.01.011.

Grosu, E., & Ichim, M. C. 2020. Turning meadow weeds into valuable species for the Romanian Ethnomedicine while complying with the environmentally friendly farming requirements of the European Union's Common Agricultural Policy. *Frontiers in Pharmacology* 11, 529. https://doi.org/10.3389/fphar.2020.00529.

Han, S. Z., Liu, H. X., Yang, L. Q., Cui, L. D., & Xu, Y. 2017. Piperine (PP) enhanced Mitomycin-C (MMC) therapy of human cervical cancer through suppressing Bcl-2 signaling pathway via inactivating STAT3/NF-Kb. *Biomedicine & Pharmacotherapy* 96: 1403–1410.

Harborne, J. B. 1982. *Introduction to Ecological Biochemistry.* 2nd ed. Academic Press, London, England, p. 278.

Holm, L., Doll, J., Holm, E., Pancho, J., & Herberger, J. 1997. *Conium maculatum* L. In: *World Weeds—Natural Histories and Distribution.* John Wiley & Sons, Inc., New York, NY; Chichester, England; Weinheim, Germany; Brisbane, Australia; Singapore; Toronto, ON, Canada, pp. 221–225.

Hotti, H., & Rischer, H. 2017. The killer of Socrates: Coniine and related alkaloids in the plant kingdom. *Molecules* 22(11): 1962. https://doi.org/10.3390/molecules22111962.

Jones, T. H., Blum, M. S., & Fales, H. M. 1982. Ant venom alkaloids from Solenopsis and Monorium species: Recent developments. *Tetrahedron* 38: 1949–1958.

Keeler, F., & Balls, D. 1978. Teratogenic effects in cattle of *Conium maculatum* and Conium alkaloids and analogues. *Clinical Toxicology* 12: 49–64.

Keeler, R. F., Balls, L. D., Shupe, J. L., & Crowe, M. W. 1980. Teratogenicity and toxicity of coniine in cows, ewes and mares. *The Cornell Veterinarian* 70: 19–26.

Kennedy, S. K., & Longnecker, D. E. 1990. History and principles of anaesthesiology. In: Gilman A. G., Rall T. W., Nies A. S., & Palmer T. (eds.), *Goodman and Gillman: The Pharmacological Basis of Therapeutics.* 8th ed. Pergamon Press, New York, NY.

Kharchoufa, L., Merrouni, I. A., Yamani, A., & Elachouri, M. 2018. Profile on medicinal plants used by the people of North Eastern Morocco: Toxicity concerns. *Toxicon* 154: 90–113. http://doi.org/10.1016/j.toxicon.2018.09.003.

Kielland, J., & Anders, O. 1998. *Conium maculatum* in Hedmark county, Eastern Norway. *Blyttia* 56: 92–93.

Kobaek-Larsen, M., El-Houri, R. B., Christensen, L. P., Al-Najami, I., Fretté, X., & Baatrup, G. 2017. Dietary polyacetylenes, falcarinol and falcarindiol, isolated from carrots prevent the formation of neoplastic lesions in the colon of azoxymethane-induced rats. *Food & Function* 8: 964–974 https://doi.org/10.1039/C7FO00110J.

Konca, C., Kahramaner, Z., Bosnak, M., & Kocamaz, H. 2014. Hemlock (*Conium maculatum*) poisoning in a child. *Turkish Journal of Emergency Medicine* 14(1): 34–36.

Lee, S. T., Green, B. T., Welch, K. D., Jordan, G. T., Zhang, Q., Panter, K. E., & Gardner, D. R. 2013. Stereoselective potencies and relative toxicities of γ-coniceine and N-methylconiine enantiomers. *Chemical Research in Toxicology* 26(4): 616–621.

Lopes, C. R., Falkowski, G. J., Brustolin, C. F., Massini, P. F., Ferreira, É. C., Moreira, N. M., Aleixo, D. L., Kaneshima, E. N., & de Araújo, S. M. 2016. Highly diluted medication reduces tissue parasitism and inflammation in mice infected by Trypanosoma cruzi. *Homeopathy: The Journal of the Faculty of Homeopathy* 105(2): 186–193. https://doi.org/10.1016/j.homp.2015.09.005.

Lopez, T. A., Cid, M. S., & Bianchini, M. L. 1999. Biochemistry of hemlock (*Conium maculatum* L.) alkaloids and their acute and chronic toxicity in livestock. A review. *Toxicon: Official Journal of the International Society on Toxinology* 37(6): 841–865. https://doi.org/10.1016/s0041-0101(98)00204-9.

Madaan, R., & Kumar, S. 2012. Screening of alkaloidal fraction of *Conium maculatum* L. aerial parts for analgesic and antiinflammatory activity. *Indian Journal of Pharmaceutical Sciences* 74(5): 457–460. https://doi.org/10.4103/0250-474X.108423.

Markham, K. 1985. Hemlock poisoning in piglets. *The Veterinary Record* 116: 27.

Martínez, P., Quintela, O., Del Valle, E., & Pérez-Gómez, B. 2021. Genetic identification and subsequent LC-QTOF MS analysis of plant remains (Oenanthe spp.) could prove the cause of an undetermined sudden death. *International Journal of Legal Medicine* 135(4): 1407–1411. https://doi.org/10.1007/s00414-020-02488-6.

Mekkonen, Y. 1994. A survey of plants (potentially) toxic to livestock in the Ethiopian flora. *Ethiopian Journal of Science* 17: 9–32.

Merriam-Webster. 2022. Pseudoconhydrine. In Merriam-Webster.com Dictionary. Available at: https://www.merriam-webster.com/dictionary/pseudoconhydrine (Assessed on March 7, 2022).

Mishra, A., Punia, J. K., Bladen, C., Zamponi, G. W., & Goel, R. K. 2015. Anticonvulsant mechanisms of piperine, a piperidine alkaloid. *Channels (Austin, Tex.)* 9(5): 317–323. https://doi.org/10.1080/19336950.2015.1092836.

Mitich, W. 1998. Poison-hemlock (*Conium maculatum* L.). *Weed Technology* 12: 194–197.

Mitra, S., Anand, U., Jha, N. K., Shekhawat, M. S., Saha, S. C., Nongdam, P., & Dey, A. 2021. Anticancer applications and pharmacological properties of piperidine and piperine: A comprehensive review on molecular mechanisms and therapeutic perspectives. *Frontiers in Pharmacology* 12: 772418. https://doi.org/10.3389/fphar.2021.772418.

Mody, N. V., Henson, R., Hedin, P. A., Kokpol, U., & Miles, D. H. 1976. Isolation of insect paralyzing agent coniine from Sarracenia flava. *Experientia* 32: 829–830.

Mondal, J., Panigrahi, A. K., & Khuda-Bukhsh, A. R. 2014. Anticancer potential of *Conium maculatum* extract against cancer cells in vitro: Drug-DNA interaction and its ability to induce apoptosis through ROS generation. *Pharmacognosy Magazine* 10(3): S524–S533. https://doi.org/10.4103/0973-1296.139792.

National Park Service. 2022. Exotic Species: Poison Hemlock—National Park Service. Available at: https://www.nps.gov/articles/poison-hemlock.htm (Accessed on March 12, 2022).

Nice, G., Johnson, B., Bauman, T., & Jordan, T. 2005. Poison Hemlock—The Toxic Parsnip; Purdue Extension—Weed Science, Purdue University: West Lafayette, IN, USA.

Nitao, J. K. 1987. Test for toxicity of coniine to a polyphagous herbivore, Heliothis zea (Lepidoptera: Noctuidae). *Environmental Entomology* 16: 656–659.

NTP. 2010. *NTP Technical Report on the Toxicology and Carcinogenesis Studies of β-Myrcene (CAS No. 123-35-3) in F344/N Rats and B6C3F1 Mice (Gavage Studies).* Research Triangle Park, National Institutes of Health, Durham, NC.

Ojima, I., & Iula, D. M. 1999. New approaches to the syntheses of piperidine, izidine, and quinazoline alkaloids by means of transition metal catalyzed carbonylations. In: Pelletier, S. W., ed. *Alkaloids: Chemical and Biological Perspectives*. Elsevier Science Ltd, Kidlington, pp. 371–410.

Panter, K. E. Gardner D.R., Stegelmeier B. L., Welch K. D., & Holstege D.2011. Water hemlock poisoning in cattle: ingestion of immature Cicuta maculata seed as the probably cause. Toxicon 57(1): 157–161.

Panter, K. E., Bunch, T. D., Keeler, R. F., & Sisson, D. V. 1988. Radio ultrasound observation of the fetotoxic effects in sheep from ingestion of Conium maculatum (poison-hemlock). *Journal of Toxicology: Clinical Toxicology* 26: 175–187.

Panter, K. E., Gardner, D. R., Lee, S. T., Pfister, J. A., Ralphs, M. H., Stegelmeier, B. L., & James, L. F. 2012. Important poisonous plants of the United States. In: Gupta, R. C., ed. *Veterinary Toxicology: Basic and Clinical Principles*. 2nd ed. Elsevier Inc., San Diego, CA, pp. 1031–1079.

Panter, K. E., James, L. F., & Gardner, D. R. 1999. Lupine, poison-hemlock, and Nicotiana spp.: Toxicity and teratogenicity in livestock. *Journal of Natural Toxins* 8: 117–134.

Panter, K. E., Welch, K. D., & Gardner, D. R. 2017. Toxic plants. In: Gupta R.C. (ed.), *Reproductive and Developmental Toxicology*. Academic Press, pp. 903–923, London, UK.

Panter, K. E., Welch, K. D., Gardner, D. R., & Green, B. T. 2013. Poisonous plants: Effects on embryo and fetal development. *Birth Defects Research Part C: Embryo Today* 99: 223–234.

Pitcher, D. 1989. The Nature Conservancy. Conium Maculatum—Bugwoodwiki. Available at: https://wiki.bugwood.org/Conium_maculatum (Accessed on 25 Feb 2022).

Radulovic, N., Đorđević, N., Denić, M., Pinheiro, M. M., Fernandes, P. D., & Boylan, F. 2012. A novel toxic alkaloid from poison hemlock (*Conium maculatum* L., Apiaceae): Identification, synthesis and antinociceptive activity. *Food and Chemical Toxicology: An International Journal Published for the British Industrial Biological Research Association* 50(2): 274–279. https://doi.org/10.1016/j.fct.2011.10.060.

Reynolds, T. 2005. Hemlock alkaloids from Socrates to poison aloes. *Phytochemistry* 66: 1399–1406. https://doi.org/10.1016/j.phytochem.2005.04.039.

Roberts, M. F. 1998. Enzymology of alkaloid biosynthesis. In: Roberts, M. F., & Wink, M., eds. *Alkaloids: Biochemistry, Ecology, and Medicinal Applications.* Springer, New York, NY, pp. 109–146.

Schep, L. J., Slaughter, R. J., & Beasley, D. M. 2009. Nicotinic plant poisoning. *Clinical Toxicology (Philadelphia, Pa.)* 47(8): 771–781. https://doi.org/10.1080/15563650903252186.

Surendran, S., Qassadi, F., Surendran, G., Lilley, D., & Heinrich, M. 2021. Myrcene—What are the potential health benefits of this flavoring and aroma agent? *Frontiers in Nutrition* 8: 400. https://doi.org/10.3389/fnut.2021.699666.

Swerczek, T. W., & Swerczek, S. J. 2012. Spotted hemlock poisoning in a herd of Angus cattle. *Journal of the American Veterinary Medical Association* 240: 1280–1281.

The British Pharmaceutical Codex, 1934. 4th ed. The Pharmaceutical Press, London, p 1768,
The Merck Index. 1976. 9th ed. Merck & Co., Inc., Rahway, NJ, p. 323.
Vetter, J. 2004. Poison hemlock (Conium maculatum L.). *Food and Chemical Toxicology: An International Journal Published for the British Industrial Biological Research Association* 42(9): 1373–1382. https://doi.org/10.1016/j.fct.2004.04.009.
Wyrembek, P., Negri, R., Kaczor, P., Czyżewska, M., Appendino, G., & Mozrzymas, J. W. 2012. Falcarindiol allosterically modulates GABAergic currents in cultured rat hippocampal neurons. *Journal of Natural Products* 75: 610–616. https://doi.org/10.1021/np2008522.

16 *Datura stramonium* (Thorn Apple or Devil's Trumpet)

Tenzin Jamtsho, Ugyen, and Phurpa Wangchuk

CONTENTS

16.1 INTRODUCTION

Datura stramonium L. is a wild-growing annual plant belonging to the Solanaceae family widely distributed and easily accessible worldwide. It is commonly known as Jimson weed, but the most common in our context is Devil snare/Thornapple (Eng), Datura (Nep), and Nyangmo-throkchang (Dz). It is believed to have originated in Central and South America, but it is found worldwide and was introduced in Bhutan in 1963 (CABI 2022). It is commonly found in warm temperate and subtropical regions growing in the abandoned farmlands, roadside, disturbed land, wasteland, and dry riverbeds/stream. It is one of the widely well-known folklore medicinal plants globally and has been widely known for its healing properties and also exploited for mystic and ceremonial purposes (Jakabová et al. 2012; Batool et al. 2020). Due to its therapeutic significance, it is widely cultivated in Europe, America, South Africa, Asia, and other tropical and subtropical regions for medicine and pharmacological purpose, despite being an invasive weed (Joshua 2019). The various parts, including flowers, leaves, roots, and seeds, are used in treating various ailments, namely rheumatism, wounds, asthma, toothache, ulcers, fever, and inflammation, and consequently, it has gained a special place in Ayurvedic medicine (Soni et al. 2012; Setshogo 2015). Additionally, seeds of *D. stramonium* are smoked to experience the hallucinogenic effect (Gaire and Subedi 2013), while leaf extract is taken orally to treat asthma, and stripped bark is applied topically to heal burns, ulcers, and swellings (Devi et al. 2011; Naik et al. 2018). Further, their medicinal values are recognized for their anti-inflammatory properties (Firdaus 2020) and antibacterial and antioxidant activities (Debnath 2017). It also exhibited antimicrobial, antidiabetic, antiasthmatic, antifungal, anti-inflammatory, antioxidant, analgesic, cytotoxic, and wound healing properties (Sharma et al.

DOI: 10.1201/b23017-18

2021). According to Soni et al. (2012), the therapeutic activities of *D. stramonium* are linked to the phytochemical constituents such as saponins, tannins, steroids, alkaloids, flavonoids, phenols, and glycosides (Soni et al. 2012). However, the entire plant is also considered as poisonous or toxic plant when taken in large doses, and it is known as a hallucinogenic plant (Vaza 2018). Moreover, the seeds are the most toxic; neither drying out nor boiling destroys the toxic properties. The ingestion of large doses of *D. stramonium* intentionally or accidentally exhibits disturbance to the central nervous system with associated symptoms like confusion, amnesia, hallucinations, and bizarre behavior (Mukhtar et al. 2019). According to Das et al. (2012), other symptoms of Datura poisoning include retention of urinary, fast heartbeat, desiccation of epidermis and lips, pupil dilation, urinary retention, impaired vision, and fast heartbeat (Das et al. 2012; Naik et al. 2018). Several incidences of accidental or intentional *D. stramonium* poisoning have been reported from different parts of the world. A study by Gaire and Subedi (2013) stated that the death from consuming *Datura stramonium* is rare; however, the recovery may take several days, hampering the health. Thus, every aspect of their therapeutic claims needs to be revealed, particularly its toxicity. Likewise, its medicinal values should be practiced only with a preceding acquaintance to avoid its adverse effects. It is vital to be well informed regarding their pharmacological and toxicological effects and the potential risks associated with their uses.

16.2 BOTANICAL DESCRIPTION

D. stramonium is an annual bushy and coarse plant growing 1–2 m in height. The roots are whitish, large, and characterized as a tap root system. The leafy erect and highly branched stems appear glabrous, green, and cylindrical. Leaves are simple, glabrous, measure 8–25 cm long, 6–17 cm wide, and appear broadly triangular-ovate with acuminate apex, dentate margin, and reticular venation. The adaxial surface is glabrous, grayish green, while the abaxial surface is slightly wrinkled and green (Radford 2010). The flowers are solitary in the axils with an upright position. It is white, funnel shaped, and measures 7–12 cm long and 2–8 cm cross with five unequal stamens attached to the corolla. Further, it is described as complete, pedicellate, ebracteate, and radially symmetrical with bilobed stigma and a superior ovary. The calyx is tubular, folded, and measures about 3.5–7 cm in length. Spiny fruits with four chamber open when matured and releases seeds. The kidney-shaped seed appears black with an irregular pitted surface and measures up to 4 mm long by 3 mm wide (Parker 1992). Seeds are elongated with a wavy structure having to bulge at either side; seed testa is formed of single-layered, lignified, and club-shaped sclerenchymatous cells; endosperm encloses curved embryo composed of thin-walled, polygonal, parenchymatous cells, usually filled with aleurone grains and abundant oil globules (Shamim and Idris 2019) (Figure 16.1).

D. stramonium is widely disturbed worldwide and possesses varying vernacular names corresponding to the regions, and it is sometimes indicated by more than one name. The genus *Datura* was derived from the ancient Hindu word for the plant, dhatura. The species name stramonium wad is formerly derived from the Greek words strychnos and manikos which means nightshade and mad, respectively. On the other hand, the species is known as thornapple in the New Latin word (Kalam and Rifat 2020). In Sanskrit (the liturgical language of Buddhism, Hinduism, and Jainism), the plants are known by several names such as dhattura, devika, dhuttura, madana, ghantika, kitava, and madakara. Similarly, the plants are named as thorn apple, hell's bells, Jimson weed, apple of peru, stink weed, devil's trumpets, moon flowers, and mad apple in English (Jarald and Jarald 2006; Khan et al. 2013); kachola in Afghani (Afghanistan); olieblaar, olieneut, pietjie, and laporte stinkblaar in Africans (Africa); jauzmasal in Arabic; pepediawu in Ashanti (Ghana); estramonio and figueirado inferno in Brazil; chan kiuetse, toukiueeul, and tsoui sin hoa in Chinese (China); doornappel in Dutch (Netherlands); chasse taupe endormeuse, endormine, estramon, and herbe in French (France); asthmakraut, botschen, dolkraut, dornapfel, nagwart in German (Germany); maszlag, tsattanto in Hungarian (Hungary); imbutoneblanco, noce spinosa, pomo spinosa, and stramonio in Italian (Italy); barbaka and shinah azghi in Loralai (Pakistan); tlaplat in Mexican (Mexico); piggeple

FIGURE 16.1 Macroscopic characteristics of *Datura stramonium* L. (a) A live flowering plant growing in wild (Bhutan); (b) dried fruits; (c) four chambers of dried fruits open to release seeds; and (d) seed.

in Norwegian (Norway); nanulah and tatulah in Persian (Iran; Tajikistan; Afghanistan); psinki in Polish (Poland); estramonia figueira do inferno in Portuguese (Portugal); durman and durnishnik in Russian (Russia); apple peru, devils apple, devils trumpet, and dewtry in South Africa; estramonio, higueraloca, and trompetilla in Spanish (Spain); spikklubbs in Swedish (Sweden) and tatule in Turkish (Turkey); dhaturo, seto dhaturo, dhattur, and madak in Nepali (Nepal); and dha-du-ra in Bhutanese traditional medicine (Bhutan) (Jarald and Jarald 2006; Phurpa Wangchuk et al. 2011; Soni et al. 2012; Gaire and Subedi 2013; Hong et al. 2015; Al-Snafi 2017; Kalam and Rifat 2020).

16.3 DISTRIBUTION

Many conservationists consider *D. stramonium* as native species to North America or Asia, although their true origin is uncertain (Clapham et al. 1987). Currently, *D. stramonium* is naturalized all over the world and consider a cosmopolitan weed that grows abundantly in the temperate and tropical

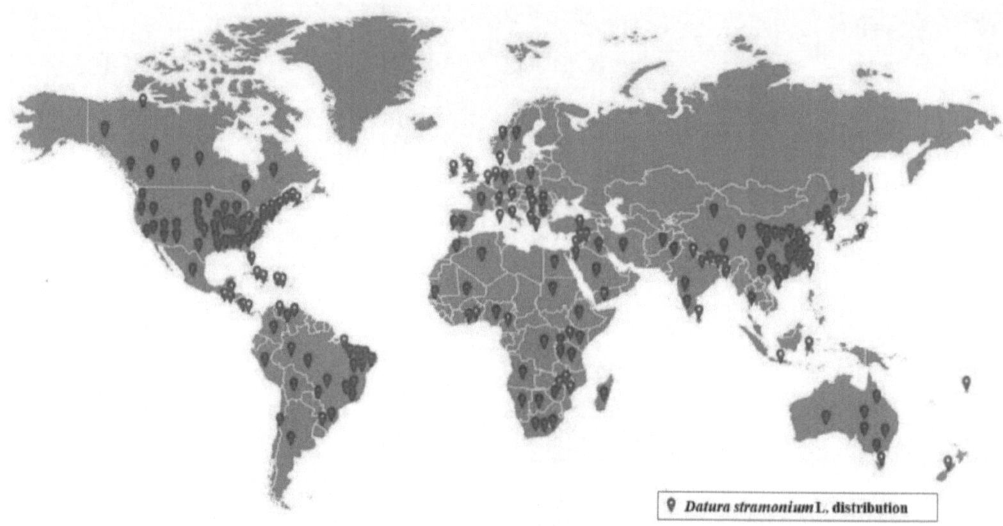

FIGURE 16.2 Distribution patterns of *Datura stramonium L.*

areas of Asia, Europe, New Zealand, South, Central, and North America, and Africa (Berkov et al. 2006; Richardson et al. 2007; Hong et al. 2015). Their habitat preference is wide with most soil types and commonly found as a weed on cultivated land, farmyard, and gardens and also in the landfills, roadsides, near a stream, river flats, waste places, and abandoned cattle yards (Weaver and Warwick 1984; Altameme et al. 2015). *D. stramonium* adapts well to open communities, mostly associated with annual plants. Their aggressive nature of growth outcompetes agricultural field crops and is considered weeds by the agriculturalist (Tutin et al. 1972; Holm et al. 1979; Chaudhary and Al-Howaishel 1980; Lorenzi 1982; Weaver and Warwick 1984; Wells et al. 1986; Kolbek and Sádlo 1996; Holm et al. 1997; Witt and Luke 2017). According to FAO and WHO (2020), *D. stramonium* is labeled as one of the world's most prevalent weeds throughout the world among the recorded *Datura* species exhibiting the growth from sea level to 2,750 m above sea level in the Himalayas (Weaver and Warwick 1984; Shamim and Idris 2019). However, *D. stramonium* has a long history of being used to treat numerous ailments. Because of this, Jimsonweed is cultivated widely, ranging from temperate to the tropical region of Asia, America, Europe, and South Africa (Jakabová et al. 2012; Shah and Seth 2013; Ally and Mohanlall 2020) (Figure 16.2).

16.4 PHYTOCHEMICAL CONSTITUENTS

The phytochemical screening of *D. stramonium* has been conducted since the early 1930s (Shamim & Idris, 2019) and subsequently, several researchers around the globe have analyzed the phytochemical constituents from the stem, leaves, flowers, fruits, and seeds using different extracts (Nain et al. 2013; Al-Snafi 2017). The preliminary qualitative phytochemical screening of the hydromethanolic root extract of *D. stramonium* conducted by Belayneh et al. (2019) revealed the presence of phenols, flavonoids, tannins, alkaloids, glycosides, and steroids, while alkaloids such as atropine, scopolamine, hyoscyamine and 7-Hydroxyhyoscyamine as major tropane alkaloids (Ally and Mohanlall 2020). Similar studies by Das et al. (2012) reported tigloidin, aposcopolamine, apoatropin, hyoscyamine N-Oxide, and scopolamine N-Oxide as minor alkaloids. Additionally, 7-Hydroxyhyoscyamine and 6a-Ditigloyloxytropane were reported in *D. stramonium* (Das et al. 2012). The distribution of hyoscyamine and scopolamine in *D. stramonium* has been explored in different plant parts. It was found that the alkaloid concentration differs with the plant parts and growth stages. For example, the leaves and stems of the young plants indicate hyoscyamine as the dominant component. However, the concentrations of atropine in different parts of both young and adult plants showed higher concentrations than the scopolamine (Devi et al. 2011). A similar finding indicates that alkaloid concentration in the leaves was maximum in the vegetative phase and decreased during maturation (Alinejad et al. 2020).

More than 67 alkaloids have been identified from different parts of *D. stramonium* by Gas Chromatography/Mass Spectrometry, where hyoscyamine and scopolamine were predominate and account for up to 70%–90% of total alkaloid that constitutes (Berkov et al. 2006; El Bazaoui et al. 2011; Li et al. 2012) isolated 12 compounds from the seeds of *D. stramonium*, namely 7-Hydroxy-Beta-Carboline-Propionic acid, 1-Acetyl-7-Hydrox-Beta-Carbol-One, scopolamine, N-Trans-Feruloyl tryptamine, umckalin, fraxetin, daturadiol, daturaolone, N-Trans-Ferulicacyltyramine, cleomiscosin A, hyoscyamilactol, and scopoletin. From the methanol extract of *D. stramonium* leaves, three alkaloids, viz., scopolamine, trigonelline, and tyramine, were isolated (Karmakar et al. 2016).

The investigation of the anti-nutrient composition of thorn apple revealed the presence of the highest phytic acid (29.19±0.40 mg/g) in the seeds, while the highest content of tannin (8.00±0.10 mg/g) and Oxalate (7.77±0.05 mg/g) was seen in the whole seeds. Similarly, the examination of mineral contents showed that seed coats constitute the highest magnesium, copper, and zinc concentration. In contrast, seeds have calcium, iron, phosphorous, and sodium and whole seeds possess the highest concentration of lead, chromium, manganese, and potassium (Oseni et al. 2011). Another study reported 26.20% of carbohydrates, 16.60% of fats, 23.70% of crude fiber, 8.70% of ash, and 8.50% of moisture in seeds of *D. stramonium* (Sharma et al. 2021). Wang and You (2012) reported sterols and their derivatives, viz., 26,26-Dimethyl-5, 24(28)-Ergostadien-3.beta.-ol, 5.Alpha.-Ergosta-7,22-dien-3.beta.-ol, and 3-Hydroxycholestan-5-yl,acetate, as chief components of the essential oil of *D. stramonium*. However, the analysis of essential oil of *D. stramonium* by GC-MS exhibited the presence of neophytadiene, β-Damascenone, and β-Eudesmol as dominant compounds (Chandan et al. 2020; Figure 16.3).

16.5 PHARMACOLOGICAL STUDIES

D. stramonium exhibits continuous biogeographical distribution with significant pharmacological potential. It is well known for its pronounced efficacy and usage in folklore medicine (Ivancheva et al. 2006; Yadav et al. 2021). The primary biologically active components in *D. stramonium* including the alkaloids atropine and scopolamine have indicated therapeutic properties (Pretorius and Marx 2006; Das et al. 2012; Batool et al. 2020). In general, different extracts from the organ of *D. stramonium* have displayed numerous biological activities such as anti-inflammatory, anticancer and cytotoxic activities, anti-insect effect, antimicrobial effect, antioxidant effect, antifungal effect, larvicidal and repellant effects, and antidiabetic, antiasthmatic and analgesic effects (Khalili Najafabadi and Atyabi 2004; Kumral et al. 2010; Belayneh et al. 2019; Alper 2019; Mir et al. 2019; Papi et al. 2021; Nasir et al. 2022; Girmay 2015; Cherie Melaku and Amare 2020; Karimzadeh and Rabiei 2020) (Table 16.1).

16.5.1 Anti-Inflammatory and Antioxidant Activity

The in vitro and in vivo anti-inflammatory activity of ethyl acetate of a powdered leaf of *D. stramonium* was investigated on Sprague-Dawley male rats using CCl4-induced (Carbon tetrachloride) liver injury model and paw and anal edema models. CCl4 (1 mL/kg of 30% CCl4 in olive oil) was administered intraperitoneally, and the anti-inflammatory potential of ethyl acetate at the dose of 150 and 300 mg/kg was evaluated. The ethyl acetate showed significant in vitro NO scavenging with an IC_{50} value of 7:625±0:51 µg/mL and also exhibited prominently in vivo anti-inflammatory activity in paw and anal edema models. The hematological examination in the CCL4 model showed vagotonic effects. Further, ethyl acetate extract-treated rats revealed significant (P<0:001–0:05) improvement in the liver functions by reducing the concentration of oxidative stress markers in liver tissues (Nasir et al. 2022).

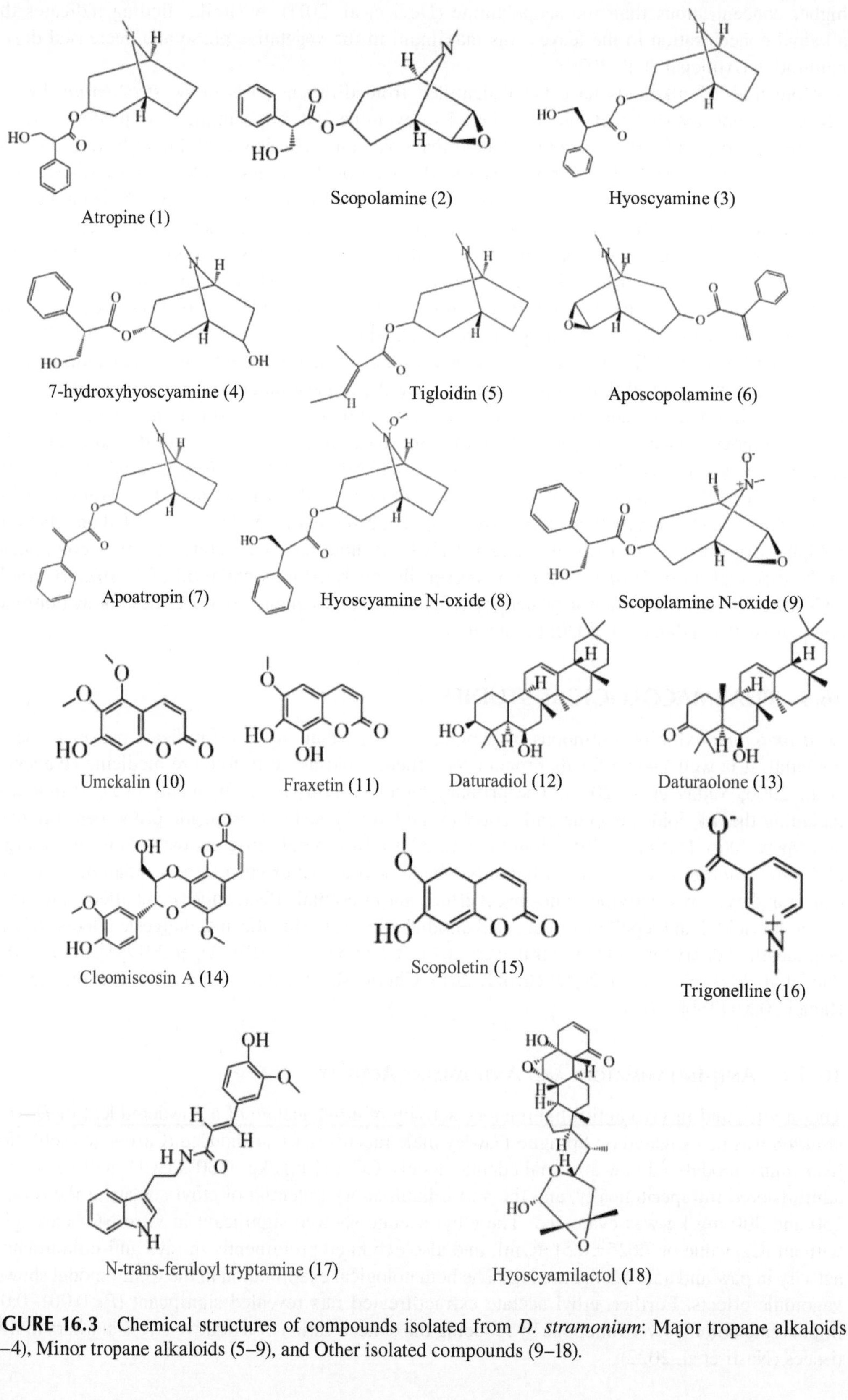

FIGURE 16.3 Chemical structures of compounds isolated from *D. stramonium*: Major tropane alkaloids (1–4), Minor tropane alkaloids (5–9), and Other isolated compounds (9–18).

TABLE 16.1
Pharmacological Activities of Different Extract and Isolated Compounds of *Datura stramonium*

Isolated Compound/ Extract	Reported Pharmacological Activities	Model Organism(s)/Cell Line(s) Used	Effective Dosage (IC_{50} Values Unless Stated)	References
Chloroform crude extract	Antimicrobial activity (in vitro)	*Bacillus subtilis*	Maximum zone of inhibition = 18.43 ± 0.57 mm	Girmay (2015)
Datura stramonium oil	Antioxidant activity	DPPH and ABTS assay	IC_{50} value of 71.35 ± 1.06 and 61.01 ± 1.07 µg/mL	Chandan et al. (2020)
Ethanol extract of stem bark	Antimicrobial activity	*K. pneumonia* *S. aureus* *S. typhi*	Diameter of the zone of inhibition of 13.0, 12.0, and 5 mm, respectively	Shagal et al. (2012)
Ethanolic extract of leaves	Antimicrobial activity (in vitro)	*Escherichia coli*, *Pseudomonas aeruginosa*, *Staphylococcus aureus*, *Proteus mirabilis*, and *Klebsiella pneumonia*	Zone of inhibition is 1.8, 1.4, 1.3, 2.01, and 1.7 mm respectively	Altameme et al. (2015)
Ethanolic extracts	Anti-larvicidal activity	*Aedes aegypti*, *Anopheles stephensi*, and *Culex quinquefasiciatus*	At a concentration of 86.25, 16.07, and 6.25 mg/L, respectively	Swathi et al. (2012)
Ethyl acetate (EA) Powdered leaf (DSL-EA)	Anti-inflammatory activity (in vivo)	Carrageenan-induced hind paw edema	Alleviated the edema caused by carrageenan in a dose- and time-dependent manner with a high dose of DSL-EA (300 mg/kg) showing more than 70% inhibitory effect	Nasir et al. (2022)
Ethyl acetate (EA) Powdered leaf (DSL-EA)	Anti-inflammatory activity (in vitro)	Nitric oxide (NO) scavenging assay using BALB/c mice	IC_{50} of 7:625 ± 0:51 µg/mL	Nasir et al. (2022)
Ethyl acetate fraction of flowers	Anti-cancer activity	Human liver cancer cell line (HePG2). Sample concentrations of 1,000, 500, 250, 125, and 62.5 µg/mL	Showed CTC_{50} values of 79.43, 64.15, 58.63, 49.52, and 37.48 µg/mL, respectively	Rajeshkanna et al. (2016)
Hexane crude extract	Antimicrobial activity (in vitro)	*Salmonella typhi*	Maximum zone of inhibition = 14.70 ± 0.32 mm	Girmay (2015)
Hexane crude extract	Antimicrobial activity (in vitro)	*Fusarium oxysporum*	Maximum zone of inhibition = 16.35 ± 0.23 mm	Girmay (2015)
Hydromethanolic root extract	Vivo antidiabetic (in vivo)	Oral glucose-loaded diabetic male Swiss albino mice	The plant extract doses of 100, 200, and 400 mg/kg significantly ($P < 0.05$) reduced blood glucose levels	Belayneh et al. (2019)

(Continued)

TABLE 16.1 (*Continued*)
Pharmacological Activities of Different Extract and Isolated Compounds of *Datura stramonium*

Isolated Compound/ Extract	Reported Pharmacological Activities	Model Organism(s)/Cell Line(s) Used	Effective Dosage (IC_{50} Values Unless Stated)	References
Hydromethanolic root extract	Antidyslipidemic	Streptozotocin (STZ)-induced diabetic male Swiss albino mice	–	Belayneh et al. (2019)
Hydromethanolic root extract	Antioxidant activities (in vitro)	2,2-Diphenyl-1-Picrylhydrazine (DPPH) free radical scavenging assay	Showed radical scavenging activity with IC_{50} value of 13.47 µg/mL	Belayneh et al. (2019)
Methanol extract of seeds	Antioxidant activity	1,1-Diphenyl-2-Picrylhydrazyl (DPPH), Superoxide radical scavenging assay, and ABTS˙ radical cation assay	IC_{50} values of 94.87, 39.59, and 41.10 µg/mL, respectively	Iqbal et al. (2017)
Methanol extract of seeds	Cytotoxic activity	Human breast cancer (MCF7) cell line (Methyl Thiazolyl Diphenyl-Tetrazolium bromide [MTT])	Showed 66.84% at 500 µg/mL concentration	Iqbal et al. (2017)
Methanolic extract of seeds	Antimicrobial activity	*E. coli*, *S. aureus*, *P. aeruginosa*, and *B. subtilis*	Zone of inhibition 20, 17.50, 16, and 15 mm, respectively	Waza et al. (2015)
Petroleum ether crude extract	Antimicrobial activity (in vitro)	*Escherichia coli*, *S. aureus*	Maximum zone of inhibition = 16.23 ± 0.62 and 19.30 ± 0.18 mm	Girmay (2015)
Petroleum ether crude extract	Antifungal activity (in vitro)	*Fusarium solani*, *Aspergillus niger*	Maximum zone of inhibition = 16.21 ± 0.64 and 17.07 ± 0.16 mm	Girmay (2015)
Scopolamine	TRAIL-resistance overcoming activity	TRAIL-resistant human gastric adenocarcinoma (AGS) cells	Treating cells with 10, 20, and 50 µM of 1 along with TRAIL (100 ng/mL) resulted in 23%, 26%, and 42% more inhibition	Karmakar et al. (2016)
Trigonelline	TRAIL-resistance overcoming activity	TRAIL-resistant human gastric adenocarcinoma (AGS) cells	Combined treatment of TRAIL 1 (100 ng/mL) and 2 (50, 100, and 150 µM) resulted in 15%, 19%, and 44% more inhibition	Karmakar et al. (2016)
Tyramine	TRAIL-resistance overcoming activity	TRAIL-resistant human gastric adenocarcinoma (AGS) cells	Compound 3 at 10, 50, and 100 µM along with TRAIL (100 ng/mL) resulted in 19%, 23%, and 32% decrease in cell viability than treatment with compound alone (without TRAIL) showing moderate TRAIL-resistance overcoming activity	Karmakar et al. (2016)

The antioxidant activity of *D. stramonium* oil was evaluated, implying Diphenyl-1-Picrylhydrazine (DPPH) and 2,2′-azino-bis(3-ethylbenzothiazoline-6-sulfonic acid) (ABTS) assay at a varying dosage ranging from 10 to 80 μg/mL. In both cases, free radical scavenging activity was marked as dose dependent, with the most significant activity at IC_{50} values of 71.35 ± 1.06 and 61.01 ± 1.07 μg/mL for DPPH and ABTS assays, respectively (Chandan et al. 2020). Investigation of in vitro anti-inflammatory activity of *D. stramonium* oil was conducted by RBC membrane stabilization and BSA denaturation method. In the study, *D. stramonium* oil exhibited significant membrane stabilization at the IC_{50} value of 68.92 ± 1.08 μg/mL and showed pronounced albumin denaturation inhibiting activity at the IC_{50} value of 91.30 ± 1.14 μg/mL (Chandan et al. 2020).

Iqbal et al. (2017) experimented with investigating the antioxidant activity of methanolic seed extracts of *D. stramonium* using DPPH and Superoxide radical assay. The result showed the highest DPPH scavenging activity at a concentration of 60 μg/mL with an IC_{50} value of 94.87 μg/mL, and the maximum superoxide radical scavenging activity was exhibited at a concentration of 30 μg/mL with an IC_{50} value of 39.59 μg/mL. A similar study by Cherie Melaku and Amare (2020) also showed in vitro antioxidant activity of the hydromethanolic seed extract with an IC_{50} value of 11.95 mg/mL.

16.5.2 Anti-Cancer and Cytotoxic Activity

Activity-guided fractionation of the MeOH extract from the leaves of *D. stramonium* led to the isolation of trigonelline, scopolamine, and tyramine alkaloids. All isolated compounds were tested against human gastric adenocarcinoma (AGS) cells to evaluate Tumor Necrosis Factor (TNF)-related apoptosis-inducing ligand (TRAIL) resistance activity. All three alkaloids, trigonelline, scopolamine, and tyramine, exhibited TRAIL-resistance overcoming activity at 50, 100, and 150 μM, respectively (Karmakar et al. 2016). In another study, the higher concentration of methanolic seed extracts of *D. stramonium* showed significant cytotoxicity on human breast adenocarcinoma (MCF-7) cells with an IC_{50} value of 113.05 μg/mL (Iqbal et al. 2017). Human cancer cells derived from the breast (DA-MB231), hypopharyngeal carcinoma cell line (FaDu), when treated with aqueous leaf extract of *D. stramonium* for 24 hours, showed significant enhancement ($P < 0.05$) of clonogenic cell killing (Ahmad et al. 2009).

Varying concentrations of Ethyl acetate fraction of flowers were examined for anti-cancer activity on the human liver cancer HePG2 cell line by MTT assay method. The finding revealed that ethyl acetate fraction concentrations of 1,000, 500, 250, 125, and 62.5 μg/mL exhibited the CTC_{50} (50% of cytotoxicity inhibition) values of 79.43, 64.15, 58.63, 49.52, and 37.48 μg/mL against the human liver cancer HePG2 cell line, respectively (Rajeshkanna et al. 2016).

16.5.3 Antimicrobial and Antifungal Activity

Several studies have shown that methanol extracts of *D. stramonium* exhibit dose-dependent antimicrobial activity against gram-positive bacteria. For instance, the antibacterial activity of methanol extract was observed highest at the concentrtion of 2.5 mg/mL against *Bacillus subtilis*, *Staphylococcus aureus*, and *Enterococous faecalis* with an inhibition zone of 11, 10, and 9 mm, respectively (Eftekhar et al. 2005). Vibriocidal effect of *D. stramonium* acetone extracts against *Vibrio cholera* and *Vibrio parahaemolyticus* strains were examined by Sharma et al. (2009). The finding showed a minimum inhibitory concentration (MIC) value in the range of 2.5–15 mg/mL, which suggests broad vibriocidal properties of *D. stramonium* (Sharma et al. 2009). Shagal et al. (2012) evaluated the antimicrobial activity of ethanolic and aqueous extracts from different parts of *D. stramonium* against various bacterial strains, and the finding suggested that ethanolic extracts had a better activity than others. In a similar study by Al-Snafi (2017), benzene, chloroform, and ethanol extract of branches and leaves of *D. stramonium* exhibited significant antibacterial activity against *Pseudomonas aeruginosa, Micrococcus luteus, Klebsiella pneumonia Escherichia coli,* and *Staphylococcus aureus.* The antimicrobial activity of the plant's stem bark extracts (ethanol) indicated inhibitory activity in the order of *K. pneumonia, S. aureus,* and *S. typhi,* with 13.0, 12.0, and 5.0 mm, respectively (Shagal et al. 2012). Furthermore, antimicrobial and antifungal activity of crude extract of *D. stramonium,* examined by paper disk diffusion, method revealed that types of solvent and parts of plants used in extraction are deciding factors in revealing the extent of activity. For example, petroleum ether, hexane, chloroform, acetone, and ethanol extract of the plants showed maximum zone of inhibition against *S. aureus* (19.30±0.18 mm), *S. aureus* (18.00±0.27 mm), *B. subtilis* (18.43±0.57 mm), *S. aureus* (15.60±0.21 mm), and *S. aureus* (15.50±0.55 mm), respectively (Girmay 2015). All the extracts showed the highest activity against *Aspergillus niger* for antifungal activity, with the maximum zone of inhibition value ranging from 13.05±0.22 to 17.07±0.16 mm (Girmay 2015).

16.5.4 Larvicidal, Repellant, and Insecticidal Activity

Studies have shown insecticidal and repellant activity of *D. stramonium* extract (Monira and Munan 2012). The ethanolic leaf extracts of *D. stramonium* were investigated for larvicidal activity and the result showed significant activity against *Aedes aegypti, Anopheles stephensi,* and *Culex quinquefasciatus* with half lethal dose (LD_{50}) values of 86.25, 16.07, and 6.25 mg/L, respectively (Swathi et al. 2012; Gaire and Subedi 2013).

Karimzadeh and Rabiei (2020) reported the larvicidal and oviposition preventive potential of the extract from different parts of *D. stramonium* against *Plutella xylostella.* The extract prepared from the roots, stem, leaves, flowers, and seeds exhibited toxicity against *P. xylostella* larvae with LC50 (median lethal concentration) values of 165.3, 841.4, 526.6, 82.3 and 146.8 µg/mL, respectively. Likewise, oviposition activity was observed highest in flower extracts, followed by root and seed extracts (Karimzadeh and Rabiei 2020). Batool et al. (2020) examined the mosquito repellent activities of ethanolic leaf extracts of *D. stramonium* against *C. quinquefasciatus, A. stephensi,* and *A. aegypti.* At higher concentrations of the extract, mosquito repellency of 2.7, 71.7, and 117.7 minutes was observed against *C. quinquefasciatus, A. stephensi,* and *A. aegypti,* respectively.

The seed extracts of *D. stramonium* were prepared using different solvents and tested for insecticidal activity against *Sitophilus oryzae* (Jawalkar et al. 2016). Findings revealed that the nature of the solvent used to prepare the extract, concentration, and exposure time were determining factors to obtain a significant level of mortality. Overall, insecticidal activity of the different extracts was in the following order: acetone extracts (LD_{50}=1.19 mL/kg), chloroform extracts (LD_{50}=7.38 mL/kg), and ethanol extract (LD_{50}=8.59 mL/kg). Further, the mortality rate increased proportionately with increasing concentration (4–16 mL/kg) of extract and exposure time (24–96 hours) (Jawalkar et al. 2016).

16.5.5 Antidiabetic

Hydromethanolic root extract of *D. stramonium* was investigated for blood sugar-lowering activity on oral glucose-loaded, normoglycemic, and streptozotocin (STZ)-induced diabetic mice models. Three doses (100, 200, and 400 mg/kg) of hydromethanolic root extract significantly ($P<0.05$, $P<0.01$) lowered blood sugar after 1 and 2 hours of treatment on oral glucose-loaded mice (Belayneh et al. 2019). In the same way, administration of 100 and 200 mg/kg of root extract significantly ($P<0.05$) improved the body weight of the diabetic mice on week 2, while oral administration of 400 mg/kg showed body weight improvement on week 1 and 2. Administration of root extracts also significantly ($P<0.05$) lowered total cholesterol and triglycerides, whereas

HDL cholesterol levels significantly ($P<0.05$) increased (Belayneh et al. 2019). A similar study by Cherie Melaku and Amare (2020) also reported that three doses of hydromethanolic seed extract exhibited significantly ($P<0.05$ at 100, $P<0.01$ at 200 and 400 mg/kg) antidiabetic activity against diabetic mice.

16.5.6 Antiasthmatic

Tropane alkaloids of *D. stramonium*, namely atropine and scopolamine, exhibit anticholinergic and broncho-dilating actions capable of blocking the muscarinic receptors (M1, M2, and M3 linked to the airway and lung tissues) and consequently dilating bronchial smooth muscles (Choudhary et al. 2021). In asthmatic patients, muscarinic receptors are highly active and stimulate bronchi constriction. However, muscarinic receptors (M2) inhibit the secretion of acetylcholine and indicate a minimal function in asthmatic patients (Papi et al. 2021). Therefore, administration of *D. stramonium* inhibits M2 function on airway smooth muscles and submucosal gland cells, eventually ensuring the release of neurotransmitters and dilation of bronchi in asthmatic patients (Kadam et al. 2018). For example, inhaling the smoke from the cigarette prepared from *D. stramonium* showed a significant decrease in airway resistance (Al-Snafi 2017).

16.6 TOXICOLOGICAL EVALUATION OF *DATURA STRAMONIUM*

D. stramonium has been globally recognized as poisonous and narcotic due to toxic compounds such as atropine, hyoscyamine, and scopolamine in all parts of the plants (Altameme et al. 2015). Subsequently, all the parts have exhibited anticholinergic effects on the central nervous system, which have ultimately induced complications and even death in severe intoxication (Sharma et al. 2021). Overall, seeds of *D. stramonium* are considered more toxic (each seed contains 0.1 mg of atropine) than other parts (Pillay and Sasidharan 2019). Despite *D. stramonium* being used extensively in folk and Ayurvedic medicine for curing various human ailments, several cases of its poisoning have been reported when used as decoction or taken accidentally or abused intentionally (Das et al., 2012; Adegoke and Alo 2013; Disel et al. 2015; Trancă et al. 2017). In any case, the symptoms of *D. stramonium* poisoning include fever, dryness of the skin, mouth, and mucous membrane, urinary retention, tachycardia, rhabdomyolysis, coma, hallucinations, mydriasis, tachypnea, rhabdomyolysis, disorientation, hallucinations, combativeness, unintelligible speech, restlessness, dizziness, abdominal pain, vomiting, and nausea (Onen et al. 2002; Lazzarini et al. 2006; Ally and Mohanlall 2020) (Table 16.1).

Many researchers have mainly studied the toxicological properties of *D. stramonium.* Dugan et al. (1989) evaluated the toxicological effects of *D. stramonium* seeds on both male and female rats. The rats were fed with a diet containing 0.5%, 1.58%, and 5.0% of *D. stramonium* seeds for 90 days. The result analysis showed a loss of body weight, the gain of serum calcium and albumin, and an increase in the weight of the testes and liver. It is concluded that a diet with 0.5% or more concentration of *D. stramonium* seeds induces adverse physiological alteration in rats (Dugan et al. 1989). In an acute oral toxicity test, female mice administered with hydroethanolic root extract at 2,000 mg/kg for 14 days showed no acute toxicity and mortality, concluding as a safe dose (Belayneh et al. 2019). The study by Ogunmoyole et al. (2019) revealed that types of solvent used to prepare *D. stramonium* seed extracts showed varying toxicity on the heart, kidney, brain, and liver. Wistar rats were injected with 2,250 mg/kg/day of *D. stramonium* seeds extract for 28 days. The results exhibited significant changes in the histological pattern, such as vacuolization, axonal atrophy, and neuronal deaths in the frontal cortices of rats. It is concluded as the effect of intoxication of *D. stramonium* seeds extract on frontal cortex neurons (Ekanem et al. 2016; Table 16.2).

TABLE 16.2
Intoxication Cases of *Datura stramonium* Reported in Different Countries

Plant Part	Country	Type of Intake	Indication of Intoxication	Consequence	References
Aqueous extract of seeds	Nigeria	Abuse	Restlessness, incoherent talk, visual hallucinations, convulsions, increased body temperature, dry mouth, dilated pupils, tachycardia	Recovered	Adegoke and Alo (2013)
Areal parts	France	Accidental	Unconsciousness, agitation, delirium, visual and tactile hallucinations M: 1.7 ng/mL atropine in blood, F: 1.4 ng/mL scopolamine in blood; 114 ng/mL atropine in urine	Recovered	Marc et al. (2007)
Flowers	Turkey	Accidental (flowers mistaken with pumpkin flowers)	Visual impairment, hallucinations, mydriasis, tachycardia, tachypnea, red and dry skin and mucous membranes, left bundle branch block, rhabdomyolysis	Recovered	Disel et al. (2015)
Flowers	Italy	Accidental (flowers mistaken with edible pumpkin)	Loss of consciousness, agitation, confusion, hallucinations, combative behavior, mydriasis, disorientation, aphasia	Recovered	Lazzarini et al. (2006)
Flowers	United States	Abuse	Hallucinations, picking at nonexistent items	Recovered	Chan (2002)
Fruit	Saudi Arabia	Accidental	Dry mouth and skin, tachycardia, dilated pupils	Recovered	Ali Taha and Mahdi (1984)
Fruit	Malaysia		Dizziness, blurred vision, restlessness	Recovered	Mohamad et al. (2009)
Seeds	Romania	Abuse	Fever, dry skin, and mucosa, tachycardia, right bundle bunch block, urinary retention, rhabdomyolysis, coma	Recovered	Trancă et al. (2017)
Seeds	United States	Abuse	Disorientation, hallucinations, combativeness, unintelligible speech, blurry vision	Recovered	Soneral and Connor (2005)
Seeds	Botswana	Accidental (sorghum porridge contaminated with *D. stramonium* seeds)	Confusion, dry mouth, restlessness, dizziness, abdominal pain, uncontrolled talking, headache, vomiting, nausea, blurred vision	Recovered	Onen et al. (2002)

(*Continued*)

TABLE 16.2 (*Continued*)
Intoxication Cases of *Datura stramonium* Reported in Different Countries

Plant Part	Country	Type of Intake	Indication of Intoxication	Consequence	References
Seeds	Canada	Abuse	Disorientation, hallucinations, combativeness	Recovered	Wiebe et al. (2008)
Seeds	Greece	Abuse	1.1 μg/mL of Hyoscyamine and 0.2 μg/mL scopolamine were determined in the blood	Died	Boumba et al. (2004)
Seeds	Hungary	Abuse (seeds used for preparing tea)	Dizziness, blurry vision, incoherent talk, dry mouth, flushed skin, mydriasis, tachycardia, coma	Recovered	Osváth et al. (2000)

16.7 TRADITIONAL AND OTHER POTENTIAL USES

D. stramonium, commonly known as Jimson weed, has been widely used in the traditional medicine of different countries from the time immemorable. In Ayurvedic and folk medicine, Jimson weeds are therapeutically indicated in treating numerous ailments such as toothache, asthma, fever, bronchitis, inflammation, and bruises (Kirtikar 1935; Ummahan 2017). The juice extracted from the leaves of Jimson weeds is taken orally with warm milk to expel the intestinal worms, and in a similar trend, seeds paste mixed with palm oil is applied externally to treat insects' bites and stings. Further, juice from the flower petals is used against ear pain and as a purgative in treating cough, fever, and asthma (Egharevba and Ikhatua 2008; Khan et al. 2013). Kadam et al. (2018) reported the efficacy of Jimson weeds in treating infertility in women. Here, 120 mg of dried powder of flowers is given with honey for a week after 10 days of menstruation. The leaves of *D. stramonium* mixed with mustard oil have proven helpful in curing skin disorders (Khan et al. 2013). A paste prepared from leaves is topically applied to treat wounds, injuries, baldness, bleedings, boils, and pains (Khan and Khatoon 2008). Juice extracted from the fruit prevents hair failure and antidandruff (Shah and Khan 2006; Rahmatullah et al. 2010). Oil from fruits helps relieve body pain when applied (Vijendra and Kumar 2010). Both seeds and dried leaves are cast off as a sedative and anticholinergic (Wazir et al. 2004). The vapors from the leaf infusion are suggested to be used to revive the pain of gout and rheumatism (Biswas et al. 2011). Savithramma et al. (2007) acclaim the inhalation of smoke from burning leaf of *D. stramonium* to reconcile asthma and bronchitis (Savithramma et al. 2007).

In different regions of Nepal, leaves of *D. stramonium* together with a stem of *Neopicrorhiza scrophulariiflora* and leaves of *Cannabis sativa* are pulverized with water and used as herbal medicine to overcome headaches. For indigestion problems, seeds and rice grains are minced and administered orally. Leaves are either warmed or roasted and applied to the inflamed region to relieve pain (Gorsi and Shahzad 2002). In the Kashmir valley of India, oil extracted after crushing leaves is prescribed for antidandruff against skin diseases, headaches, and its stalks and stems are masticated as a treatment for toothache and respiratory problems (Khan et al. 2016). Besides the medicinal potential of *D. stramonium* in treating various ailments, its usage for non-medicinal purposes is also recorded. For instance, its usage among the students is recorded in Ethiopia to enhance imagination power and infuse a receptive learning mindset. In many European countries, the plant is used for brewing beer and in liniments as a craft witch (Khan et al. 2016). The traditional uses of *D. stramonium* in various parts of the world are listed in Table 16.3.

TABLE 16.3
The Traditional Uses of *Datura stramonium* in Various Parts of the World

Ailments	Country	Part Used	Preparation Form/Traditional Uses	References
	Mexico	Leaves	Cataplasm	BDMTM (2009)
Anesthetic	Spain	Leaves	Cataplasm	Velasco (2009)
Antidandruff	Pakistan	Fruit	Juice extracted is applied to the scalp for falling hairs	Khan and Khatoon (2008)
Asthma	Spain	Leaves, roots and flowers	Cigarettes, decoction	Carrió and Vallès (2012)
Asthma and tonsil	Pakistan	Seeds	Chewed	Khan and Khatoon (2008)
Asthma, Bronchitis, Cough, Headache, Joint pain, Rheumatism, and Tonic	India (Kashmir)	Leaves	Crushed and boiled and taken orally with water	Khan et al. (2016)
Boils, Headache, Skin disease, Dandruff	India (Kashmir)	Seeds	Its seeds are crushed and oil is obtained which acts as an ointment Tropical application of ointment	Khan et al. (2016)
Bronchitis	Spain	Flowers	Fumigated	Carrió (2013)
Bronchitis and asthma	India	Fruits	Burnt and ash are given orally with honey	Belayneh et al. (2019)
Burns	Mexico Spain	Leaves	Direct application, Cataplasm	Salas and Cáceres (2003)
Cold and catarrh	Spain	Indet., leaves	Indet., Cigarettes	Carrió (2013)
Cough	Mexico Spain	Leaves	Infusion Cigarettes	Abigail et al. (1994), Velasco (2009)
Cough, fever, and asthma	Pakistan	Seeds	Used as purgative, smoked for its narcotic action	Khan et al. (2013)
Diabetes mellitus	Ethiopia	Root extract	The root of the plant is taken orally	Belayneh et al. (2019)
Earache	Pakistan	Flower	Extract the juice and apply it externally	Khan et al. (2013)
Earache	Pakistan	Flower	Extract the juice and apply it externally	Shah and Khan (2006)
Eczema	Spain	Aerial parts	Cataplasm	Akerreta et al. (2013)
Erysipelas	Mexico	Leaves	Direct application (externally)	BDMTM (2009)
Asthma and Bronchitis	India	Leaf	The smoke from the burning leaf is inhaled	Savithramma et al. (2007)
Furuncle and Traumatic injury	China	Leaf	Pound fresh part applied on the affected area, treating for	Hong et al. (2015)
Gout	Mexico	Indet.	Indet.	BDMTM (2009)
Headache	Mexico	Flor	Decoction	BDMTM (2009)

(*Continued*)

TABLE 16.3 (*Continued*)
The Traditional Uses of *Datura stramonium* in Various Parts of the World

Ailments	Country	Part Used	Preparation Form/Traditional Uses	References
Headache	Nepal	Leaves	Leaves along with the leaves of *Cannabis sativa* and stem of *Neopicrorhiza scrophulariiflora*, are pounded with water and applied topically	Rajbhandari (2001)
Headache	Pakistan	Seeds	Seeds (2–5) are added to a cup of green tea	Hussain et al. (2006)
Helminthiasis	Mexico	Leaves	Decoction	BDMTM (2009)
Hemorrhoids	Mexico Spain	Leaves	Decoction	Abigail et al. (1994), Salas and Cáceres (2003)
Impotency	–	Seeds	Boil 15 ripe fruits in 8 L of cow milks and extract butter churning. Use butter to message penis and spin every morning and evening	Prasad (2008)
Indigestion	Nepal	Seeds	Seeds with grains of rice are crushed and taken orally	Rajbhandari (2001)
Infertility in women	India	Dried flower powder	Administered at the dose of 120 mg is given with honey 10 days after menstruation. It is given for 5–7 days	Kadam et al. (2018)
Inflammation	Mexico, Spain	Leaves	Direct application, infusion, Cataplasm	Abigail et al. (1994)
Inflammation of the uterus	Mexico	Leaves	Direct application (externally)	BDMTM (2009)
Intestinal worms (tapeworm)	Nepal	Leaf	Juice extracted from the leaf is given with mil to expel intestinal worms	Rajbhandari (2001)
Joint pain	Mexico	Leaves, Seeds	Decoction and added to a bath, or spread. Direct application	BDMTM (2009)
Labor pain	Mexico	Indet., Seeds	Indet., Cataplasm, Decoction	BDMTM (2009), Morton (1981)
Lactation	Pakistan	Leaves	Warm up 5–8 leaves in low fire and then tie on flagged breasts to bust them up. This treatment is continued for 15 days	Hussain et al. (2006)
Mumps	Mexico	Leaves	Decoction and externally applied	BDMTM (2009)
Neuralgia	Mexico	Seeds	Tincture	Morton (1981)
Parasites ("bichos")	Mexico	Leaves	With butter	Abigail et al. (1994)
Pertussis	Mexico	Leaves	Decoction and added to a bath	BDMTM (2009)
Pimples, grains	Mexico	Leaves	Decoction and externally applied	Abigail et al. (1994)
Rheumatism and gout	Bangladesh	Leaf	Vapors of leaf infusion are used to relieve the pain of	Biswas et al. (2011)
Rheumatism	Spain	Roots, Leaves	Macerated in water or alcohol, Cataplasm	Abigail et al. (1994), Velasco Santos et al. (2010)

(*Continued*)

TABLE 16.3 (*Continued*)
The Traditional Uses of *Datura stramonium* in Various Parts of the World

Ailments	Country	Part Used	Preparation Form/Traditional Uses	References
Scorpion stings	Saudi Arabia	Whole plants	Leaves are dried, crushed, heated, and applied topically to the sting point	Tounekti et al. (2019)
Skin diseases	Bangladesh	Leaf	Paste of leaves is topically applied	Rahmatullah et al. (2010)
Skin disorders	Pakistan	Leaf	Mixed with mustard oil and apply externally	Khan et al. (2013)
Softening the boils	Pakistan	Green leaves	Used externally for softening the boils	Shah and Khan (2006)
Sores and stings	Nigeria	Leaf and seeds	The crushed leaves and seeds are mixed with palm oil and applied to severe cases of insects bites and stings	Egharevba and Ikhatua (2008)
Sores and stings	Nigeria	Leaves and seeds	The crushed leaves and seeds are mixed with palm oil and applied topically	Egharevba and Ikhatua (2008)
Sudorific	Spain	Seeds	Infusion	Salas and Cáceres (2003)
Toothache	Mexico	Seeds	Direct application	BDMTM (2009)
Toothache, Aphrodisiac, Intestinal worms, and Respiratory disorder		Stems and stalk	Chewed	Khan et al. (2016)
Tranquilizer – Sedative	Mexico, Spain	Indet. Leaves	Indet., Cigarettes, Decoction	BDMTM (2009) Garrido (2008)
Vomiting inducer	Spain	Seeds	Infusion	Salas and Cáceres (2003)
Wounds, bleedings, and baldness	Pakistan	Leaf	Paste and extract are applied externally	Khan and Khatoon (2008)

16.8 CONCLUSION AND FUTURE REMARKS

D. stramonium is a cosmopolitan species. It has aggressive nature of growth by releasing allelopathic chemicals and outcompetes many associated plant species. It is commonly known as Jimson weed, devil's trumpet, and evil seeds and has both toxic and medicinal properties. All parts of a plant are toxic if ingested and smoked, often resulting in delirium, photophobia, hyperthermia, neurotoxicity, severe mydriasis, fever, dryness of the skin, mouth and mucous membrane, rhabdomyolysis, hallucinations, mydriasis, tachypnea, rhabdomyolysis, disorientation, combativeness, unintelligible speech, restlessness, dizziness, vomiting, and nausea. In this view, it is indispensable to be mindful of toxicity and likely risk linked with the therapeutic or recreational use of *D. stramonium*. Traditionally, the plant is used for curing various human ailments such as skin disorders, inflammation, toothache, cough, fever, bronchitis, convulsion, asthma, burn, ulcer, wound, rheumatism, sciatica, and gout. However, one has to be cautious when administering orally. Pharmacologically, based on the type of solvent used for extraction and parts of plant use, plant extract has exhibited varying biological activities, including antifungal, antiasthmatic, larvicidal, antibacterial, anti-inflammatory, and antioxidant, antispasmodic, anti-rheumatoid, anti-ulcer, and antinociceptive activities. However, biological activities are limited to plant extract. *D. stramonium* encompasses several phytochemicals, including tropane alkaloids, flavonoids, phenolic compounds, carbohydrates, cardiac glycosides, tannins, amino acids, and minerals. However, most compounds are identified or detected, and isolated compounds are very few. Hence, reported biological activities of *D. stramonium* are limited to crude extracts of different solvents. For example, about 67 alkaloids have been identified, but they are least studied for their pharmacological properties. We recommend that future studies focus on compound isolation and studying the pharmacological activities of those compounds that are already isolated, like alkaloids.

NOTES

Tenzin Jamtsho, Yangchenphugh Higher Secondary School, Ministry of Education, Thimphu, Bhutan

Ugyen, Jigme Khesar Strict Nature Reserve, Department of Forests & Park Services, MoAF, Haa, Bhutan

Phurpa Wangchuk, Center for Molecular Therapeutics, Australian Institute of Tropical Health and Medicine (AITHM), James Cook University, Smithfield, Australia

REFERENCES

Abigail, A., Raúl, C. J., Soledad, C., Patricia, J., & Edith, L. M. 1994. *Herbario medicinal del Instituto Mexicano del Seguro Social: informacion etnobotanica*. 1st ed. Juárez: Información Etnobotánica.

Adegoke, S., & Alo, L. 2013. Datura stramonium poisoning in children. *Nigerian Journal of Clinical Practice* 16 (1); 116.

Ahmad, I. M., Abdalla, M. Y., Mustafa, N. H, & Qnais, E. Y. 2009. Datura aqueous leaf extract enhances cytotoxicity via metabolic oxidative stress on different human cancer cells. *Jordan Journal of Biological Sciences* 2 (1); 9–15.

Akerreta, S., Calvo, M. I., & Cavero, R. Y. 2013. *Sabiduría popular y plantas curativas. Recopilación Extraída De un Estudio Etnobotánico En Navarra*. Madrid, Spain: Ediciones I.

Ali Taha, S., & Mahdi, A. H. 1984. *Datura* intoxication in Riyadh. *Transactions of the Royal Society of Tropical Medicine and Hygiene* 78 (1); 134–135.

Alinejad, S., Sarabi, V., Bakhtvari, A. R. S.,& Hashempour, H. 2020. Variation in physiological traits, yield and secondary metabolites of jimsonweed (*Datura stramonium* L.) under different irrigation regimes and nutrition systems. *Industrial Crops and Products* 143; 111916.

Ally, F., & Mohanlall, V. 2020. An overview of tropane alkaloids from *Datura stramonium* L. *Journal of Pharmacognosy and Phytochemistry* 9 (3); 05–13.

Alper, M. 2019. Investigation of potential anti-cancer and anti-inflammatory effects of *Datura stramonium* ethanolic extracts against selected human cancer cell lines. *Parlar Scientific Publications* 28 (12); 8993–9003.

Al-Snafi, A. E. 2017. Medical importance of *Datura fastuosa* (syn: *Datura metel*) and *Datura stramonium*—A review. *IOSR Journal of Pharmacy* 7 (2); 43–58.

Altameme, H. J., Hameed, I. H., & Kareem, M. A. 2015. Analysis of alkaloid phytochemical compounds in the ethanolic extract of *Datura stramonium* and evaluation of antimicrobial activity. *African Journal of Biotechnology* 14 (19); 1668–1674.

Batool, A., Batool, Z., Qureshi, R., & Iqbal Raja, N. 2020. Phytochemicals, pharmacological properties and biotechnological aspects of highly medicinal plant. *Journal of Plant Sciences* 8 (2); 29.

Belayneh, Y. M., Birhanu, Z., Birru, E. M., & Getenet, G. 2019. Evaluation of in vivo antidiabetic, antidyslipidemic and in vitro antioxidant activities of hydromethanolic root extract of Datura stramonium L. (Solanaceae). *Journal of Experimental Pharmacology* 11; 29–38.

Berkov, S., Zayed, R., & Doncheva, T. 2006. Alkaloid patterns in some varieties of Datura stramonium. *Fitoterapia* 77 (3); 179–182.

Biblioteca Digital de la Medicina Tradicional Mexicana (BDMTM). 2009. http://www. medicinatradicionalmexicana.unam.mx/index.php (Accessed on 20 April, 2022).

Biswas, K. R., Khan, T., Monalisa, M. N., Swarna, A., Ishika, T., Rahman, M., & Rahmatullah, M. 2011. Medicinal plants used by folk medicinal practitioners of four adjoining villages of Narail and Jessore Districts, Bangladesh. *Advances in Natural and Applied Sciences* 5 (1); 23.

Boumba, V. A., Mitselou, A., & Vougiouklakis, T. 2004. Fatal poisoning from ingestion of *Datura stramonium* seeds. *Veterinary and Human Toxicology* 46 (2); 81–82.

CABI. 2022. *Invasive Species Compendium*. Wallingford, England: CAB International. www.cabi.org/isc (Accessed on 4 May, 2022).

Carrió, E. 2013. *Contribució al coneixement etnobotànic de Mallorca. La biodiversitat vegetal i la seva gestió en una illa mediterrània (Tesis doctoral)*. Barcelona, Spain: Facultat de Farmàcia, Universitat de Barcelona.

Carrió, E., & Vallès, J. 2012. Ethnobotany of medicinal plants used in Eastern Mallorca (Balearic Islands, Mediterranean Sea). *Journal of Ethnopharmacology* 141 (3); 1021–1040.

Chan, K. 2002. Jimson weed poisoning—A case report. *The Permanente Journal* 6 (4); 28–34.

Chandan, G., Kumar, C., Verma, M. K., Satti, N. K., Saini, A. K., & Saini, R. V. 2020. *Datura stramonium* essential oil composition and it's immunostimulatory potential against colon cancer cells. *3 Biotech* 10 (10); 451.

Chaudhary, S. A., & Al-Howaishel, Z. 1980. Weeds of Southwestern Arabian Peninsula. *Tropical Pest Management* 26 (3); 299–302.

Cherie Melaku, B., & Amare, G. G. 2020. Evaluation of antidiabetic and antioxidant potential of hydromethanolic seed extract of *Datura stramonium* Linn (Solanaceae). *Journal of Experimental Pharmacology* 12; 181–189.

Choudhary, M., Sharma, I., Agrawal, D. C., Dhar, M. K., & Kaul, S. 2021. Neurotoxic potential of alkaloids from thorn apple (*Datura stramonium* L.): A commonly used Indian folk medicinal herb. In: Agrawal, D. C., & Dhanasekaran, M., eds. *Medicinal Herbs and Fungi*. Singapore: Springer, pp. 391–420.

Clapham, A. R., Tutin, T. G., & Moore, D. M. 1987. *Flora of the British Isles* (3rd ed). Cambridge University Press, New York, United States.

Das, S., Kumar, P., & Basu, S. P. 2012. Phytoconstituents and therapeutic potentials of *Datura stramonium* linn. *Journal of Drug Delivery and Therapeutics* 2 (3); 4–7.

Debnath, T. 2017. Newer insights into the psychoactive and pharmacological properties of *Datura stramonium* Linn. *Global Journal of Addiction & Rehabilitation Medicine* 4 (3); 555638. https://doi.org/10.19080/GJARM.2017.04.555638

Devi, M. R., Bawari, M., Paul, S. B., & Sharma, G. D. 2011. Neurotoxic and medicinal properties of *Datura stramonium* L.—Review. *Assam University Journal of Science & Technology* 7; 139–144.

Disel, N. R., Yilmaz, M., Kekec, Z., & Karanlik, M. 2015. Poisoned after dinner: Dolma with *Datura stramonium*. *Turkish Journal of Emergency Medicine* 15 (1); 51–55.

Dugan, G. M., Gumbmann, M. R., & Friedman, M. 1989. Toxicological evaluation of jimson weed (*Datura stramonium*) seed. *Food and Chemical Toxicology* 27 (8); 501–510.

Eftekhar, F., Yousefzadi, M., &Tafakori, V. 2005. Antimicrobial activity of *Datura innoxia* and *Datura stramonium*. *Fitoterapia* 76 (1); 118–120.

Egharevba, R. K. A., & Ikhatua, M. I. 2008. Ethno- medical uses of plants in the treatment of various skin diseases in Ovia North East, Edo State, Nigeria. *Research Journal of Agriculture and Biological Sciences* 4 (1); 58–64.

Ekanem, P. E., Ekanem, R., & Gaim, K. 2016. Histological patterns of neurodegeneration of frontal cortex neurons in *Datura stramonium* treated wistar rats. *Journal of Behavioral and Brain Science* 06 (02); 85–92.

El Bazaoui, A., Bellimam, M. A., & Soulaymani, A. 2011. Nine new tropane alkaloids from *Datura stramonium* L. identified by GC/MS. *Fitoterapia* 82 (2); 193–197.

FAO & WHO. 2020. *Guidance Document on Physical Datura Stramonium Seed Contamination*. Rome, Italy: FAO.

Firdaus, N. V. 2020. Potenial and pharmacological ations of dhatura safed (*Datura metel* L.): As a deadely poision and as drug: An overview. *International Journal of Pharmaceutical Science and Research* 11 (7); 3123–3137.

Gaire, B. P., & Subedi, L. 2013. A review on the pharmacological and toxicological aspects of *Datura stramonium* L. *Journal of Integrative Medicine* 11 (2); 73–79.

Garrido, G. 2008. *Gran diccionario de las plantas medicinales: propiedades curativas*. Madrid, Spain: Libro Hobby Club.

Girmay, S. 2015. Preliminary phytochemical screening and in vitro antimicrobial activity of *Datura stramonium* leaves extracts collected from eastern Ethiopia. *International Research Journal of Biological Sciences* 4; 1–5.

Gorsi, M. S., & Shahzad, R. 2002. Medicinal uses of plants with particular reference to the people of Dhirkot. Azad Jammu and Kashmir. *Asian Journal of Plant Sciences* 1 (3); 222–223.

Holm, L., Doll, J., Holm, E., Pancho, J & Herberger, I., eds. 1997. *World Weeds: Natural Histories and Distribution*. New York, NY: Wiley.

Holm, L., pancho, J. V., herberger, J. P., & Plucknett, D. L. eds. 1979. *A Geographical Atlas of World Weeds*. New York: Wiley.

Hong, L., Guo, Z., Huang, K., Wei, S., Liu, B., Meng, S., & Long, C. 2015. Ethnobotanical study on medicinal plants used by Maonan people in China. *Journal of Ethnobiology and Ethnomedicine* 11 (1); 32.

Hussain, F., Badshah, L., & Dastagir, G. 2006. Folk medicinal uses of some plants of South Waziristan, Pakistan. *Pakistan Journal of Plant Sciences* 12; 27–39.

Iqbal, S., Sivaraj, C., & Gunasekaran, K. 2017. Antioxidant and anticancer activities of methanol extract of seeds of *Datura stramonium* L. *Free Radicals and Antioxidants* 7 (2); 184–189.

Ivancheva, S., Nikolova, M., & Svetkova, R. 2006. Pharmacological activities and 4 biologically active compounds of Bulgarian medicinal plants. *Phytochemistry: Advances in Research* 37 (1); 87–103.

Jakabová, S., Vincze, L., Farkas, Á., Kilár, F., Boros, B., & Felinger, A. 2012. Determination of tropane alkaloids atropine and scopolamine by liquid chromatography–mass spectrometry in plant organs of Datura species. *Journal of Chromatography A* 1232; 295–301.

Jarald, E. E., & Jarald, S. E. 2006. *Colour Atlas of Medicinal Plants (Common Names & Classifications)*. 1st ed. New Delhi, India: CBS Publishers & Distributors.

Jawalkar, N., Zambare, S., & Zanke, S. 2016. Insecticidal property of *Datura stramonium* L. seed extracts against Sitophilus oryzae L. (Coleoptera: Curculionidae) in stored wheat grains. *Journal of Entomology and Zoology Studies* 4 (6); 92–96.

Joshua, P. E. 2019. Comparative effects of *Datura stramonium* leaf and seed extracts on membrane stabilization and platelet aggregation in-vitro. *African Journal of Biomedical Research* 22; 363–368.

Kadam, S. D., Chavhan, S. A., Shinde, S. A., & Sapkal, P. N. 2018. Pharmacognostic review on Datura. *Research Journal of Pharmacology and Pharmacodynamics* 10 (4); 171.

Kalam, M. A., & Rifat, I. 2020. Datura species (Dhatura Safed and Dhatura Seyah): A review with special emphasis on single-use and compound formulations and pharmacological studies relevant to Unani System of Medicine. *Indian Journal of Integrative Medicine* 2 (1); 9.

Karimzadeh, J., & Rabiei, A. 2020. Larvicidal and oviposition deterrent effects of the jimsonweed (*Datura stramonium* L.) extracts on the Diamondback Moth, Plutella xylostella (L.). *Journal of Agricultural Science and Technology* 22 (5); 1279–1293.

Karmakar, U. K., Toume, K., Ishikawa, N., Arai, M. A., Sadhu, S. K., Ahmed, F., & Ishibashi, M. 2016. Bioassay-guided isolation of compounds from *Datura Stramonium* with TRAIL-resistance overcoming activity. *Natural Product Communications* 11 (2); 185–187.

Khalili Najafabadi, M., & Atyabi, S. 2004. Evaluation of analgesic effect of *Datura stramonium* seed extract in hot plate and formalin tested on male rats. *Iranian Journal of Medicinal and Aromatic Plant* 20 (3); 309–322.

Khan, J., Khan, R., & Qureshi, R. A. 2013. Ethnobotanical study of commonly used weeds of district Bannu, Khyber Pakhtunkhwa (Pakistan). *Journal of Medicinal Plants Studies* 1 (2); 1–6.

Khan, S., Kamili, A., & Gupta, R. C. 2016. Economic and medicinal properties of some medicinal plants found in Kashmir Himalaya. *Journal of Medicinal Plants Studies* 4; 38–44.

Khan, S. W., & Khatoon, S. 2008. Ethnobotanical studies on some useful herbs of Haramosh and Bugrote valleys in Gilgit, northern areas of Pakistan. *Pakistan Journal of Botany* 40 (1); 43–58.

Kirtikar, K. R. 1935. *Indian Medicinal Plants: By K.R. Kirtikar, B.D. Basu, and An I.C.S. In 4 Volumes*. 2nd ed. Allahabad, India: Lalit Mohan Basu.

Kolbek, J., & Sádlo, J. 1996. Some short-lived ruderal plant communities of non-trampled habitats in North Korea. *Folia Geobotanica et Phytotaxonomica*, 31 (2); 207–217. https://doi.org/10.1007/BF02812064

Kumral, N. A., Çobanoğlu, S., & Yalcin, C. 2010. Acaricidal, repellent and oviposition deterrent activities of *Datura stramonium* L. against adult Tetranychus urticae (Koch). *Journal of Pest Science* 83 (2); 173–180.

Lazzarini, D., Baffoni, M. T., Cangiotti, C., Di Fronzo, G., Gerboni, S., Micheli, R., Morelli, S., Morolli, L., & Ioli, G. 2006. Food poisoning by *Datura stramonium*: An unusual case report. *Internal and Emergency Medicine* 1 (1); 88–90.

Li, J., Lin, B., Wang, G., Gao, H., & Qin, M. 2012. Chemical constituents of *Datura stramonium* seeds. *Zhongguo Zhong Yao Za Zhi = Zhongguo Zhongyao Zazhi = China Journal of Chinese Materia Medica* 37 (3); 319–322.

Lorenzi, H. 1982. Weeds of Brazil, terrestrial and aquatic, parasitic, poisonous and medicinal. In: *Plantas daninhas de Brasil, terrestres, aquaticas, parasitas, toxicas e medicinais*. Sao Paulo, Brazil: Nova Odessa, p. 425.

Marc, B., Martis, A., Moreau, C., Arlie, G., Kintz, P., & Leclerc, J. 2007. Intoxications aiguës à *Datura stramonium* aux urgences. *La Presse Médicale* 36 (10); 1399–1403.

Mir, M. A., Hamdani, S. S., Sheikh, B. A., & Mehraj, U. 2019. Recent advances in metabolites from medicinal plants in cancer prevention and treatment. *Current Immunology Reviews* 15 (2); 185–201.

Mohamad, N., Baharuddin, K. A., & Ahmad, R. 2009. A traditional Malay myth leading to unintentional self intoxication with kecubung fruit. *The Southeast Asian Journal of Tropical Medicine and Public Health* 40 (6); 1331–1334.

Monira, K. M., & Munan, S. M. 2012. Review on *Datura metel*: A potential medicinal plant. *Global Journal of Research on Medicinal Plants & Indigenous Medicine* 1 (4); 123–132.

Morton, J. F. 1981. *Atlas of Medicinal Plants of Middle America: Bahamas to Yucatan*. Springfield, IL: C.C. Thomas.

Mukhtar, Y., Tukur, S., & Bashir, R. A. 2019. An overview on *Datura stramonium* L. (Jimson weed): A notable psychoactive drug plant. *American Journal of Natural Sciences* 2 (1); 1–9.

Naik, V. K. M., Babu, K. S., Latha, J., & Kolluru, B. 2018. A review on phytochemical and pharmacological activity of *Daturas stramonium* (medicinal plant). *Research Journal of Pharmacognosy and Phytochemistry* 10 (1); 77.

Nain, J., Bhatt, S., Dhyani, D. S., & Joshi, N. 2013. Phytochemical screening of secondary metabolites of *Datura stramonium* L. *International Journal of Current Pharmaceutical Research* 5 (2); 151–153.

Nasir, B., Khan, A. U., Baig, M. W., Althobaiti, Y. S., Faheem, M., & Haq, I.-U. 2022. *Datura stramonium* leaf extract exhibits anti-inflammatory activity in CCL4-induced hepatic injury model by modulating oxidative stress markers and iNOS/Nrf2 expression. *BioMed Research International* 2022; 1–20.

Ogunmoyole, T., Adeyeye, R. I., Olatilu, B. O., Akande, O. A., & Agunbiade, O. J. 2019. Multiple organ toxicity of *Datura stramonium* seeds extracts. *Toxicology Reports* 6; 983–989.

Onen, C. L., Othol, D., Mbwana, S. K., & Manuel, I. L. 2002. *Datura stramonium* mass poisoning in Botswana. *South African Medical Journal* 92 (3); 213–214.

Oseni, O. A., Olarinoye, C. O., & Amoo, I. A. 2011. Studies on chemical compositions and functional properties of thorn apple (*Datura stramonium* L) solanaceae. *African Journal of Food Science* 5 (2); 40–44.

Osváth, P., Nagy, A., Fekete, S., Tényi, T., Trixler, M., & Radnai, I. 2000. A case of datura stramonium poisoning—General problems of differential diagnosis. *Orvosi Hetilap* 141 (3); 133–136.

Papi, A., Fabbri, L. M., Kerstjens, H. A. M., Rogliani, P., Watz, H., & Singh, D. 2021. Inhaled long-acting muscarinic antagonists in asthma—A narrative review. *European Journal of Internal Medicine* 85; 14–22.

Parker, C. 1992. *Weeds of Bhutan. National Plant Protection Centre, Simtokha. Royal Government of Bhutan*. Thimphu, Bhutan: National Plant Protection Centre.

Pillay, V. V., & Sasidharan, A. 2019. Oleander and *Datura* poisoning: An update. *Indian Journal of Critical Care Medicine* 23 (S4); S250–S255.

Prasad, B. 2008. *Monographs on Datura Stramonium L.* Kaski, Nepal: The School of Pharmaceutical and Biomedical Sciences Pokhara University, pp. 1–114.

Pretorius, E., & Marx, J. 2006. *Datura stramonium* in asthma treatment and possible effects on prenatal development. *Environmental Toxicology and Pharmacology* 21 (3); 331–337.

Radford, A. E. 2010. *Manual of the Vascular Flora of the Carolinas*. Chapel Hill: The University of North Carolina Press.

Rahmatullah, M., Islam, R., Kabir, Z., Jahan, R., Begum, R., Seraj, S., Khatun, M. A., & Chowdhury, A. R. 2010. Folk medicinal practices in Vasu Bihar Village, Bogra District, Bangladesh. *American-Eurasian Journal of Sustainable Agriculture* 4 (1); 86–93.

Rajbhandari, K. R. 2001. *Ethnobotany of Nepal*. Kathmandu, Nepal: Kishor Offset Press Private Limited, pp. 142–143.

Rajeshkanna, A., Prabhakaran, D., Senthamilselvi, M. M., & Muruganantham, N. 2016. Anti cancer activity of *Datura stramonium* (flowers) against human liver cancer. *Indo American Journal of Pharmaceutical Sciences* 3 (6); 582–585.

Richardson, W. H., Slone, C. M., & Michels, J. E. 2007. Herbal drugs of abuse: an emerging problem. *Emergency Medicine Clinics of North America* 25 (2); 435–457.

Salas, M., & Cáceres, M. T. 2003. Notas Históricas y estudios de algunas plantas mesoamericanas en Canarias: Piteras, tuneras y estramonio. *Vegueta* 7; 255–263.

Savithramma, N., Sulochana, C. H., & Rao, K. N. 2007. Ethnobotanical survey of plants used to treat asthma in Andhra Pradesh, India. *Journal of Ethnopharmacology* 113 (1); 54–61.

Setshogo, M. P. 2015. A review of some medicinal and or hallucinogenic solanaceous plants of Botswana: The genus Datura L. *International Journal of Medicinal Plants and Natural Products* 1 (2); 15–23.

Shagal, M. H., Modibbo, U. U., & Liman, A. B. 2012. Pharmacological justification for the ethnomedical use of *Datura Stramonium* stem-bark extract in treatment of diseases caused by some pathogenic bacteria. *International Research of Pharmacy and Pharmacology* 2 (1); 16–19.

Shah, B. N., & Seth, A. K. 2013. *Textbook of Pharmacognosy and Phytochemistry*. 1st ed. New Delhi, India: CBS Publisher and Distributors.

Shah, G. M., & Khan, M. A. 2006. Common medicinal folk recipes of Siran Valley, Mansehra, Pakistan. *Ethnobotanical Leaflets* 10; 49–62.

Shamim, & Idris, M. 2019. Datura Stramonium (Dhatura). Pharmacological Actions, therapeutic uses and phytochemistry. A review. *International Journal of Pharmacy and Biological Sciences*, 9 (1); 622–631.

Sharma, A., Patel, V., & Chaturvedi, A. 2009. Vibriocidal activity of certain medicinal plants used in Indian folklore medicine by tribals of Mahakoshal region of central India. *Indian Journal of Pharmacology* 41 (3); 129.

Sharma, M., Dhaliwal, I., Rana, K., Delta, A. K., & Kaushik, P. 2021. Phytochemistry, pharmacology, and toxicology of Datura Species—A review. *Antioxidants* 10 (8); 1291.

Soneral, S. N., & Connor, N. P. 2005. Jimson weed intoxication in five adolescents. *WMJ: Official Publication of the State Medical Society of Wisconsin* 104 (7); 70–72.

Soni, P., Siddiqui, A. A., Dwivedi, J., & Soni, V. 2012. Pharmacological properties of *Datura stramonium* L. as a potential medicinal tree: An overview. *Asian Pacific Journal of Tropical Biomedicine* 2 (12); 1002–1008.

Swathi, S., Murugananthan, G., Ghosh, S. K., & Pradeep, A. S. 2012. Larvicidal and repellent activities of ethanolic extract of datura stramonium leaves against mosquitoes. *International Journal of Pharmacognosy and Phytochemical Research* 4 (1); 25–27.

Tounekti, T., Mahdhi, M., & Khemira, H. 2019. Ethnobotanical study of indigenous medicinal plants of Jazan Region, Saudi Arabia. *Evidence-Based Complementary and Alternative Medicine* 2019; 1–45.

Trancă, S. D., Szabo, R., & Cociş, M. 2017. Acute poisoning due to ingestion of *Datura stramonium*—A case report. *Romanian Journal of Anaesthesia and Intensive Care* 24 (1); 65–68.

Tutin, T. G., Heywood, V. H., Burges, N. A., Moore, D. M., Valentine, D. H., Walters, S. M., & Webb, D. A. 1972. *Flora Europaea, Diapensiacea To Myoporaceae*. Zenodo. Cambridge University Press. https://doi.org/10.5281/ZENODO.305475.

Ummahan, O. 2017. The antifungal effects of *Datura stramonium* L., *D. metel* l., *D. innoxia* mill. in flora of Turkey. *Mugla Journal of Science and Technology* 3 (2); 96–103.

Vaza, J. S. 2018. Phytochenistry, medicinal value and pharamacological potential of *Datura stramonium* L. review. *International Journal of Pharamaceutics and Drug Analysis* 6 (7); 540–544.

Velasco, J. M. 2009. *Guía de plantas útiles y perjudiciales en Castilla y León.* 1ª ed. Salamanca, Spain: Caja Duero.

Velasco Santos, J. M., Criado Coca, J., & Blanco Castro, E. 2010. *Uso tradicional de las plantas en la provincia de Salamanca: una aproximaci??n al estudio de las relaciones de las plantas y los pueblos de Salamanca.* Salamanca, Spain: Instituto de las Identidades, Diputación de Salamanca.

Vijendra, N., & Kumar, K. P. 2010. Traditional knowledge on ethno-medicinal uses prevailing in tribal pockets of Chhindwara and Betul Districts, Madhya Pradesh, India. *African Journal of Pharmacy and Pharmacology* 4 (9); 662–670.

Wang, S., & You, L. 2012. Allelopathic potential in different polarity phases of *Datura stramonium* L. *Advance in Biomedical Engineering* 21 (3–4); 165–170.

Wangchuk, P., Pyne, S. G., & Keller, P. A. 2011. Ethnobotanical authentication and identification of Khrogsman (lower elevation medicinal plants) of Bhutan. *Journal of Ethnopharmacology* 134 (3); 813–823. https://doi.org/10.1016/j.jep.2011.01.034

Waza, S. A., Anthony, P., & Dar, S. 2015. Phytochemical analysis, antioxidant and antimicrobial activities of methanolic extract of datura stramonium seeds. *International Journal of Pharmaceutical Sciences and Research* 6 (7); 3021–3026.

Wazir, S. M., Dasti, A. A., & Shah, J. 2004. Common medicinal plants of Chapursan valley, Gojal ii, Gilgit-Pakistan. *Journal of Research (Science)* 15 (1); 41–43.

Weaver, S. E., & Warwick, S. I. 1984. The biology of canadian weeds.: 64. *Datura stramonium* L. *Canadian Journal of Plant Science* 64 (4); 979–991.

Wells, M. J., Balsinhas, A. A., Joffe, H., Engelbrecht, V. M., Harding, G., & Stirton, C. H., eds. 1986. *A Catalogue of Problem Plants in Southern Africa: Incorporating the National Weed List of Southern Africa*. Pretoria, Republic of South Africa: Botanical Research Institute.

Wiebe, T. H., Sigurdson, E. S., & Katz, L. Y. 2008. Angel's Trumpet (*Datura stramonium*) poisoning and delirium in adolescents in Winnipeg, Manitoba. *Paediatrics & Child Health* 13 (3); 193–196.

Witt, A., & Luke, Q. 2017. *Guide to the Naturalized and Invasive Plants of Eastern Africa*. Wallingford, England: CABI.

Yadav, B., Singla, A., Srivastava, N., & Gupta, P. 2021. Pharmacognostic and phytochemical screening of *Datura stramonium* by TLC and GC-MS: A forensic approach. *Biomedical and Pharmacology Journal* 14 (4); 2221–2226.

17 *Digitalis purpurea* (Foxglove Plant)

Tina, Vedanshi Pal, and Manu Pant

CONTENTS

17.1 INTRODUCTION

Digitalis purpurea is a very popular herbaceous ornamental plant. This plant, commonly known as 'Foxglove', is native to Europe and African islands. *D. purpurea* was initially placed under the family Scrophulariaceae, but was subsequently shifted to the family Plantaginaceae after elaborate phylogenetic studies. The plant was first described in 1753 by Carl Linnaeus in *Species Plantarum*. It derives its name from the Latin word *Digitalis* meaning 'finger' as the flowers have a finger-like appearance. The species name *purpurea* refers to the flower pigments which give them the characteristic purple color, and sometimes yellow or even white hues.

Besides being an ornamental plant used for decorative purposes, *D. purpurea* is widely known for its curative properties in heart-related ailments since ancient times. This is owing to the presence of cardiac and steroidal glycosides like Digitoxin ($C_{41}H_{64}O_{13}$) and Digoxin ($C_{41}H_{64}O_{14}$) which are found in abundance in the entire plant body (Lungeanu et al. 1963). Digitoxin is the principal active ingredient of the species and accounts for several therapeutic properties of the plant. Besides the glycosides, the plant is a rich source of microelements like boron, copper, and manganese. However, the *Digitalis* plant also can uptake potentially toxic elements like lead, chromium, nickel, and arsenic calling for focused attention on the analysis of mineral content in the plant species (Roca-Pérez et al. 2005 Negi et al. 2012).

Digitalis is a term used for inotropic drugs that induce myocardial contractility. Digitalis administration plays a very important role in the treatment of congestive heart failure, atrial fibrillation, and cardiac dilation for more uptake of blood in the heart and chronic kidney diseases (Sessa et al.

DOI: 10.1201/b23017-19

2016). Digitalis is also extensively used by herbal practitioners because of its natural plant source and less clinical index (Reddy 2010). However, the toxicity of the Digitalis drug is a debatable topic. Repeated or above optimal dosage of the drug leads to serious impacts on the patients' health. Atrial tachycardias, atrioventricular blockage, and fatal ventricular arrhythmias are listed as the major effects of digitalis intoxication (Hauptman and Kelly 1999). In severe cases, it may even lead to death. General sickness, antiperistalsis, eating disorders, dizziness, confused eyesight, objects appearing greenish or yellowish, polyuria chilling perspiration, syncope, and slow pulse which can be 35 beats/minute are all symptoms of Digitalis toxicity (Withering 1937).

This chapter discusses the botanical description, distribution, phytochemicals, traditional and therapeutic uses, cultivation methods, drug extraction process, mechanism of action of the drug, pharmacokinetics, and its toxic responses.

17.2 BOTANICAL DESCRIPTION

17.2.1 Habit and Habitat

D. purpurea is a biennial or perennial herb that attains a height of about 45–150 cm with around 60–100 cm in width. The leaves are simple, coarsely reticulate-veined, around 9–35 cm long and 5–13 cm wide, ovate, petiolated, and arranged spirally. They are dark green and bitter due to the presence of glycosides. The leaves possess several glandular trichomes on both surfaces giving them a woolly appearance. Flowering takes place in early summers, from mid-June to mid-July where each flower remains open for 10 days. The flowers are tubular and drooping down. They are arranged as clusters at the terminal ends (Figure 17.1). The petals are elongated, clubber together, and heavily

FIGURE 17.1 *Digitalis purpurea* plant.

spotted from inside. The main pollinators of Foxglove are Bumblebees (Best and Bierzychudek 1982). The reproductive part consists of two long and two short stamens, and the flower type is hypogynous. The fruiting period of the Foxglove plant is from July to August and the seeds ripen from August to October (Al-Snafi 2017). The fruit is a round and glandular capsule. It splits open and releases numerous seeds when ripe. The seeds are brown and rectangular with a mesh of ridges across the surface.

Cold to a moderate temperature range of 25°C and nutrient-rich, well-drained slightly acidic soil (pH 5.5–6.5) are the most preferable conditions for plant growth (Clapham et al. 1962; Phillips and Rix 1991). The plants require partial shade conditions and are reported to grow poorly in complete darkness. They are photosensitive along with a high germination rate when sown at a depth of 10 mm (Bliss and Smith 1985; Huxley 1992).

17.2.2 Propagation Methods

17.2.2.1 Conventional Method of Propagation

D. purpurea is cultivated for its beautiful flowers and medicinal purposes. The most common method of propagation is seeds. Sowing is done during spring or early summers. Germination occurs within 2–3 weeks and quick watering is required to avoid drying of the seedlings. The light-weighted seeds are mixed with sieved sand to facilitate proper seed scattering. In general, sowing is done in damp soil at around 20°C and in rows that are at 10–11 cm intervals. Once sown, the seeds are covered with a fine layer of soil and sand. The seedlings are ready for transplantation within 40–45 days after sowing (Reddy 2010).

In the first year of growth, the stem grows with spirally arranged leaves in a rosette form. By the second year of growth, the plants are mature and flowering starts in April (in the first year of growth) followed by fruiting and seed development in the second year of growth. Different combinations of Farmyard manure and fertilizers (NPK) are applied at different intervals to ensure healthy plant propagation (Rowson 1955). Healthy leaves from plants in the second year of growth and with more than half fully developed flowers are most suitable for extraction purposes. The leaves can be harvested for up to 2 years from a single plant. In the first year, leaves of 8–10 cm in length can be collected in the months of July and August, while the second harvest is done after around one and a half to 2 months of the first harvest. During the first year, harvesting can be done 2–3 times and 1–2 harvests can be done during the second year. Once harvested, the leaves are stored in airtight containers with drying substances like silica gel beads and leaves are dried properly in hot air up to 60°C with occasional stirring. The standard green color of the leaves is preserved by rapid drying. The process of drying is carried out till the moisture content drops down to less than 5% since it causes hydrolysis of cardiac glycosides which leads to loss of cardiac activity. After drying, the leaves are packed under pressure in airtight containers to avoid any kind of moisture and environmental interactions. It is estimated that around 2–2.5 tonnes of dried leaves can be obtained from *D. purpurea* plantations in one acre of land (Reddy 2010). Bright purple/magenta colored, heavily spotted flowers are also collected for medicinal purposes. Ripe seeds are collected and stored at ~5% moisture content (fresh weight basis) and at 50°C for further sowing (Hay et al. 2008).

The plants are known to be sufficiently disease resistant, though they may suffer from powdery mildew and crown rot due to humid weather (Wallace and Crouch 2018). To avoid contamination of the planting stock, the infected stalk is excised and removed before the infection reaches the seeds (Garibaldi et al. 2022).

17.2.2.2 In Vitro Propagation

To facilitate large-scale propagation of disease-free plant material of *D. purpurea* in a small space, plant tissue culture technology offers the best option. Plant cells, tissue, organs, protoplast, and seeds can be grown artificially on nutrient medium under aseptic and regulated physicochemical conditions. Table 17.1 summarizes the various micropropagation procedures used to propagate *D. purpurea* from various tissues (explants).

TABLE 17.1
In Vitro Propagation Strategies for *D. purpurea*

Explant	Response	Medium and Plant Growth Regulators	References
Anthers	Haploid callus culture	Nitsch and Nitsch+2,4-D	Corduan and Spix (1975)
Callus	Shoot derived culture	Liquid based medium+BAP+IAA	Hagimori et al. (1982)
Callus or leaf tissue	Root culture	Torrey+Saccharose+IAA	Rucker et al. (1983)
Leaf segments	Shoot organogenesis	MS Medium+NAA+BAP	Pèrez-Alonso et al. (2018)
Leaf, petiole, root	To be asked	MS_{30}+2,4-D+BAP+NAA	Rad et al. (2021)
Leaves	Root differentiation	Heller and MS Media+IAA +NAA+2,4-D+GA3	Rücker et al. (1981)
Leaves	Root system and little callus development	Heller medium+IAA	Rücker et al. (1976)
Leaves and roots of in vitro germinated seeds	Shoot derived culture	MS Medium+thiamine+myo -inositol+6- BAP+IAA+Sucrose	Pèrez-Alonso et al. (2014)
Nodal, intermodal, leaf	Multiple shoot formation	MS medium+BA+Kin+TDZ +IAA+NAA+2,4-D	Patil et al. (2013)
Seedlings	Callus induction	Basal medium+IAA+Agar	Hagimori et al. (1982)
Seedlings	Callus tissue derived	MS Medium+2,4-D+Kin	Furuya et al. (1970)
Seeds	Seedlings development	MS Medium+BA+Kin ++ TDZ +NAA+IAA+2,4-D	Patil et al. (2012)
In vitro germinated seeds	Shoot derived culture	MS Medium+thiamine HCl+6-BAP +IAA+Myoinositol+Sucrose	Pèrez-Alonso et al. (2009)
2-year-old culture of proliferating shoots	Shoot multiplication	MS Medium+IBA+Adenine sulfate	Chaturvedi and Jain (1994)

17.2.2.3 Propagation for Enhanced Glycoside Production

Since the Foxgloves are valued for their glycoside content, there have been studies on standardizing propagation methods for the production of quality crops with enhanced glycosides. It was shown that chloroplasts are not necessary for the synthesis of digitoxin in *Digitalis* cells (Hagimori et al. 1982). Thereafter, the studies were focused on different physical factors for plant and phytochemical production. A study showed that treatment of potassium chloride and progesterone can enhance cardenolide production (Patil et al. 2013). Exposure to red and blue lights in the ratio of 2:8 has been reported to induce maximum production of cardenolides in the leaf tissues of the plant (Verma et al. 2018). Besides, optimum light and temperature conditions have been proven to be essential for Foxglove development as they directly influence photosynthetic rate, nutrient absorption, hormone and enzymatic activities in the plant body. For *D. purpurea*, a greenhouse temperature of 22°C during day and 18°C at night is most preferable for plants. Besides, periodic watering and supply of nutrients is a must for the development of healthy plants with optimal phytoingredients (Verma et al. 2018).

17.3 DISTRIBUTION

D. purpurea is found in a wide range of climatic conditions varying from oceanic to suboceanic areas to Mediterranean areas (Meusel 1954; Hinz 1990; Bruelheide and Heinemeyer 2002). Despite appropriate climatic and soil conditions, Digitalis occurrence reduces as one moves from west to east (Werner 1964) earning them the name 'ephemerophyte'. In 1980s, it was reported from the British Columbia and western Washington state (Best and Bierzychudek 1982). The species is known to be indigenous to the European regions (Albania, Belgium, Finland, Sweden, United Kingdom, Italy,

Norway, France, Portugal, Spain, Denmark, Germany, Croatia, Canary Island), New Zealand, and Africa (Madeira Islands, Morocco, Cape Verde) (; Bräuchler and Heubl 2004; Sletvold and Rydgren 2007; Kreis 2017). In the eastern Germany boundaries (like Hainleite Mountains in Southern Lower Saxony and Thuringia) and in Poland, *D. purpurea* is mainly affected by frosts in winters and drought periods in summers limiting the growth of adult plants (Bruelheide and Heinemeyer 2002). The species is subsequently introduced in other parts of the world and now grows widely in parts of North America, Iraq, Japan, Kurdistan, Mexico, Turkey, Nepal, and India (Verma et al. 2016). In India, the Nilgiri mountains, Kashmir, and Darjeeling are the hotspots for the cultivation of *D. purpurea*, while some cultivation is also done in parts of Uttarakhand and Himachal Pradesh (Chopra and De 1926; Negi et al. 2012). In Japan, England, Hungary, and Germany, it is primarily cultivated for drug production (Rana et al. 2022).

17.4 PHYTOCHEMICAL CONSTITUENTS

D. purpurea is a chemically active species. It is rich in glycosides, saponins, carbohydrates, volatile oil, gums, fatty matter, and sugars and glycosides. Glycosides like verodoxin, odoroside-H, digitalose, gitaloxin, gitaloxigenin, and glucoverodoxin are mostly found in the leaves (Ali Esmaile et al. 2017). Essential saponins like digitonin, digitoxigenin and gitonin, luteolin, and flavones are known to impart color to the drug made from *D. purpurea* (Sharma and Purkait 2012).

17.4.1 CARDIAC GLYCOSIDES IN *D. PURPUREA*

The glycosides found in *D. purpurea* can be categorized as primary and secondary, 0.2%–0.45% of which are present in the leaves. Primary glycosides include purpurea glycosides A and B and glucogitaloxin. The secondary glycosides (e.g., digitoxin, digoxin) are synthesized from the primary ones and are far more absorbed, stable, and therefore, more widely used. Cardiac glycosides are the secondary glycosides composed of a carbohydrate moiety and an aglycone (steroid nucleus with unsaturated lactone) (Mishra et al. 2020). These are naturally occurring steroids with a high impact stimulating action on the cardiac muscles. The therapeutic property of *D. purpurea* is mainly attributed to the cardiac glycosides – digoxin and digitoxin which are found in all the parts of the plant like roots, leaves, stem, and flowers.

Digoxin ($C_{41}H_{64}O_{14}$, molecular weight 780.95 Da) has a white crystalline appearance, is odorless, and exhibits complete solubility in pyridine, partial solubility in alcohol and complete insolubility in water. Digitoxin ($C_{41}H_{64}O_{13}$, molecular weight 764.939) (Figure 17.2) has a similar structure and

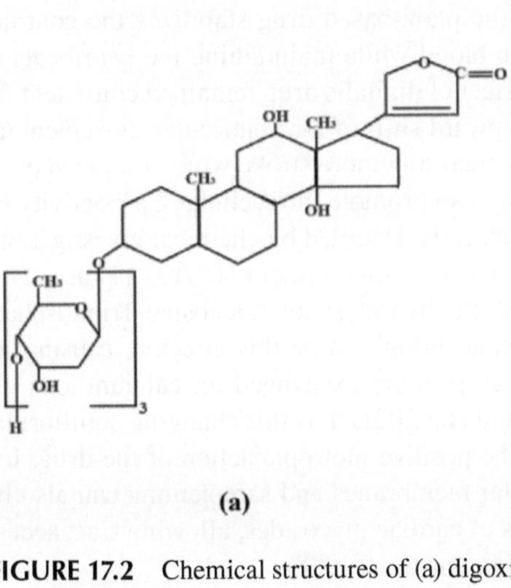

(a)

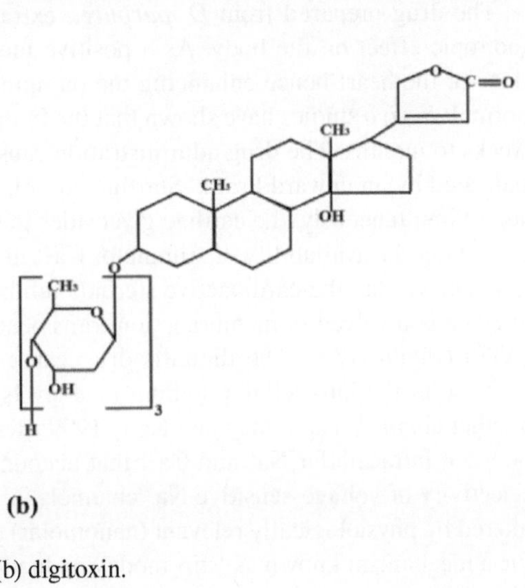

(b)

FIGURE 17.2 Chemical structures of (a) digoxin and (b) digitoxin.

phytochemical effects as digoxin (except that digoxin has an extra hydroxyl [–OH] group), but has a prolonged half-life and long-lasting effects in the body. It is insoluble in water; soluble in alcohol, chloroform, and acetone; and partially soluble in methanol. Synthesis of digitoxin takes place in the fresh leaves by the hydrolytic action of enzyme digipuridase on purpurea glycosides A and B.

17.5 PHARMACOLOGICAL STUDIES

Out of different organic compounds present in *D. purpurea*, only a few compounds (basically glycosides and saponins) are pharmacologically active and exploited for treatment of disorders like heart failure, arrhythmia, neurological diseases, and cancer (Lindholm et al. 2002). Foxglove is highly valued for its glycoside content which stimulates the cardiac muscles, and it has antioxidative, hepatoprotective, and cytotoxic properties (Zahid and Rizwani 2016). Drugs prepared through the plant extracts are widely used in the treatment of heart diseases, improving functioning of cardiac muscles without the need for excess oxygen and regulating heartbeat regulation (Navarro et al. 2000). Edema, which is a common result of heart failure, is also known to be effectively treated through plant-derived drug (Rosen et al. 1985). Foxglove's cardiac glycosides have been considered to be very safe inotropes for patients suffering from serious heart-related disorders (Schwinger et al. 2003).

Leaves are the major source for the extraction of cardiac glycosides digitoxin and digoxin (Roca-Pérez et al. 2004). Digoxin is rapidly absorbed by the cells, undergoes minimal hepatic metabolism, and finally excreted out completely through the kidneys. The major mechanism of digoxin is to suppress the DNA topoisomerases I and II and to increase the intracellular Ca^{2+} concentration. It also upregulates HIF-1 thereby inducing the cell cycle arrest (Yeh et al. 2001; Winnicka et al. 2008; Elbaz et al. 2012). Digoxin is known to have a vagomimetic action on the atrioventricular node. The parasympathetic nervous system is activated which reduces heart rate by slowing electrical conduction at the atrioventricular node. As calcium concentration rises, phase 4 and phase 0 of action potential extend, increasing the atrioventricular node's refractory time hence improving the ventricular response (Ren et al. 2020).

Unlike digoxin, which is not extensively metabolized in humans but is excreted mostly unchanged by the kidneys, digitoxin is eliminated via the liver and can be used in patients with poor or erratic kidney function (Haux 1999). The mechanism of action of digitoxin is to increase the intracellular Ca^{2+} concentration and downregulate cyclin B1 and Cdc2, hence inducing cell cycle arrest in G2/M phase (Hina Zahid and Rizwani 2016). Digitoxin affects the function of calcium metabolism genes as well as IAP (Inhibitors of Apoptosis) genes that control cell death and caspases (Riedl and Shi 2004).

The drug prepared from *D. purpurea* extract (commonly called digitalis drug) has a positive inotropic effect on the body. As a positive inotrope, the plant-based drug stabilizes the contractions of the heart hence enhancing the pumping of the blood while maintaining the heartbeats to normal. *In vivo* studies have shown that the inotropic effect of digitalis drug remained consistent for weeks to months. The drug administration causes an upward shift of the ventricular movement (as indicated by an upward Frank–Starling curve), hence generating more stroke work at a given pressure. Simultaneously, the cardiac glycosides in the drug also promote intracellular Ca^{2+} activity by improving the availability of stimulator Ca^{2+} in the heart cells. Detailed biochemical investigations have shown that the cardioactive steroids inhibit the intermembrane protein Na^+/K^+-ATPase. This enzyme is involved in membrane ion translocation with the hydrolysis of Adenosine Triphosphate (ATP) (Figure 17.3). The digitalis drug causes reversible inhibition of this enzyme, causing an increase in the intracellular sodium ions levels which in turn are exchanged for calcium ions via another channel (Hauptman and Kelly 1999; Rehman and Hai 2021). It is this changing equilibrium between intracellular Na^+ and Ca^{2+} that accounts for the positive inotropic action of the drug. Ion selectivity of voltage-sensitive Na^+ channels in T-tubular membranes and sarcolemma can also be altered by physiologically relevant (nanomolar) dosages of cardiac glycosides, allowing Ca^{2+} access via a mechanism known as 'slip-mode conductance' (Santana et al. 1998).

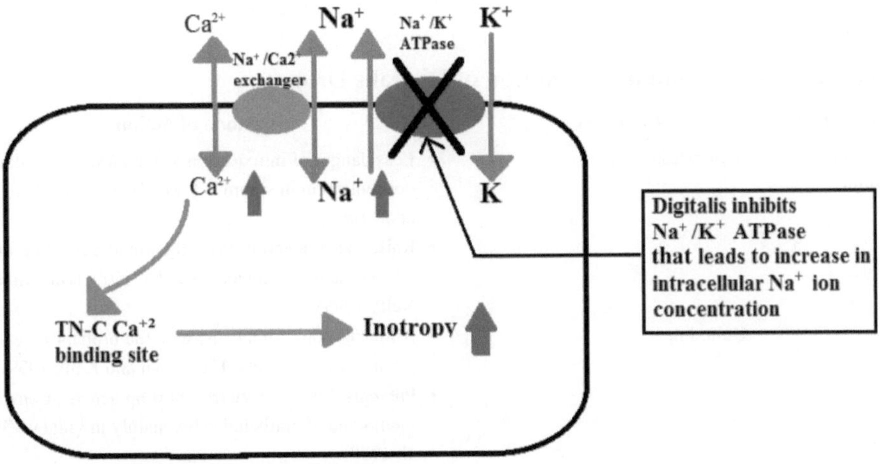

FIGURE 17.3 Mechanism of action of digitalis drug.

17.5.1 Pharmacokinetics

Digitalis drug absorption and metabolism is a crucial part. Digitoxin (90%) has higher oral availability than Digoxin (75%) and has a much improved half-life (160 days) as compared to digoxin (40 days). However, the prolonged half-life of these cardiac glycosides means that a consistent dosage is needed before steady-state therapeutic plasma levels are achieved. Therefore, the procedure of 'digitalization' is done before initiating digitalis drug treatment. This involves special dosing patterns to instantly increase digoxin plasma levels. The safe levels of therapeutic plasma concentration for digoxin are between 0.5 and 1.5 ng/mL. A higher concentration than the upper limit could be fatal for the patient and can lead to digitalis toxicity as digitalis drugs have a comparatively narrow therapeutic safety index (Rehman et al. 2022).

The absorption of digoxin and digitoxin from the gastrointestinal tract can be altered by manipulating the tract motility, adsorption of drugs, modifying intestinal wall features, or altering the intestinal microflora by the use of different chemical agents that interact with the glycoside compounds (Table 17.2).

17.6 TOXIC RESPONSE

Toxicity refers to an unfavorable outcome of medication caused by an amplification of the effect that causes the therapeutic response. Digitalis drug has been used across the world for centuries, but its use has been restricted due to its toxicity, digoxin adverse effects (AEs), and multiple drug interactions. Besides, the narrow therapeutic index of digoxin calls for close and constant monitoring after drug administration. However, since digoxin plays an important role in the treatment of heart failure, atrial fibrillations, etc., it has been opted as adjunctive therapy, rather than baseline therapy (Yancy et al. 2013).

The toxic effects of digitalis can be grouped as cardiac AEs and non-cardiac AEs. The cardiac AEs include ventricular arrhythmias, atrioventricular block, atrial arrhythmias, and sinus bradycardia. The non-cardiac AEs are manifested as anorexia, nausea, vomiting, stomach ache, headache, eating disorders, visual impairment, objects appearing green or yellow, increased urine secretion, psychosis, etc. (Wade 1986). In severe cases, chilling perspiration, convulsions, syncope, slow pulse (as slow as 35 beats/minute), and death have been associated with digitalis poisoning (Withering 1937). Accidental digitalis toxicity is rare and exposure is more likely in infants and children under the age of six than in adults (Watson et al. 2002). Elderly patients having improper muscle mass and

TABLE 17.2
Chemical Agents That Modify the Action of Digitalis Drug

Category of Drugs	Examples	Mode of Action
Antiarrhythmic medications	Amiodarone	• Less danger of intoxication as increase of digitoxin concentration in serum is lower than that of digoxin (Peters et al. 1980). • Raise the concentration of digoxin at a constant rate and lowers the maintenance doses by ≥50% (Hauptman and Kelly 1999).
1.	Quinidine	• Antiarrhythmic agents increase the probability of proarrhythmic events (Hauptman and Kelly 1999). • Prevents digoxin from transporting across plasma membrane of epithelial cells, mainly in kidneys (Fromm et al. 1999). • Competes for binding sites with digoxin, controls the renal elimination of digoxin (Peters et al. 1980; Hauptman and Kelly 1999). • May raise the concentration of digitoxin in serum (Wagner et al. 2007).
Antibiotics	Aminosalicylic acid, Bleomycin, Neomycin and Sulfasalazine	• Reduce digoxin concentration in plasma by inhibiting absorption and elevating clearance from the body (Kuhlmann 1982).
Antithyroid drug	Propylthiouracil	• Inhibits secretion from thyroid gland by suppressing the thyroid peroxidase enzyme (Amisha and Rehman 2021).
Benzodiazepines	• Alprazolam, Chlordiazepoxide	• Bind to gamma amino butyric acid (GABA) receptors and opens the chloride channels (Kang et al. 2021).
Beta blockers	Catecholamines, Epinephrine, and Norepinephrine	• Bind to B1 and B2 receptors thus inhibiting chronotropic and inotropic effects and relaxation of smooth muscles (Farzam and Jan 2021).
Calcium channel blocking drugs	Verapamil	• Reduce the rate of absorption and increase the rate of elimination of digoxin (Kuhlmann 1982).
Cytostatic and chemotherapeutic drugs	Bleomycin, Erythromycin, Tetracycline, and Vincristine	• Increase the elimination of digoxin from the body and limits its absorption (Frank et al. 1985)
Cytotoxic agents	Acarbose, Sucralfate	• Reduce digoxin concentration in the plasma (Frank et al. 1985).
Diuretics	Spironolactone	• Downregulates the plasma potassium levels (hypokalemia), thus affecting the action of digitalis drug (Frank et al. 1985).
	Thiazide diuretics	• Checks filtration through glomerulus (lowering volume) and initiates changes in various ions or electrolytes (Hauptman and Kelly 1999).
Gastrointestinal medications	Antacid gels, Cholestyramine, Kaolin-pectate	• Decrease digoxin bioavailability (Rodin 1988).
Specific Fab fragments	Digibind	• Exhibit a strong affinity with digoxin stronger than any cell membrane thus affecting the laying down of digoxin. • Seize cytoplasmic and histo-bound digoxin into the extracellular matrix leading to a rise in the concentration of digoxin in serum (Ujhelyi and Robert 1995).
Vasodilators	• Alprostadil IV, Deponit, Hydralazine	• Increase renal clearance of digoxin (Hooymans and Merkus 1985).

renal functions are more prone to plasma digoxin intoxication because Na^+/K^+-ATPase proteins in muscles act as a large binding reservoir for the digitalis drug.

The cardiac AEs of digitalis poisoning are the most dreaded ones. Arrhythmias caused by digitalis poisoning may be life-threatening (Eisner and Smith 1991; Kelly and Smith 1996). With digoxin, once the plasma concentration level exceeds the highest range and toxicity occurs, it may take a very long time for the plasma concentrations to fall to safe levels because of the long half-life of digoxin. Tachycardias caused by enhanced atrial automaticity are a common supraventricular arrhythmia linked with digitalis poisoning. Intoxication is even observed in patients whose serum levels are within the normal clinical range (0.5–0.9 ng/mL) such as the combination of decreased conduction and increased automaticity, for instance, atrioventricular block with an accelerated junctional pacemaker (Hauptman and Kelly 1999). Frequent junctional pacemakers may begin to discharge at a higher dosage which results in nonparoxysmal atrioventricular junctional tachycardia. This is manifested as premature ventricular contractions in an electrocardiogram.

Another effect of digitalis poisoning is the automaticity of cardiac tissues due to increased amounts of cardiac glycosides in the body. This ventricular automaticity is aggravated by increased Ca^{2+} concentration in the serum. Patients with digitalis toxicity are known to develop fatal ventricular arrhythmias when given intravenous calcium (Hauptman and Kelly 1999). One major toxic effect of digitalis poisoning is on the kidneys. This is because the SDC (Serum Digoxin Concentration) increases due to renal insufficiency which limits the extravascular volume of distribution of digoxin. Consequently, digitalis drug dosage needs to be closely monitored in the case of kidney disease (Sessa et al. 2016; Ziff and Kotecha 2016).

Many cases of digitalis poisoning have been reported from different parts of the world. In one case, digitoxin poisoning in a 77-year-old patient with heart disease was reported (Hess et al. 1983). The patient had arrhythmias, extracardiac intoxication symptoms, and severe thrombocytopenia. Antibody treatment resulted in increase in platelet count from 26,000/mm^3 to 47,000/mm^3 within 12 hours and to over 60,000/mm^3 within 16 hours. In 2009, a case was reported in Italy (Maffè et al. 2009) where three people of one family suffered from digitalis poisoning after consuming potato dumplings flavored with *Borago officinalis* (borage) leaves mixed with Foxglove leaves. The patients suffered from nausea, vomiting, symptoms of bradyarrhythmia, and elevation in hepatic enzyme levels. In a more severe case, one of the patients had a full atrioventricular block with a 10-second asystolic stop, requiring the insertion of a temporary pacemaker into the femoral vein. Treatment with digoxin-specific antibody Fab fragments helped in overcoming the digitalis poisoning. Janssen et al. (2016) reported a case of digitalis poisoning in a 67-year-old Chinese woman. The patient had uneasiness in the chest, raised heartbeats, vomiting, and nausea in the initial stages. Subsequently, she complained of abdominal cramps and fatigue. The clinical investigations showed a high total leukocyte count and the electrocardiogram showed an atrioventricular block of the first degree, which eventually moved to the third degree. Alongside, dysrhythmias developed accompanied by premature ventricular systoles.

The problem with the treatment of digitalis poisoning is the difficulty in its identification due to the nonspecificity of the related signs and symptoms. Fatigue, weakness, nausea, and anorexia are the most typical symptoms that are very nonspecific. Similarly, cardiac glycoside-induced electrocardiographic abnormalities may be nondiagnostic. Arrhythmias associated with digitalis poisoning are frequently nonspecific and can reflect the patient's underlying heart condition (Bayer 1991). In case of digoxin toxicity, the following measures are prescribed (Mauskopf and Wenger 1991; Marie Nicole et al. 2021):

- Checking medication profile of a patient.
- Administration of Digoxin immune Fab [ovine]-Fab fragments (Digibind).
- Maintenance of electrolyte repletion and intravenous hydration.
- Referring patient to Intensive Care Unit.
- Proper consideration of nephrologists inputs in case of hemodialysis.

Nevertheless, the risk versus advantages of digitalis drug remains debatable.

17.7 TRADITIONAL AND OTHER POTENTIAL USES

The health benefits of *Digitalis* spp. were first described in 1785 by William Withering (AlSnafi 2017). In ancient times, the plant was used to prepare medicine for the treatment of edema, boils, ulcers, cysts, headaches, paralysis, wounds, cuts, scratches, etc. With time, Foxglove was increasingly utilized for several other purposes like treatment of congestive heart failure (CHF), atrial fibrillation, atrial tachycardia, inducing contractions of the heart and increasing the excitability of cardiac muscles, reducing swelling in arms and ankles as well as improving blood flow throughout the body (Lunney et al. 2020). The digitalis drug is also known to have antioxidant, anti-diabetic, anticancer, antiviral, insecticidal, neurological, and immunological effects (Stenkvist et al. 1980; Ebaid et al. 2006; Pierre et al. 2007; Soto-Blanco 2022).

Cardiotonic properties: Digitalis compounds have anciently been used in the treatment of chronic heart failure due to their ability to impart effective action on the cardiac muscles. Scientific studies have shown that conjunction with vasodilators and diuretics with digoxin improves ejection fraction and cardiac output and reduces heart pressures and pulmonary capillary wedge pressure (which reduces pulmonary congestion and edema) in patients with little fluctuation of heart rate (Ribner et al. 1987; Campbell and MacDonald 2003). The main impact that drug has on the patients is increased inotropy. However, the immediate effect of digoxin on blood vessels is narrowing of blood vessels (vasoconstriction) and a fall in the systemic vascular resistance. This leads to improvement in cardiac output and removal of compensatory vasoconstrictor mechanisms. Drug digitalis has a less yet straight diuretic impact on the kidneys, which proves to be quite beneficial in heart failure patients.

Atrial fibrillation and atrial flutter is an abnormal state of the heart characterized by irregular and rapid heartbeats as well as disruption of the coordination between the atria and the ventricle of the heart, which results in low cardiac output. This eventually leads to low blood supply to the organs which could be a cause of organ failure. Complications with this kind of abnormality can be fatal yet can be cured and treated with proper medications. Glycosides such as digoxin are preferred for lowering the ventricular rate. The mechanism of digoxin is to stimulate the parasympathetic nervous system and subsequently activate vagal efferent nerves; this lowers the conduction of electrical impulses within the AV node and hence decreases heartbeat rate. Along with this mechanism, Digoxin also follows the path of reversibly inhibiting the Na/K-ATPase enzyme, and improving the heart rate (Fauchier et al. 2016).

Anticancer properties: Several interesting anticancer properties have been observed in digitalis after 1960 when studies reported the effective inhibition of tumor cells in vitro due to glycosides. Research studies have analyzed the in vitro anti-cancerous properties of 'cardenolides', extracted from digitalis, suggesting its possible therapeutic use in cancer treatment (Verma et al. 2018). Digoxin generally uses the mechanism of inhibiting the activity of HIF (Hypoxia inducing factor), FGF, and Nuclear transcription factor (NTF), hence initiating cell death (apoptosis). It also regulates the tumor cell suppression by binding with Na/K-ATPase receptor which controls tumor cell multiplication.

Antiviral properties: Cardiac glycosides from *Digitalis* have shown potential lethal effects on a wide range of DNA viruses like dengue virus, human papilloma virus, and Ebola. The mode of action includes impairment of viral particle replication, transcription and translation process of the virus (Blanco 2022).

17.8 FUTURE PROSPECTS

The significance of Foxglove-derived drug has been recognized in the treatment of ailments like congestive heart failures, atrial fibrillation, atrial arrhythmias, tachycardia (atrial and supraventricular), cancer, and virus infections. Despite these significant therapeutic applications, it is not applied as a preferential therapy due to its restricted therapeutic window, need for close monitoring, and possible harmful effects. Unprescribed medication and drug hypersensitivity may have lethal effects on the individuals. Such observations suggest the need for more concerted studies on drug development from *D. purpurea*. These include standardization of dosage parameters, countering intoxication due to multiple drug interactions, predicting AEs in patients with different physiological status, etc. Digibind has been shown to significantly minimize the risk of Foxglove toxicity. However, many more of these medications, with no or minimum side effects in patients, must be developed to reverse Digitalis toxicity. As cardiac glycosides are of special interest in Foxglove drugs, there is always a need of large quantities of these compounds. It is suggested that plant tissue culture technology can be utilized for mass-scale production and extraction of cardiotonic glycoside and its derivatives for drug development. Moreover, use of recombinant DNA technology can be utilized for enhanced production by hairy root culture or gene transfer into the host from which glycoside purification is easier. Another important aspect is the consideration of the phytoremediation potential of the plant. Since Foxglove can remove heavy metals from the substratum, due care is needed for drug preparation from the plant growing in heavy metal contaminated soil to avoid side effects on the body.

ACKNOWLEDGMENT

The authors are thankful to Ms. Kim Schrik, the Netherlands for providing the photograph of *Digitalis purpurea* plant.

NOTES

Tina, Department of Life Sciences, Graphic Era Deemed to be University, Dehradun, India
Vedanshi Pal, Department of Life Sciences, Graphic Era Deemed to be University, Dehradun, India
Manu Pant, Department of Life Sciences, Graphic Era Deemed to be University, Dehradun, India

REFERENCES

Al-Snafi, A. 2017. Phytochemical constituents and medicinal properties of *Digitalis Lanata* and *Digitalis purpurea*-a review. *Indo American Journal of Pharmaceutical Sciences* 4: 225–234.

Amisha, F., & Rehman, A. 2021. *Propylthiouracil (PTU). StatPearls. PDR for Herbal Medicines.* 2nd ed. Montvale, NJ: Medical Economics Company, pp. 248–252.

Bayer, M. J. 1991. Recognition and management of digitalis intoxication: Implications for emergency medicine. *American Journal of Emergency Medicine* 9(2): 29–32.

Best, L. S., & Bierzychudek, P. 1982. Pollinator foraging on foxglove (*Digitalis purpurea*): A test of as new model. *Evolution* 36(1): 70–79.

Blanco, B. S. 2022. Herbal glycosides in healthcare. In: Mandal, S. C., Nayak, A. K., & Dhara, A. K. (Eds.), *Herbal Biomolecules in Healthcare Applications.* Cambridge, MA: Elsevier Inc., pp. 239–282.

Bliss, D., & Smith, H. 1985. Penetration of light into soil and its role in the control of seed germination. *Plant, Cell & Environment* 8(7): 475–483.

Bräuchler, H. M., & Heubl, G. 2004. Molecular phylogeny of the genera Digitalis L. and Isoplexis (Lindley) Loudon (Veronicaceae) based on ITS- and trnL-F sequences. *Plant Systematics and Evolution* 248: 111–128.

Bruelheide, H., & Heinemeyer, A. 2002. Climatic factors controlling the eastern and altitudinal distribution boundary of *Digitalis purpurea* L. in Germany. *Flora—Morphology, Distribution, Functional Ecology of Plants* 197(6): 475–490.

Campbell, T. J., & MacDonald, P. S. 2003. Digoxin in heart failure and cardiac arrhythmias. *The Medical Journal of Australia* 179(2): 98–102.

Chaturvedi, H. C., & Jain, M. 1994. Restoration of regeneration potentiality in prolonged culture of *Digitalis purpurea*. *Plant Cell, Tissue and Organ Culture* 38: 73–75.

Chopra, R. N., & De, P. 1926. Indian digitalis. *The Indian Medical Gazette* 61(3): 117–120.

Clapham, A. R., Tutin, T. G., & Warburg, E. F. 1962. *Flora of the British Isles*. Cambridge, England: Cambridge University Press.

Corduan, G., & Spix, C. 1975. Haploid callus and regeneration of plants from anthers of *Digitalis purpurea* L. *Planta* 124(1): 1–11.

Ebaid, G. M., Faine, L. A., Diniz, Y. S., Rodrigues, H. G., Galhardi, C. M., Ribas, B. O., Fernandes, A. A., & Novelli, E. L. 2006. Effects of digitonin on hyperglycaemia and dyslipidemia induced by high-sucrose intake. *Food and Chemical Toxicology* 44(2): 293–299.

Eisner, D. A., & Smith, T. W. 1991. The Na-K pump and its effectors in cardiac muscle. In: Fozzard, H. A., Haber, E., Katz, A. M., & Morgan, H. E. (Eds.), *The Heart and Cardiovascular System*. New York, NY: Raven Press, pp. 863–902.

Elbaz, H. A., Stueckle, T. A., Tse, W., Rojanasakul, Y., & Dinu, C.Z. 2012. Digitoxin and its analogs as novel cancer therapeutics. *Experimental Hematology & Oncology* 1(4): 1–10.

Farzam, K., & Jan, A. 2021. Beta Blockers. StatPearls. https://www.ncbi.nlm.nih.gov/books/NBK532906/.

Fauchier, L., Laborie, G., & Babuty, D. 2016. Beta –blocker or digoxin for atrial fibrillation and heart failure? *Cardiac Failure Review* 2(1): 35–39.

Fromm, M. F., Kim, R. B., Stein, C. M., Wilkinson, G. R., & Roden, D. M. 1999. Inhibition of P-glycoprotein-mediated drug transport: A unifying mechanism to explain the interaction between digoxin and quinidine. *Circulation* 99(4): 552–557.

Furuya, R., Hirotani, M., & Shinohara, T. 1970. Biotransformation of digitoxin by suspension callus culture of *Digitalis purpurea. Chemical and Pharmaceutical Bulletin* 18(5): 1080–1081.

Garibaldi, A., Tabone, G., & Gullinoa, M. L. 2022. First report of root rot caused by *Pythium oopapillum* on *Digitalis purpurea* in Italy. *Plant Disease* 106(4): 1309.

Hagimori, M., Matsumoto, T., & Obi, Y. 1982. Studies on the production of *Digitalis* cardenolides by plant tissue culture. *Plant Physiology* 69(3): 653–656.

Hauptman, P. J., & Kelly, R. A. 1999. Digitalis. *Circulation* 99(9): 1265–1270.

Haux, J. 1999. Digitoxin is a potential anticancer agent for several types of cancer. *Medical Hypotheses* 53(6): 543–548.

Hay, F. R., Probert, R. J., & Smith, R. D. 2008. The effect of maturity on the moisture relations of seed longevity in foxglove (Digitalis purpurea L.). *Seed Science Research* 7(4): 341–350.

Hess, T., Riesen, W., Scholtysik, G., & Stucki, P. 1983. Digitoxin intoxication with severe thrombocytopenia: Reversal by digoxin-specific antibodies. *European Journal of Clinical Investigation* 13: 159–163.

Hinz, P.-A. 1990. E´ tudebiosyste´matique de l'agre´gat *Digitalis purpurea* L. (Scrophulariaceae) enMe´diterrane´e occidentale. XII. Synthese. *Candollea* 44: 181–199.

Hooymans, P. M., & Merkus, F. W. 1985. Current status of cardiac glycoside drug interactions. *Clinical Pharmacology* 4(4): 404–413.

Huxley, A. 1992. *The New RHS Dictionary of Gardening*. London, England: MacMillan. ISBN 0-333-47494-5.

Janssen, R. M., Berg, M., & Ovakim, D. H. 2016. Two cases of cardiac glycoside poisoning from accidental foxglove ingestion. *Canadian Medical Association Journal* 188(10): 747–750.

Kang, M., Galuska, M. A., & Ghassemzadeh, S. 2021. Benzodiazepine Toxicity. StatPearls. https://www.ncbi.nlm.nih.gov/books/NBK482238/.

Kelly, R. A., & Smith, T. W. 1996. Pharmacological treatment of heart failure. In: Hardman, J.G., Limbird, L.E., Molinoff, P.B., Ruddon, R.W., Gilman, A.G. (Eds.), *Goodman & Gilman's The Pharmacological Basis of Therapeutics*. 9th ed. New York, NY: McGraw-Hill, pp. 809–838.

Kuhlmann, J. 1982. Inhibition of digoxin absorption but not of digitoxin during cytostatic drug therapy. *ArzneimittelForschung* 32(6): 698–704.

Kreis, W. 2017. The Foxgloves (Digitalis) revisited. *Planta Medica*. 83(12–13): 962–976.

Lindholm, P., Gullbo, J., Claeson, P., Goransson, U., Johansson, S., Backlund, A., Larsson, R., & Bohlin, L. 2002. Selective cytotoxicity evaluation in anticancer drug screening of fractionated plant extracts. *Journal of Biomolecular Screening* 7(4): 333–340.

Lungeanu, I., Calcandi, V. & Calcandi, I. 1963. Kariologische und phytocemische untersuchungen uber *Digitalis purpurea* L. ssp. heywoodii. *Die Naturwissenschaften*. 21: 673–674.

Lunney, M., Ruospo, M., Natale, P., Quinn, R. R., Ronksley, P. E., Konstantinidis, I., Palmer, S. C., Tonelli, M., Strippoli, G. F., & Ravani, P. 2020. Pharmacological interventions for heart failure in people with chronic kidney disease. *Database of Systematic Review* 2(2): CD012466.

Maffè, S., Cucchi, L., Zenone, F., Bertoncelli, C., Beldì, F., Colombo, M. L., Bielli, M., Paino, A. M., Parravicini, U., Paffoni, P., Dellavesa, P., Perucca, A., Pardo, N. F., Signorotti, F., Didino, C., & Zanetta, M. 2009. *Digitalis* must be banished from the table: A rare case of acute accidental *Digitalis* intoxication of a whole family. *Journal of Cardiovascular Medicine (Hagerstown, Md.)* 10(9):727–732.

Mauskopf, J. A., & Wenger, T. L. 1991. Cost-effectiveness analysis of the use of digoxin immune Fab (ovine) for treatment of digoxin toxicity. *American Journal of Cardiology* 68(17): 1709–1714.

Meusel, H. 1954. Distribution maps of central German key plants, 7th - Math.-Nat. Series. *Martin Luther University Halle Wittenberg* 3(1): 10–17.

Mishra, N., Tiwari, V. K., & Schmidt, R. R. 2020. Recent trends and challenges on carbohydrate- based molecular scaffolding: general consideration towards impact of carbohydrates in drug discovery and development. In: Tiwari, V. K. (Eds.), *Carbohydrates in Drug Discovery and Development*. Cambridge, MA: Elsevier Inc., pp. 1–69.

Navarro, E., Alonso, P., Alonso, S., Trujillo, J., Pèrez, C., Toro, M. V., & Ayuso, M. J. 2000.Cardiovascular activity of a methanolic extract of *Digitalis purpurea* spp. heywoodii. *Journal of Ethnopharmacology* 71: 437–442.

Negi, J. S., Bisht, V. K., Bhandari, A. K., & Sundriyal, R. C. 2012. Determination of mineral contents of Digitalis purpurea L. and Digitalis lanata Ehrh. *Journal of Soil Science and Plant Nutrition* 12(3): 463–470.

Patil, J. G., Ahire, M. L., & Nikam, T. D. 2012. Influence of plant growth regulators on *in vitro* seed germination and seedling development of *Digitalis purpurea* L. *The Asian and Australian Journal of Plant Science and Biotechnology* 6(1): 12–18.

Patil, J. G., Ahire, M. L., Nitnaware, K. M., Panda, S., Bhatt, V. P., Kishor, P. B. K., & Nikam, T. D. 2013. *In vitro* propagation and production of cardiotonic glycosides in shoot cultures of *Digitalis purpurea* L. by elicitation and precursor feeding. *Applied Microbiology and Biotechnology* 97(6): 2379–2393.

Pérez-Alonso, N., Martín, R., Capote, A., Pérez, A., Hernández-Díaz, E. K., Rojas, L., Jiménez, E., Quiala, E., Angenon, G., Garcia-Gonzales, R., & Chong-Pérez, B. 2018. Efficient direct shoot organogenesis, genetic stability and secondary metabolite production of micropropagated *Digitalis purpurea* L. *Industrial Crops and Products* 116: 259–266.

Pérez-Alonso, N. L., Labrada, F. A., Pérez, A. C., Pérez, A. P., Sosa, R., Mollineda, A., & González, E. J. 2014. Stimulation of cardenolides in *Digitalis purpurea* L. shoots cultivated in vitro by means of elicitors. *Revista Colombiana de Biotecnología* 16(1): 51–61.

Pèrez-Alonso, N. L., Wilken, D., Gerth, A., Jähn, A., Nitzsche, H. M., Kerns, G., Pèrez, A. C., & Jimènez, E. 2009. Cardiotonic glycosides from biomass of *Digitalis purpurea* L. cultured in temporary immersion systems. *Plant Cell, Tissue and Organ Culture* 99: 151–156.

Peters, U., Risler, T., Grabensee, B., Falkenstein, U., & Kroukou, J. 1980. Interaction of quinidine and digitoxin in the human. *Deutsche MedizinischeWochenschrift* 105(13): 438–442.

Phillips, R. & Rix, M. 1991. *Perennials Volumes 1 and 2*. London, UK: Pan Books. ISBN 0-330-30936-9.

Pierre, S. V., Yang, C., Yuan, Z., Seminerio, J., Mouas, C., Garlid, K. D., Dos-Santos, P., & Xie, Z. 2007. Ouabain triggers preconditioning through activation of the Na+, K+-ATPase signaling cascade in rat hearts. *Cardiovascular Research* 73(3): 488–496.

Rad, M. M., Abdossi, V., Moradi, P., Rakhshandehroo, F., & Mehrafarin, A. 2021. Phytochemical changes of *Digitalis purpurea* L. in response to polyamines and methyl jasmonate application in callus culture. *Journal of Plant Biochemistry and Biotechnology* 31: 310–319. DOI:10.1007/s13562-021-00678-w.

Rana, S., Singh, S., & Bisht, A. S. 2022. *Digitalis pupurea* L.: An overview on important medicinal plant. In: *Advances in Agriculture For Doubling Farmers' Income*. Lucknow, India: BFC Publications Private Limited, pp. 190–194.

Reddy, B. A. 2010. Digitalis therapy in patients with congestive heart failure. *International Journal of Pharmaceutical Sciences Review and Research* 3(2): 90–95.

Rehman, R., Dawson, A. H., & Hai, O. 2022. Digitalis toxicity. In *StatPearls*. StatPearls Publishing. https://pubmed.ncbi.nlm.nih.gov/29083729/

Ren, Y., Ribas, H. T., Heath, K., Wu, S., Ren, J., Shriwas, P., Chen, X., Johnson, M. E., Cheng, X., Burdette, J. E., & Kinghorn, A. D. 2020. Na+/K+-ATPase-targeted cytotoxicity of (+)-digoxin and several semisynthetic derivatives. *The Journal of Natural Products* 83(3): 638–648.

Ribner, H. S., Zucker, M. J., Stasior, C., Talentowski, D., Stadnicki, R., & Lesch, M. 1987. Vasodilators as first-line therapy for congestive heart failure: a comparative hemodynamic study of hydralazine, digoxin, and their combination. *American Heart Journal* 114: 91–96.

Riedl, S. J. & Shi, Y. 2004. Molecular mechanisms of caspase regulation during apoptosis. *Nature Reviews Molecular Cell Biology* 5: 897–907.

Roca-Pérez, L., Boluda, R., Gavidia, I., & Pérez-Bermúdez, P. 2004. Seasonal cardenolide production and Dop5br gene expression in natural populations of *Digitalis obscura*. *Phytochemistry* 65: 1869–1878.

Roca-Pérez, L., Pérez-Bermúdez, P., Gavidia, I., & Boluda, R. 2005. Relationships among soil characteristics, plant macronutrients, and cardenolide accumulation in natural populations of *Digitalis obscura*. *Journal of Plant Nutrition and Soil Science* 168: 774–780

Rodin, S. M., & Johnson, B. F. 1988. Pharmacokinetic interactions with digoxin. *Clinical Pharmacokinetic* 15(4): 227–244.

Rosen, M. R. 1985. Cellular electrophysiology of digitalis toxicity. *Journal of the American College of Cardiology* 5: 22A–34A.

Rowson, J. M. 1955. Studies in the genus Digitalis. *Journal of Pharmacy and Pharmacology* 7(1): 932–941.

Rücker, W., Jentzsch, K., & Wichtl, M. 1976. Root differentiation and glycoside formation in tissues of *Digitalis purpurea* L. cultured *in vitro. Journal of Plant Physiology* 80(4): 323–335.

Rücker, W., Jentzsch, K., & Wichtl, M. 1981. Organ differentiation and glycoside-formation in tissues of *Digitalis purpurea* L. cultured *in vitro*; the effect of various growth substances, basal media and light conditions. *Journal of Plant Physiology* 102(3): 207–222.

Rucker, W., Jentzsch, K., & Wichtl, M. 1983. Studies on growth, morphogenesis and glycoside formation in roots organ cultures of *Digitalis purpurea* L. growth, morphology and glycosides formation in root organs cultures of *Digitalis purpurea* L. *Bichemie und physiologie der*

Santana, L. F., Gomez, A. M., & Lederer, W. J. 1998. Ca2+ flux through promiscuous cardiac Na+ channels: Slip-mode conductance. *Science* 279: 1027–1033.

Schwinger, R. H. G., Bundgaard, H., Muller-Ehmsen, J., & Kjeldsen, K. 2003. The Na, K-ATPase in the failing human heart. *Cardiovascular Research* 57: 913–920.

Sessa, M., Mascolo, A., Andersen, M. P., Rosano, G., Rossi, F., & Capuano, A. 2016. Effect of chronic kidney diseases on mortality among digoxin users treated for non-valvular atrial fibrillation: A nationwide register-based retrospective cohort study. *PLOS One* 11(7): e0160337.

Sharma, A., & Purkait, B. 2012. Identification of medicinally active ingredient in ultradiluted *Digitalis purpurea*: Fluorescence spectroscopic and cyclic-voltammetric. *Journal of Analytical Methods in Chemistry* 2012: 109058. DOI:10.1155/2012/109058.

Sletvold, N., & Rydgren, K. 2007. Population dynamics in *Digitalis purpurea*: The interaction of disturbance and seed bank dynamics. *Journal of Ecology* 95: 1346–1359.

Stenkvist, B., Bengtsson, E., Eklund, G., Eriksson, O., Holmquist, J., Nordin, B., & West-man-Naeser, S. 1980. Evidence of a modifying influence of heart glucosides on the development of breast cancer. *Analytical and Quantitative Cytology and Histology* 2(1): 49–54.

Ujhelyi, M. R., & Robert, S. 1995. Pharmacokinetic aspects of digoxin-specific Fab therapy in the management of digitalis toxicity. *Clinical Pharmacokinetics* 28(6): 483–493.

Verma, S.K., Das, A.K., Cingoz, G.S., & Gürel, E. 2016. *In vitro* culture of Digitalis L. (Foxglove) and the production of cardenolides: An up-to-date review. *Industrial Crops and Products* 94 : 20–51.

Verma, S. K., Gantait, S., Jeong, B. R., & Hwang, S. J. 2018. Enhanced growth and cardenolides production in *Digitalis purpurea* under the influence of different LED exposures in the plant factory. *Scientific Reports* 8(1): 1–12.

Wade, O. L. 1986. Review articles digoxin 1785–1985 I. Two hundred years of *Digitalis. Journal of Clinical and Hospital Pharmacy* 11: 3–9.

Wagner, L. & Kenreigh, C. 2007. Digitoxin. In: Enna, S. J. & Bylund, D. B. (Eds.). *xPharm: The Comprehensive Pharmacology Reference.* Elsevier.

Wallace, E. C., & Crouch, J. A. 2018. First report of *Peronospora digitalidis* causing downy mildew disease on foxglove in Oregon. *Plant Disease* 102(4): 827.

Watson, W. A., Litovitsìz, T. L., Rodgers, G. C., Jr., Klein-Schwartz, W., Youniss, J., Rose, S. R., Borys, D., & May, M. E. 2002. Annual report of the American Association of Poison Control Center toxic exposure surveillance system. *The American Journal of Emergency Medicine* 21: 353–421.

Werner, K. 1964. Die Verbreitung der Digitalis-Arten. Wiss Z Univ Halle-Wittenberg. *Math-Naturwiss Reihe* 13: 453–486

Winnicka, K., Bielawski, K., Bielawska, A., & Surazyński, A. 2008. Antiproliferative activity of derivatives of ouabain, digoxin and proscillaridin A in human MCF-7 and MDA-MB-231 breast cancer cells. *Biological & Pharmaceutical Bulletin* 31(6): 1131–1140.

Withering, W. 1937 (Ed.). An Account of the Foxglove, and Some of Its Medical Uses; With Practical Remarks on Dropsy, and Other disease. Birmingham, London: G.G.J. and J. Robinson.

Yancy, C. W., Jessup, M., Bozkur, B., Butler, J., Casey, D. E., Drazner, M. H., Fonarow, G. C., Geraci, S. A., Horwich, T., Januzzi, J. L., Johnson, M. R., Kasper, E. K., Levy, W. C., Masoudi, F. A., McBride, P. E., McMurray, J. J. V., Mitchell, J. E., Sam, F., Stevenson, L. W., Tang, W. H. W., Tsai, E. M., & Wilkoff, B. L. 2013. ACCF/AHA guideline for the management of heart failure: A report of the American College of Cardiology Foundation/ American Heart Association Task Force on practice guidelines. *Circulation* 128: e240–e327.

Yeh, J-Y. Y., Huang, W. J., Kan, S-F., & Wang, P. S. 2001. Inhibitory effects of *Digitalis* on the proliferation of androgen-dependent and independent prostate cancer cells. *The Journal of Urology* 166(5): 1937–1942.

Zahid, H., & Rizwani, H. G. 2016. *Digitalis purpurea* L.: A concise drug review with probable clinical uses. *HamdardMedicus* 59(2): 25–32.

Ziff, O., & Kotecha, D. 2016. Digoxin: The good and the bad. *Trends in Cardiovascular Medicine* 26(7): 585–595.

18 *Gloriosa superba* (Glory Lily)

Umakanta Dash, Ankita Bhardwaj, Ashwath M. N., Preeti Vats, Sourav Ranjan Mohapatra, and Kajal Samantara

CONTENTS

18.1 INTRODUCTION

Plants do play a crucial role as a fulcrum in any ecosystem and also contribute to the welfare of mankind. Since antiquity, they have been chief source of therapeutic agents. Increasing recognition of conventional herbal systems of medicine, like *Ayurveda*, within India and outside has resulted in the resumption of ancient traditions of medicine (Bala, 1982). Traditional medicine is defined

DOI: 10.1201/b23017-20

as 'the sum total of knowledge, skills, and practises based on theories, beliefs, and experiences indigenous to various cultures, whether explicable or not, used in the maintenance of health, as well as the prevention, diagnosis, improvement, or treatment of physical and mental illnesses' (WHO, 2000). Medicinal plants account for 10% of all vascular plants (Badwaik et al., 2011). They are non-sedative and have fewer side effects, thus looked upon as a source of affordable healthcare cum important item of international trade and commerce, i.e., global market of worth US$800 billion a year (Rajasekharan and Ganeshan, 2002). Allopathic medicine, too, owes a huge debt to medicinal plants; in the United States, one out of every four prescriptions is either a synthetic form of or derived from plant ingredients (Srivastava et al., 1995). The secondary metabolites produced by the plants are usually responsible for the curative characteristics of plant species used throughout the world (Figure 18.1; Dar et al., 2017). Traditional medicine is widely used in China, India, Japan, Pakistan, Sri Lanka, and Thailand; in China, traditional tribal remedies account for around 40% of overall medicinal usage (Singh, 2015). Conventional medicines, largely plant pharmaceuticals, are estimated to cater the health needs of 80% of the world's population or 4.3 billion people. According to current assessments, around 9,000 plants have therapeutic uses in diverse cultures and countries, and this is without conducting extensive research among indigenous and other populations (Farnsworth and Soejarto, 1991). While it is true that man has reaped the greatest advantage from the plants available to him, this knowledge pales in contrast to the world's plant wealth. Only in the last century, systematic explorations of the plant wealth are being carried out for documenting their reliable usages (Holley and Cherla, 1998). Several medico-potential drugs have been discovered from various plant sources, *viz.* Ajmaline and Reserpine from *Rauvolfia serpentina* (antihypertensive), Vinblastine and Vincristine from *Catharanthus roseus* (for Hodgkin's disease and paediatric leukaemia), Ephedrine from *Ephedra sinica* (bronchodilator), Quinine from *Cinchona pubescens* (antimalarial), etc. (Batugal et al., 2004).

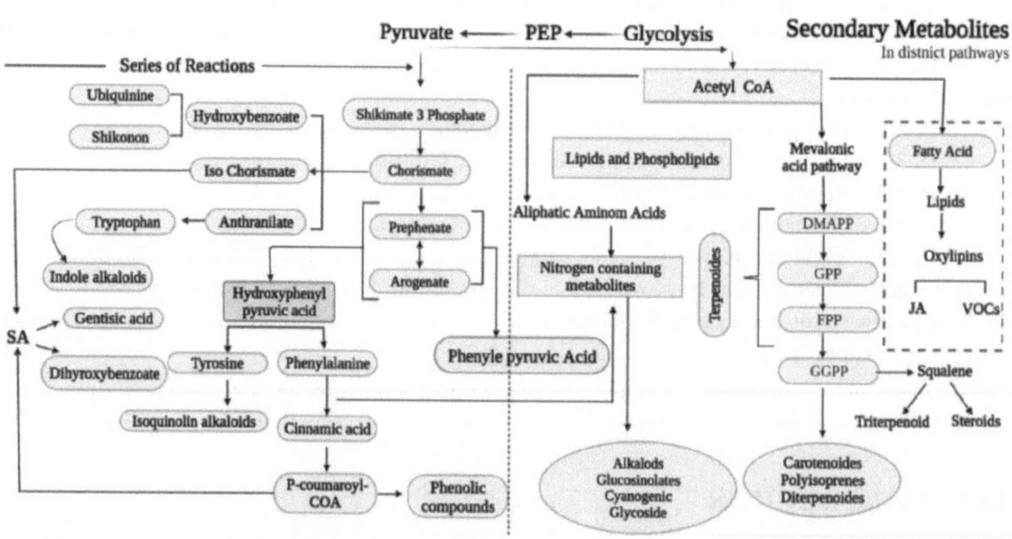

FIGURE 18.1 Pathways of secondary metabolite synthesis in plants.

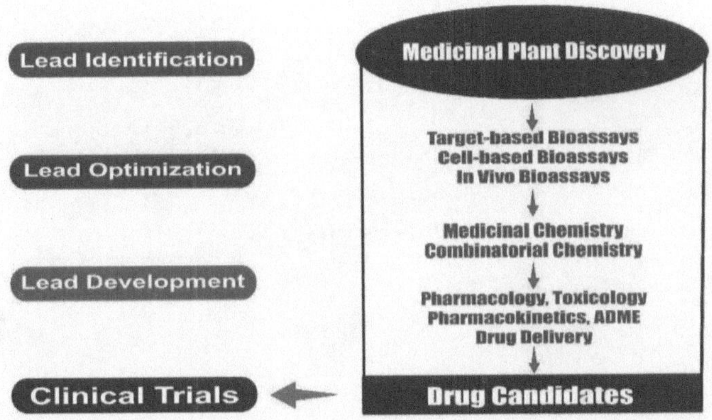

FIGURE 18.2 Schematic representation of a typical medicinal plant drug discovery process and development.

Likewise, *Gloriosa superba* Linn. is such an aesthetically pleasing and medicinally potent plant species belonging to the family Liliaceae. It is commonly known as 'malabar glory lily, climbing lily, tiger claw or flame lily' in English, '*Agnimukhi*' in Sanskrit, and '*Kalihari*' in Hindi (Kavina et al., 2011a). *Gloriosa* gets its name from the Latin words 'gloriosus,' which means handsome, and *superba*, which means splendid or majestic (Jagtap and Satpute, 2014). It is one of the *seven upavishas* (semi-poisonous drugs) in Indian medicine, which cures many disorders but can be fatal if misused (Jain et al., 2021). Despite the toxic response, tribes have utilized *G. superba* as a traditional remedy since 2000 BC (Jana and Shekhawat, 2011). Due to their extensive usage, the plant has been entered into threatened (endangered) category (Lal and Mishra, 2011). It is included in the top ten prioritized medicinal plants by the National Medicinal Plants Board (NMPB) of India (Mishra and Sharma, 2020).

18.2 GLOBAL DISTRIBUTION

G. superba is native to the tropical region of Africa and South East Asia (Yadav et al., 2012a). Its range in Africa extends from Senegal east to Ethiopia and Somalia, as well as to South Africa (Jana and Shekhawat, 2011). Its natural distribution spreads mainly in tropical Asia *viz*. India, Sri Lanka, Malaysia, and Myanmar (Phatak and Hegde, 2014). The species' altitudinal range is up to 2,100 m above sea level, and it is widely distributed across India's Himalayan regions and southern states (Chopra et al., 1956). The herbaceous plant's flower is the state flower of Tamil Nadu, India, as well as Zimbabwe's national flower (Ashokkumar, 2015; Chandel et al., 1996). It is currently grown in Tamil Nadu (especially in the Karur and Moolanur regions) and other parts of south India (Gopinath et al., 2014). This vital plant is also propagated by a number of botanical research institutes and nurseries. Because of its floral beauty, it is also being used as an indoor plant in gardens.

18.3 PHYTOGRAPHY

G. superba (Figure 18.4) is a remarkably beautiful, greenish, perennial climbing herb with hollow, slender stem of about 6 m long, arising from a fleshy tuberous rhizome (Figure 18.4). Rhizome is cylindrical, up to 12 inches long and 1.5 inches in diameter, bifurcated usually V- or L-shaped,

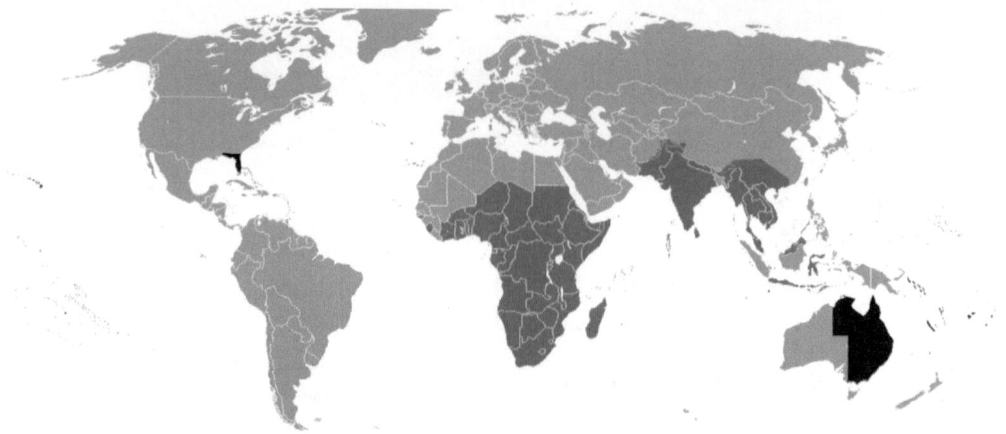

FIGURE 18.3 Distribution of *Gloriosa superba*. Natural distribution (dark grey) and area of introduction (black).

FIGURE 18.4 *Gloriosa superba*.

yellowish green colour, bilobed with two limbs equal or unequal in length and pointed ends (Basak et al., 2012). Separation of bilobed tubers results in higher percentage of plants as compared to undivided ones (Mishra and Sharma, 2020). Young tubers are shorter, but grow in V-shape when mature. One to four stems emerge from a single tuber per year in rainy season (June–July). It sprouts well in warm, humid, and tropical conditions. It is conventionally propagated through corms. Daughter corms flower only 2–3 years once they have achieved a critical size (Jana and Shekhawat, 2011). Economical tuber harvest is after 3–4 years (Sivakumar et al., 2003a, b). Although it is conventionally propagated through corms, the proliferation rate is still very low (Anandhi et al., 2013). Ideal weight of tubers has to be about 50–60 g.

Leaves are alternate, opposite or 3-nately whorled, subsessile, ovate, linear-lanceolate with acuminate tips spirally twisted to serve as tendrils. Colour of leaf lamina is mainly pale green or dark green, yet a few accessions were observed with pale green streaks on dark green coloured lamina (Patel et al., 2020). Flowers are showy, large, 4.5–7 cm in diameter, solitary or corymbose, 6-Merous with perianth segments having wavy margins, greenish at first later becoming yellow and finally scarlet or crimson (Smith, 1979; Floridata, 2004; Esther et al., 2020). Small insects cannot pollinate the huge blooms because of their unusual features, which include six perianth lobes twisted backwards, six radiating anthers, and a style bent nearly 90° at the site of attachment to the ovary (Horioka et al., 2020). Only huge insects such as bumble bees and long-beaked birds such as *Nectarinia zeylonica* and *Nectarinia asiatica* have been observed visiting these blooms (Subramanya and Radhamani, 1992). Although wind is another aspect that would aid in pollination, the flower structure reduces the chance of good cross-pollination. Fruits or capsules are 3 grooved, oblong, ellipsoid, 5 cm long containing many rounded seeds. Seeds are smooth and around 4–5 mm in diameter, with a spongy red testa surrounding them (Maori and Van der Maesen, 2011).

18.4 TRADITIONAL AND OTHER POTENTIAL USES

Among practitioners of traditional medicine, *G. superba* is a widely known NTFP that has long been in regular demand, especially in India. In *Ayurvedic* as well as *Unani* systems of medicine, it can be taken alone or in combination with other medications (Raghavendra, 1996). The local people used to exploit the plant as barter from foreign traders (Jana and Shekhawat, 2011). Glory lily has been proved effective in more than 30 disease conditions and approximately 158 formulations are being prepared from this drug (Bhide and Acharya, 2012). Five different pharmaceutical formulations commonly used are paste, powder, maceration (soaking in cold water), decoction (preparation in hot water) and using the whole plant without specific preparation (Maroyi and Maeson, 2011). All its parts are diversely used, and it treats a wide range of human ailments.

In a total of 12 states in India, 36 tribes have been reported to make use of this plant ethnobotanically. Main tribes include Baiga, Bhil, Gond, Irular, Kokani, Kuruma, Mullu, Mavachi, Munda, Oraon, Pawara, and Santal. The fact that its tuber paste is applied externally to ease labour pain after childbirth implies its ethnogynaecomedicinal use (Kal-Kutti as its local name). The paste of tubers which are grown in the northern direction with distil portion downwards is popular among Kani tribes with the name 'Karal Vatti' for accelerating childbirth (Jana and Shekhawat, 2010). The Gond tribe of Madhya Pradesh grinds its tuber, mixes it with ghee, and applies it orally, in the case of induced abortion (Vaishnavi et al., 2019). Jharkhand's Birhors apply rhizome extract to the navel and vaginal area to lessen labour discomfort and ensure a normal delivery (Tiwari and Yadav, 2003; Padamapriya et al., 2015). Vaidhyas (local healers of Jharkhand) usually give a dosage of 250–500 mg of the rhizome. If given to a woman who is 1 or 2 months pregnant, this dose may cause abortion. Deogarh tribes deploy *G. superba* to cure piles. Tubers are extremely poisonous if collected at the time of flowering of the plant, i.e., from July to October (Suri et al., 2001). It is collected and carefully cooked for a long time like 14–20 hours as it is poisonous after which its thin slices are kept in curd for a night and sundried for 4–5 days before being consumed . During food scarcity, tribals use them. Unfortunately, if it is eaten uncooked, it will cause major harm to the throat or may kill the person (Fatima and Jadhav, 2017). Jharkhand tribals use rhizome powder mixed with coconut oil for 5 days to treat skin eruptions and illnesses. Despite its toxic nature, Indian system of medicine recommends this as an antidote against snake and scorpion bites (Joshi, 1993; Kavina et al., 2011b). Similarly, Maroyi and Maeson (2011) purported that glory lily is snake and scorpion repellant.

Other uses of rhizome include poultices to relieve neuralgia, arthritic conditions, swellings of the joints, sprains, and dislocations (Kande et al., 2016; Patro, 2016). The tuber is conventionally exploited as a suitable drug for the treatment of cancer, colic, chronic ulcers, haemorrhoids, and leprosy (Nadkarni, 1978; Eddleston, 2000; Neuwinger, 1994). External use of tuber paste is used to treat parasitic skin disorders (Sivakumar, 2002). A similar study in Bidar demonstrates an equal portion of ginger paste mixed with the tuber and administered to the concerned area of the skin. Its pods, tuber, and leaves are used to treat the infection caused by liver fluke, guinea worms, roundworm, tapeworm, schistosomes (causing bilharzia) as well as filarial (Kirtikar and Basu, 1935). However, all portions of the plants are exceedingly poisonous and can be lethal if consumed (Lal and Mishra, 2011).

In *Charaka Samhita*, the plant is quoted as *Shakavarga*, which is an essential drug to be kept in sutikagara (labour room) and the main content of tiktekshvakvadi taila (Joshi and Rao, 2011). In *Mulaja visha*, Langalaki is the name mentioned by Acharya Charaka (Bhide and Acharya, 2012). Various medical conditions such as Shotha (inflammation/oedema), Vrana (wound), Gandamala (lymphadenitis), Charmaroga (skin diseases), Khalitya (hair loss), Agnimandya (loss of appetite), Arsha (piles), and Vatavyadhi (joint pain/arthritis) are mentioned in Ayurvedic classical formulations. It is well known for its anthelmintic, laxative, uterotonic, and alexiteric properties in Ayurveda. Mishra and Sharma (2020) highlighted the Rasayana properties (tonic and rejuvenator) of glory lily. Soup prepared from leaves or sap of tuber after due processing is administered to cure sterility, delayed puberty, delayed childbirth, and menstrual issues in women. Juice of leaves, unripe fruits along with butter, and macerated tuber has been reported to kill lice in the scalp (Burkill, 1995; Ghatapanadi et al., 2011). The sap from the leaf tip is used for pimples and skin eruptions. Rajith et al. (2010) reported that in South Kerala, tuber paste is applied on eyebrows to cure conjunctivitis. Tribals crush roots of the plant in water and apply them to the head for curing baldness. Likewise, to remove dandruff, dried powdered tubers are mixed with *Pongamia pinnata* oil and applied to the scalp. Flowers are used at Dongarya Dev's religious festivities in Nandurbar, Maharashtra. Glory lily has applications as a repellent for snakes and scorpions (Maroyi and Maeson, 2011).

Glory lily has a wide range of traditional uses in Africa. In the case of fainting, a leaf decoction used as a liniment soothes cough and general pain, and leaf juice is sprayed into the nose (Badwaik et al., 2011). Faso leaves are used as an enema decongestant in Cote d'Ivoire and Burkina Faso. In the Congo, crushed leaves are applied to the chest to treat asthma. Scrotal dropsy is treated with a leaf decoction in Burundi, whereas rheumatism is treated with leaf pulp (Rahmatullah et al., 2009). The Ulanga people of Tanzania burn the herb and apply the ash to wounds to aid healing. The plant juice is also used as an antimalarial. Tuber sap is used to make a sleep-inducing drink in Sudan. A paste made from the tuber is applied externally to promote parturition. Tanzanians utilize tuber juice as ear drops to treat earaches (Burkill, 1995), whereas Zimbabweans use tuber juice to treat toothaches (Burkill, 1995). Glory lily is also used in the manufacture of an impotence medication in Zambia. In Ethiopia, the powdered dry root of *Gloriosa superba* L. was used to treat a tumour and was taken orally with tea and coffee (Grubben and Denton, 2004; Yineger and Yewhalaw, 2007).

18.5 PHYTOCHEMICAL CONSTITUENTS

This charismatic plant species contains a variety of bioactive chemicals, including phenolics, vitamins, glycosides, alkaloids, tannins, steroids, flavonoids, saponins, terpenoids, and minerals. Plant tubers and seeds contain colchicine and its derivatives, a pharmaceutically important alkaloid; the amount of colchicine in tubers is 2–5 times lower than in seeds. Colchicine and colchicoside can be found in up to 0.9% and 0.8% of a single plant, respectively. The majority of these alkaloids are found in the plant's stem tubers. Because of the larger amount of colchicine in the seeds and tubers, they are extremely poisonous (Senthilkumar, 2013). Other sections of the plant with high colchicine concentrations (%) include the stem (0.33–0.41), ovary (0.08), corms (0.3), seeds (0.06), and flower (1.18) (Kavina et al., 2011a; Umavathi et al., 2020; Gurung et al., 2021). Phytochemicals present in different plant parts are illustrated in Figures 18.5 and 18.6. Studies also showed that methanol extract of leaves from mother plants contains higher phenols and flavonoids concentration than in vitro propagation of plants. In contrast, stem acetone extract of the mother plant contains higher amounts of tannins (Sharma et al., 2021). To study the biochemical attributes of *G. superba*, a number of solvents like ethanol, methanol, acetone, acetonitrile, and distilled water have been tried. Methanol and acetone show the best results for all bioactive compounds: phenolics, flavonoids, flavonols, tannins, and proanthocyanidins (Bahukhandi et al., 2021).

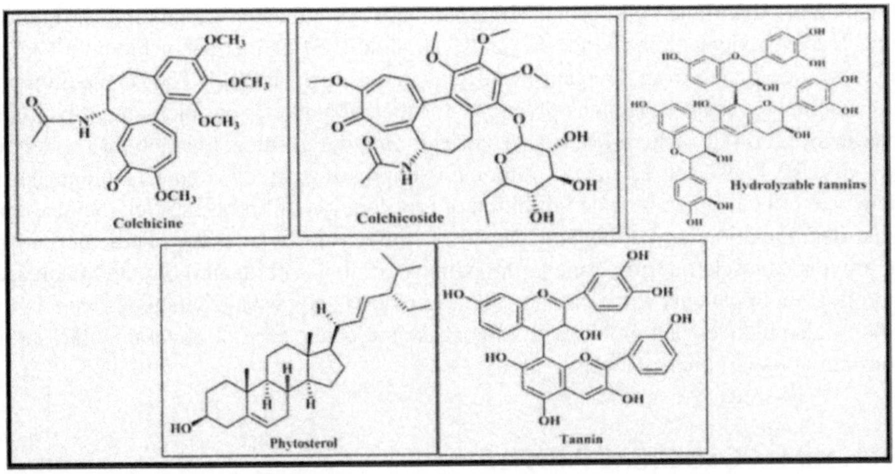

FIGURE 18.5 Phytocompounds of *Gloriosa superba*.

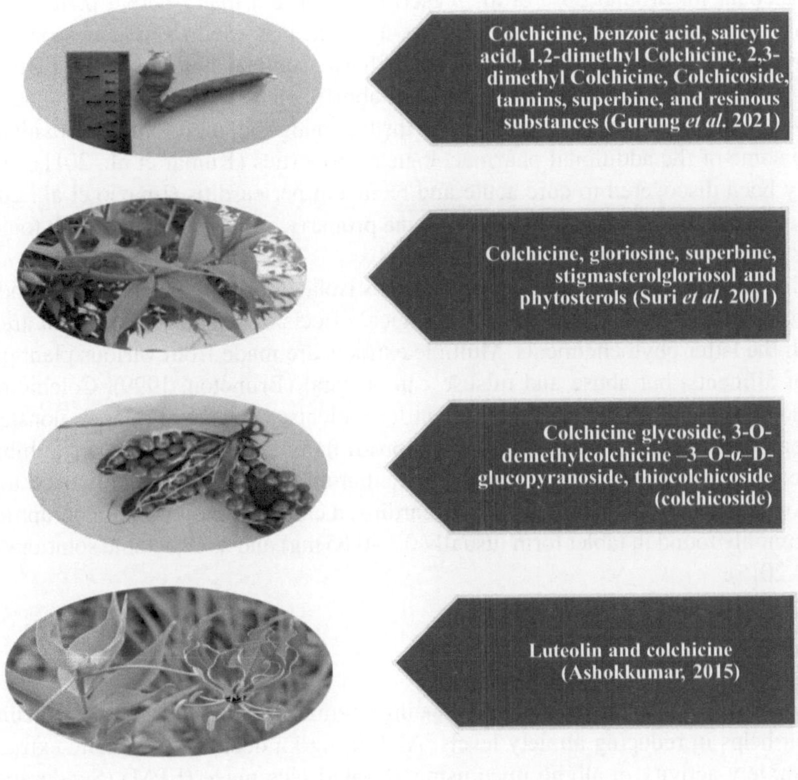

FIGURE 18.6 Phytochemicals in different plant parts of *Gloriosa superba*.

18.5.1 Colchicine

It is an amino alkaloid that is derived from two amino acids, i.e., phenylalanine and tyrosine, and chemically known as N-[(7S)-1,2,3,10-Tetra methoxy-9-oxo-5,6,7,9 – Tetrahydrobenzo[a] Heptalen-7-yl]acetamide. The other compounds in the plant that also possess some medicinal properties like colchicine are 3-Dimethyl colchicine, 1,2-Dimethyl colchicine, Gloriosine,

Lumicolchicine, superbine, N-Formyl International 75 diacetyl colchicine, 3-Dimethyl-N-Deformyl-N-Deacetylcolchicine, and the latest alkaloid 3-O-Demethylcolchicine-3-O-Alpha-D-Glucopyranoside isolated from the plant was discovered as early as 2001. The B5 medium prepared from *G. superba* act as an exogenous precursor for the biosynthesis of colchicine (Sivakumar and Krishnamurthy, 2004). Colchicine acts as a superior antigout agent and is therefore widely used to treat gout disease. It also acts as an anti-mitotic for suppressing cell division. Colchicine has significant biological effects which include inhibition of proliferation of fibroblasts, leukocyte diapedesis, lysosomal degranulation, and collagen transport to the extracellular space. These actions prevent or treat amyloidosis scleroderma, relieve the pain associated with acute gout, and decrease interleukin production in patients with primary biliary cirrhosis and chronic cutaneous leukocytoclastic vasculitis. It also induces polyploidy with the occasional occurrence of mutations like chlorophyll mutations but at low frequency levels.

18.6 PHARMACOLOGICAL PROFILE

According to the World Health Organization, around 11% of the current 252 medications generated only from flowering plants are vital for human health (Raskin et al., 2002). In developed countries, at least one ingredient in medicines is derived directly or indirectly from nature, and natural prescriptions account for around 25% of all prescriptions. It is a major herbal plant with numerous applications in the pharmaceutical business. As a result, it is used to treat a variety of illnesses both historically and in modern pharmacology. Pharmaceutical businesses produce anti-inflammatory, anthelmintic, antibacterial, analgesic, and abortifacient pharmaceuticals under therapeutic purposes. Anti-cancerous, anti-diabetic, anti-arthritic, analgesic, uterotonic, antioxidant, and anti-anxiety are some of the additional pharmacological properties (Kumar et al., 2011), and its utility has recently been discovered to cure acute and recurrent pericarditis (Imazio et al., 2009; Mishra and Sharma, 2020). The plant also holds Rasayana property, i.e., it behaves as both tonic as well as rejuvenator (Bhide and Acharya, 2012).

Colchicine and Gloriosine, two valuable alkaloids isolated from the plant's seeds and tubers, are known for their ethnobotanical and pharmacological effects. Rheumatism and gout are commonly treated with the latter phytochemicals. Multiple extracts are made from various plant parts to treat a variety of ailments, but abuse and misuse can be fatal (Bruneton, 1999). Colchicine has been widely utilized as a gout treatment replacement for patients who cannot tolerate nonsteroidal anti-inflammatory medications (NSAIDs). Multiple pro-inflammatory processes are inhibited, resulting in increased amounts of anti-inflammatory mediators. It is also commonly used to relieve the symptoms of stomach cancer and to avoid pericarditis, a cardiac condition (monograph). Colchicine is most commonly found in tablet form (usually 0.5–0.65 mg) and as injectable solutions (Chatterjee and Ghosh, 2015).

18.6.1 Anti-Anxiety

The aerial part of the plant is known for producing chemical compounds, i.e., petroleum ether and ethanol, that helps in reducing anxiety levels. At 300 mg/kg dose, the ethanolic extract exhibited good anti-anxiety activity in albino mice using elevated plus maze (EPM) (Sundaraganapathy et al., 2013).

18.6.2 Antimicrobial

The antimicrobial potential of *G. superba* extracts documented by Khan et al. (2008) for the superb antifungal activity confirmed against *Microsporum canis*, *Candida albicans*, *Staphylococcus aureus*, *Candida glabrata*, and *Trichophyton longifusus*.

18.6.3 Anti-Nemic

The rind of *G. superba* which has nematicidal properties usually goes to waste. Its usefulness is comparable and equal to the seeds, leaves, flowers, and rhizomes, as realized through evidence by its effectiveness in inhibiting the hatching of *Meloidogyne incognita* eggs and juvenile mortality. All the parts of *G. superba* exhibited anti-nemic activity against *M. incognita* and their antifungal and antibacterial actions. Due to nematicidal activities, FYM-enriched rinds can be applied to soil as an amendment that enhances nutrient availability. In integrated nematode management programmes, it can be used as one of the low-cost technologies (Ashokkumar et al., 2017).

18.6.4 Antibacterial Activity

The ethanolic tuber extract of *G. superba* is reported for high antibacterial activity against bacteria like *Escherichia coli*, *Staphylococcus aureus*, *Micrococcus luteus*, *Pseudomonas aeruginosa*, and *Salmonella abony* (Uchimahali et al., 2019). Acetone extract of *G. superba* tuber showed a more antagonistic effect on the growth of pathogenic bacteria. They have the highest negative impact on *E. coli* and the lowest impact on *Bacillus subtilis* (Kola et al., 2015). The methanol extract from seeds and tubers showed maximum antibacterial potential activity against gram-positive than gram-negative organisms, while chloroform extract showed moderate (Senthilkumar, 2013).

18.6.5 Antifungal Activity

The ethanolic extract from tubers of *G. superba* is reported to have an effective antifungal response against fungi like *Candida krusei*, *Aspergillus niger*, and *Candida albicans* (Uchimahali et al., 2019).

18.6.6 Anthelmintic Activity

Another property of the plant is its anthelmintic property, which helps treat parasitic worms such as *Eisenia fetida* and possesses an antimicrobial effect that is more effective for gram-positive bacteria than for gram-negative bacteria. Significant anthelmintic activity is reported against Indian earthworms (*Pheretima posthuma*) with aqueous and ethanolic whole plant extract of *G. superba* compared to standard drug piperazine citrate (Pawar et al., 2010). *G. superba* also contains a rich source of antioxidants in the leaves, seeds, and tubers and shows resistance against venom due to its anti-coagulant property by inhibiting clotting induced by thrombin.

18.6.7 Anti-Inflammatory

Thiocolchicoside (TCC) is semi-synthetic and derived from naturally occurring colchicoside. It is characterized by anti-inflammatory and analgesic effects (Padamapriya et al., 2015). TCC may thus represent an attractive therapeutic option for managing bone metastatic disease (Micheau et al., 2012). Flavonoids are ketonic compounds that can induce anti-inflammatory activity, as they inhibit prostaglandin synthesis and oxygen compounds, enzyme cyclooxygenase-dependent pro-inflammation activity (Gopinath and Arumugam, 2012).

18.6.8 Antioxidants

Polyphenols and phenols found in plants are two secondary metabolites considered as natural antioxidants (Davis, 1964).

18.6.9 Anti-Cancerous

This role is due to the alkaloid colchicine and its anti-mitotic activity as it binds to the α- and β-Tubulin heterodimer and is also used as a drug for chemotherapy (Jothi et al., 2019).

18.6.10 Anti-Coagulant/Anti-Thrombotic

The methanolic aqueous extract from the roots of the plant has exhibited anti-thrombotic properties, which may be due to the inhibition of thrombin-induced clotting (Jain and Suryavanshi, 2010). Thus, it cures thromboembolic diseases (Kumar et al., 2011).

18.7 TOXIC RESPONSE

The untrue and misleading belief about herbal drugs is that they are perfectly safe and devoid of adverse effects (Mordeniz, 2019; Kharchoufa et al., 2020). Due to insignificant knowledge, there lies a huge gap regarding their safety (Kharchoufa et al., 2021). India ranks first in recording the highest incidence of acute poisoning in the world; every year, more than 50,000 people die from toxic exposure (Peranantham et al., 2014). Therefore, despite being medicinal, the toxicity of *G. superba* cannot be disproved or neglected. The whole plant parts, especially the tubers, are highly poisonous. In rural areas, the tubers have normally been used as a suicidal agent among women and been used for homicide (Ellenhorn et al., 1996). Toxicity of this plant is attributed to the presence of colchicine, the biological hallmark of the family to which *G. superba* belongs.

18.7.1 Physico-Chemical Characteristics of Colchicine

Colchicine comes in the form of odourless crystals or amorphous scales/powder that ranges in colour from pale yellow to greenish yellow. When exposed to light, it darkens. It has a melting point of 157°C and a solubility of around 1/20 in water; it is easily soluble in chloroform and alcohol (Aleem, 1992; Windholz, 1983).

18.7.2 Mechanism of Colchicine Fatal Action

Primary mechanism is tubulin (structural protein of microtubules) disruption thereby inhibits its alpha and beta forms from polymerization to form microtubules (key component of cytoskeleton). At lower concentration of colchicine, microtubule growth arrested and its higher concentration promotes microtubule depolymerization (Leung et al., 2015). Microtubules have a role in a variety of cellular functions, including cell migration and division, cell shape regulation, intracellular trafficking, and ion channel control; the termination of these mechanisms may lead to multi-organ dysfunction and failure (Massarotti et al., 2012). Colchicine is a conventional anti-mitotic drug that disrupts spindle apparatus during metaphase, thereby leading to separation of chromosomes (Nuki, 2008). The arrest of mitosis is most dangerous to organs with a high rate of cell turnover (e.g., the intestinal epithelium, hair follicles, and bone marrow) as explicated by Gautam et al. (2017). Colchicine is also a phosphatase inhibitor (Craker and Simson, 1986). Its fatal dose for humans is thought to be around 60 mg, while lesser doses also have been seen to kill people (Angunawela and Fernando, 1971). Colchicine's LD_{50} in rats is 5 mg/kg, according to relevant animal research (Dunuwille et al., 1968).

On an average, cell cycle inhibition and apoptosis in the gastrointestinal region are the key events in colchicine poisoning, which results in deformation of tight junction proteins in intestinal barrier/villi. Furthermore, due to dilation of intercellular spaces and change in intestinal bacteria population, endotoxin enters into blood flow, thereby causing massive cytokine production.

18.7.3 Clinical Effects

Acute colchicine poisoning can be caused by overuse of medication or accidental intake of Colchicine-containing plants, and it manifests itself in three phases that are generally overlapping: (1) Early gastrointestinal symptoms, blood volume depletion, hypotension due to severe vomiting and diarrhoea, and peripheral leukocytosis; 10–24 hours after intake. (2) Alteration in mental status, oliguric renal failure, rhabdomyolysis, haematopoietic inhibition, and cardiac arrest; 2–7 days after consumption. Rapidly increasing multi-organ failure, sepsis, or shock cause death. (3) Rebound leukocytosis and alopecia more than 7 days after consumption (Finkelstein et al., 2010).

18.7.4 Illustrative Cases

A 35-year-old female developed symptoms of nausea, bloody diarrhoea, and dehydration. Her blood pressure was 100/70 mm Hg, pulse rate was 112/minute, and the respiratory rate was 20/minute. After 24 hours of hospital admission, serum urea and creatinine levels were found to be markedly elevated (urea: 70 mg/dL, creatinine: 3.5 mg/dL) and her gastrointestinal symptoms worsened. The patient ultimately died due to shock and multi-organ failure after 30 hours of hospitalization. History revealed that she had consumed 3–5 tubers of *G. superba* (Peranantham et al., 2014). Similar case was reported from a remote village in the Western Ghats of Karnataka, where a 54-year-old housewife died of endotoxin shock and multi-organ failure after 40 hours of intake of *G. superba* tubers (George, 2011). A non-fatal case of a 27-year-old man who suffered from profuse vomiting, diarrhoea, respiratory distress, mild renal impairment, thrombocytopenia, and localized alopecia after drinking herbal tea has been reported where *G. superba* seeds were mistakenly used in place of coriander. He was fully recovered after 10–15 days hospital care. Another non-fatal case was an 18-year-old girl who had developed harsh gastrointestinal symptoms, vaginal bleeding, acute renal failure, 6 hours after eating uncooked tubers of *G. superba* (Gautam et al., 2017).

18.8 CULTIVATION

The high market demand for glory lily has put further strain on the natural resources used to produce this plant. Because these species are not commercially cultivated, they are more reliant on wild collections, which have exacerbated the issue (Badola, 2002). As a result, immediate conservation measures are required to save these species in the wild by sustaining wild populations and expanding cultivation practices in many nations, without which the species may become extinct (Chatterjee and Ghosh, 2015). As a result, numerous propagation methods have been tried to solve the problem, including seed propagation, tuber multiplication, tissue culture, and micropropagation.

18.8.1 Propagation through Tubers and Seeds

The propagation of glory is generally done through specialized vegetative structures, i.e., tubers. Seeds are rarely used for plant production, as it requires more time than vegetative propagation. The matured seeds need to be pre-treated for better germination. Germination is intermittent and may take 3 weeks to 3 months time. Removal of the sarcotesta reduces the seed dormancy and allows seeds to germinate in 11–15 days (Rajamani and Padmapriya, 2017). Storage of seeds at ambient temperature affects germination and viability as an act of dormancy release. Le Roux and Robbertse (1994) reported 97% germination on incubation of seeds at 20°C–25°C for 30–31 days. Over the chemical treatment, glory lily showed maximum germination of 32.75% in hot-water treatment for 1 hour (Anandhi et al., 2016). Hot-water treatment (100°C boiled and flame released) for 40 minutes was reported to maximize germination up to 62% (Venudevan et al., 2010). Some seeds may exhibit the characters of producing twin seedlings from one seed, which may be considered

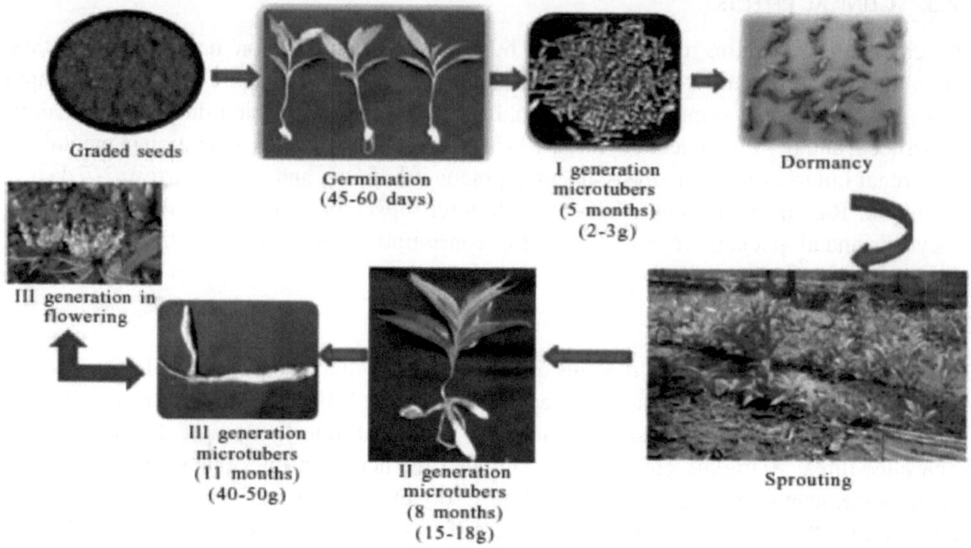

FIGURE 18.7 Cost-effective protocol for microtuber production technology. (Adapted from Padmapriya and Rajamani, 2021.)

abnormal as they compete and affect growth (Rane et al., 2012). Standardized methodology for plant development from seed implies that a medium containing decomposed coir with red soil and sand promotes germination in microtubers. Seed-origin seedlings generally start producing tubers in the second year, and flowering can be seen from the third or fourth year.

Vegetative propagation of glory lily using tubers is expected for high seed yield. The size and time of tuber and collection will decide the success rate of propagation. However, propagation is sluggish since each plant produces just one tuber in the next generation. Separating two-lobed tubers yields more blooming plants than keeping the tubers undivided. Sprouting is uneven, reaching around 60% in 30 days (Rajamani and Padmapriya, 2017). Tubers weighing 40–60 g are chosen for planting because they are vigorous and provide a maximal field stand. Tubers can be stored for a short period in the sand. The preservation of tubers in leaf mould was deemed the optimum storage option considering decay and sprouting percentage (Rajagopalan, 1994).

The growth promoters showed increased sprouting capacity of tubers. Treating tubers with ethrel @ 500 ppm was reported for early and peak sprouting percentages (Anandhi et al., 2016). Flower, pod yield, and quality were found to vary with varying concentrations and weight of seed tuber used for planting. Sivasankar et al. (2017) reported improved gain by using tuber weighing above 250 g treated with GA_3 @ 250 ppm. Padmapriya and Rajamani (2017) induced mutagenesis using Ethyl Methane Sulfonate (EMS) and Diethyl Sulphate (DES) to seed tubers. Dry seed yield was enhanced in plants treated with EMS 2% and succeeded in producing short-statured plants with DES 1.5% (Figure 18.7).

18.8.2 In Vitro Micropropagation and Tissue Culture

Several attempts were made to generate plants by tissue culture and micropropagation technologies due to poor germination, late seed productivity, and challenges in tuber development. There have been reports on micropropagation, which has been performed by multiple shoot formation or in vitro tuber synthesis using various plant hormones. After 6 weeks of inoculation,

Ghosh et al. (2007) produced one tuber per explant and three tubers per explant after 12 weeks of culture. The root explants on MS media treated with 2,4-Dichlorophenoxyacetic acid (2,4-D), Indole-3-Acetic Acid (IAA), and Napthelene Acetic Acid (NAA) had the highest percentage of callus. Gopinath et al. (2014) reported an 80% plant survival rate on planting rooted shoots in media containing sterilized cow dung powder, sand, and red soil in 1:2:3 ratios and stored in a mist house.

Invitro callus induction and proliferation were achieved by different researchers (Manju et al., 2010; Gopinath and Arumugam, 2012; Mahajan et al., 2016; Mahendran et al., 2018; Sivakumar et al., 2019; Mishra and Sharma, 2020). Mishra and Sharma (2020) reported proficient callus production in MS media with 1.5–3 mg/L of Benzyl Amino Purine (BAP) and 0.2–0.5 mg/L of NAA. Callus initiation from rhizome explants in MS media with different concentrations of 2,4-D (1.0–5.0 mg L^{-1}), α-NAA (0.5 mg L^{-1}), *Ulva lactuca* extracts (ULE, 10%–50%), and ULAgNPs (0.1–0.5 mg L^{-1}) was reported by Mahendran et al. (2018).

In vitro shoot induction is also practiced for quality plant production (Sivakumar and Krishnamurthy, 2000; Ade and Rai, 2009; Manju et al., 2010; Mahendran et al., 2016; Balamurugan et al., 2019). The tuber explants immersed in MS media with 2.0 mg/L of BAP, 0.5 mg/L of Kin, and 1.0 mg/L of Gibberellin showed proficient shoot multiplication (Mahajan et al., 2016). High shoot multiplication was observed in MS media with 1.5 mg/L of BAP, 0.6 mg/L of NAA, and 15 mg/L of putrescine (Sivakumar et al., 2019).

Root induction (Nikhila et al., 2014; Mahajan et al., 2016; Ghosh et al., 2015; Subiramani et al., 2019; Sivakumar et al., 2019) and Somatic embryogenesis (Nikhila et al., 2014; Mahendran et al., 2018; Balamurugan et al., 2019) through different techniques were achieved from intermodal, stem, seed, and other explants and successfully produced roots and plants. Sivakumar et al. (2019) reported in vitro root proliferation on MS media with 1.0 mg/L of IBA, 0.6 mg/L of IAA, and coconut water (15%). Microtuber roots on callus were observed in elicitor treatment with $AlCl_3$ at 125 and 150 µM (Subiramani et al., 2019). Cell suspension in a liquid medium supplemented with 2 mg/L of BAP showed high somatic embryogenesis (Balamurugan et al., 2019). Somatic embryogenesis was achieved at 17 days of internode implant with MS fresh medium with 0.5 mg/L 2,4-D and 0.25 mg/L Kinetin (Nikhila et al., 2014). Casein hydrolysate and salicylic acid were the most efficient elicitors for generating increased synthesis of colchicine and TCC content levels in cell suspension cultures (Mahendran et al., 2018).

18.8.3 Planting Management

Quality planting material can be planted directly in July and August on broad ridges with adequate organic matter in an east-west orientation with 1.5-m spacing. Application of FYM @ 20 t/ha or vermicompost @ 12.5 t/ha is recommended for improving soil organic content. One to two tonnes of quality tubers are required for one hectare based on the spacing. Closer spacing encourages cross-pollination and flowering, although there is a higher chance of disease incidence. TNAU recommends using *Trichoderma*, *Pseudomonas fluorescens*, and carbendazim for better management against disease and yield without loss. Glory lily, as a climber, requires support in the form of standards, either dead or live poles. The crop requires consistent soil moisture for tuber sprouting, vine development, primary stem branching, blooming, pod set, and maturity. Irrigation must be ensured uniformly (Premaratna et al., 2015).

18.8.4 Nutrient Management

G. superba requires high nutrients for better development and yield. Application of NPK in 120:50:75 kg/ha and micronutrients such as $ZnSO_4$ and $MgSO_4$ in two split doses are recommended (Kavithamani et al., 2013). Phatak et al. (2016) recommend fertigation of 150:50:100 kg of NPK

ha^{-1}, along with vermicompost @ 5 t/ha (as basal), $ZnSO_4$ @ 25 kg/ha, $FeSO_4$ @ 50 kg/ha, Borax @ 10 kg/ha, and Sodium molybdate @ 0.5 kg/ha. Foliar application of $FeSO_4$ (1%), $ZnSO_4$, (0.5%), Boric acid (0.2%), and gibberellic acid as foliar spray twice @ 200 mg/kg was considered to be the best for crop growth. Foliar treatment with humic acid 0.1%+panchagavya 4%+vermiwash 20% coupled with foliar application of *Pseudomonas fluorescens*+*Trichoderma viride* enhanced seed production and colchicine content (Balakumbahan and Rajamani, 2012).The combined application of vermicompost 4 t/ha with 40 kg N, 16.67 kg P, and 25 kg K/ha was reported to achieve high growth and seed yield (Gupta et al., 2013).

18.8.5 Yield and Harvest

Pod and flower yield depends on the management of plants. For final seed harvest, the crop requires 160–180 days. The collected pods are dried in the shade for 2–3 days to allow the capsules to open up and expose bright orange-yellow seeds. Maximized pod set is encouraged by artificial pollination than open pollination owing to typical flowers shape during the flower development and anthesis (Selvarasu and Kandhasamy, 2012). Different drying techniques of seeds were investigated for colchicine content by Balakumbahan and Rajamani (2008), and they reported that drying seeds in a mechanical drier had the greatest colchicine than shade drying. The average yield reported is 400–500 kg dry seeds/ha and 1,000 kg tubers/ha/year (Rajamani and Padmapriya, 2017). The colchicine yield is significantly increased with arbuscular mycorrhizal (AM) inoculation, *Glomus mosseae, Glomus fasciculatum, Glomus margarita,* and *Glomus gilmorei.* Among the four AM fungi, *G. mosseae* was found to be the best (Pandey et al., 2014). Yadav et al. (2012b) reported improved survival rate, growth, and colchicine content in AMF inoculated plantlets. Colchicine content was highest in tubers extracted from the tubers treated with *Acaulospora laevis* + *G. mosseae.* Intercropping glory lily with cashew was reported to have double the farmer's income in the Tamil Nadu region (Deivasigamani, 2016).

18.9 GENETIC DIVERSITY AND IMPROVEMENT

In different regions of India, significant heterogeneity in *G. superba* was reported in terms of environment, colchicine concentration, and morphological features of plants. Inter-Simple Sequence Repeat (ISSR), Random Amplified Polymorphic DNA (RAPD), Simple Sequence Repeat (SSR) or microsatellites are the most often utilized PCR-based DNA markers for genetic diversity investigations.

Padmalatha and Prasad (2006) optimized DNA isolation and PCR reaction protocol for genetic analysis using RAPD primers. CTAB method of DNA extraction was modified and standardized for quality DNA extraction. Seventy RAPD makers were employed to assess the genetic diversity among the 18 accessions selected from Tamil Nadu and Andhra Pradesh (Rajagopal and Rajamani, 2013). Fifty-eight primers showed polymorphism with a total of 413 amplicons. The unweighted pair group method using arithmetic averages showed considerable variation among the accessions. Ghosh et al. (2009) explored diversity among five populations with the help of RAPD markers and showed 76% polymorphism. Genetic diversity between the micropropagated, cultivated, and wild genotypes from Tamil Nadu, Madhya Pradesh, Chhattisgarh, and Andhra Pradesh was assessed with 10 RAPD and 6 ISSR markers (Tilwari and Sharma, 2021). 96.43% and 91.83% of polymorphism was shown in RAPD and ISSR, respectively. Based on a dendrogram formed using UPGMA and a dispersed plot derived from PCA analysis, high genetic variation was identified among all genotypes. The genetic fidelity of micropropagated plants of glory lily was assessed with the help of RAPD and ISSR markers (Yadav et al., 2013). The selected genotypes were propagated through standardized tissue culture protocol, and plantlets produced were used to analyse genetic stability using 50 RAPD and 30 ISSR markers. Out of which 10 RAPD and 7 ISSR primers produced reproducible amplicons.

Selvarasu and Kandhasamy (2017) reported DNA marker and agro-morphological genetic diversity among the glory lily mutants in VM 2 generations. Plant, Flower, and Pod morphological characters were analysed, and GCV, PCV, and heritability were estimated. 45 ISSR markers were used for maker-based diversity analysis. Characters showed high diversity coupled with polymorphism of ISSR amplification. High heritability and genetic advance were observed for fruit and plant characters among 20 accessions from Gujarat (Patel et al., 2021). High PCV and GCV were recorded for colchicine content; tuber and flower characteristics suggest high variability, which can be improved through phenotypic selection.

Five accessions were selected based on Genetic Variability, Heritability, and Genetic Advance (Rajagopal and Kandhasamy, 2009) and hybridization (Selvarasu and Kandhasamy, 2013) was carried out with higher improved colchicine content. Intraspecific hybridization studies were carried out to enhance the pod and seed production in the F1 generation. The elite chemotypes were identified based on colchicine content and evaluated for in vitro anti-arthritic potential from Eastern Himalayas (Misra et al., 2017) and for in vitro antigout activity from Gangetic Plains (Misra et al., 2020). Mahajan et al. (2022) explored diversity among 50 established accessions collected from Maharashtra region of Western Ghats based on morphological traits, ISSR markers, and metabolite content. The populations are grouped into three classes based on genetic markers. Significant morphometric variation for seed traits and vegetative traits was observed among the accessions. Metabolic examination showed high colchicine content in unexplored tissue, such as pod shell (Acharya and Anshu, 2008). The study recommends selecting high-yielding accessions as not many verities are developed in the particular medicinal crop.

18.10 CURRENT STATUS AND FUTURE REMARKS

G. superba is believed as the most important medicinal plant which is used in many medicinal and pharmaceutical industries. However, its overexploitation and illimitable collection of seeds and roots for the foreign market is creating a shortage of raw materials for the local drug industries in India. If it is allowed likewise, then endangered species like *G. superba* will be extinct in India. It is therefore a need of the hour to come forward and protect this important glorious herb from extinction. Therefore, simultaneous strenuous efforts of botanists, biotechnologists, policy makers, and conservationists are required. It is really a matter of great concern of conservation to conserve this plant; otherwise, we will be losing it in upcoming years (Balunas and Kinghorn, 2005).

Furthermore, there is a crucial need to preserve this plant using biological tactics. Emphasis should be given to enhance productivity and increase colchicine content in plant to fulfil the industrial demands. Even though various biotechnological techniques are done to improve the productivity and chemical constituent of *G. superba*, but a satisfactory improvement wasn't reported about colchicine content. Hence, great attention must be paid for overall improvement of plant that will benefit the global population and can encounter the everlasting bids of pharmaceutical industries.

18.11 CONCLUSION

G. superba is a commercially potent medicinal plant which possesses diverse medicinal applications. A number of pharmacologically important phytochemicals such as gloriosine and colchicines have been isolated from this plant. Moreover, it has been widely used for a number of diseases, and eventually, the plant is moving towards local extinction due to its overexploitation. Hence, it has been affirmed as endangered plant by IUCN. So, there is an immediate need of conservation to fulfil the ever-increasing gap of demand and supply. Plant biotechnological tools, especially plant tissue culture technique including micropropagation, somatic hybridization, cybridization, protoplast culture, embryo rescue, and mutational breeding, could also be applied to assist and advance our knowledge on all aspects of plant biology. With this technique, we not only cultivate good quality of plant but also enhance its valuable component and reduce the overharvesting of plant. Furthermore, responsiveness should be generated among the common people concerning the importance of *G. superba* and its overexploitation by the people. This is needed for informed management strategies to minimize the projected extinction of such commercially important medicinal plant. In this chapter, plant morphology, phytochemical constituents, pharmacological values, propagation methodology, and all other important aspects of glory lily were explained. So, the present chapter can help researchers to explore this plant to a further extent.

NOTES

Umakanta Dash, Department of Silviculture and Agroforestry, Kerala Agricultural University, Thrissur, India
Ankita Bhardwaj, Department of Silviculture and Agroforestry, Kerala Agricultural University, Thrissur, India
Ashwath M. N., Department of Forest Biology and Tree Improvement, Kerala Agricultural University, Thrissur, India
Preeti Vats, Department of Forest Products and Utilization, Kerala Agricultural University, Thrissur, India
Sourav Ranjan Mohapatra, Department of Forest Biology and Tree Improvement, Kerala Agricultural University, Thrissur, India
Kajal Samantara, Department of Genetics and Plant Breeding, Institute of Agricultural Sciences, Siksha 'O' Anusandhan (Deemed to be University), Bhubaneswar, India

REFERENCES

Acharya, D., & Anshu, S. 2008. *Indigenous Herbal Medicines: Tribal Formulations and Traditional Herbal Practices*. Aavishkar Publishers, Rajasthan, India.
Ade, R., & Rai, M. K. 2009. Current advances in *Gloriosa superba* L. *Biodiversitas Journal of Biological Diversity* 10(4): 210–214.
Aleem, H. M. 1992. *Gloriosa superba* poisoning. *Journal of Association of Physicians of India* 40(1): 541–542.
Anandhi, S., Rajamani, K., & Jawaharlal, M. 2013. Propagation studies on *Gloriosa superba. African Journal of Agricultural Research* 1(1): 1–4.
Anandhi, S., Rajamani, K., & Jawaharlal, M. 2016. Propagation studies in *Gloriosa superba. African Journal of Agricultural Research* 11(4): 217–220.
Angunawela, R. M., & Fernando, H. A. 1971. Acute ascending polyneuropathy & dermatitis following poisoning by tubers of *G. Superba. Ceylon Medical Journal* 16: 233–235.
Ashokkumar, K. 2015. *Gloriosa superba* L.: A brief review of its phytochemical properties and pharmacology. *International Journal of Pharmacognosy & Phytochemical Research* 7(6): 1190–1193.
Ashokkumar, K., Shanthi, A., Sivakumar, M., & Rajamani, K. 2017. Studies on the nutrient enhancement of Glory lily (*Gloriosa superba* L.), *Journal of Pharmacognosy and Phytochemistry* 6(6): 1410–1412.
Badola, H. K. 2002. Endangered medicinal plant species in Himachal Pradesh. A report on the international workshop on 'Endangered Medicinal Plant species in Himachal Pradesh' organized by G.B. Plant Institute of Himalayan Environment and Development at Himachal Unit, Mohal-Kullu. *Current Science* 83: 797–798.
Badwaik, H., Giri, T. K., Tripathi, D. K., Singh, M., & Khan, A. H. 2011. A review on pharmacological profile for phytomedicine known as *Gloriosa superba* Linn. *Research Journal of Pharmacognosy and Phytochemistry* 3(3): 103–107.
Bahukhandi, A., Barola, A., & Bhatt, I. D. 2021. Impact of solvent system on polyphenolics and antioxidant activity of *Gloriosa superba* L.: Herbaceous species of Western Himalaya. *National Academy Science Letters* 44(1): 9–12.
Bala, P. 1982. The Ayurvedic and Unani systems of medicine in medieval India. *Studies in the History of Medicine* 6(4): 282–289.
Balakumbahan, R., Rajamani, K., & Sivakumar, V. 2012. Effect of pre-harvest treatments on growth and yield of *Gloriosa superba. Tropical Ecosystems: Structure, Function, and Services* 8: 240–246.

Balamurugan, V., Amal, T. C., Karthika, P., Selvakumar, S., & Vasanth, K. 2019. Somatic embryogenesis and plant regeneration in *Gloriosa superba* L.: an endangered medicinal plant. In: Kumar, M., Muthusamy, A., Kumar, V., Bhalla-Sarin, N. (eds). *In vitro plant breeding towards novel agronomic traits, biotic and abiotic stress tolerance*. Springer, Singapore, 27–42.

Balunas, M. J., & Kinghorn, D. A. 2005. Drug discovery from medicinal plants review article. *Life Sciences* 78(5): 431–441.

Basak, U. C., Dash, D., & Mahapatra, A. K. 2012. Estimation of colchicine in tubers of *Gloriosa superba* L. originated from different agroclimatic zones of Odisha, India International. *Journal of Pharmacognosy and Phytochemical Research* 4(3): 157–161.

Batugal, P. A., Kanniah, J., Lee, S. Y., & Jeffrey, T. O. (eds). 2004. *Medicinal Plants Research in Asia, Volume 1: The Framework and Project Workplans*. International Plant Genetic Resources Institute—Regional Office for Asia, the Pacific and Oceania (IPGRI-APO), Selangor DE, Malaysia.

Bhide, B., & Acharya, R. 2012. *Gloriosa superba* Linn. and its therapeutic importance in Ayurveda—A review. *International Journal of Ayurvedic Medicine* 3(2): 58–67.

Bhide, B., & Acharya, R. 2012. Uses of Langali *Gloriosa superba* Linn.: An ethnomedicinal perspective. *International Journal of Ayurveda and Allied Sciences* 1(3): 65–72.

Bruneton, J. 1999. *Toxic Plants Dangerous to Humans and Animals*. Intercept, Andover, Hampshire.

Burkill, H. M. 1995. *The Useful Plants of West Tropical Africa*. Vol. 3. Royal Botanic Gardens, Kew, UK.

Chandel, K. P. S., Shukla, G., & Sharma, N. 1996. *Biodiversity in Medicinal and Aromatic Plants in India: Conservation and Utilization*. NBPGR, New Delhi, India.

Chatterjee, T., & Ghosh, B. 2015. Gloriosa: A high demanding pharmaceutical plant in near future. *International Journal of Economic Plants* 2(2): 089–102.

Craker, L. E., & Simson, J. C. 1986. *Recent Advances in Horticulture & Pharmacology Botany*. Oryx Press, Phoenix, AZ.

Dar, A. R., Shahnawaz, M., & Hassan, Q. R. 2017. General overview of medicinal plants: A review. *The Journal of Phytopharmacology* 6(6): 349–351.

Davis, B. J. 1964. Disc electrophoresis. II. Method and application to human serum proteins. *Annals of the New York Academy of Sciences* 121: 404–427.

Deivasigamani, S. 2016. Economics of intercropping systems on Glori Lily (*Gloriosa superba* l.) in cashewnut (*Anacardium occidentale*) plantations. *International Journal of Research in Management, Economics, and Commerce* 6(5): 12–14.

Dunuwille, R., Balasubramanium, K., & Bible, S. W. 1968. Toxic principles of *Gloriosa superba*. *Ceylon Journal of Medical Science* 17(2): 1–6.

Eddleston, M. 2000. Patterns and problems of deliberate self-poisoning in the developing world. *QJM* 93: 715–731.

Ellenhorn, M. J., Schonwald, S., Ordog, G., Wasser, B., & Ellenhorn, J. 1996. *Medical Toxicology Diagnosis & Treatment of Human Poisoning*. Williams & Wilkins, Baltimore, MD.

Esther, S. M., Jose, A. G. C., & Francisco, M. A. 2020. Worldwide research trends on medicinal plants. *International Journal of Environmental Research and Public Health* 17: 11–20.

Farnsworth, N. R., & Soejarto, D. D. 1991. Global importance of medicinal plants. In: Heywood, V., & Synge, H. (eds). Conservation of Medicinal Plants. Cambridge University Press, Cambridge, UK.

Fatima, S., & Jadhav, R. G. 2017. Review of ethnomedicinal uses of *Gloriosa superba* plant of Nandurbar district, Maharashtra. *International Journal of Multidisciplinary Research* 3(7): 1–6.

Finkelstein, Y., Aks, S. E., Hutson, J. R., Juurlink, D. N., Nguyen, P., Dubnov-Raz, G., Pollak, U., Koren, G., & Bentur, Y. 2010. Colchicine poisoning: The dark side of an ancient drug. *Clinical Toxicology* 48(5): 407–414.

Floridata. 2004. A Floridata plant profile. Retrieved April, 2015 from http:www.floridata.com/ref/g/glor_rot.cfm.

Gautam, A., Rao, C. H., Paswan, S. K., & Verma, P. 2017. Poisonous plant (*Gloriosa superbA L.*): Its pharmacological and therapeutic profile. *Elixir Applied Botany* 104: 45969–45972.

Ghatapanadi, S. R., Johnson, N., & Rajasab, A. H. 2011. Documentation of folk knowledge on medicinal plants of Gulbarga district, Karnataka. Indian Journal of Traditional Knowledge 10(2): 349–353.

Ghosh, S., Ghosh, B., & Jha, S. 2007. In vitro tuberisation of *Gloriosa superba* L. on basal medium. *Scientia Horticulturae* 114(3): 220–223.

Ghosh, S., Ghosh, B., & Jha, S. 2009. Polymorphism in *Gloriosa superba*. *Plant Genetic Resources* 7(1): 9–15.

Ghosh, S., Ghosh, B., & Jha, S. 2015. Role of exogenous carbohydrate and amino acid sources on biomass and colchicine production in nontrans formed root cultures of Gloriosa superba. *Plant Tissue Culture and Biotechnology* 25(2): 247–256.

Gopinath, K., & Arumugam, A. 2012. In vitro micropropagation using corm bud explants: An endangered medicinal plant of *Gloriosa superba* L. *Asian Journal of Biotechnology* 4: 120–128.

Gopinath, K., Karthika, V., Gowri, S., & Arumugam, A. 2014. In vitro morphogenetic regeneration from root explant of *Gloriosa superba* L. for enhanced crop production. *Science & Technology Journal* 1: 27–30.

Grubben, G. J. H., & Denton, O. A. 2004. *Plant Resources of Tropical Africa*. PROTA Foundation, Wageningen, Netherlands.

Gupta, L. M., Gupta, M., & Sharma, V. 2013. Integrated nutrient management for growth and yield in Glory Lily (*Gloriosa superba* L.). *Journal of Medicinal Plants Research* 7(43): 3197–3201.

Gurung, R., Sharma, S., Sharma, S., & Sharma, V. 2021. *Gloriosa superba*: Its properties and *in vitro* production methods. *International Journal of Botany Studies* 6(3): 74–77.

Holley, J., & Cherla, K. 1998. *The Medicinal Plants Sector in India: A Review*. South Asia Regional Office, IDRC, Canada Medicinal and Aromatic Plants Program in Asia (MAPPA), New Delhi, India, p. 117.

Horioka, K., Tanaka, H., Isozaki, S., Konishi, H., Fujiya, M., Okuda, K., Asari, M., Shiono, H., Ogawa, K., & Shimizu, K. 2020. Acute colchicine poisoning causes Endotoxemia *via* the destruction of intestinal barrier function: The curative effect of endotoxin prevention in a murine model. *Digestive Diseases and Sciences* 65(1): 132–140.

Imazio, M., Brucato, A., & Trinchero, R. 2009. Colchicine for pericarditis: Hype or hope. *European Heart Journal* 30: 532–539.

Jagtap, S., & Satpute, R. 2014. Phytochemical screening, antioxidant, antimicrobial and flavonoid analysis of *Gloriosa superba* Linn. rhizome extracts. *Journal of Academia and Industrial Research* 3(6): 247–254.

Jain, A., & Suryavanshi, S. 2010. *Gloriosa superba* Linn: A pharmacological review. *IJPRD* 2: 24–28.

Jain, R., Asaf, S., Numan, M. L., & Kim, K. M. 2021. Plant secondary metabolite biosynthesis and transcriptional regulation in response to biotic and abiotic stress conditions. *Agronomy* 11: 968.

Jana, S., & Shekhawat, G. S. 2011. Critical review on medicinally potent plant species: *Gloriosa superba*. *Fitoterapia* 82(3): 293–301.

Joshi, P. 1993. Tribal remedies against Snakebites and Scorpion Stings in Rajasthan. *Glimpses in Plant Research* 10: 23–25.

Joshi, P., & Rao, N. 2011. *Role of Indigenous People in Conservation of Biodiversity of Medicinal Plants: An Indian Case Study*. Survival and Sustainability, Environmental Earth Sciences, Delhi, India.

Jothi, U., Jebamalar, J. A., & Sivakumar, T. 2019. Study on estimation and antioxidant activity of *Gloriosa superba* L. whole plant extract. *International Journal of Scientific Research. Biological Sciences* 6(3): 50–55.

Kande, V., Ekanayeka, C. J. R., & Wijewardane, D. K. 2016. Case report: A rare case of attempted homicide with *Gloriosa superba* seeds. *BMC Pharmacology and Toxicology* 17: 26.

Kavina, J., Gopi, R., & Panneerselvam, R. 2011a. G. *superba* Linn.—A medicinally important plant. *Drug Invention Today* 3(6): 69–71.

Kavina, J., Gopi, R., & Panneerselvam, R. 2011b. Quantification of Colchicine in seed and tuber samples of *Gloriosa superba* by high-performance liquid chromatography method. *Journal of Applied Pharmaceutical Science* 01(07): 116–119.

Kavithamani, D., Umadevi, M., & Geetha, S. 2013. A review on *Gloriosa superba* L as a medicinal plant. *Indian Journal of Research in Pharmacy and Biotechnology* 1(4): 554.

Khan, H., Khan, M. A., & Mahmood, T. 2008. Antimicrobial activities of *Gloriosa superba* Linn extract. *Journal of Enzyme Inhibition and Medicinal Chemistry* 6: 855–859.

Kharchoufa, L., Bouhrim, M., Bencheikh, N., Addi, M., Hano, C., Mechchate, H., & Elachouri, M. 2021. Potential toxicity of medicinal plants inventoried in Northeastern Morocco: An ethnobotanical approach. *Plants* 10: 1108.

Kharchoufa, L., Bouhrim, M., Bencheikh, N., El Assri, S., Amirou, A., Yamani, A., Choukri, M., Mekhfi, H., & Elachouri, M. 2020. Acute and subacute toxicity studies of the aqueous extract from Haloxylon scoparium pomel (Hammada scoparia (pomel)) by oral administration in Rodents. *BioMed Research International* 6: 4020647.

Kirtikar, K. R., & Basu, B. D. 1935. Indian medicinal plants. *International Book Distributor, Dehradun, India* 1(5): 209.

Kola, S. G., Gajula, R. G., Thurpu, S. R., & Chiliveri, P. 2015. Anti-bacterial activity of *Gloriosa superba*. *World Journal of Pharmacy and Pharmaceutical Sciences* 4(3): 1398–1401.

Kumar, S., Joseph, L., George, M., & Sharma, A. 2011. A review on anticoagulant/antithrombotic activity of natural plants used in traditional medicine. *International Journal of Pharmaceutical* 8(1): 70–74.

Lal, H. S., & Mishra, P. K. 2011. *Gloriosa superba*—An endangered plant spotted for the first time in the forest of Tpchanchi, Hazaribagh (Jharkhand), India. *Science Research Reporter* 1(2): 61–64.

Le Roux, L. G., & Robbertse, P. J. 1994. Tuber ontogeny, morphology, and vegetative reproduction of *Gloriosa superba* L. *South African Journal of Botany* 60(6): 321–324.

Leung, Y. Y., Yao, H. L. L., & Kraus, V. B. 2015. Colchicine-Update on mechanisms of action and therapeutic uses. *Seminars in Arthritis and Rheumatism* 45(3): 341–350.

Mahajan, R., Kapoor, N., & Billowria, P. 2016. Callus proliferation and in vitro organogenesis of *Gloriosa superba*: An endangered medicinal plant. *Annals of Plant Sciences* 5: 1466–1471.

Mahajan, Y.A., Shinde, B.A., Mulani, F.A., Gade, A.B., Kasodekar, A.K., Thulasiram, H.V., Kadoo, N.Y., & Nikam, T.D. 2022. Diversity assessment of *Gloriosa superba* accessions from Western Ghats of India based on morphological traits, ISSR markers and metabolite content. *Journal of Applied Research on Medicinal and Aromatic Plants*, 100388.

Mahendran, D., Kavi Kishor, P. B., Sreeramanan, S., & Venkatachalam, P. 2018. Enhanced biosynthesis of colchicine and thiocolchicoside contents in cell suspension cultures of *Gloriosa superba* L. exposed to ethylene inhibitor and elicitors. *Industrial Crops and Products* 120: 123–130.

Mahendran, D., Thiyagarajan, M., Kavi Kishore, P. B., & Venkatachalam, P. 2016. Efficient plant regeneration from shoot tip explants of *Gloriosa superba*-an endangered anti-cancer medicinal plant. *Plant Cell Biotechnology and Molecular Biology*, 17(1–2): 31–38.

Manju, M., & Joseph, J. P. 2010. Histological marker to differentiate organogenic calli from non-organogenic calli of *Gloriosa superba* L. *Plant Tissue Culture and Biotechnology* 20(1): 1–5.

Massarotti, A., Coluccia, A., Silvestri, R., Sorba, G., & Brancale, A. 2012. The tubulin colchicine domain: a molecular modeling perspective. *ChemMedChem* 7(1): 33–42.

Micheau, O., Dufourr, F., & Walczak, H. 2012. Thiocolchicoside a semi-synthetic derivative of the Glory Lily: a new weapon to fight metastatic bone resorption? *British Journal of Pharmacology* 165: 2124–2126.

Mishra, T., & Sharma, P. 2020. A critical review of glory lily: a rare medicinal plant. *World Journal of Pharmacy and Pharmaceutical Sciences* 9(10): 1123–1133.

Misra, A., Srivastava, A., Khalid, M., Kushwaha, P., & Srivastava, S. 2017. Evaluation of anti arthritic potential of *Gloriosa superba* (L.) elite germplasm collected from Eastern Himalayas, India. *Pharmacognosy Journal* 9(1–6): s87–s92.

Mordeniz, C. 2019. Integration of traditional and complementary medicine into evidence-based clinical practice. In *Traditional and Complementary Medicine*; IntechOpen, London, pp. 1–8, 13.

Nadkarni, K. M. 1978. *Indian Materia Medica*, Vol I. Popular Prakashan, Mumbai.

Neuwinger, H. D. 1994. *African Ethnobotany: Poisons and Drugs: Chemistry, Pharmacology, Toxicology*. CRC Press.

Nikhila, G. S., Sangeetha, G., Nair, A. G., Pradeesh, S., & Swapna, T. S. 2014. High-frequency embryogenesis and organogenesis in *Gloriosa superba* L.—A plant in need of conservation. *Journal of Aquatic Biology and Fisheries* 2: 398–402.

Nuki, G. 2008. Colchicine: Its mechanism of action and efficacy in crystal-induced inflammation. *Current Rheumatology Reports* 10(3): 218–227.

Padamapriya, S., Rajamani, K., & Sathiyamurthy, V. A. 2015. Glory lily (*Gloriosa superba* L.)—A review. *International Journal of Current Pharmaceutical Review and Research* 7(1): 43–49.

Padmalatha, K., & Prasad, M. N. V. 2006. Optimization of DNA isolation and PCR protocol for RAPD analysis of selected medicinal and aromatic plants of conservation concern from Peninsular India. *African Journal of Biotechnology* 5(3): 230–234.

Padmapriya, S., & Rajamani, K. 2017. Induced mutagenesis in glory lily (*Gloriosa superba* L.) for economic variability. *Medicinal Plants-International Journal of Phytomedicines and Related Industries* 9(4): 231–236.

Padmapriya, S., & Rajamani, K. 2021. Protocol for cost-effective and rapid propagation of quality planting material of *Gloriosa superba* L. *Medicinal Plants-International Journal of Phytomedicines and Related Industries* 13(3): 460–466.

Pandey, D. K., Malik, T., Dey, A., Singh, J., & Banik, R. M. 2014. Improved growth and colchicine concentration in *Gloriosa superba* on mycorrhizal inoculation supplemented with phosphorus-fertilizer. *African Journal of Traditional, Complementary and Alternative Medicines* 11(2): 439–446.

Patel, A. I., Desai, B. S., Chaudhari, B. N., & Vashi, J. M. 2020. Genetic improvement in glory lily (*Gloriosa superba* L.): A review. *International Journal of Chemical Studies* 8(4): 255–260.

Patel, M. H., Patel, A. I., Desai, B. S., Jha, S. K., Patel, D. P., & Tandel, M. B. 2021. Genetic variability, heritability and genetic advance studies in *Gloriosa superba* Linn. *International Journal of Current Microbiology and Applied Sciences* 10(02): 3278–3282.

Patro, L. 2016. Medicinal plants of India: With special reference to Odisha. *International Journal of Advance Research and Innovative Ideas in Education* 2(5): 2395–4396.

Pawar, B. M., Wavhal, V. P., Pawar, N. D., Agarwal, M. R., Shinde, P. B., & Kamble, H. V. 2010. Anthelmintic activity of *Gloriosa superba* Linn (Liliaceae). *International Journal of Pharm Tech Research* 2(2): 1483–1487.

Peranantham, S., Manigandan, G., & Shanmugam, K. 2014. Fatal *Gloriosa superba* poisoning—A case report. *International Journal of Medical Research and Pharmaceutical Sciences* 4(10): 21–24.

Peter George, M. D. 2011. Death related to herbal therapy for joint pains—A rare case of Gloriossa Superba poisoning. *Journal of Clinical and Diagnostic Research* 5(2): 379–380.

Phatak, R. S., & Hegde, L. N. 2014. Glory lily (*Gloriosa superba* L.): An important medici nal crop—A review. *HortFlora Research Spectrum* 3(3): 282–287.

Phatak, R. S., Hegde, L., Narayanpur, V., & Hegde, N. K. 2016. Effect of nutrient doses on growth, seed yield, and tuber yield of glory lily (*Gloriosa superba* L.). *Research in Environment and Life Sciences* 9(5): 634–636.

Premaratna, R., Weerasinghe, M. S., Premawardana, N. P., & Silva, H. J. 2015. *Gloriosa superba* poisoning mimicking an acute infection—A case report. *BMC Pharmacology and Toxicology* 16(1): 1–4.

Raghavendra, R. R. 1996. Traditional knowledge and sustainable development, key role of ethnobiologist. *Journal of Ethnobotany* 8: 14–24.

Rahmatullah, M., Noman, A., Hossan, M. S., Harun-Or-Rashid, M., Rahman, T., Chowdhury, M. H., & Jahan, R. 2009. A survey of medicinal plants in two areas of Dinajpur district, Bangladesh including plants that can be used as functional foods. *Journal of Sustainable Agriculture and Environment* 3(4): 862–876.

Rajagopal, C., & Kandhasamy, R. 2009. Genetic variability, heritability and scope of improvement for yield components in glory lily (*Gloriosa superba* L.). *International Journal of Plant Breeding* 3(2): 139–143.

Rajagopal, C., & Rajamani, K. 2013. Assessment of genetic diversity of *Gloriosa superba* L. accessions detected by random amplified polymorphic DNA analysis. *Journal of Medicinal Plants Research* 7(28): 2122–2127.

Rajagopalan, A. 1994. Investigation on certain aspects of growth, development, crop production and quality in glory lilly (*Gloriosa Superba* L.) (Doctoral dissertation, Tamil Nadu Agricultural University, Coimbatore).

Rajak, R. C., & Rai, M. K. 1990. *Herbal Medicines, Biodiversity and Conservation Strategies*. International Book Distributors, New Delhi, India, pp. 75–79.

Rajamani, K., & Padmapriya, S. 2017. Glory lily (*Gloriosa superba*): A phytochemical crop of commercial significance. *Medicinal Plants-International Journal of Phytomedicines and Related Industries* 9(1): 12–19.

Rajasekharan, P. E., & Ganeshan, S. 2002. Conservation of medicinal plant biodiversity in Indian perspective. *Journal of Medicinal and Aromatic Plant Sciences* 24(1): 132–147.

Rajith, N. P., Navas, M., Thaha, A. M., Manju, M. J., & Anish, N. 2010. A study on traditional mother care plants of rural communities of south Kerala. *Indian J Traditional Knowledge* 9(1): 203–208.

Rane, A. D., Boat, P. G. R., & Patil, Y. 2012. A note on twin seedling in *Gloriosa superba* L. An important medicinal plant. *Life Sciences Leaflets* 23: 68.

Raskin, I., Ribnicky, D. M., & Komarnytsky, S. 2002. Plants and human health in the twenty-first century. *Trends in Biotechnology* 20: 522–531.

Rastogi, R. P., & Mehrotra, B. N. 1993. *Compendium of Indian Medicinal Plants*. CDRI and Publication, and Information Directorate, New Delhi, India, p. 5.

Selvarasu, A., & Kandhasamy, R. 2012. Reproductive biology of *Gloriosa superba*. *Journal of Medicinal and Aromatic Plants* 3(2): 1–5.

Selvarasu, A., & Kandhasamy, R. 2013. Intraspecific hybridization in glory lily. *Floriculture Ornamental Biotech* 7: 99–102.

Selvarasu, A., & Kandhasamy, R. 2017. Molecular and agro-morphological genetic diversity assessment of *Gloriosa superba* mutants. *European Journal of Medicinal Plants* 21(1): 1–13.

Senthilkumar, M. 2013. Phytochemical screening and antibacterial activity of *Gloriosa superba* Linn. *International Journal of Pharmacognosy and Phytochemical Research* 5(1): 31–36.

Sharma, S., Devi, B., & Sharma, V. 2021. Estimation of phytochemicals from mother plants and in vitro raised plants of *Gloriosa superba*. *International Journal of Economic Plants* 8(3): 127–130.

Singh, R. 2015. Medicinal plants: A review. *Journal of Plant Sciences* 3(1): 50–55.

Sivakumar, G. 2002. *Gloriosa superba* L.—A very useful medicinal plant. *Ethnomedicine and Pharmacognosy*, 40: 465–482.

Sivakumar, G., & Krishnamurthy, K. V. 2000. Micropropagation of *Gloriosa superba* L.—An endangered species of Asia and Africa. *Current Science* 78(1): 30–32.

Sivakumar, G., & Krishnamurthy, K. V. 2004. In vitro organo genetic responses of *Gloriosa superba*. *Russian Journal of Plant Physiology* 51: 713–721.

Sivakumar, G., Krishnamurthy, K. V., & Rajendran, T. D. 2003a. Embryoidogenesis and plant regeneration form leaf tissue of *Gloriosa superb*. *Planta Medica* 69: 479–481.

Sivakumar, G., Krishnamurthy, K. V., & Rajendran, T. D. 2003b. In vitro corm production in *Gloriosa superba* L., an Ayurvedic medicinal plant. *Journal of Horticultural Science and Biotechnology* 78(4): 450–453.

Sivakumar, S., Siva, G., Sathish, S., Kumar, G. P., Vigneswaran, M., Vinoth, S., Kumar, T. S., Sathishkumar, R., & Jayabalan, N. 2019. Influence of exogenous polyamines and plant growth regulators on high frequency in vitro mass propagation of *Gloriosa superba* L. and its colchicine content. *Biocatalysis and Agricultural Biotechnology* 18: 101–103.

Sivasankar, S., Manivannan, K., & Muruganandam, C. 2017. Effect of growth regulators treatments with different weight of tubers on pod yield and quality of Glory lily (*Gloriosa superba* L.). *International Journal of Current Research in Life Sciences* 6(7): 603–605.

Smith, A. C. 1979. Flora vitiensis nova: A new flora of Fiji. Lawai, Kauai, Hawaii. *National Tropical Botanical Garden* 1: 141–142.

Srivastava, J., Lambert, J., & Vietmeyer, N. 1995. *Medicinal Plants: A Growing Role in Development*. World Bank, Washington, DC.

Subiramani, S., Sundararajan, S., Govindarajan, S., Sadasivam, V., Ganesan, P. K., Packiaraj, G., Manickam, V., Thiruppathi, S. K., Ramalingam, S., & Narayanasamy, J. 2019. Optimized in vitro micro-tuber production for colchicine biosynthesis in *Gloriosa superba* L. and its anti-microbial activity against Candida albicans. *Plant Cell, Tissue and Organ Culture (PCTOC)* 139(1): 177–190.

Subramanya, S., & Radhamani, T. R. 1992. Pollination by birds and bats. *Current Science* 65: 201–209.

Sundaraganapathy, R., Niraimathi, V., Thangadurai, A., Kamalakannan, D., Narasimhan, B., & Deep, A. 2013. Antianxiety activity of *Gloriosa superba* Linn. *Hygeia—Journal for Drugs and Medicines* 5(1): 148–151.

Suri, O. P., Gupta, B. D., & Suri, K. A. 2001. A new glycoside, 3- O-demethylcolchicine-3-O-alpha-d-glucopyranoside from Gloriosa seeds. *Natural Product Research* 15: 217–219.

Tilwari, A., & Sharma, R. 2021. Random amplified polymorphic DNA and inter simple sequence repeat markers reveals genetic diversity between micro propagated, wild and field cultivated genotypes of *Gloriosa superba*: An endangered medicinal plant. *Molecular Biology Reports* 48(3): 2437–2452.

Tiwari, D. S., & Yadav, A. 2003. Ethnobotanical investigation of some medicinal plants availed by Gond Tribe of Naoradehi Wildlife Sanctuary, Madhya Pradesh. *The Anthropologist* 5(3): 201–202.

Uchimahali, J., Jebamalar, A., Duraikannu, G., & Thirumal, S. 2019. Phytochemical analysis and evaluation of antimicrobial activity in the whole plant extracts of *Gloriosa superba*. *Phytochemical Analysis* 12(6): 245–249.

Umavathi, S., Gopinath, K., Manjula, M. S., Balalakshmi, C., & Arumugam, A. 2020. *Gloriosa superba* L: A critical review of recent advances. *Abasyn Journal of Life Sciences* 3(2): 48–65.

Vaishnavi, B. A., Khan, H., & Bhoomika, H. R. 2019. Review on pharmacological properties of glory lily (*Gloriosa superba* Linn.)—An endangered medicinal plant. *International Journal of Current Microbiology and Applied Sciences* 8(2): 1359–1364.

Venudevan, B., Sundareswaran, S., & Vijayakumar, A. 2010. Optimization of dormancy breaking treatments for germination improvement of Glory lily (*Gloriosa superba* L.) seeds. *Madras Agricultural Journal* 97: 31–32.

Windholz, M. (ed). 1983. *The Merck Index. An Encyclopedia of Chemicals, Drugs and Biological*. 10th ed. Merck & Co. Inc, Rahway, NJ.

World Health Organization (WHO). 2000. General guidelines for methodologies on research and evaluation of traditional medicine. Retrieved 12 May, 2022 from https://apps.who.int/iris/bitstream/handle/10665/66783/WHO_EDM_TRM_2000.1.pdf.

Yadav, K., Agarwal, A., & Singh, N. 2012a. Arbuscular mycorrhizal fungi (AMF) induced acclimatization, growth enhancement and colchicine content of micropropagated *Gloriosa superba* L. plantlets. *Industrial Crops and Products* 45: 88–93.

Yadav, K., Aggarwal, A., & Singh, N. 2012b. Actions for *ex situ* conservation of *Gloriosa superba* L.: An endangered ornamental cum medicinal plant. *Journal of Crop Science and Biotechnology* 15: 297–303.

Yadav, K., Agarwal, A., & Singh, N. 2013. Evaluation of genetic fidelity among micropropagated plants of *Gloriosa superba* L. using DNA-based markers—A potential medicinal plant. *Fitoterapia* 89: 265–270.

Yineger, H., & Yewhalaw, D. 2007. Traditional medicinal plant knowledge and use by local healers in Sekoru District, Jimma Zone, Southwestern Ethiopia. *Journal of Ethnobiology and Ethnomedicine* 3(1): 1–7.

19 *Nerium oleander* (Oleander or Nerium)

Kolagani Chandramohan and Salman Khan

CONTENTS

19.1 INTRODUCTION

Nerium/oleander/Kaner (*Nerium oleander* L.) is an important plant species belonging to the genus *Nerium*, family Apocynaceae and sub-family Apocynoideae. The major characteristics of any Apocynaceae member plants include the presence of poisonous milky substance that is secreted from stem and twigs when cut, trees or shrubs with significant amount of foliage, a simple arrangement of opposite leaf pattern, dry pods or follicle pair consisting of fruits and aesthetically colourful flowers. Linnaeus in 1753 reported and described *N. oleander*. The genus name *Nerium* is taken from the Greek word Nerion, which means water, as this species is present near flowing water bodies. The oleander is an evergreen poisonous shrub or plant of caution due to the presence of poisonous natural compounds that contain potent cardiac glycosides (Gupta and Mittal 2010; Ebrahimi et al. 2018). The oleander is commonly available in the region of sub-tropical climate, and due to its aesthetic importance, the plantation of this species is usually carried out for roadside avenue beautification, near parks & gardens and other places (Culotta et al. 2005; Al-Lawati and Al-Lawati 2020). This species has white, red and pink flowers. White-flowered oleander plants contain less toxic compounds in comparison to the red- and pink-flowered plants (Karawya et al. 1973; Derwich et al. 2010). The oleander plant is an easy growing species due to its acceptance to the soil with low nutrients concentration; it grows under a wide range of pH (5–8.3) and is tolerant to harsh climatic conditions such as heat, drought and salt richness (Popenoe 2008; Popenoe et al. 2019). The flowers grow more vigorously with regular watering at certain interval. The oleander plant does not require

DOI: 10.1201/b23017-21

pruning activities as it can bloom and sustain in harsh condition on its own, but as it grows, the mature branches become lanky and grow unevenly by forming thin sub-branches. Due to this reason, it is recommended to carry out pruning or thinning operation in mature and aged individuals during autumn to provide form and to help developing new growth and to induce timely flowering after winter season (Laughlin 1997; Popenoe 2008). The seeds of the oleander plants are stored only if harvester needs it; otherwise, the seed pods are pruned away (Kathleen 2007).

19.2 TAXONOMY

N. oleander is an evergreen small tree or shrub species which can grow up to 4 m in height; branches produce milky latex (Laughlin 1997). Leaves whorled, 10–20 × 1.5–2.0 cm, lanceolate or linear to lanceolate with tapering ends; distinct nerves (Zibbu and Batra 2010). The flowers are white, pink or red having mild to high fragrance, flowers bracteate, bracts small, green, 3.5–4 cm across, arranged in terminal cymes; peduncles short. Calyx green, five lobed, divided at the base, linear, acute at apex. Corolla campanulate, white or red and pink (in some cases) in colour; rounded, overlapping, corona having five scales exists on the inner wall of corolla, each scales are divided into 2–7 free segments, ca. 2.5 cm long. Stamens are five, including short filaments, anthers connivent into a cone and adherent to the stigma, connective tissues formed into a feathery appendage. Carpels two. Stigma is dumb-bell shaped having distinct and long style. The root system of *N. oleander* is extensive and is primarily helpful in stabilizing the soils in tropical areas (Sinha and Biswas 2016). It comprises one pair of follicles, 7.5–17.5 cm long, lanceolate, split along single side to discharge the seeds (Sinha and Biswas 2016). The seeds are oblong, with a plume of hairs at one end (Choudhary and Prasad 2014; Figure 19.1).

FIGURE 19.1 Flower variations in *Nerium oleander.*

19.3 DISTRIBUTION

It is distributed throughout the India and is also available in Sri Lanka (Ghangale et al. 2021). Other distribution records of *N. oleander* include Afghanistan, Albania, Algeria, Australia, China, Croatia, Cyprus, France, Greece, Iran, Iraq, Italy, Japan, Jordan, Lebanon, Libya, Malta, Morocco, Nepal, Niger, Pakistan, Palestine, Portugal, Spain, Syria, Tunisia, Turkey, UAE, USA, middle-east countries and other areas (Derwich et al. 2010; Rahman and Akter 2015; Al-Snafi 2020; Farkhondeh et al. 2020).

19.4 CHEMICAL CONSTITUENTS

N. oleander contains cardiac glycosides that are similar to the active compounds present in *Digitalis* spp. (Radford et al. 1986). The major and important glycosides present in *N. oleander* include Oleandrin, Neriine, Cardenolides, Gentiobiosyl and Odoroside (Figure 19.2) (Duke 1985; Begum et al. 1999). The secondary metabolites formed by any plant can be different in quality and quantity depending upon the environmental conditions and stress level (Liu et al. 2015; Saeedfar and Miri 2011). Additionally, Folinerin, Rosagenin, Rutin and Oleandomycin, which are pharmacologically significant, are also identified. Two new cardenolides, 3-Beta-Hydroxy-8,14-Epoxy-5 Beta-Carda-16,20(22)-Dienolide (2) and 3-Beta-O-(D-2-O-Methyl-Digitalosyl)-14-Beta-Hydroxy-5-Beta-CARDA-16,20(22)-Dienolide (1), and two known cardenolides, 3-Beta-O-(D-Digitalosyl)-14-Beta-Hydroxy-5 Beta-Card-20 (22)-Enolide (4) and 3-Beta-O-(D-Digitalosyl)-14-Beta-Hydroxy-16-Beta-Acetoxy-5 Beta-Card-20 (22)-Enolide (3), have been reported from the leaves of *N. oleander* (Siddiqui et al. 1997; Ebrahimi et al. 2018).

Zhao et al. (2006) reported that two new taraxasterane-type triterpenes, 20-Beta, 28-Epoxytaraxaster-21-en-3beta-ol and 20-Beta,28-Epoxy-28 Alpha- Methoxytaraxasteran-3beta-ol, are present in the leaves of *N. oleander* which can be isolated by using ethyl acetate extract along with ursane-type triterpenes, 28-nor-Urs-12-ENE-3beta, 17beta-diol and 3beta-Hydroxyurs-12-en-28-Aldehyde. The spectroscopic analysis confirms their structural component (Ebrahimi et al. 2018; Farooqui and Tyagi 2018). Ishikawa et al. (2007) reported that the hot aqueous extract of *N. oleander* leaves yields phenolic compounds, 3-O-Caffeoylquinic acid, which is also known as chlorogenic acid and its structural isomer, 5-O-Caffeoylquinic acid (Cinnamate easter), which express alpha-glucosidases inhibition. Due to this inhibitory property in a non-competitive style, they are called as anti-hyperglycemias (Ishikawa et al. 2007).

Wang et al. (2009) described that three glycosides steroids can be prepared from the ethyl acetate extract of *N. oleander* using fractionation process. These cardenolides include 3

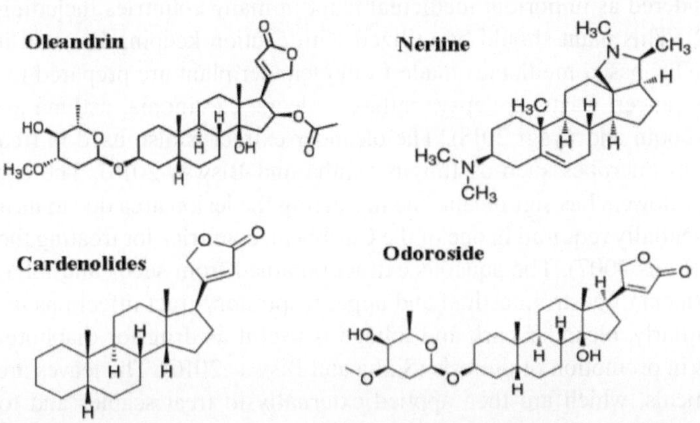

FIGURE 19.2 Major and important glycosides present in *Nerium oleander.*

Beta-O-(Beta-D-Diginosyl)-14, 15alpha-Dihydroxy-5alpha-Card-20(22)-Enolide, Cardenolide N-1 and uzarigenin (Wang et al. 2009). The leaves contain high content of polyphenols of which cinnamic acid is reported to be of maximum quantity followed by catechin, epicatechin and chlorogenic acid (Ebrahimi et al. 2018). Sinha and Biswas (2016) obtained a novel Labdanes and Oleanderen from the newly emerged foliage of *N. oleander* (Sinha and Biswas 2016). Zibbu and Batra (2010) reported the extraction of many volatile compounds, sugars and Prussic acid from the leaves of oleander plant. The seeds, bark and root of *N. oleander* contain glycosides of different groups and properties (Kumar et al. 2013).

19.5 ETHNO-MEDICINAL VALUE OF *NERIUM OLEANDER*

The Indian traditional medical systems represent Naturopathy, Ayurveda and many others including Yoga which are indigenous and are parts of Indian traditional knowledge (TK). These TKs are the information that is forwarded from one generation to another in the form of documents, faith and culture. Use of these information for treatment and cure of any disease is still prevalent. Many ailments and diseases can be successfully cured with the proper use of TK. Different indigenous communities and tribes use plant-, animal- and mineral-based medicines as their TK in their medical treatment and diagnosis, spiritual therapies and manual techniques formed to treat different prolonged and fatal diseases (Dey and Chaudhuri 2014; Keshavarzi et al. 2015).

Phytomedicines are meant to the traditional medicinal or folk medicines which are practiced by the utilization of plants and plant extracts. The plants are important components in any traditional medicines. Plants with medicinal properties have been vital constituents of both curative and preventive medical therapy preparations for human being. The preparation of phytomedicines requires the extraction of medicinal component/complex from the flowers, leaves, seeds, roots, bark, fruits and sometimes whole plants. Earlier, people chiefly relied on the raw plant material for their medical needs to retain vitality and to treat various diseases. Despite the fact that phytochemicals are important as medicines, there are some poisonous plant-based compounds too. Poisonous plants have always been part of our everyday life and few of them are so common that we do not even doubt their toxicity. Indigenous tribal people all over the world use toxic plants for various purposes such as hunting, fishing, fighting and treating diseases. The poisonous compounds are usually applied on the arrow which helps in killing the prey by the activity of toxins. The toxins from plants are much related to health aspects of human being and animals. These toxic compounds are sometimes fatal, but they are also used for refractory symptoms and effective cure of some chronic diseases. These secondary metabolites are chemicals that are produced by any plant under stress and have lethal or therapeutic characteristics (Liu et al. 2015).

When we discuss about the medicinal or poisonous plants, *N. oleander* is not lagging behind. *N. oleander* is considered as important medicinal plant in many countries including China and India (Dey et al. 2012). This plant should be utilized with caution keeping in mind its toxicity (Zibbu and Batra 2010). TK-based medicines made from oleander plant are prepared to cure and manage diseases such as cancer, warts, epilepsy, calluses, ulcers, carcinoma, asthma and heart disorders (Duke 1985; Farooqui and Tyagi 2018). The oleander extract is also used in treating various skin diseases related to microbes such as fungus (Sinha and Biswas 2016). The compound obtained from the oleander flowers has significant role in treating the lesion area due to its anti-inflammatory property. It is essentially required in one of the Caribbean countries for treating the disorders related to reproduction (Lans 2007). The aqueous extract obtained from seeds and foliage of *N. oleander* helps in the treatment of gastrointestinal and upper respiratory tract infections in Kenya (Nanyingi et al. 2008). Similarly, oleander bark and foliage is useful as drug for diaphoretic, cardiac tonic, emetic and helps in promotion of diuresis (Sinha and Biswas 2016). The leaves are boiled to extract the herbal chemicals, which are then applied externally to treat scabies and to reduce swelling (Farooqui and Tyagi 2018). Root bark oil collected from *N. oleander* is used to treat the Leprosy and dermal infections (Ebrahimi et al. 2018). As per TK-based medicinal system in Italy, the oleander plant part is required for treating Malaria (Tagarelli et al. 2010). Similarly, as per the TKs of South Indian state, the sap obtained from the bark of oleander plant is useful in providing relief from ear pain (Muthu et al. 2006). Ebrahimi et al. (2018) mentioned that small quantity of leaf latex is used to treat snake and other venomous bites.

19.6 BIOLOGICAL ACTIVITIES

N. oleander exhibits high medicinal properties. The botanical compounds derived from different parts of the plant are important and used as antioxidant, antiviral, anti-fungal, anti-bacterial, anti-cancerous, antihistamine, anti-inflammatory, antifeedant and larvicidal agents (Zibbu and Batra 2010; Dey and Chaudhuri 2014; Sinha and Biswas 2016; Ebrahimi et al. 2018; Farooqui and Tyagi 2018; Table 19.1).

19.6.1 Antioxidant Activity

N. oleander is a well-known plant for its antioxidant property which provides protection and support to cell from injuries triggered by unstable molecule (Mohadjerani 2012). Almost all plant parts are considered as vital and important for production of antioxidants, which eventually help in maintaining good health. Aruna et al. (2020) reported that the antioxidant potency of Crude *N. oleander* leaves extract is greater than that of flowers. The antioxidant property is interrelated to the concentration of benzenol (Mohadjerani 2012). The oleander leaf exhibits significant hydroxyl radical, hypochlorous acid, peroxy-nitrite scavenging and iron chelation activity (Ebrahimi et al. 2018).

TABLE 19.1
Biological Uses of Different Plant Parts of *Nerium oleander*

Plant Part	Medicinal Properties	References
Leaves	Diaphoretic, cardiotonic, diuretic, sternutatory, expectorant, antimicrobial, larvicidal, antifeedant, anti-diabetic, anti-cancerous, anti-proliferative, antioxidant, anti-depressant, analgesic, herbicidal, anti-fungal, anti-bacterial, insecticidal, headache and snakebite	Sinha and Biswas (2016), Malik et al. (2015), Gupta and Thorsteinson (1960), Jacobson (1975), Grainge (1985), Kumar et al. (2012), Roni et al. (2013), Sikarwar et al. (2009), Pathak et al. (2000), Choudhary and Prasad (2014), Wong et al. (2011), Begum et al. (1999), Zia et al. (1995), Hinai et al. (2020), Uygur et al. (1997), Rao and Narayanasamy (1990), Mallets et al. (1994), Mulas et al. (2002), Bhuvaneshwari et al. (2007), Joshi et al. (2009), Al-Lawati and Al-Lawati (2020)
Flower	Diaphoretic, cardiotonic, diuretic, sternutatory, expectorant, antimicrobial, larvicidal, antifeedant, anti-cancerous, antioxidant, anti-inflammatory and anti-scabies	Sinha and Biswas (2016), Hadizadeh et al. (2009), Gupta and Thorsteinson (1960), Jacobson (1975), Grainge et al. (1985), Raveen et al. (2014), Ali et al. (2009), Singhal and Gupta (2012), Nurgun et al. (2003), Choudhary and Prasad (2014), Derwich et al. (2010), Islam and Lucky (2019)
Root	Antimicrobial, larvicidal, antifeedant, active metabolites, anti-dermatitis, diuretic, cardiac tonic, resolvent, attenuant, joint and abdominal pain	Huq et al. (1999), Hussain and Gorsi (2004), Gupta and Thorsteinson (1960), Jacobson (1975), Grainge et al. (1985), Sinha and Biswas (2016), Kawalekar et al. (2012), Rahman and Akter (2015)
Sap/Juice	Larvicidal, antifeedant, headache, snakebite, anti-inflammatory, joint pain and dermatitis	Gupta and Thorsteinson (1960), Jacobson (1975), Grainge et al. (1985), Sinha and Biswas (2016), Al-Lawati and Al-Lawati (2020), Hinai et al. (2020)
Bark/stem	Antimicrobial, antifeedant, headache, larvicidal and snakebite	Tannu et al. (2011), Gupta and Thorsteinson (1960), Jacobson (1975), Grainge et al. (1985), Hinai et al. (2020), Sinha and Biswas (2016), Al-Lawati and Al-Lawati (2020)

Similarly, stem displayed sufficient Nitric oxide and DDPH unstable molecules scavenger potency, while root shows oxidative degradation of lipids, hydrogen peroxide and superoxide anions (Dey and Chaudhuri 2014; Ebrahimi et al. 2018).

19.6.2 Anti-Inflammatory Activity

Inflammation is auto-immune process of the body to defend the harmful foreign agents including bacteria, viruses and other pathogens (Dey and Chaudhuri 2014; Felman 2020). Dried leaves and fresh flower ethanolic extracts of *N. oleander* possess anti-inflammatory property in mice with no side effect (Farooqui and Tyagi 2018). Zibbu and Batra (2010) reported that lipid soluble constituent of flower may possess anti-inflammatory property.

19.6.3 Antimicrobial Activity

The oleander plant is toxic but at the same time has antimicrobial properties (Farooqui and Tyagi 2018) due to the presence of benzenol which reduces the availability of free radicals. Sinha and Biswas (2016) reported that the roots of *N. oleander* contain a novel cardenolide, which exhibits the antimicrobial properties with additional activity similar to digoxin (Sinha and Biswas 2016). The extract of leaf possesses anti-bacterial properties against some bacteria, e.g., *Bacillus* and *Nyctanthes* spp. (Hussain and Gorsi 2004; Ebrahimi et al. 2018). Dey and Chaudhuri (2014) also provide information on the anti-bacterial activities of oleander plant extracted mostly from the root and bark. Similarly, essential oil obtained from the flower of oleander plant shows anti-bacterial characteristic towards *Escherichia coli* and some bacteria (Hussain and Gorsi 2004; Sinha and Biswas 2016). Farooqui and Tyagi (2018) reported that the ethanol extract of oleander flower has antimicrobial activities towards many species including *Alternaria*, *Fusarium* and *Rhizoctonia* spp.

19.6.4 Larvicidal Activity

Kumar et al. (2012) mentioned that the oleander plant exhibits potential to control the insect larva and eggs. Similarly, Roni et al. (2013) reported the adulticidal and ovicidal properties of oleander plant against *Anopheles stephensi*. Similarly, the oleander plant is said to have larvicidal and ovicidal effects on important mosquito species that transmits diseases like Dengue and Japanese encephalitis (Lokesh et al. 2010; Raveen et al. 2014; Al-Hakimi et al. 2022). Most of the plant parts (chiefly leaf and flower) of *N. oleander* show insecticidal and antifeedant activity against Diamondback moth (Sinha and Biswas 2016).

19.6.5 Anti-cancer Activity

Ali et al. (2009) extracted essential oil from oleander flowers and reported that it shows an anti-tumour property on Ehrlich cells. The Oleandrin and its aglycone oleandrigenin are reported to possess anti-cancerous properties (Zibbu and Batra 2010; Sinha and Biswas 2016). Pathak et al. (2000) analysed different concentrations of Anvirzel and Oleandrin against cancerous cell. They concluded that both the products are having potential to initiate cell mortality, with the exception of mice tumour cells that exhibit signs of apoptosis (Zibbu and Batra 2010).

19.6.6 Cellular and Humoral Immune Responses

The drug prepared from plant parts of *N. oleander* has shown very positive and encouraging immune-stimulating properties (Farooqui and Tyagi 2018). *N. oleander* has a significant immuno-modulatory effect on rabbit. The oleander leaf extract application on rabbits reduces the formation

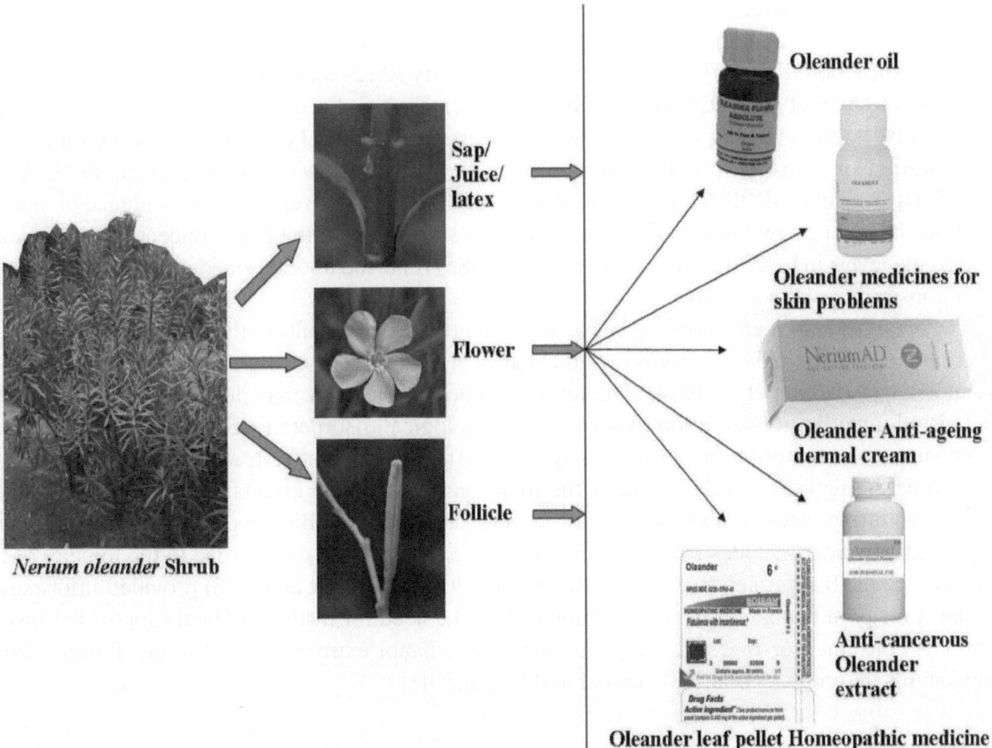

FIGURE 19.3 *Nerium oleander* plant parts and their useful products.

of antibodies. Apart from this, the phagocytosis and improper immune-related stimulus can be treated with small dosage (Farooqui and Tyagi 2018; Ebrahimi et al. 2018).

19.6.7 Antiviral Activity

Oleanderin, a cardiac glycoside found in oleander leaves, suppresses the response to the stimulus of Human Immunodeficiency Virus protein (Sanna et al. 2021). Small amount of Anvirzel was potentially sufficient to break the HIV infection potential rate (Dey and Chaudhuri 2014; Ebrahimi et al. 2018). Further, the medicinal uses of different plant parts of *N. oleander* are depicted in Figure 19.3.

19.7 TOXICITY

Farkhondeh et al. (2020) have concluded that *N. oleander* (Kaner) is a poisonous plant which contains toxins for animal when ingested in large concentration. Most of the plant parts exhibit poisonous compounds including oleandrin, neriifolin, and thevetins A and B (Bandara et al. 2010). The maximum concentration of toxicity is found in the seeds and roots of this plant (Ebrahimi et al. 2018). Even smoke originating or interacting with the foliage of this shrub is lethal for some animals. Oleander leaves have a lethal dose range of 15–30 g for horses, more than 10 g for cows and approx. 5 g for sheep, according to reports (Popenoe et al. 2019). The signs of toxicity by Kaner plant include abdominal ache, oral fissure pain, vomiting and its sensation, diarrhoea and muscle contraction. For Avifauna, even smaller quantity (0.12–0.7 g) of plant has been reported to cause death (Arai et al. 1992). Researchers have identified and reported that most of the animals are sensitive to the oleander toxicity related to heart, and the level of toxicity is based on amount and duration of ingestion except in mice (Szabuniewicz et al. 1972; Hougen et al. 1979; Adam et al. 2001; Aslani et al. 2007). Blum and Reiders (1983) have also reported an incidence of human fatalities brought on by Oleander intake. The autopsy confirmed the high concentration of oleandrin tissue levels in her. Similarly, Haynes et al. (1985) reported another incident of a woman dying after consuming oleander 'tea'.

The oleander leaves are known to be harmful to cattle, which results in their death within 36 hours after intake. The toxicity symptoms include heartbeat related issues, blockage of heart AV node and haemorrhages (Mahin et al. 1984). The juicy secretion of oleander may develop dermatitis, uveitis and visual uneasiness in addition to dermal immunological disorders (Zibbu and Batra 2010). The responses to consumption of oleander may bring effects on digestive tract, alimentary canal and heart functioning in the body. Some of the major gastrointestinal effects that have been reported include vomiting, nausea, abdominal pain, excess salivation and diarrhoea, while heart functioning related disorder includes irregular heartbeat. The children are more prone to the toxicity of the oleander plant (Bandara et al. 2010; Derwich et al. 2010). Goetz et al. (1998) provided information on the severity of oleander toxicity by emphasizing the need for instant medical support for anyone who has consumed or ingested the plant part. The ethanol extract of *N. oleander* flowers shows anticonvulsant property in mice (Singhal and Gupta 2011).

19.8 CONCLUSION AND FUTURE REMARKS

After critical review, it was concluded that *N. oleander* can be utilized in treating different ailments and has been therapeutically proven to prevent cancer, tumour formation and control sexually transmitted diseases. TKs and healthcare systems of tribal are parts of their culture and belief. The various plant parts are important ingredients in the preparation of traditional medicines. This makes the world to shift towards plant-based natural medicines and utilize them rather than depending solely on allopathic drugs. From this plant (of different parts), various bioactive compounds have been reported which are lethal. Thus, there is a need to discover the potentiality and applicability of natural plant-based medicines which help in treating various incurable diseases & disorders and support pharmacognosy-based medical system. *N. oleander* is most curative drug which has toxicity as well as medicinal importance. The drug obtained from this plant is being used as antioxidant, anti-proliferative, anti-cancerous, anti-diabetic, antimicrobial, anti-depressant, in skin diseases and for other illnesses. Therefore, further research on therapeutic potential of *N. oleander* is needed for better health system.

ACKNOWLEDGEMENT

The authors are greatly indebted to Shri. Anoop Singh IFS, Director General, Forest Survey of India, Dehradun and Shri. Chaturbhuja Behera IFS, Regional Director, Forest Survey of India Central Zone, Nagpur for their support and facilities.

NOTES

Kolagani Chandramohan, National Forest Inventory, Forest Survey of India, Ministry of Environment, Central Zone, Seminary Hills, Nagpur, India

Salman Khan, National Forest Inventory, Forest Survey of India, Ministry of Environment, Central Zone, Seminary Hills, Nagpur, India

REFERENCES

Adam, S. E., Al-Yahya, M. A., & Al-Farhan, A. H. 2001. Acute toxicity of various oral doses of dried *Nerium oleander* leaves in Sheep. *The American Journal of Chinese Medicine* 29(3–4): 525–532.

Al-Hakimi, A. N., Abdulghani, M. A., Alhag, S. K., Aroua, L. M., & Mahyoub, J. A. 2022. Larvicidal activity of leaf extract of *Nerium oleander* L. and its synthesized metallic nanomaterials on dengue vector, Aedes aegypti. *Entomological Research* 52(3): 148–158.

Ali, H. F. M., El-Ella, F. M. A., & Nasr, N. F. 2009. Screening of chemical analysis, antioxidant antimicrobial and antitumor activities of essential oil of oleander (*Nerium oleander*) flower. *International Journal of Biological Chemistry* 4(4): 190–202.

Al-Lawati, Z., & Al-Lawati, A. 2020. Uses, local practices and side effects of six medicinal plants in the sultanate of Oman: A review article. *Journal of Medicinal Plants* 8(5): 5–13.

Al-Snafi, A. E. 2020. Bioactive ingredients and pharmacological effects of *Nerium oleander. IOSR Journal of Pharmacy* 10(9): 19–32.

Arai, M., Stauber, E., & Shropshire, C. M. 1992. Evaluation of selected plants for their toxic effects in canaries. *Journal of the American Veterinary Medical Association* 200(9): 1329–1331.

Aruna, P., Mangaiyarkarasi, R., Pavithra, V., Rathnaswathy, S., & Renisha, S. 2020. Quantification of biopigments from *Nerium oleander* L. and assessing their antioxidant potential. *Journal of Pharmaceutical Innovation* 9(3): 331–333.

Aslani, M. R., Movassaghi, A. R., Janati-Pirouz, H., & Karazma, M. 2007. Experimental oleander (*Nerium oleander*) poisoning in goats: A clinical and pathological study. *Iranian Journal of Veterinary Research* 8: 58–63.

Bandara, V., Weinstein, S. A., White, J., & Eddleston, M. 2010. A review of the natural history, toxinology, diagnosis and clinical management of *Nerium oleander* (common oleander) and *Thevetia peruviana* (yellow oleander) poisoning. *Toxicon* 56(3): 273–281.

Begum, S., Siddiqui, B. S., Sultana, R., Zia, A., & Suria, A. 1999. Bio-active cardenolides from the leaves of *Nerium oleander. Phytochemistry* 50(3): 435–438.

Bhuvaneshwari, L., Arthy, E., Anitha, C., Dhanabalan, K., & Meena, M. 2007. Phytochemical analysis & antibacterial activity of *Nerium oleander. Ancient Science of Life* 26(4): 24–28.

Blum, L. M., & Reiders, F. 1983. Oleandrin distribution in a fatality from rectal and oral Nerium oleander extract administration. *Journal of Analytical Toxicology* 11: 219–221.

Choudhary, K., & Prasad, D. N. 2014. A review on: *Nerium oleander* Linn. (Kaner). *International Journal of Pharmacognosy and Phytochemical Research* 6(3): 593–597.

Culotta, L., Gianguzza, A., & Orecchio, S. 2005. Leaves of *Nerium oleander* L. as bioaccumulators of polycyclic aromatic hydrocarbons (PAH) in the air of Palermo (Italy): Extraction and GC-MS analysis, distribution and sources. *Polycyclic Aromatic Compounds* 25(4): 327–344.

Derwich, E., Benziane, Z., & Boukir, A. 2010. Antibacterial activity and chemical composition of the essential oil from flowers of *Nerium oleander. Electronic Journal of Environmental, Agricultural and Food Chemistry* 9(6): 1074–1084.

Dey, P., & Chaudhuri, T. K. 2014. Pharmacological aspects of *Nerium indicum* Mill: A comprehensive review. *Pharmacognosy Review* 8(16): 156–162.

Dey, P., Roy, S., & Chaudhuri, T. K. 2012. A quantitative assessment of bioactive phytochemicals of *Nerium indicum*: An ethnopharmacological herb. *International Journal of Research in Pharmaceutical Sciences* 3: 579–587.

Duke, J. A. 1985. Rose Bay, Rose Laurel (*Nerium oleander* L.). In: *Handbook of Medicinal Herbs*. Edited by J.A. Duke, Boca Raton, FL, Washington, DC, CRC Press, pp. 627–628.

Ebrahimi, F., Nohooji, M. G., & Miri, S. M. 2018. Agronomic and pharmacological aspects of *Nerium oleander*: An important medicinal plant. In proceedings of *The First National Congress and International Fair of Medicinal Plants and Strategies for Persian Medicine That Affect Diabetes*. Mashhad University of Medical Sciences, Iraninan Society of Medicinal Plant,Mashhad, Iran, pp. 9–11.

Farkhondeh, T., Kianmehr, M., Kazemi, T., Samarghandian, S., & Khazdair, M. R. 2020. Toxicity effects of *Nerium oleander*, basic and clinical evidences: A comprehensive review. *Human & Experimental Toxicology* 39(6): 773–784.

Farooqui, S., & Tyagi, T. 2018. *Nerium oleander*: Its application in basic and applied science: A review. *International Journal of Pharmacy and Pharmaceutical Sciences* 10(3): 1–4.

Felman, A. 2020. Everything you need to know about inflammation. https://www.medicalnewstoday.com/articles/248423 (Assessed on 22/5/2022).

Ghangale, A. L., Ghangale, N. A., Band, R. M., Sapate, A. B., & Gaiwale, S. 2021. Brief review of update in oleander plant poisoning, treatment & medico-legal aspects. *International Journal of Medical Toxicology & Legal Medicine* 24(1&2): 234–237.

Goetz, R. J., Jordan, T. N., McCain, J. W., & Su, N. Y. 1998. Oleander. In: *Indiana Plants Poisonous to Livestock and Pets*. Cooperative Extension Service, Purdue University, West Lafayette, IN, pp. 56–57.

Grainge, M. 1985. *Plant Species Reportedly Possessing Pest-Control Properties: An EWC/UH Database*. Resource Systems Institute, East-West Center (EWC), Honolulu, HI.

Gupta, P. D., & Thorsteinson, A. J. 1960. Food plant relationships of the diamondback moth (*Plutella maculipennis* (Curt)). I. Gustation and olfaction in relation to botanical specificity of the larva. *Entomologia Experimentalis et Applicata* 3: 241–250.

Gupta, V., & Mittal, P. 2010. Phytochemical and pharmacological potential of *Nerium oleander*: A review. *International Journal of Pharmaceutical Sciences and Research* 1(3): 21–27.

Hadizadeh, I., Peivastegan, B., & Kolahi, M. 2009. Antifungal activity of nettle (*Urtica dioica* L.), colocynth (*Citrullus colocynthis* L. Schrad), oleander (*Nerium oleander* L.) and konar (*Ziziphus spinachristi* L.) extracts on plants pathogenic fungi. *Pakistan Journal of Biological Sciences* 12(1): 58–63.

Haynes, B. E., Bessen, H. A., & Wightman, W. D. 1985. Oleander tea: Herbal draught of death. *Annals of Emergency Medicine* 14: 350–353.

Hinai, A. A., Lupton, D. A., & Al-Issai, G. 2020. Indigenous knowledge and folk use of medicinal plants in the Eastern Hajar Mountains, Oman. *Journal of Medicinal Plants Research* 8: 104–110.

Hougen, T., Lloyd, B., & Smith, T. 1979. Effects of inotropic and arrhythmogenic digoxin doses and of digoxin-specific antibody on myocardial monovalent cation transport in the dog. *Circulation Research* 44: 23–31.

Huq, M. M., Jabbar, A., Rashid, M. A., & Hasan, C. M. 1999. A novel antibacterial and cardiac steroid from the roots of *Nerium oleander*. *Fitoterapia* 70(1): 5–9.

Hussain, M. A., & Gorsi, M. S. 2004. Antimicrobial activity of *Nerium oleander* Linn. *Asian Journal of Plant Sciences* 3(2): 177–180.

Ishikawa, A., Yamashita, H., Hiemori, M., Inagaki, E., Kimoto, M., Okamoto, M., Tsuji, H., Memon, A. N., Mohammadio, A., & Natori, Y. 2007. Characterization of inhibitors of postprandial hyperglycemia from the leaves of *Nerium indicum*. *Journal of Nutritional Science and Vitaminology* 53(2): 166–173.

Islam, M. S., & Lucky, R. A. 2019. A study on different plants of Apocynaceae family and their medicinal uses. *Journal of Pharmacy Research* 4(1):40–44.

Jacobson, M. 1975. Nerium oleander. In: *Insecticides from Plants: A Review of the Literature, 1954–1971*. Washington, DC, Agricultural Research Service, United States, Department of Agriculture, Government Printing Office, p. 8.

Joshi, B., Sunil, L., & Anuja, S. 2009. Antibacterial property of different medicinal plants: *Ocimum sanctum*, *Cinnamomum zeylanicum*, *Xanthoxylum armatum* and *Origanum majorana*. *Kathmandu University Journal of Science, Engineering and Technology* 5(1): 143–150.

Karawya, M., Balbaa, S., & Khayyal, S. 1973. Estimation of cardenolides in *Nerium oleander*. *Plant Medicine* 23(01): 70–73.

Kathleen, N. B. 2007. *Western Garden Book*. Manlo Park, California, Sunset Publishing.

Kawalekar, J. S., Varsha, P., & Vijayalakshmi, N. 2012. Preliminary phytochemical investigations on roots of *Nerium Oleander*, Linn. *International Journal of Pharmacognosy and Phytochemical Research* 4(3): 134–138.

Keshavarzi, M., Rezaie, M. B., & Miri, S. M. 2015. *Evaluation of Morphological and Phytochemical of Stachys lavandulifolia Vahl in Field Conditions*. Tehran, Iran, 4th National Congress on Medicinal Plants.

Kumar, A., De, T., Mishra, A., & Mishra, A. K. 2013. Oleandrin: A cardiac glycosides with potent cytotoxicity. *Pharmacognosy Reviews* 7(14): 131–139.

Kumar, G., Karthik, L., Rao, K. V. B., Kirthi, A. V., & Rahuman, A. A. 2012. Phytochemical composition and mosquito controlling property of *Nerium oleander* leaves (Apocynaceae) against *Culex tritaeniorhynchus* and *Culex gelidus* (Diptera: Culicidae). *Asian Pacific Journal of Tropical Biomedicine* 2: 1–6.

Lans, C. 2007. Ethnomedicines used in Trinidad and Tobago for reproductive problems. *Journal of Ethnobiology and Ethnomedicine* 3: 13.

Laughlin, C. W. 1997. Oleander. *Ornamentals & Flowers* 4: 1–3.

Liu, W., Liu, J., Yin, D., & Zhao, X. 2015. Influence of ecological factors on the production of active substances in the anti-cancer plant *Sinopodophyllum hexandrum* (Royle) TS Ying. *PLoS One* 10(4): 1–22.

Lokesh, R., Barnabas, E. L., Madhuri, P., Saurav, K., & Sundar, K. 2010. Larvicidal activity of *Trigonella foenum* and *Nerium oleander* leaves against mosquito larvae found in Vellore City, India. *Current Research Journal of Biological Sciences* 2(3): 154–160.

Mahin, L., Marzou, A., & Huart, A. 1984. A case report of *Nerium oleander* poisoning in cattle. *Veterinary and Human Toxicology* 26(4): 303–304.

Malik, R., Bokhari, T. Z., Siddiqui, M. F., Younis, U., Hussain, M. I., & Khan, I. A. 2015. Antimicrobial activity of *Nerium oleander* L. and *Nicotiana tabacum* L.: A comparative study. *Pakistan Journal of Botany* 47(4): 1587–1592.

Mallet, J. F., Cerrati, C., Ucciani, E., Gamisans, J., & Gruber, M. 1994. Antioxidant activity of plant leaves in relation to their alpha-tocopherol content. *Food Chemistry* 49(1): 61–65.

Mohadjerani, M. 2012. Antioxidant activity and total phenolic content of *Nerium oleander* L. grown in North of Iran. *Iranian Journal of Pharmaceutical Research* 11(4): 1121–1126.

Mulas, M., Perinu, B., & Francesconi, A. H. D. 2002. Evaluation of spontaneous oleander (*Nerium oleander* L.) as a medicinal plant. *Journal of Herbs, Spices & Medicinal Plants* 9(2–3): 121–125.

Muthu, C., Ayyanar, M., Raja, N., & Ignacimuthu, S. 2006. Medicinal plants used by traditional healers in Kancheepuram district of Tamil Nadu, India. *Journal of Ethnobiology and Ethnomedicine* 2: 43.

Nanyingi, M. O., Mbaria, J. M., Lanyasunya, A. L., Wagate, C. G., Koros, K. B., Kaburia, H. F., Munenge, R. W., & Ogara, W. O. 2008. Ethnopharmacological survey of Samburu district, Kenya. *Journal of Ethnobiology and Ethnomedicine* 4: 14.

Nurgun, E., Esru, K., & Erdem, Y. 2003. Anti-inflammatory and Antinociceptive activity assessment of plants used as remedy in Turkish folk medicine. *Journal of Ethnopharmacology* 89: 123–129.

Pathak, S., Multani, A. S., Narayan, S., Kumar, V., & Newman, R. A. 2000. Anvirzel, an extract of *Nerium oleander*, induces cell death in human but not murine cancer cells. *Anti-Cancer Drugs* 11(6): 455–463.

Popenoe, J. 2008. *Oleander*. Gainesville, FL, The Institute of Food and Agricultural Science Extension, University of Florida.

Popenoe, J., Warwick, C. R., Bourdon, J., & Chen, J. 2019. Key plant, key pest: Oleander (*Nerium oleander*): ENH310/EP574, 8/2019. *EDIS* 2019(4): 5.

Radford, D. J., Gillies, A. D., Hinds, J. A., & Duffy, P. 1986. Naturally occurring cardiac glycosides. *The Medical Journal of Australia* 144: 540–544.

Rahman, A. H., & Akter, M. 2015. Taxonomy and traditional medicinal uses of Apocynaceae (Dogbane) family of Rajshahi district, Bangladesh. *Research & Reviews: Journal of Botanical Sciences* 4(4): 1–2.

Rao, G. N., & Narayanasamy, P. 1990. Effect of plant extracts and oils on rice yellow dwarf infection. *Madras Agricultural Journal* 77(5–6): 197–201.

Raveen, R., Kamakshi, K. T., Deepa, M., Arivoli, S., & Tennyson, S. 2014. Larvicidal activity of *Nerium oleander* L. (Apocynaceae) flower extracts against *Culex quinquefasciatus* Say (Diptera: Culicidae). *International Journal of Mosquito Research* 1(1): 38–42.

Roni, M., Murugan, K., Panneerselvam, C., Subramaniam, J., & Hwang, J. S. 2013. Evaluation of leaf aqueous extract and synthesized silver nanoparticles using *Nerium oleander* against *Anopheles stephensi* (Diptera: Culicidae). *Parasitology Research* 112: 981–990.

Saeedfar, S., & Miri, S. M. 2011. Production of plant metabolites via bioreactor. *Zeitoon* 31(221): 32–38 (in Farsi).

Sanna, G., Madeddu, S., Serra, A., Collu, D., Efferth, T., Hakkim, F. L., & Rashan, L. 2021. Anti-poliovirus activity of *Nerium oleander* aqueous extract. *Natural Product Research* 35(4): 633–636.

Siddiqui, B. S., Sultana, R., Begum, S., Zia, A., & Suria, A. 1997. Cardenolides from the methanolic extract of *Nerium oleander* leaves possessing central nervous system depressant activity in mice. *Journal of Natural Products* 60: 540–544.

Sikarwar, M. S., Patil, M. B., Kokate, C. K., Sharma, S., & Bhat, V. 2009. Antidiabetic activity of *Nerium indicum* leaf extract in alloxan-induced diabetic rats. *Journal of Young Pharmacists* 1(4): 333–335.

Singhal, K. G., & Gupta, G. D. 2011. Some central nervous system activities of *Nerium oleander* Linn (Kaner) flower extract. *Tropical Journal of Pharmaceutical Research* 10(4): 455–461.

Singhal, K. G., & Gupta, G. D. 2012. Hepatoprotective and antioxidant activity of methanolic extract of flowers of *Nerium oleander* against CCl4 induced liver injury in rats. *Asian Pacific Journal of Tropical Medicine* 5(9): 677–685.

Sinha, S. N., & Biswas, K. 2016. A concise review on *Nerium oleander* L.—An important medicinal plant. *Tropical Plant Research* 3(2): 408–412.

Szabuniewicz, M., Schwartz, W. L., McCrady, J. D., Russell, L. H., & Camp, B. J. 1971. Treatment of experimentally induced oleander poisoning. *Archives Internationales de Pharmacodynamie et de Therapie* 189: 12–21.

Szabuniewicz, M., Schwartz, W. L., McCrady, J. D., Russell, L. H., & Camp, B. J. 1972. Experimental oleander poisoning and treatment. *Southwestern Veterinarian* 25: 105–114.

Tagarelli, G., Tagarelli, A., & Piro, A. 2010. Folk medicine used to heal malaria in Calabria (southern Italy). *Journal of Ethnobiology and Ethnomedicine* 6: 27.

Uygur, F. N., & Iskenderoglu, S. N. 1997. Allelopathic and bioherbicide effects of plant extracts on germination of some weed species. *Turkish Journal of Agriculture and Forestry* 21(2): 177–180.

Wang, X. B., Li, G. H., Zheng, L. J., Ji, K. Y., Lü, H., Liu, F. F., Dang, L. Z., Mo, M. H., & Zhang, K. Q. 2009. Nematicidal cardenolides from *Nerium indicum* Mill. *Chemistry and Biodiversity* 6(3): 431–436.

Wong, S. K., Lim, Y. Y., Abdullah, N. R., & Nordin, F. J. 2011. Antiproliferative and Phytochemical analysis of leaf extracts of ten Apocyanaceae species. *Pharmacognosy Research* 3(2): 100–106.
Zhao, M., Zhang, S., Fu, L., Li, N., Bai, J., Sakai, J., Wang, L., Tang, W., Hasegawa, T., Ogura, H., Kataoka, T., Oka, S., Kiuch, M., Hirose, K., & Ando, M. 2006. Taraxasterane- and ursane-type triterpenes from *Nerium oleander* and their biological activities. *Journal of Natural Products* 69(8): 1164–1167.
Zia, A., Siddiqui, B. S., Begum, S., Siddiqui, S., & Suria, A. 1995. Studies on the constituents of the leaves of *Nerium oleander* on behavior pattern in mice. *Journal of Ethnopharmacology* 49: 33–39.
Zibbu, G., & Batra, A. 2010. A review on chemistry and pharmacological activity of *Nerium oleander* L. *Journal of Chemical and Pharmaceutical Research* 2(6): 351–358.

20 *Ricinus communis* (Castor Oil Plant)

Ambreen Bano, Adria Hasan, Swati Sharma, and Snober S. Mir

CONTENTS

DOI: 10.1201/b23017-22

20.1 INTRODUCTION

Plants are man's best companion, providing him with food, medicine, and fuel since the dawn of civilization (Garekae et al. 2022). Plants are still a key basis of medicine. Throughout history, societies all across the globe have discovered how to employ medicinal plants to treat sickness (Odieka et al. 2022). Their application is more relevant in present day scenario as they have comparably less side effects. This has led researchers to focus on plants that have medicinal efficacy. Plants have a variety of bioactive chemicals that have the potential to be developed as therapeutic medicines (El-Saadony et al. 2021; Husen et al. 2021; Husen 2021, 2022). The global demand for herbal medicines is increasing (Liu et al. 2021). Hence, uncontrolled practices with a lack of concern on conservation efforts are now endangering medicinal plants (Sofi et al. 2022) through the validation of traditional medicines, ethnopharmacological practices, and indigenous therapies (Xue et al. 2022).

There are several natural crude drugs in traditional medicine that have the efficiency to cure a variety of ailments and problems, one of them is *Ricinus communis* L. (family: Euphorbiaceae) commonly known as Castor bean (Kebede and Shibeshi 2022). It is a monotypic species having 60% distinctive oil, 90% of which is ricinoleic acid. As it has low solidification point (12°C–18°C), high molecular weight, very low melting point (5°C) and is stable having high viscosity, this acid has a lot of industrial importance (Pham et al. 2022). Ricinoleic acid is accountable for castor bean oil's popularity, since it has the greatest and high stable viscosity index of any vegetable oil, as well as excellent lubricity, particularly at low temperatures (Rodríguez-Cabal et al. 2020; Porokhovinova et al. 2022).

Castor is a non-oilseed plant with significant nutritional potential due to its 90% monounsaturated fatty acids concentration along with bioactive compounds like phenolics, vitamin E (tocotrienols or tocopherols), and phospholipids (Yeboah et al. 2020; Sbihi et al. 2018). These biocompounds contribute to castor oil's taste as well as stability, making it appropriate for a wide range of applications (Sedeek et al. 2012). In castor oil, tocopherol is a natural antioxidant that has anti-inflammatory and anti-proliferative activities. The major tocopherol isomers, 30.89–52.7 mg/g for γ and 43.1–96.62 mg/g for δ, are greater than those found in sunflower, hazelnuts, and olives (Said et al. 2016; Sbihi et al. 2018). Castor oil also has good oil quality due to its physicochemical qualities like high thiobarbituric acid and saponification value and low iodine and acid value (Omari et al. 2015).

On the other hand, Castor bean cultivation is complicated due to the presence of ricinine alkaloid, toxic ricin, as well as allergenic protein polysaccharide Castor bean allergenic fraction (CB-1A) in the plant. Ricin inhibits the synthesis of protein by inactivating ribosomes which is harmful in any form. Ricin which is one of the most powerful plant poisons is also found in castor beans.

Fritz Hauschild considers ricin in the list of five most dangerous compounds known: gramicidin, diphtheria toxin, tetanus toxin, botulinum toxin, and ricin (Hauschild 1960). As a result, it is not surprising that castor plant has been designated as a "poisonous plant 2018" in Germany.

20.2 BOTANICAL DESCRIPTION

Castor bean is a 5 m tall, semi-woody lager shrub or small tree and evergreen herbaceous plant (Figure 20.1). This plant is rapid growing that at first it grows straight up and then branches later. The stem varies in colour from green to reddish purple and has hollow internodes. Flowers are greenish yellow.

20.3 DISTRIBUTION

Although the castor bean is native to the Eastern Africa, Southern Mediterranean Basin, and India, it is thought to have originated from tropical Africa. Castor bean is extensively cultivated in subtropical, tropical, and temperate regions; it optimally grows in summer rainfall areas notably India, Brazil, and China due to its significant economic worth (Kamusoko and Jingura 2022). Breeders are

FIGURE 20.1 *Ricinus communis.*

paying close attention to breeding and variety enhancement because of rising castor bean demand in various countries (Fatimah et al. 2022). In tropical or subtropical locations, most varieties are big perennials that frequently mature into small trees; however, in frost-prone places, it is generally shorter and smaller and grows yearly. It is clear that castor bean has a high level of phenotypic variation and phenotypic plasticity in response to environmental influences.

It grows well at temperatures ranging from 20°C to 25°C. The seeds' higher oil content may be related to the warmer climate areas, but temperatures above 38°C can cause deprived seed condition. Furthermore, it is known that plants are harmed by temperatures that are low enough to result in frost formation. Because production is concentrated primarily in India, China, and Brazil, its total output varies considerably season to season due to variations in rainfall and planting area size. As a result, this concentration has resulted in the production of cyclic castor. Thus, diversifying production regions and increasing irrigation production should help to reduce the climatic impact on castor supplies. According to Union Minister of State for Agriculture and Farmers Welfare, India is the world's leading producer of castor oil. Gujarat is the leading producer of castor in India. The major castor growing states in the country are Gujarat, Tamil Nadu, Rajasthan, Andhra Pradesh, and Odisha. It is grown in Maharashtra, Karnataka, Uttar Pradesh, Bihar, and Madhya Pradesh, among other places (Senthilvel et al. 2022; Ramoliya 2022). As in India, Gujarat accounts for around 86% of this seed production, followed by Andhra Pradesh and Rajasthan. In India, Castor oil production is concerted in the Gujarat districts of Banaskantha, Mehsana, and Kutch/Saurashtra, as well as the Andhra Pradesh districts of Mahabubnagar and Nalgonda. Castor crop economic success in Gujarat in 1980s and later may be attributed to combination of a good breeding programme as well as an excellent extension model, leading to national and international market (Patel et al. 2016). India is the leading exporter of castor oil. China, the Netherlands, the United States, France, and Japan are the top export destinations. In 2020–2021, France received Rs. 641.03 crore from oil exports, while China received Rs. 3,144.08 crore. Export to the Netherlands totaled 97,288.73 MT worth Rs. 882.12 crore the United States received 70,770.76 MT worth Rs. 678.14 Crore, and Japan received 23,164.38 MT worth Rs. 222.95 crore. In 2020, the castor oil production in India amounted to just over 1.84 million tons. In comparison, castor oil production in the Philippines was approximately 40 tons in 2020 (www.Triged.com).

20.4 POISONOUS COMPOUNDS FOUND IN *RICINUS COMMUNIS*

Several compounds have been identified and isolated as toxic from *R. communis*, including the extremely toxic ricin (RCA-II, RCA60, ricin D), highly homologous but least toxic, *R. communis* agglutinin, allergenic compounds, and the nudiflorin and ricinine alkaloids (Napagoda 2022) (Figure 20.2).

20.4.1 Ricin

Ricin is present in various isoforms, depending on the variety of plant and type of seed from which it is extracted (Sehgal et al. 2010; Kumaraswamy et al. 2022).

20.4.1.1 Ricin (RCA-II, RCA60, Ricin D)

Ricin D is stored in mature ricinus seeds within endosperm cells in vacuoles, where it is susceptible to rapid degradation by hydrolysis during the post-germination, early growth stages (Tyagi et al. 2015; Lopez Nunez et al. 2017). While the ricin endogenous property within the plant is unknown, it is hypothesized that its potent cytotoxicity may act as a defence against plant-damaging organisms or various herbivores (Worbs et al. 2011).

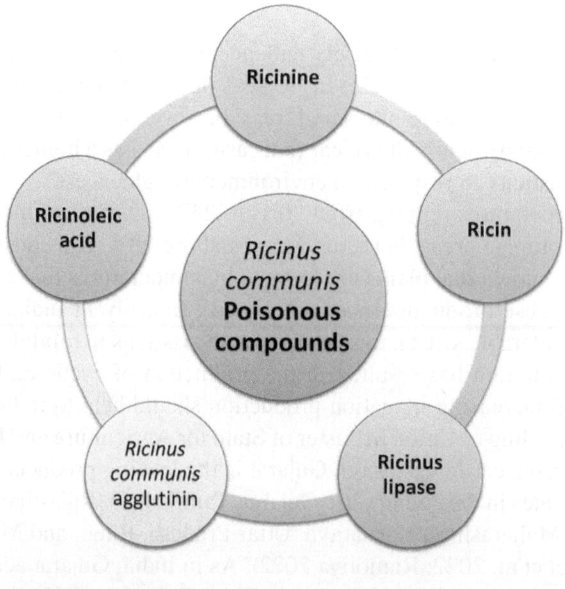

FIGURE 20.2 Poisonous compounds found in *Ricinus communis.*

20.4.2 Ricinoleic Acid

90% Ricinoleic acid as a hydroxylated unsaturated fatty triglyceride is present in cold-pressed oleum ricini. Its oil has been used as a purgative and it is said to be a good laxative.

Following ingestion of castor oil, intestinal lipases hydrolyze the triglyceride into ricinoleic and glycerol acid. It affects water flux and ions transport in intestine mucosa as well as intestinal motility. It acts as a laxative and stimulates labour in pregnant women (Teuscher and Lindequist 2010; Tunaru et al. 2012).

20.4.3 *Ricinus communis* Agglutinin

Ricinus communis agglutinin (RCA, comprising two subunits including A and B subunits) is present in the seeds. The A chains homology between ricin (RCA60) and RCA120 is over 93% and the B chains homology is 84% (Franke et al. 2019). Despite being closely related to ricin, it is classified as protein having significantly lower toxicity. RCA120 is dissimilar to ricin; it is not effectively cytotoxic, although it has an affinity for red blood cell, causing subsequent haemolysis as well as agglutination (Audi et al. 2005; Severino et al. 2012). RCA is poorly absorbed in the gastrointestinal tract, resulting in clinically considerable haemolysis after intravenous infusion (Audi et al. 2005).

20.4.4 Ricinine

It is an alkaloidal toxin that belongs to the piperidine alkaloids group. It is present in trace amounts in every parts of plant, but it also translocates throughout the plant based on its age (Audi et al. 2005; Worbs et al. 2011). Besides that, because of its higher resistance to temperature, ricinine, unlike ricin, cannot be inactivated by conventional heat treatment (Worbs et al. 2011). As ricinine and ricin can be co-extracted, they can be used as potential markers for tracking overdosage caused by plant crude extracts (Lopez Nunez et al. 2017).

20.4.5 Allergenic Compounds

R. communis pollen and seed components can be allergenic, causing urticaria and rhinitis or conjunctivitis (Bradberry et al. 2003; Teuscher and Lindequist 20/10; Deus-de-Oliveira et al. 2011; Severino et al. 2012). Castor bean allergies have been discovered in a variety of industrial sectors, including workers in fertilizer retail, the furnishing industry, and oil processing mill (Deus-de-Oliveira et al. 2011). It has been observed that castor bean dust causes endemic asthma near a castor oil mill and people exposed to this develop the allergic syndrome (Bradberry et al. 2003; Severino et al. 2012; Patel et al. 2016). Both type I and IV allergenic responses to castor bean dust have been described, as well as the involvement of eosinophils and IgE receptors (Bradberry et al. 2003). It has been proposed that the castor allergens Ric c 3 and Ric c 1 are a new class of amylase inhibitor and may have an important role in plant defence based on their ability to inhibit insect amylase (Nascimento et al. 2011).

20.4.6 Ricinus Lipase

R. communis has high lipase activity that is accountable for triglyceride hydrolysis (Srivastava et al. 2016). In spherosomes derived from the seeds' endosperm tissue, Lipase activity is found (Srivastava et al. 2016). Furthermore, Morlon-Guyot and colleagues demonstrated that ricin contains active site for functional lipase. The structural analysis of type II ribosome-inactivating proteins revealed that lipase site is present in toxic members of this family (ricin, abrin), but it is transformed into nontoxic or less toxic members (ebulin 1, mistletoe lectin I) (Morlon-Guyot et al. 2003). This finding suggests that lipolytic activity is crucial in ribosome-inactivating protein cytotoxicity.

20.5 PHYTOCHEMICAL OF *RICINUS COMMUNIS*

Several bioactive compounds are identified from *R. communis* leaves, seed, stem, and root extracts including flavonoids, coumarins, alkaloids, benzoic acid derivatives, terpenoids, tocopherols, fatty acids, and terpenoid. Some of these compounds are discussed below along with their pharmacological activities in the following sections.

20.5.1 ALKALOIDS

These belong to a class of chemically diverse secondary metabolites that are used for their medicinal properties (Facchini 2001; Ndagijimana et al. 2013). From *R. communis*, five alkaloids have been isolated.

Ricinine alkaloid is the most likely compound produced by *R. communis* that has received the most attention. Wachira et al. (2014) investigated the toxicity and larvicidal activity of Ricinine and its carboxylic acid derivative 3-Carboxy-4-Methoxy-N-Methyl-2-Pyridone, methyl-5-(3-Cyano-1-Methyl-2-Oxo-1,2-Dihydropyridine-4-il) pentanoate, 1-Methyl-4-(4-Metilpentiloksi) Pyridine-2 (1H)-on, and N-Demethyl ricinine.

20.5.2 FLAVONOIDS

Flavonoids are a considerable group of secondary metabolites found in plants that can be classified into several classes (Falcone Ferreyra et al. 2012). One molecule of cinnamoyl-CoA is combined with three molecules of malonyl-CoA to produce these chemicals (Falcone Ferreyra et al. 2012), Quercetin, Quercetin-3-O-d-Glucopyranoside, Quercetin-3-O-d-Galactoside, Quercetin-3-O-d-Xylopyranoside, Quercetin-3-O-d-Rutinoside, and Quercetin-3-O-d-xylopyranoside, from the leaves of *R. communis*. Three kaempferol glycosyl derivatives were obtained: kaempferol-3-O-d-Rutinoside, kaempferol-3-O-d-Glucopyranoside, and kaempferol-3-O-d-Xylopyranoside. From the roots and leaves of *R. communis*, Epicatechin and catechin were extracted (Singh and Chauhan 2009; Wafa et al. 2014; Zahir et al. 2012). Leaves and roots of *R. communis* were used to isolate the flavones vitexin and luteolin.

20.5.3 BENZOIC ACID DERIVATIVES

From *R. communis* roots and leaves, ellagic, gentisic, gallic, and vanillic acids were isolated. The presence of the o- and p-Dihydroxybenzene system, which is an antioxidant enhancer, is responsible for increased activity (Singh and Chauhan 2009).

20.5.4 COUMARINS

Three coumarins were isolated from *R. communis* – 8-Dihydroxy coumarin, 3,4-Dimethoxy-6 from the flowers (Khafagy et al. 1979) and scopoletin and isofraxidine from the leaves (Bigi et al. 2004) – despite the fact that these compounds were not tested for pharmacological activity.

20.5.5 TOCOPHEROLS

From *R. communis* cotyledons and roots, α-, β-, δ-, and γ-Tocopherols were identified (Ribeiro et al. 2014). In the cotyledons, at high temperatures (35°C), α-Tocopherol levels increased. In contrast, the roots showed no temperature-related changes. δ- and γ-Tocopherols followed a similar pattern to a much lesser extent. These chemicals protect cells from the oxidative stress caused by high temperatures.

20.5.6 Terpenoids

20.5.6.1 Monoterpenoids

From *R. communis* leaves, 1,8-Cineole, α-Thujone, α-Pinene, camphene, and camphor were identified in the essential oil (Kadri et al. 2011), whereas fusidic acid was extracted from *R. communis* leaves in methanol (Li et al. 2014).

20.5.6.2 Diterpenoids

From *R. communis* cotyledons, roots, and leaves, phytol was detected (Li et al. 2014; Ribeiro et al. 2014). Callypsinol was isolated from *R. communis* leaves methanol extract (along with 36 other compounds) and it was observe to show significant in vitro inhibitory activity and selectivity against mouse and human 11-HSD (Li et al. 2014). In a pioneer study conducted by Ribeiro et al. (2014) Casbene, (+)-Cembrene, (+)-Beyerene, (+)-Sandaracopimaradiene, (–)-Kaurene, and (–)-Trachylobane were identified and were produced from mevalonic acid by 2.5-day-old seedlings exposed to cultures of *Aspergillus niger*, *Rhizopus stolonifer*, and *Fusurium moniliforme*. Casbene is reported to act as a phytoalexin.

20.5.6.3 Sterols

In the cotyledons and roots of *R. communis* seedlings grown at various temperatures, Stigmasterol, Campesterol, and Beta-Sitosterol were detected (Ribeiro et al. 2014). 7-Oxo-Sitosterol and 3-O-d-Glycosylsitosterol were isolated from *R. communis* stem and leaves (Bigi et al. 2004; Ohishi et al. 2014). Stigmasterol oleate, Stigmasterol arachidate, and stigmasterol stearate were isolated from *R. communis* roots (Mittal and Ali 2012), whereas Stigmast-4-en-6-ol-3-one, Stigmast-4-en-3-one, 3-O-d-Glycosylstigmasterol, and Stigmast4-en-3,6-dione were isolated from the leaves (Bigi et al. 2004; Li et al. 2014).

20.5.6.4 Triterpenoids

Srivastava et al. (2014) isolated and identified two triterpenes from *R. communis* root: a new diketone pentacyclic triterpene and lupeol, a known compound. The compounds lup-20(29)-en-3, 30-nor-Lupan-3-ol-20-one, 15-Diol, and acetylaleuritolic acid were isolated from the leaves of *R. communis* and demonstrated in vitro inhibitory activity against mouse and human 11-HSD (Li et al. 2014). Lup-20(29)-en-15-ol-3-one was isolated from *R. communis* leaves (Tan et al. 2009).

20.5.7 Fatty Acids

In the chemical industry, Ricinoleic acid (12-Hydroxyoctadeca-9-Enoic acid) is an important hydroxylated fatty acid (Bafor et al. 1991). *R. communis* endosperm contains more than 80% of this fatty acid. Although it was isolated primarily from *R. communis* endosperm, it was also isolated from its leaves and roots (Oloyede 2012; Wafa et al. 2014). GC–MS analysis revealed four fatty acids in *R. communis* hexane leaf and seed extracts: linolenic acid, linoleic acid, stearic acid, and palmitic acid (Ramos-López et al. 2012; Ribeiro et al. 2015).

Oloyede (2012) identified methyl linoleate and methyl ricinoleate from seeds, but it is most likely that these compounds are a marker of the extraction procedure, which involved the use of methanol. Gondoic acid and oleic acid were found in the leaves and roots (Wafa et al. 2014). 1-Palmitic acid Glycerol-2,3-Dimethylketal ester and 1-Palmitic acid glycerol ester were isolated from leaves (Bigi et al. 2004). Triricinolein and n-Heptacosanyl oleate were isolated from roots (Mittal and Ali 2012; Tang et al. 2012).

20.5.8 Other Compounds

Other compounds including n-Butyl Ricicommunioate, Arachidoyl arabinoside, methyl communisoate, ethyl brevifolincarboxylate, and n-Hexatriacont-14-ene were isolated from the roots of *R. communis* (Mittal and Ali 2012; Tang et al. 2012), although no pharmacological efficacy was reported.

TABLE 20.1
Pharmacological Activities of Different Parts of *Ricinus communis*

Plant Parts	Activity	Solvent	Extraction Method	Compound Detected	References
Roots and leaves	Antioxidant	Methanol	Maceration	Flavonoids, phenols, and condensed tannins	Wafa et al. (2014)
Leaves		-	Hydrodistillation	α-Pinene, α-Thujone, 1,8-Cineole, camphene, and camphor	Kadri et al. (2011)
		Methanol	Soxhlet	Steroids, saponins, and alkaloids	Taur et al. (2011)
Roots, Leaves, stems, capsules, and seeds		Methanol	Maceration	Alkaloids and cardiac glycosides	Ibraheem and Maimako (2014)
Root	Anti-inflammatory	Ethanol	Soxhlet	–	Lomash et al. (2010)
Leaves	Antidiabetic (hypoglycemic)	Ethanol	Maceration	–	Ndagijimana et al. (2013)
Leaves	Antimicrobial	Methanol and ethyl acetate	Maceration	Coumarins, alkaloids, tannins, flavonoids, and phenols	Rampadarath et al. (2014)
		Methanol, ethanol, and water	Maceration	–	Naz and Bano (2012)
		Ethanol and methanol	Soxhlet	Cardiac glycosides, saponins, flavonoids, tannins, and terpenoids	Jeyaseelan and Jashothan (2012)
Roots, leaves, stems, capsules, and seeds		Methanol	Maceration	Cardiac glycosides and alkaloids	Ibraheem and Maimako (2014)

20.6 PHARMACOLOGICAL EFFICACY OF *RICINUS COMMUNIS*

R. communis being a poisonous plant has several therapeutic potential. Various pharmacological activities of *R. communis* are shown in Table 20.1. Various studies on biological activities of *R. communis* plants have been conducted and published (Khalid et al. 2022). Such activities are present in the crude extract because of its bioactive compounds. Presently, *R. communis* properties mentioned in this chapter include antimicrobial, antidiabetic, anti-cancer, antioxidant, and anti-inflammatory (Figure 20.3).

20.6.1 Antimicrobial Activity

Because of the rise in the number of human infections caused by a variety of clinical strains of bacteria, as well as the development of antibiotic resistance, there is an urgent need to identify consistent alternative bases to fight against antibiotic resistance (Friedman and Rasooly 2013; Murray et al. 2022). Castor and its phytochemicals have antimicrobial properties against a variety of microorganisms as depicted in Table 20.2. The crude extract has been shown to inhibit the growth of bacteria

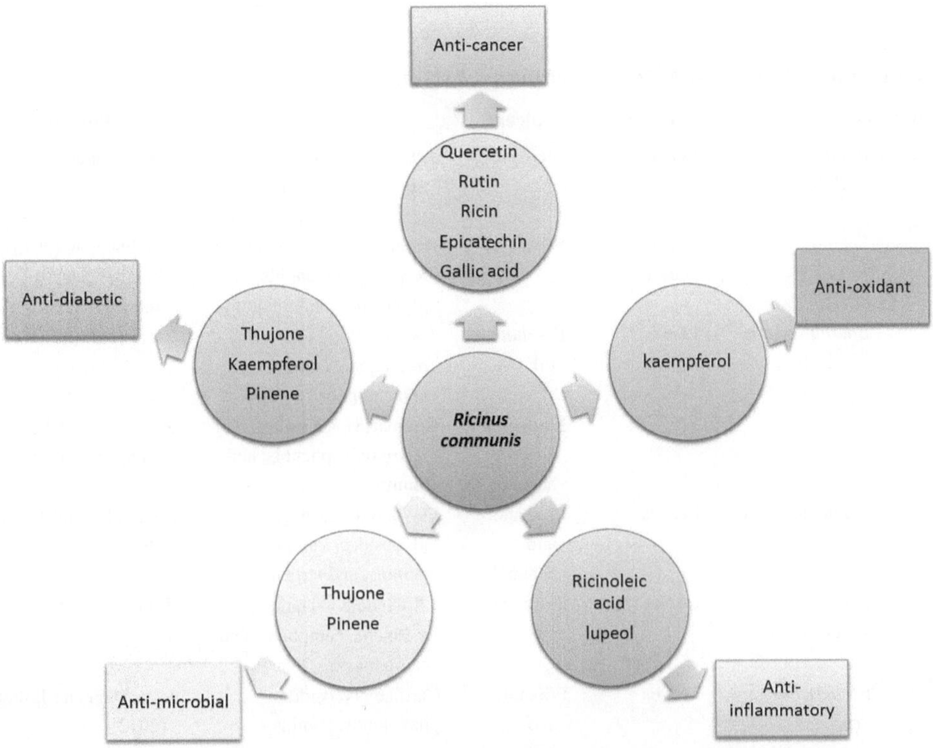

FIGURE 20.3 Pharmacological efficacy of *Ricinus communis* with their specific compounds.

such as *Enterococcus faecalis*, *Staphylococcus aureus*, *Escherichia coli*, *Streptococcus mutans*, as well as methicillin-resistant *S. aureus* (Abew et al. 2014; Alli et al. 2022). Various aqueous and solvent-based extracts of *R. communis* were used to test its activity. The various solvent systems employed include methanolic extract, ethanolic extract, butanolic extract, and ethyl acetate extract. The highest activity against *S. aureus* with a MIC of 5 mg/mL was found in ethanolic extract and it was found to be most effective in many cases (Jeyaseelan and Jashothan 2012). In another study, the lowest activity against *Bacillus subtilis* and the highest activity against *E. coli* were found in methanolic extract (Abd-Ulgadir et al. 2015). While the lowest activity against *Klebsiella pneumonia* and the highest activity against *S. aureus* were found in aqueous extract (Naz and Bano 2012). Castor activity as a complete cleanser solution was demonstrated in a randomized clinical study against bacterial pathogens (Sandford et al. 2021). In *R. communis* oil, the presence of sodium ricinoleate inhibits biofilm formation, causing cell wall damage and cell death due to cytoplasmic component loss (Salles et al. 2015). Based on the findings, it is possible to conclude that *R. communis* could be a valuable source of an antibacterial drug against a variety of bacterial pathogens.

R. communis has antifungal activity in several parts, including the root, leaf, and stem. *R. communis* methanolic and aqueous extracts have been found to be effective against a wide range of fungal species. A study was conducted to test the castor extract antifungal activity; the least activity was found against *Alternaria solani* and the highest antifungal activity was found against *Candida albicans* (Vandita et al. 2013). In one of the studies, methanolic extract showed less activity against *Aspergillus flavus* and prominent inhibitory activity against *Aspergillus fumigatus* and *Aspergillus niger* (Jeyaseelan and Jashothan 2012). In another study, aqueous leaf extract showed the least activity against *Aspergillus flavus* and *Aspergillus fumigatus* (Abd-Ulgadir et al. 2015). The research

TABLE 20.2
Antibacterial Activities of *Ricinus communis* Extracts

Microorganisms		Tissue	Solvent	Compound	References
Bacteria	*Bacillus subtilis*	Leaves	Methanol	Alkaloids, tannins, cardiac glycosides, flavonoids, terpenoids, and steroids	Vandita et al. (2013)
		Callus cultures	Ethanol	Anthocyanin, alkaloids, terpenes, flavonoids, glycosides, and tannins	Vandita et al. (2013), Zarai et al. (2012), Naz and Bano (2012)
	Bacillus cereus	Leaves	Essential oil	1,8-Cineole, α -Pinene, α-Thujone, camphene, and camphor	Zarai et al. (2012)
			Methanol and ethyl acetate	Coumarins, alkaloids, flavonoids, phenols, and tannins	Zarai et al. (2012), Rampadarath et al. (2014)
	Staphylococcus aureus	Leaves	Methanol and Ethanol	Flavonoids, cardiac glycosides, saponins, tannins, and terpenoid	Jeyaseelan and Jashothan (2012)
	Klebsiella pneumoniae	Leaves	Essential oil	1,8-Cineole, αThujone, α-Pinene, camphene, and camphor	Zarai et al. (2012)
	Escherichia coli	Leaves	Ethanol and methanol	Cardiac glycosides, flavonoids, tannins, saponins, and terpenoid	Jeyaseelan and Jashothan (2012)
			Methanol and ethyl acetate	Coumarins, alkaloid, tannins, flavonoid, and phenol	Rampadarath et al. (2014)
			Methanol	Alkaloids, flavonoids, tannins, cardiac glycosides, terpenoids, and steroids	Vandita et al. (2013)
Fungi	*Aspergillus niger*	Stem	Water	-	Vandita et al. (2013), Naqvi et al. (2011)
	Aspergillus fumigatus	Leaves and stems	Water	Sugars and free amino acids	Naz and Bano (2012), Habib Naqvi et al. (2011)
	Aspergillus flavus	Leaves	Methanol	-	Naz and Bano (2012)

findings above confirm *R. communis* extract's antifungal activity, and it could be a good source for finding a new drug candidate to inhibit the fungi.

Kebede and Shibeshi (2022) reported that castor leaf extract and fraction were tested for antimicrobial activity. At 400 mg/mL, the maximum antimicrobial activity for ethyl acetate fraction ranged from 20.33 mm (*S. aureus*) to 14.67 mm (clinical *E. coli*), while n-Hexane had least antimicrobial activity. The MIC values for the tested fractions ranged from 16.67 mg/mL (ethyl acetate) (*Candida albicans*) to 1.5625 mg/mL (*S. aureus*). The ethyl acetate fraction demonstrated bactericidal activity against all microorganisms tested.

20.6.2 Antidiabetic Activity

R. communis seed has antidiabetic activity (Cheikhyoussef and Cheikhyoussef 2022). Ethanol roots extract has potential effects on lipid profile, fasting blood glucose, liver and kidney functions and no significant difference on alkaline phosphatase, serum bilirubin, serum glutamate oxaloacetate transaminases, serum glutamate pyruvate transaminases, creatinine, as well as total protein. Roots ethanolic extract antidiabetic activity was tested in a study and it was discovered to be effective against hypoglycemic rats. After treating alloxan diabetic rats with a single dose of 500 mg/kg body weight for 20 days, blood was collected on day 0, day 10, and day 20 of the experiment to examine the lipid profile. Their investigation showed a significant reduction in fasting blood sugar to near-normal levels and enhance in insulin levels, resulting in an improvement in body weight and lipid profile. Blood glucose levels decreased from 149 ± 11 mg/dL (control) to 379 ± 72 mg/dL (diabetic rat) (Shokeen et al. 2008). The other in vivo study on alloxan-induced diabetic rats found that after treatment with *Ricinus* extract, blood glucose levels decreased from 390.0 to 148.5 mg/dL, representing a 61.97% decrease in blood glucose over a 7-day period (Matthew et al. 2012). These findings point to *R. communis* potential as a therapeutic agent for diabetic control.

20.6.3 Anti-Cancer Activity

Several research have been conducted using fractions with 100% methanol, ethanol, and an aqueous phase that have demonstrated activity against particular cancer cell lines like melanoma, hepatic cancer (HepG2), breast cancer (MCF7), cervical cancer, and pancreatic cancer (PC3) (Prakash and Gupta 2014). Castor has been shown to have anti-cancer activity in both in vitro and in vivo studies using different parts of plant and ricin lectins (Ravishankar et al. 2012). Numerous researchers studied and reported on anti-cancer activity of *R. communis*. The cytotoxic effect of castor lectins on three diverse cell lines, human erythrocytes, sarcoma 180, and HeLa cells, was first discovered. *R. communis* aqueous extracts were found to have cytotoxic effect on human melanoma (A375 cancer cell lines) with an IC_{50} value of 48 g/mL (Shah et al. 2015). According to You et al. (2010), Ricinus agglutinin I caused rapid down-regulation of endothelial cell apoptosis as well as vascular endothelial growth factor-2 (VEGFR-2) in tumor blood vessels.

Other in vitro studies were conducted to test cytotoxic effects of *R. communis* on different cell lines, including liver cancer, breast cancer, colon cancer, cervix cancer, skin melanoma (B16F10), prostate cancer, and ovarian cancer (OVCAR-5). The extract from *R. communis* was tested against these cancer cell lines at lower concentrations of about 100 g/mL and found to be effective (Prakash and Gupta 2014). Several studies have shown that *R. communis* has anti-cancer activity, implying that it could be a better source of anti-cancer therapeutic.

R. communis fruit extract (RCFE) also shown to have potential anti-cancer activity, and this was investigated using estrogen positive MCF-7 and highly aggressive, triple negative MDA-MB-231 breast cancer cells. In these cells, RCFE induces cytotoxicity in a time- and dose-dependent manner. It also inhibited migration, adhesion, invasion, and the expression of matrix metalloproteinases (MMPs) 9 as well as 2 in both cell lines, indicating potent anti-metastatic activity. The anti-cancer efficacy of a common castor plant fruit extract can be potentially a promising source for the treatment of breast cancer (Majumder et al. 2019).

20.6.4 Anti-Inflammatory Activity

Castor has been reported to a have strong anti-inflammatory properties. Diverse fractions such as methanolic, ethanolic, and hexane have been used to assess anti-inflammatory potential of *R. communis*. Its extract's anti-inflammatory activity was tested in one study using the acetone, hexane, and methanol fractions, in which significant anti-inflammatory activity was demonstrated in methanolic extract, which was attributed to the flavonoid found in it. Raw 264.7 macrophage cell lines were used in this study to test anti-inflammatory activity that leads to 95% scavenging activity at 2.5 g/mL with methanolic extract (Nemudzivhadi and Masoko 2014).

Lindauer et al. (2010) found that ricinolein mediated anti-inflammatory and proinflammatory activity after repeated application. Anti-inflammatory activity was tested in two animal models, namely the cotton pellet granuloma model and carrageenan-induced paw edema model. A dose of 100 mg/kg was found to inhibit edema by 26.47% in a carrageenan model. At doses of 250 and 500 g/kg, *Ricinus* methanolic extract was reported to inhibit edema by 43.28% in cotton pellet granuloma models (Srivastava et al. 2014). The activation of leucine-rich protein (NALP3) inflammasome and nucleotide-binding domain was responsible for this activity.

Another in vivo study demonstrated anti-inflammatory activity of *R. communis* in histamine or carrageenan-induced edema in guinea pigs or mice, respectively. For 8 days (0.9 mg/mouse), ricinoleic acid was applied topically and was found to inhibit edema by 58%. This demonstrates that ricinoleic acid has the efficacy to be a new capsaicin-like substance that belongs to a class of potent anti-inflammatory compounds (Vieira et al. 2000).

20.6.5 Antioxidant Activity

The first line of defence are Antioxidants against the damage caused by free radical, and they play an important role in preserving viability of cells and overall health.

Numerous neurodegenerative diseases are linked to free radicals in the cell. The antioxidant activity of *R. communis* is revealed in vitro by the presence of compounds such as quercitin, gallic acid, epicatechin, rutin, ellagic acid, and gentisic acid in the leaves methanolic extract (Ahmed et al. 2015).

R. communis methanolic extract antioxidant potential was observed to scavenge free radicals such as 2,2-Diphenyl-1-Picrylhydrazyl (DPPH) and 2,2′-Azino-bis(3-Ethylbenzothiazoline-6-Sulfonic acid) (ABTS) by Hussain et al. (2022). Another study demonstrated the ability of *R. communis* to scavenge DPDH, NO, and superoxide radicals, proving it to be a novel antioxidant. Butanol fraction of *Ricinus* showed the highest antioxidant activity (Iqbal et al. 2012; Höferl et al. 2014). *R. communis* ethyl acetate extract was also discovered to be a powerful antioxidant. The antioxidant activity of *R. communis* is attributed to flavonoids (Nath et al. 2015).

20.6.6 Analgesic Activity

Several researches have been carried out to demonstrate the analgesic activity of *R. communis* extract. *R. communis* extract has been studied and proven to have typical central nervous system stimulant as well as neuroleptic effects (Franke et al. 2019). The alkaloid ricinine in *R. communis* causes stimulant effects like hyperactivity, clonic seizures, and memory improvement. Ricinine is not anxiogenic because it does not inhibit brain exploration (Ferraz et al. 1999).

Another study compared *R. communis* aqueous extract of root bark to the standard drug diclofenac for the analgesic activity at a dose of 50 mg/kg. In Albino mice, *R. communis* extract *doses* of 100 and 200 mg/kg were used. The analgesic activity was determined using two methods: Eddy's hot plate method and tail immersion method (Almeida et al. 2001). The effect of methanolic extract of *Ricinus* leaves on antinociceptive activity was also demonstrated in one study. A 150 mg/kg dose resulted in a gradient increase in the tail flick of mice from (2.9000.194) in control mice at 0 minute to (6.300.110) in *Ricinus* treated mice, indicating the analgesic activity of *R. communis* (Taur et al. 2011).

20.6.7 Anthelmintic Activity

R. communis was studied for its anthelmintic activity in inducing paralysis of worms as well as the time required to kill the worm. Anthelmintic activity was assessed using both ethanolic and aqueous extracts. When compared to ethanolic extract, aqueous *Ricinus* extract demonstrated higher activity at 100 mg/mL in less time, i.e., 8.500.64 (paralysis) and 31.501.25 (death) at a concentration of 100 mg/mL. As a result, the aqueous extract of *R. communis* was found to have higher anthelmintic activity (Rana and Dhamija 2012).

20.6.8 Anti-Fertility Activity

R. communis was studied for its anti-fertility effects in male rats and they showed a decrease in epididymal sperm count. The main effects of treatment with 50% ethanolic extracts included changes in morphology and motility of sperms (Rocha et al. 2022). In another study, semen parameters were observed, and testicular function was suppressed in male Wistar rats treated with castor (10 mg/kg; Nath et al. 2015). Isichei et al. (2000) conducted a clinical study for females which showed antifertility effects with a single dose that prevented pregnancy for 12 months.

20.6.9 Antiulcer Activity

Castor was observed to have considerable antiulcer properties. The initial dose of 500 mg/kg revealed that *R. communis* has antiulcer properties. The mechanism of basic *R. communis* cytoprotective action and strengthening of the gastric mucosa are examples of its antiulcer activity, which also results in improved mucosal defence (Rachhadiya et al. 2011).

20.6.10 Antiasthmatic Activity

In this study, *R. communis* demonstrated significant antiasthmatic activity conducted by Taur and Patil (2011). The saponin content of *R. communis* in its roots was reported to cause mast cell stabilization, Flavonoids, on the other hand, are responsible for smooth muscle relaxation and bronchodilation activity. In vivo studies, which are critical in the treatment of asthma, were used to determine the anti-allergic activity. Because of the presence of saponins and flavonoids, the ethanolic extract was effective in reducing milk-induced eosinophilia and leukocytosis.

20.7 CONCLUSIONS

The medicinal effect of *R. communis* plant has occupied a distinct place as a medicinal plant from ancient time. It has a variety of medicinal properties, some of which are discussed here, but this plant still has a lot of untapped potential. The medicinal properties reported in this chapter confirm that *R. communis* has a high therapeutic value and the potential to develop a new, safe, potential, and less expensive drug in the future. However, more research, pharmacological studies, exploration, clinical trials, and public awareness are required for the better utilization of its medicinal properties. As a whole, *R. communis* phytochemical constituents and pharmacological activities have great potential and significance in the field of medicinal plant research. As a result, industrial entrepreneurs should also come up with new ideas and steps to make the better use of this potentially medicinal plant. *R. communis* extract can also be used to synthesize nanoparticles for testing its activity against microbial pathogens as well as cancer cell lines; these nanoparticles will also be of great interest in target drug delivery. Hence, more in vitro and animal model studies of these phytoconstituents as well as nanoparticles can be used to develop new drugs to tap the plant's full bioactive potential.

NOTES

Ambreen Bano, Department of Biosciences, Faculty of Sciences, Integral University, Lucknow, India
Adria Hasan, Department of Bioengineering, Faculty of Engineering, Integral University, Lucknow, India
Swati Sharma, Department of Biosciences, Faculty of Sciences, Integral University, Lucknow, India
Snober S. Mir, Department of Biosciences, Faculty of Sciences, Integral University, Lucknow, India

REFERENCES

Abd-Ulgadir, K. S., Suliman, S. I., Zakria, I. A., and Hassan, N. E. A. 2015. Antimicrobial potential of methanolic extracts of Hibiscus sabdariffa and Ricinus communis. *Advancement in Medicinal Plant Research, 3*(1), pp. 18–22.

Abew, B., Sahile, S., and Moges, F. 2014. In vitro antibacterial activity of leaf extracts of Zehneria scabra and Ricinus communis against Escherichia coli and methicillin resistance Staphylococcus aureus. *Asian Pacific Journal of Tropical Biomedicine, 4*(10), pp. 816–820.

Ahmed, D., Khan, M. M., and Saeed, R. 2015. Comparative analysis of phenolics, flavonoids, and antioxidant and antibacterial potential of methanolic, hexanic and aqueous extracts from Adiantum caudatum leaves. *Antioxidants, 4*(2), pp. 394–409.

Alli, S., Dulger, G., Kilicciogl u, I., Alli, A., and Dulger, B. 2022. Castor oil—Based graft copolymers: Synthesis, characterization antimicrobial activity and antiproliferative effects against breast cancer cell lines. *Polymer Bulletin, 1*, p. 23.

Almeida, R. N., Navarro, D. S., and Barbosa-Filho, J. M. 2001. Plants with central analgesic activity. *Phytomedicine, 8*(4), pp. 310–322.

Audi, J., Belson, M., Patel, M., Schier, J., and Osterloh, J. 2005. Ricin poisoning: A comprehensive review. *JAMA, 294*(18), pp. 2342–2351.

Bafor, M., Smith, M. A., Jonsson, L., Stobart, K., and Stymne, S. 1991. Ricinoleic acid biosynthesis and triacylglycerol assembly in microsomal preparations from developing castor-bean (Ricinus communis) endosperm. *The Biochemical Journal, 280* (Pt 2), 507–514.

Bigi, M. F. M., Torkomian, V. L., De Groote, S. T., Hebling, M. J. A., Bueno, O. C., Pagnocca, F. C., Fernandes, J. B., Vieira, P. C., and Da Silva, M. F. G. 2004. Activity of Ricinus communis (Euphorbiaceae) and ricinine against the leaf-cutting ant Atta sexdens rubropilosa (Hymenoptera: Formicidae) and the symbiotic fungus Leucoagaricus gongylophorus. *Pest Management Science, 60*(9), pp. 933–938.

Bradberry, S. M., Dickers, K. J., Rice, P., Griffiths, G. D., and Vale, J. A. 2003. Ricin poisoning. *Toxicological Reviews, 22*(1), pp. 65–70.

Cheikhyoussef, N., and Cheikhyoussef, A. 2022. Bioactive phytochemicals from castor (Ricinus communis Linneo) seed oil processing by-products. In: Ramadan Hassanien, M.F. (eds.), *Bioactive Phytochemicals from Vegetable Oil and Oilseed Processing By-products* (pp. 1–20). Cham, Switzerland: Springer International Publishing.

Deus-de-Oliveira, N., Felix, S. P., Carrielo-Gama, C., Fernandes, K. V., DaMatta, R. A., and Machado, O. L. 2011. Identification of critical amino acids in the IgE epitopes of Ric c 1 and Ric c 3 and the application of glutamic acid as an IgE blocker. *PLoS One, 6*(6), p. e21455.

El-Saadony, M. T., Zabermawi, N. M., Zabermawi, N. M., Burollus, M. A., Shafi, M. E., Alagawany, M., Yehia, N., Askar, A. M., Alsafy, S. A., Noreldin, A. E., and Khafaga, A. F. 2021. Nutritional aspects and health benefits of bioactive plant compounds against infectious diseases: A review. *Food Reviews International, 2*, pp. 1–23.

Facchini, P. J. 2001. Alkaloid biosynthesis in plants: Biochemistry, cell biology, molecular regulation, and metabolic engineering applications. *Annual Review of Plant Biology, 52*(1), pp. 29–66.

Falcone Ferreyra, M. L., Rius, S., and Casati, P. 2012. Flavonoids: Biosynthesis, biological functions, and biotechnological applications. *Frontiers in Plant Science, 3*, p. 222.

Fatimah, I., Sagadevan, S., Murugan, B., and Muraza, O. 2022. Castor oil (Ricinus communis). *Biorefinery of Oil Producing Plants for Value-Added Products, 1*, pp. 51–78.

Ferraz, A. C., Angelucci, M. E. M., Da Costa, M. L., Batista, I. R., De Oliveira, B. H., and Da Cunha, C. 1999. Pharmacological evaluation of ricinine, a central nervous system stimulant isolated from Ricinus communis. *Pharmacology Biochemistry and Behavior, 63*(3), pp. 367–375.

Franke, H., Scholl, R. and Aigner, A. 2019. Ricin and *Ricinus communis* in pharmacology and toxicology-from ancient use and "Papyrus Ebers" to modern perspectives and "poisonous plant of the year 2018". *Naunyn-Schmiedeberg's Archives of Pharmacology 392*, 1181–1208.

Friedman, M., and Rasooly, R. 2013. Review of the inhibition of biological activities of food-related selected toxins by natural compounds. *Toxins, 5*(4), pp. 743–775.

Garekae, H., Shackleton, C. M., and Tsheboeng, G. 2022. The prevalence, composition and distribution of forageable plant species in different urban spaces in two medium-sized towns in South Africa. *Global Ecology and Conservation, 33*, p. e01972.

Hauschild, F. 1960. *Pharmakologie und Grundlagen der Toxikologie.* Leipzig, Germany: VEB Georg Thieme.

Höferl, M., Stoilova, I., Schmidt, E., Wanner, J., Jirovetz, L., Trifonova, D., Krastev, L., and Krastanov, A. 2014. Chemical composition and antioxidant properties of Juniper berry (Juniperus communis L.) essential oil. Action of the essential oil on the antioxidant protection of *Saccharomyces cerevisiae* model organism. *Antioxidants, 3*(1), pp. 81–98.

Husen, A. 2021. *Traditional Herbal Therapy for the Human Immune System*. Boca Raton, FL: CRC Press. https://doi.org/10.1201/9781003137955

Husen, A. 2022. *Herbs, Shrubs and Trees of Potential Medicinal Benefits*. Boca Raton, FL: Taylor & Francis Group. https://doi.org/10.1201/9781003205067

Husen, A., Bachheti, R. K., and Bachheti, A. 2021. *Non-Timber Forest Products (Food, Healthcare and Industrial Applications)*. Cham, Switzerland: Springer Nature Switzerland AG. https://doi.org/10.1007/978-3-030-73077-2

Hussain, A., Aslam, B., Muhammad, F., and Faisal, M. N. 2022. In vitro antioxidant activity and in vivo anti-inflammatory effect of *Ricinus communis* (L.) and *Withania somnifera* (L.) Hydroalcoholic extracts in rats. *Brazilian Archives of Biology and Technology*, *64*(2).

Ibraheem, O., and Maimako, R. F. 2014. Evaluation of alkaloids and cardiac glycosides contents of Ricinus communis Linn (Castor) whole plant parts and determination of their biological properties. *International Journal of Toxicological and Pharmacological Research, 6*(3), pp. 34–42.

Iqbal, J., Zaib, S., Farooq, U., Khan, A., Bibi, I., and Suleman, S. 2012. Antioxidant, antimicrobial, and free radical scavenging potential of aerial parts of Periploca aphylla and Ricinus communis. *International Scholarly Research Notices*, *2012*, p. 563267.

Isichei, C. O., Das, S. C., Ogunkeye, O. O., Okwuasaba, F. K., Uguru, V. E., Onoruvwe, O., Olayinka, A. O., Dafur, S. J., Ekwere, E. O., and Parry, O. 2000. Preliminary clinical investigation of the contraceptive efficacy and chemical pathological effects of RICOM-1013-J of Ricinus communis var Minor on women volunteers. *Phytotherapy Research: An International Journal Devoted to Pharmacological and Toxicological Evaluation of Natural Product Derivatives, 14*(1), pp. 40–42.

Jeyaseelan, E. C., and Jashothan, P. J. 2012. In vitro control of Staphylococcus aureus (NCTC 6571) and Escherichia coli (ATCC 25922) by Ricinus communis L. *Asian Pacific Journal of Tropical Biomedicine, 2*(9), pp. 717–721.

Kadri, A., Gharsallah, N., Damak, M., and Gdoura, R. 2011. Chemical composition and in vitro antioxidant properties of essential oil of Ricinus communis L. *Journal of Medicinal Plants Research, 5*(8), pp. 1466–1470.

Kamusoko, R., and Jingura, R. M. 2022. Phytoremediation potential of non-edible biofuel plants: Physic nut (Jatropha curcas), castor bean (Ricinus communis) and water hyacinth (Eichhornia crassipes). In: Puthur T. J., and Dhankher P. O. (eds.), *Bioenergy Crops* (pp. 71–93). Boca Raton, FL: CRC Press.

Kebede, B., and Shibeshi, W. 2022. In vitro antibacterial and antifungal activities of extracts and fractions of leaves of Ricinus communis Linn against selected pathogens. *Veterinary Medicine and Science*, *8*, pp. 1802–1815.

Khafagy, S., Mahmoud, Z., and Salam, N.E. 1979. Coumarins and flavonoids of Ricinus communis growing in Egypt. *Planta Medica*, *37*(10), 191–191.

Khalid, A., Algarni, A. S., Homeida, H. E., Sultana, S., Javed, S. A., Abdalla, H., Alhazmi, H. A., Albratty, M., and Abdalla, A. N. 2022. Phytochemical, cytotoxic, and antimicrobial evaluation of Tribulus terrestris L., Typha domingensis Pers., and Ricinus communis L.: Scientific evidences for folkloric uses. *Evidence-Based Complementary and Alternative Medicine*, *2022*, p. 6519712.

Kumaraswamy, H. H., Kumar, V. D., Lavanya, C., Ushakiran, B., Senthilvel, S., Sujatha, M., Durga Bhavani, K. B., Konda, A. K., Lakshmidevi, G., Ashfaq, A. M., and Kumar, B. S. 2022. Biotechnological approaches for genetic improvement of castor bean (Ricinus communis L.). In: Gosal, S. S., and Wani, H. S. (eds.), *Accelerated Plant Breeding, Vol. 4* (pp. 359–418). Cham, Switzerland: Springer.

Li, S. H., Deng, Q., Zhu, L., Lai, C. H., Wang, H. S., and Tan, Q. G. 2014. Terpenoids and sterols from Ricinus communis and their activities against diabetes. *Zhongguo Zhong yao za zhi = Zhongguo Zhongyao Zazhi= China Journal of Chinese Materia Medica, 39*(3), pp. 448–452.

Lindauer, M., Wong, J., and Magun, B. 2010. Ricin toxin activates the NALP3 inflammasome. *Toxins, 2*(6), pp. 1500–1514.

Liu, M., Gao, Y., Yuan, Y., Yang, K., Shi, S., Tian, J., and Zhang, J. 2021. Efficacy and safety of herbal medicine (Lianhuaqingwen) for treating COVID-19: A systematic review and meta-analysis. *Integrative Medicine Research, 10*(1), p. 100644.

Lomash, V., Parihar, S. K., Jain, N. K., and Katiyar, A. K. 2010. Effect of Solanum nigrum and Ricinus communis extracts on histamine and carrageenan-induced inflammation in the chicken skin. *Cellular and Molecular Biology, 56*(3), pp. 1239–1251.

Majumder, M., Debnath, S., Gajbhiye, R. L., Saikia, R., Gogoi, B., Samanta, S. K., Das, D. K., Biswas, K., Jaisankar, P., and Mukhopadhyay, R. 2019. Ricinus communis L. fruit extract inhibits migration/invasion, induces apoptosis in breast cancer cells and arrests tumor progression in vivo. *Scientific Reports, 9*(1), pp. 1–14.

Matthew, O. O., Olusola, L., and Matthew, A. O. 2012. Preliminary study of hypoglycaemic and hypolipidemic activity of aqueous root extract of Ricinus communis in alloxan-induced diabetic rats. *Journal of Physiology and Pharmacology Advances, 2*(10), pp. 354–359.

Mittal, A., and Ali, M. 2012. New phenolic methyl esters and arachidonyl arabinoside from the roots of Ricinus communis L. *Der Pharmacia Lettre, 4*, pp. 1461–1466.

Morlon-Guyot, J., Helmy, M., Lombard-Frasca, S., Pignol, D., Piéroni, G., and Beaumelle, B. 2003. Identification of the ricin lipase site and implication in cytotoxicity. *Journal of Biological Chemistry, 278*(19), pp. 17006–17011.

Murray, C. J., Ikuta, K. S., Sharara, F., Swetschinski, L., Aguilar, G. R., Gray, A., ... and Naghavi, M. 2022. Global burden of bacterial antimicrobial resistance in 2019: A systematic analysis. *The Lancet, 399*, pp. 629–655.

Napagoda, M. 2022. 9 Poisonous plants and their toxic metabolites. *Chemistry of Natural Products: Phytochemistry and Pharmacognosy of Medicinal Plants, 12*, p. 173.

Naqvi, S. H., Dahot, M. U., Rafiq, M., Khan, M. Y., Ibrahim, I., Lashari, K. H., Ali, A., and Korai, A. L. 2011. Anti-microbial efficacy and biochemical analysis from different parts of Acacia nilotica L. and Ricinus communis L. extracts. *Journal of Medicinal Plants Research, 5*(27), pp. 6299–6308.

Nascimento, V. V. D., Castro, H. C., Abreu, P. A., Oliveira, A. E. A., Fernandez, J. H., Araújo, J. D. S., and Machado, O. L. T. 2011. In silico structural characteristics and α-amylase inhibitory properties of Ric c 1 and Ric c 3, allergenic 2S albumins from Ricinus communis seeds. *Journal of Agricultural and Food Chemistry, 59*(9), pp. 4814–4821.

Nath, S., Kadasi, A., Grossmann, R., Sirotkin, A. V., Kolesarova, A., Talukdar, A. D., and Choudhury, M. D. 2015. Ricinus communis L. stem bark extracts regulate ovarian cell functions and secretory activity and their response to Luteinising hormone. *International Journal of Impotence Research, 27*(6), pp. 215–220.

Naz, R., and Bano, A. 2012. Antimicrobial potential of Ricinus communis leaf extracts in different solvents against pathogenic bacterial and fungal strains. *Asian Pacific Journal of Tropical Biomedicine, 2*(12), pp. 944–947.

Ndagijimana, A., Wang, X., Pan, G., Zhang, F., Feng, H., and Olaleye, O. 2013. A review on indole alkaloids isolated from Uncaria rhynchophylla and their pharmacological studies. *Fitoterapia, 86*, pp. 35–47.

Nemudzivhadi, V., and Masoko, P. 2014. In vitro assessment of cytotoxicity, antioxidant, and anti-inflammatory activities of Ricinus communis (Euphorbiaceae) leaf extracts. *Evidence-Based Complementary and Alternative Medicine, 2014*, p. 625961.

Nunez, O. F. L., Pizon, A. F., and Tamama, K. 2017. Ricin poisoning after oral ingestion of castor beans: A case report and review of the literature and laboratory testing. *The Journal of Emergency Medicine, 53*(5), pp. e67–e71.

Odieka, A. E., Obuzor, G. U., Oyedeji, O. O., Gondwe, M., Hosu, Y. S., and Oyedeji, A. O. 2022. The medicinal natural products of Cannabis sativa Linn.: A review. *Molecules, 27*(5), p. 1689.

Ohishi, K., Toume, K., Arai, M. A., Sadhu, S. K., Ahmed, F., Mizoguchi, T., Itoh, M., and Ishibashi, M. 2014. Ricinine: A pyridone alkaloid from Ricinus communis that activates the Wnt signaling pathway through casein kinase 1α. *Bioorganic & Medicinal Chemistry, 22*(17), pp. 4597–4601.

Oloyede, G. K. 2012. Antioxidant activities of methyl ricinoleate and ricinoleic acid dominated Ricinus communis seeds extract using lipid peroxidation and free radical scavenging methods. *Research Journal of Medicinal Plants, 6*(7), pp. 511–520.

Omari, A., Mgani, Q. A., and Mubofu, E. B. 2015. Fatty acid profile and physico-chemical parameters of castor oils in Tanzania. *Green and Sustainable Chemistry, 5*(04), p. 154.

Patel, V. R., Dumancas, G. G., Viswanath, L. C. K., Maples, R., and Subong, B. J. J. 2016. Castor oil: Properties, uses, and optimization of processing parameters in commercial production. *Lipid Insights, 9*, p. LPI-S40233.

Pham, N. K. T., Tran, T. T. L., Duong, T. H., Trung, N. T., Phan, D. C. T., Mai, D. T., Nguyen, V. K., Huynh, B. L. C., Nguyen, T. A. T., Tran, T. D., and Tran, T. N. M. 2022. Ricicomin A, a new alkaloid from the leaves of Ricinus communis Linn. *Natural Product Research, 36*(8), pp. 1973–1979.

Porokhovinova, E. A., Matveeva, T. V., Khafizova, G. V., Bemova, V. D., Doubovskaya, A. G., Kishlyan, N. V., Podolnaya, L. P., and Gavrilova, V. A. 2022. Fatty acid composition of oil crops: Genetics and genetic engineering. *Genetic Resources and Crop Evolution, 69*, pp. 1–17.

Prakash, E., and Gupta, D. K. 2014. In vitro study of extracts of Ricinus communis Linn on human cancer cell lines. *Journal of Medical Sciences and Public Health, 2*(1), pp. 15–20.

Rachhadiya R M., Prassad, and Shetty, R. V. 2011. Evaluation of antiulcer activity of castor oil in rats. *International Journal of Research in Ayurveda and Pharmacy* IJRAP, 2(4), pp. 1349–1353.

Ramoliya, R. K. 2022. Growth and instability of major oilseed crops in Gujarat. *International Journal of Agriculture Sciences, ISSN, 14*, pp. 0975–3710.

Ramos-López, M. A., González-Chávez, M. M., Cárdenas-Orteg, N. C., and Zavala-Sánchez, M. A. 2012. Activity of the main fatty acid components of the hexane leaf extract of Ricinus communis against Spodoptera frugiperda. *African Journal of Biotechnology, 11*(18), pp. 4274–4278.

Rampadarath, S., Puchooa, D., and Ranghoo-Sanmukhiya, V. M. 2014. A comparison of polyphenolic content, antioxidant activity and insecticidal properties of Jatropha species and wild Ricinus communis L. found in Mauritius. *Asian Pacific Journal of Tropical Medicine, 7*, pp. S384–S390.

Rana, M., and Dhamija, M. 2012. Ricinus communis L. – A Review. *International Journal of PharmTech Research, 4*(4), pp. 1706–1711.

Ravishankar, K., Indira, N., and Vijay Bhaskar, R. 2012. In vivo hepatoprotective activity of Ricinus communis. linn leaf extract against ccl4 induced hepatic damage in albino rats. *International Journal of Biological & Pharmaceutical Research, 3*(3), pp. 444–449.

Ribeiro, P. R., Dekkers, B. J., Fernandez, L. G., de Castro, R. D., Ligterink, W., and Hilhorst, H. W. 2014. Identification of reference genes for gene expression studies during seed germination and seedling establishment in Ricinus communis L. *Seed Science Research, 24*(4), pp. 341–352.

Ribeiro, P. R., Willems, L. A., Mudde, E., Fernandez, L. G., de Castro, R. D., Ligterink, W., and Hilhorst, H. W. 2015. Metabolite profiling of the oilseed crop Ricinus communis during early seed imbibition reveals a specific metabolic signature in response to temperature. *Industrial Crops and Products, 67*, pp. 305–309.

Rocha, A. C., da Silveira Alves, F. G., Salles, H. O., Pompeu, R. C. F. F., Ludke, J. V., Severino, L. S., and Cândido, M. J. D. 2022. The industrial process of solvent extraction of castor oil reduces the toxicity of the meal. *Industrial Crops and Products, 181*, p. 114800.

Rodríguez-Cabal, H. A., Jaramillo-Mazo, C. Y., Franco-Sierra, N. D., Villanueva-Mejía, D. F., and Alvarez, J. C. 2020. A differentially expressed gene from a high oil producer cultivar of castor bean (Ricinus communis) is involved in the biosynthesis of ricinoleic acid. *American Journal of Plant Sciences, 11*(3), pp. 393–412.

Said, G., Daniel, P., Badr, K., Mohamed, I., and Zoubida, C. 2016. Chemical characterization and oxidative stability of castor oil grown in Morocco. *Moroccan Journal of Chemistry, 4*(2), pp. 4–12.

Salles, M. M., Badaro, M. M., Arruda, C. N. F. D., Leite, V. M. F., Silva, C. H. L. D., Watanabe, E., Oliveira, V. D. C., and Paranhos, H. D. F. O. 2015. Antimicrobial activity of complete denture cleanser solutions based on sodium hypochlorite and Ricinus communis—A randomized clinical study. *Journal of Applied Oral Science, 23*, pp. 637–642.

Sandford, E. C., Muntz, A., and Craig, J. P. 2021. Therapeutic potential of castor oil in managing blepharitis, meibomian gland dysfunction and dry eye. *Clinical and Experimental Optometry, 104*(3), pp. 315–322.

Sbihi, H. M., Nehdi, I. A., Mokbli, S., Romdhani-Younes, M., and Al-Resayes, S. I. 2018. Hexane and ethanol extracted seed oils and leaf essential compositions from two castor plant (Ricinus communis L.) varieties. *Industrial Crops and Products, 122*, pp. 174–181.

Sedeek, S., El-Ghobashy, R., and Tawfik, M. 2012. Thermal stability of cottonseed oil mixed with jojoba or castor oil during frying process. *Journal of Biological Chemistry and Environmental Science, 7*(2), pp. 39–56.

Sehgal, P., Khan, M., Kumar, O., and Vijayaraghavan, R. 2010. Purification, characterization and toxicity profile of ricin isoforms from castor beans. *Food and Chemical Toxicology, 48*(11), pp. 3171–3176.

Senthilvel, S., Manjunatha, T., and Lavanya, C. 2022. Castor breeding. In: Yadava, D. K., Dikshit, H. K., Mishra, G. P., and Tripathi, S. (eds.), *Fundamentals of Field Crop Breeding* (pp. 945–970). Springer Nature, Singapore.

Severino, L. S., Auld, D. L., Baldanzi, M., Cândido, M. J., Chen, G., Crosby, W., Tan, D., He, X., Lakshmamma, P., Lavanya, C., and Machado, O. L. 2012. A review on the challenges for increased production of castor. *Agronomy Journal, 104*(4), pp. 853–880.

Shah, T. I., Sharma, E., and Shah, G. A. 2015. Inhibitory property of aqueous extract of Ricinus communis leaves on proliferation of melanoma treated against A375 cell lines. *World Journal of Pharmaceutical Sciences, 3*, pp. 758–761.

Shen, N., Wang, T., Gan, Q., Liu, S., Wang, L., and Jin, B. 2022. Plant flavonoids: Classification, distribution, biosynthesis, and antioxidant activity. *Food Chemistry, 383*, p. 132531.

Shokeen, P., Anand, P., Murali, Y. K., and Tandon, V. 2008. Antidiabetic activity of 50% ethanolic extract of Ricinus communis and its purified fractions. *Food and Chemical Toxicology, 46*(11), pp. 3458–3466.

Singh, P. P., and Chauhan, S. M. S. 2009. Activity guided isolation of antioxidants from the leaves of Ricinus communis L. *Food Chemistry, 114*(3), pp. 1069–1072.

Sofi, I. I., Ganie, A. H., Shah, M. A., Verma, S., and Sharma, N. 2022. Threat status of three important medicinal Himalayan plant species and conservation implications. *Nature Conservation Research, 7*(1).

Srivastava, A., Mathur, R. M., Prakash, R., and Agrawal, S. 2016. Studies on Ricinus lipase enzyme isolated from castor seeds. *Oriental Journal of Chemistry, 32*(2), p. 1235.

Srivastava, P., Jyotshna, Gupta, N., Maurya, A. K., and Shanker, K. 2014. New anti-inflammatory triterpene from the root of Ricinus communis. *Natural Product Research, 28*(5), pp. 306–311.

Tan, Q. G., Cai, X. H., Du, Z. Z., and Luo, X. D. 2009. Three terpenoids and a tocopherol-related compound from Ricinus communis. *Helvetica Chimica Acta, 92*(12), pp. 2762–2768.

Taur, D. J., and Patil, R. Y. 2011. Antiasthmatic activity of Ricinus communis L. roots. *Asian Pacific Journal of Tropical Biomedicine, 1*(1), pp. S13–S16.

Taur, D. J., Waghmare, M. G., Bandal, R. S., and Patil, R. Y. 2011. Antinociceptive activity of Ricinus communis L. leaves. *Asian Pacific Journal of Tropical Biomedicine, 1*(2), pp. 139–141.

Teuscher, E., and Lindequist, U. 2010. *Biogene Gifte: Biologie-Chemie-Pharmakologie-Toxikologie.* Wissenschaftl. Verlag-Gesellschaft, GmbH, Stuttgart, 3 Auflage.

Tunaru, S., Althoff, T. F., Nüsing, R. M., Diener, M., and Offermanns, S. 2012. Castor oil induces laxation and uterus contraction via ricinoleic acid activating prostaglandin EP3 receptors. *Proceedings of the National Academy of Sciences, 109*(23), pp. 9179–9184.

Tyagi, N., Tyagi, M., Pachauri, M., and Ghosh, P. C. 2015. Potential therapeutic applications of plant toxin-ricin in cancer: Challenges and advances. *Tumor Biology, 36*(11), pp. 8239–8246.

Vandita, P., Amin, N., Khyati, P., and Monisha, K. 2013. Effect of phytochemical constituents of Ricinus communis, Pterocarpus santalinus, Terminalia belerica on antibacterial, antifungal and cytotoxic activity. *Steroids, 12*(13), p. 14.

Vieira, C., Evangelista, S., Cirillo, R., Lippi, A., Maggi, C. A., and Manzini, S. 2000. Effect of ricinoleic acid in acute and subchronic experimental models of inflammation. *Mediators of Inflammation, 9*(5), pp. 223–228.

Wachira, S. W., Omar, S., Jacob, J. W., Wahome, M., Alborn, H. T., Spring, D. R., Masiga, D. K., and Torto, B. 2014. Toxicity of six plant extracts and two pyridone alkaloids from Ricinus communis against the malaria vector Anopheles gambiae. *Parasites & Vectors, 7*(1), pp. 1–8.

Wafa, G., Amadou, D., and Larbi, K. M. 2014. Larvicidal activity, phytochemical composition, and antioxidant properties of different parts of five populations of Ricinus communis L. *Industrial Crops and Products, 56*, pp. 43–51.

Worbs, S., Köhler, K., Pauly, D., Avondet, M. A., Schaer, M., Dorner, M. B., and Dorner, B. G. 2011. Ricinus communis intoxications in human and veterinary medicine—A summary of real cases. *Toxins, 3*(10), pp. 1332–1372.

Xue, H. T., Stanley-Baker, M., Kong, A. W. K., Li, H. L., and Goh, W. W. B. 2022. Data considerations for predictive modeling applied to the discovery of bioactive natural products. *Drug Discovery Today, 27*, pp. 2235–2243.

Yeboah, A., Ying, S., Lu, J., Xie, Y., Amoanimaa-Dede, H., Boateng, K. G. A., Chen, M., and Yin, X. 2020. Castor oil (Ricinus communis): A review on the chemical composition and physicochemical properties. *Food Science and Technology, 41*, p. 19620.

You, W. K., Kasman, I., Hu-Lowe, D. D., and McDonald, D. M. 2010. Ricinus communis agglutinin I leads to rapid down-regulation of VEGFR-2 and endothelial cell apoptosis in tumor blood vessels. *The American Journal of Pathology, 176*(4), pp. 1927–1940.

Zahir, A. A., Rahuman, A. A., Bagavan, A., Geetha, K., Kamaraj, C., and Elango, G. 2012. Evaluation of medicinal plant extracts and isolated compound epicatechin from Ricinus communis against Paramphistomum cervi. *Parasitology Research, 111*(4), pp. 1629–1635.

Zarai, Z., Chobba, I. B., Mansour, R. B., Békir, A., Gharsallah, N., and Kadri, A. 2012. Essential oil of the leaves of Ricinus communis L.: In vitro cytotoxicity and antimicrobial properties. *Lipids in Health and Disease, 11*(1), pp. 1–7.

21 *Thevetia peruviana* (Yellow Oleander)

Subhrajit Banerjee, Senjuti Banerjee, Kasturi Sarkar, and Parames C. Sil

CONTENTS

DOI: 10.1201/b23017-23

21.1 INTRODUCTION

Yellow oleander is a drought resistant and a high temperature tolerant tropical plant. It bears a good amount of foliage and have vibrant yellow flowers for which it is preferred as an ornamental plant in gardens. It finds different uses in human urban establishments. It is used as screens in highway medians, in demarcation of land boundaries, and as hedges. Its toxicity can be found in the literature as early as 1863 (Chen and Chen, 1934a). The family Apocynaceae was named by Carl Linnaeus. The word "Apocynaceae" is derived from Asclepias who was the son of Apollo, the Greek God of healing. The family is also called Dogbane and Linnaeus named some 130 milk weed species under this family. Andre Thevet (1502–92), a priest and cosmologist, described this plant. He found the plant in South American Peru. The plant, *Thevetia peruviana*, is named after him and the country Peru, where he found the plant. DeVry in 1863 named the water-soluble cardiac glycoside isolated from the plant to be Thevetin. Blas in 1868 gave the formula to thevetin and crystallized it. The digitalis like actions of Thevetin on heart was reported by Husemann and König in 1876. Dumontier 1856 (Chen and Chen, 1934a) gave an account of accidental poisoning of a 3-year-old child, ingesting the plant. Bhattacharya and Ayyar (1927) reported cattle poisonings by seeds of the plant. Chen and Chen (1934a) isolated thevetin from the kernels of the seeds. Occasional accidental poisonings were reported from various parts of the world in humans, livestock, and animals throughout the nineteenth century. The plant came into focus when its rampant use as a suicidal agent started in South-east Asia in the early 1980s. Poisoning cases of oleander (303 cases) were reported in mostly young children, and among them, the highest number of cases (12 cases) was of below 1-year old in Texas, United States during 1994 (Langford and Boor, 1996). All parts of the plants, leaves, barks, flowers, roots, and seeds are reported to be toxic due to the presence of cardiac glycosides present in the plant (Ghatak et al., 1933; Bardosi, 1963; Romano and Mombelli, 1990; Oji and Okafor, 2000). The kernels of the fruit contain the highest concentration of cardiotoxic glycosides called cardenolides (Saravanapavananthan and Ganeshmoorthy, 1988). Saravanapavananthan (1985) first reported widespread suicidal cases in Sri Lanka. Such suicidal cases were reported in India in the adjacent southern states later. Samal (1989) reported suicidal cases in Orissa.

The plant has uses in ethnomedicinal practices like treatment of nervous shock in Haiti and the uses of the plant extracts as emetocathartic and febrifuge in Antilles as reported by Weitz and Boulay (1923). Chen and Chen (1934b) reported abortifacient use of the plant. The reports of the traditional use of the plant for generating fish poison in Brazil (Chen and Chen, 1934b) and arrow poisoning in Malay (Weitz and Boulay, 1923) are found. Before 1980s, toxicity reports were mostly due to unintentional cases of poisonings, consuming leaves and fruits. All plant-derived cardiac glycosides like digitalis found in foxglove bind to Na^+/K^+-ATPase pump and inactivate this enzyme. This is thought to be a co-evolutionary defence against herbivory. Many different plant families produce these cardenolides. Features of toxicity in Thevetia poisonings resemble toxicity of digitalis overdosing.

21.2 BOTANICAL DESCRIPTION

21.2.1 Taxonomic Description

Yellow oleander belongs to the botanical family Apocynaceae, which includes various poisonous species, including common Oleander or Nerium oleander (González-Stuart and Rivera, 2017). Its accepted botanical name is *Thevetia peruviana* (Alvarado-Cárdenas and Ochoterena, 2007).

According to the Spanish word, "Cascabela" stands for a small bell shape, resembling the bell of a venomous rattlesnake from which the plant derives its name Cascabela. The Plant List 2013 recognizes Cascabela as a separate accepted genus. Cascabela is a Neotropical genus of Apocynaceae with six species *Cascabela balsaensis*, *Cascabela gaumeri*, *Cascabela ovata*, *Cascabela pinifolia*, *Cascabela thevetia*, and *Cascabela thevetioides* (Alvarado-Cárdenas et al., 2017). Common names of the tree are Lucky nut, Be-still Exile oleander, Bastard oleander, Trumpet flower, and so on (Table 21.1). The colloquial names of the plant in Guianas and the Antilles are "Ahouai," "Bagage a collier," "Snake Nut," and "Milk Tree" (Weitz and Boulay, 1923). In the Mexican ethnomedicinal terms, its names are "Yollotl", "Codo de fraile" and Friar's elbow (González-Stuart and Rivera, 2017).

21.2.2 Plant Description

It is a drought-resistant, high temperature tolerant tree or shrub, with highly branched, dense crown and a poisonous milky sap (Bandara et al, 2010). All of these characteristics are of the Dogbane/Apocynaceae family. In some instances, the old trees grow to a height of about 20–30 feet (Langford and Boor, 1996). The leaves are green, glossy and linear, alternately arranged, narrow, with a prominent mid-rib, about 13–15 cm in length, narrow less than 1 cm in width, and closely arranged in spirals at the end of branches (Swaminathan and Kochhar, 2019). Bright yellow flowers at the tip of twigs scattered in diffuse branches are seen throughout the plant (Picture 21.1).

Flower colours varying from dull orange to peach are also common. The flower is funnel shaped or tubular (Bandara et al., 2010), present in few terminal clusters, measuring about 5–6 cm in length and 1–2 cm in diameter with five petals twisted spirally (Saravanapavananthan and Ganeshamoorthy, 1988). Yellow oleanders produce flowers all year round with subsequent production of seed pods (Langford and Boor, 1996). The fruit is globular, with diameter being 4–5 cm, changing colour from light green to black as it ripens. Each fruit contains 2–4 kernels (Saravanapavananthan and Ganeshamoorthy, 1988). When older, the trees are capable of producing large amounts of seeds ranging from 400 to 800 throughout the year, depending on the age of the plant (Bora et al., 2014). Few distinctive features differentiate the yellow oleander from *Nerium oleander. N. oleander* has three, dark green, leathery and lanceolate leaves arranged in whorls, while yellow oleanders have alternate leaves. Narrower, lighter green leaves with a long distinct mid-rib and non-lanceolate in shape are the characteristic features of Yellow Oleander distinguishing it from common oleanders (*N. oleander*). Nerium bears pink and white flowers.

TABLE 21.1
Common Names in Different Languages

Gele oleander	Dutch
Oléandre jaune	French
Gelber Schellenbaum	German
Adelfa amarilla	Spanish
Codo de Fraile	Spanish
Oleandro-amarelo	Portuguese
Thong thien/rumpei	Thailand
Geel oleander	Africans
Captain cook	Australia
Ayoyote	Mexican
In Different Indian Languages	
Pili kaner	Hindi
Kolke phul	Bengali
Bitti	Marathi
Manjal	Tamil

PICTURE 21.1 Picture of the plant with sparsely distributed flowers at the crown of each branch. Colours from bright yellow (b) to peach (c), the fruit is a green drupe (c), fruit peeled of exposes the seed (d), the seed coat is hard and measures 3–4 cm (e), the kernels of the seed are shown in (f), and the plant can grow to a tree with good amount of foliage (g).

21.3 DISTRIBUTION

T. peruviana is native to tropical and subtropical parts of South and Central America in countries like Brazil, Mexico, Caribbean, and Bahamas. It is extensively found throughout the tropical parts of Asia significantly in South-east Asia, especially India and Sri Lanka and in various parts of Africa and Australia. The trees are also distributed in South-eastern USA from Florida to California and on several Pacific islands (Langford and Boor, 1996). In Australia, this plant is considered as a noxious weed.

21.4 PHYTOCHEMICAL CONSTITUENTS

The plant is rich in various phytochemical constituents. Seeds have a diverse content of cardiac glycosides (CG; refer Table 21.2). Kernels of the seed have highest concentration of cardenolides

TABLE 21.2
Compounds Isolated from Seed Extract of Yellow Oleander

Serial Number	Name of the Compound	R1	R2	Molecular Formula	References
1	19-nor-Neriifolin	H	*α*-L Thevetose	$C_{29}H_{44}O_8$	Tian et al. (2016)
2	19-nor-Digitoxigenin 3-O-*β*-D-Glucosyl-(1→4)-*α*-L-Thevetoside	H	*β*-D-Glucosyl-(1→4)-*α*-L Thevetose	$C_{35}H_{54}O_{13}$	Tian et al. (2016)
3	19-nor-Digitoxigenin 3-O-*β*-Gentiobiosyl-(1→4)-*α*-L-Thevetoside	H	*β*-Gentiobiosyl-(1→4)-*α*-L-Thevetose	$C_{41}H_{64}O_{18}$	Tian et al. (2016)
4	19-nor-10-Hydroxydigitoxigenin 3-O-*α*-L-Thevetoside	OH	*α*-L-Thevetose	$C_{29}H_{44}O_9$	Tian et al. (2016)
5	19-nor-10-Hydroxydigitoxigenin 3-O-*β*-D-Glucosyl-(1→4)-*α*-L-Thevetoside	OH	*β*-D-Glucosyl-(1→4)-*α*-L-Thevetose	$C_{35}H_{54}O_{14}$	Tian et al. (2016)
6	19-nor-10-Hydroxydigitoxigenin 3-O-*β*-Gentiobiosyl-(1→4)-*α*-L-Thevetoside	OH	*β*-Gentiobiosyl-(1→4)-*α*-L-Thevetose	$C_{41}H_{64}O_{19}$	Tian et al. (2016)
7	19-nor-10-Hydroperoxydigitoxigenin 3-O-*β*-D-Glucosyl-(1→4)-*α*-L-Thevetoside	OOH	*β*-D-Glucosyl-(1→4)-*α*-L-Thevetose	$C_{35}H_{54}O_{15}$	Tian et al. (2016)
8	19-nor-10-Hydroperoxydigitoxigenin 3-O-*β*-Gentiobiosyl-(1→4)-*α*-L-Thevetoside	OOH	*β*-Gentiobiosyl-(1→4)-*α*-L-Thevetose	$C_{41}H_{64}O_{20}$	Tian et al. (2016)
9	Digitoxigenin 3-O-*β*-Gentiotriosyl-(1→4)-2′-O-Acetyl-*α*-L-Thevetoside	CH_3	*β*-Gentiobiosyl-(1→4)-2′-O-Acetyl-*α*-L-Thevetose	$C_{50}H_{78}O_{24}$	Tian et al. (2016)
10	Uzarigenin 3-O-*β*-Gentiobiosyl-(1→4)-2′-O-Acetyl-*α*-L-Thevetoside	CH_3	*β*-Gentiobiosyl-(1→4)-2′-O-Acetyl-*α*-L-Thevetose	$C_{44}H_{68}O_{19}$	Tian et al. (2016)
11	Cannogenol 3-O-*β*-D-Glucosyl-(1→4)-*α*-L-Thevetoside	CH_2OH	*β*-D-Glucosyl-(1→4)-*α*-L-Thevetose	$C_{36}H_{56}O_{14}$	Tian et al. (2016)
12	Cannogenol 3-O-*β*-Gentiobiosyl-(1→4)-*α*-L-Thevetoside	CH_2OH	*β*-Gentiobiosyl-(1→4)-*α*-L-Thevetose	$C_{42}H_{66}O_{19}$	Tian et al. (2016)
13	Cannogenol 3-O-*β*-D-Glucosyl-(1→4)-*α*-Acofrioside	CH_2OH	*β*-D-Glucosyl-(1→4)-*α*-L-Acofriose	$C_{36}H_{56}O_{14}$	Tian et al. (2016)
14	Cannogenin 3-O-*β*-D-Glucosyl-(1→4)-2′-O-Acetyl-*α*-L-Thevetoside	CHO	*β*-D-Glucosyl-(1→4)-2′-O-Acetyl-*α*-L-Thevetose	$C_{38}H_{56}O_{15}$	Tian et al. (2016)
15	Cannogenin 3-O-*β*-D-Glucosyl-(1→4)-*α*-L-Thevetoside	CHO	*β*-D-Glucosyl-(1→4)-*α*-L-Thevetose	$C_{36}H_{54}O_{14}$	Tian et al. (2016)
16	19-Carboxydigitoxigenin 3-O-*β*-D-Glucosyl-(1→4)-*α*-L-Thevetoside	COOH	*β*-D-Glucosyl-(1→4)-*α*-L-Thevetose	$C_{36}H_{54}O_{15}$	Tian et al. (2016)

(Continued)

TABLE 21.2 (*Continued*)
Compounds Isolated from Seed Extract of Yellow Oleander

Serial Number	Name of the Compound	R1	R2	Molecular Formula	References
17	2′-O-Acetylthevetin B	CH_3	β-Gentiobiosyl-(1→4)-2′-O-Acetyl-α-L-Thevetose	$C_{44}H_{68}O_{19}$	Kohls et al. (2012)
18	Digitoxigenin 3-O-β-D-Glucopyranosyl-(1→4)-α-L-aco-Friopyranoside	CH_3	β-D-Glucopyranosyl-(1→4)-α-L-aco-Friose	$C_{36}H_{56}O_{13}$	Tian et al. (2016)
19	Digitoxigenin β-D-Glucosyl-(1→4)-α-L-Thevetoside	CH_3	β-D-Glucosyl-(1→4)-α-L-Thevetose		Abe et al. (1992)
20	Thevetin B	CH_3	β-Gentiobiosyl-(1→4)-α-L-Thevetose	$C_{42}H_{66}O_{18}$	Tian et al. (2016)
21	Digitoxigenin β-Gentiobiosyl-(1→4)-α-L-Acofrioside	CH_3	β-Gentiobiosyl-(1→4)-α-L-aco-Friose		Abe et al. (1992)
22	Neriifolin	CH_3	α-L-Thevetose	$C_{30}H_{46}O_8$	Yamauchi et al. (1987)
23	Digitoxigenin 3-O-β-D-Glucopyranosyl-(1→4)-2′-O-Acetyl-α-L-Thevetopyranoside	CH_3	β-D-Glucosyl-(1→4)-2′-O-Acetyl-α-L-Thevetose		Endo et al. (1997)
24	Uzarigenin β-Gentiobiosyl-(1→4)-α-L-Thevetoside	CH_3	β-Gentiobiosyl-(1→4)-α-L-Thevetose		Abe et al. (1992)
25	Uzarigenin β-Gentiobiosyl-(1→4)-α-L-Acofrioside	CH_3	β-Gentiobiosyl-(1→4)-α-L-aco-Friose		Abe et al. (1992)
26	Thevefolin	CH_3	α-L-Thevetose	$C_{30}H_{46}O_8$	Miyagawa et al. (2009)
27	Uzarigenin β-D-Glucosyl-(1→4)-α-L-Thevetoside	CH_3	β-D-Glucosyl-(1→4)-α-L-Thevetose		Abe et al. (1992)
28	Ruvoside	CH_2OH	α-L-Thevetose	$C_{30}H_{46}O_9$	Tian et al. (2016)
29	Peruvoside	CHO	α-L-Thevetose	$C_{30}H_{44}O_9$	Voigtländer et al. (1969)
30	2′-O-Acetylthevetin A	CHO	β-Gentiobiosyl-(1→4)-2′-O-Acetyl-α-L-Thevetose	$C_{44}H_{66}O_{20}$	Kohls et al. (2012)
31	Thevetin A	CHO	β-Gentiobiosyl-(1→4)-α-L-Thevetose	$C_{42}H_{64}O_{19}$	Kohls et al. (2012)
32	Perusitin	COOH	α-L-Thevetose		Tian et al. (2016)
33	19-Carboxydigitoxigenin β-Gentiobiosyl-(1→4)-α-L-Thevetoside	COOH	β-Gentiobiosyl-(1→4)-2′-O-Acetyl-α-L-Thevetose		Abe et al. (1994)
34	Thevetin C	CH_3	β-Gentiobiosyl-(1→4)-α-L-Thevetose	$C_{42}H_{66}O_{19}$	Kohls et al. (2012)
35	Acyetylthevetin C	CH_3	β-Gentiobiosyl-(1→4)-2′-O-Acetyl-α-L-Thevetose	$C_{44}H_{68}O_{20}$	Kohls et al. (2012)

(4.8%) with a drastic decrease in leaves (0.07%), fruit (0.045%), and milk sap (0.036%; Kyerematen et al., 1985; Saravanapavananthan and Ganeshamoorthy, 1988).

21.4.1 Cardiac Glycosides

21.4.1.1 Early Studies on Cardiac Glycosides

Chen and Chen (1934a) reported the presence of thevetin, crystallized it, and derived the empirical formula of thevetin as $C_{29}H_{46}O_{13}.2H_2O$ which was a larger molecule than that reported by Blas et al. (1868) earlier. Jacobs (1933) mentioned the presence of lactones attached to steroid rings of thevetin. Thevetin has toxic effects in organs which have smooth muscles like the gastrointestinal (GI) tract and blood vessels and causes nausea and vomiting. The digitalis and ouabain like action of the cardiotoxic glycosides present in the plant were demonstrated through animal experiments by Chen and Chen (1934b) and they measured the emetic and lethal doses of thevetin. Thevetin is pharmacologically the most active substance for the heart. Chen compared the cardiac actions of Thevetin to an equal weight of ouabain and arrived at a factor of 1/8 times potency to that of ouabain.

The purified, crystallized glycosides from the plant mostly studied for their chemical actions are Thevetin A/B/C, peruvoside, ruvoside, and neriifolin. Structurally, cardenolides extracted from yellow oleander are very similar to the digitalis cardenolides like digoxin and digitoxin derived from the foxglove plant. In spite of the similarity, plant-derived cardenolides differ in their pharmacokinetic properties and in terms of therapeutic window and toxicity (Roberts et al., 2006).

21.4.1.2 Structure of Cardiac Glycosides/Cardenolides

The plant CGs are composed of a hydrophobic steroid ring (aglycone part) composed of 23 carbon structure having four fused rings (A, B, C, D rings). Attached to steroid ring is a five- or six-membered unsaturated lactone ring at Carbon-17 position of steroid backbone at the D ring, and a hydrophilic sugar chain (Glycone part) linked through a glycosidic linkage to the steroid backbone at Carbon-3 of the A ring (Figure 21.1). The steroid nucleus with the lactone ring is referred to as "genin"/"aglycone" portion of the CG. To give cardio-active effect, the unsaturated lactone ring in the structural backbone is necessary. Butenolide is an unsaturated five-membered lactone ring with a 20(22)-enolide. CGs are thus considered as amphiphilic molecules (Prassas and Diamandis, 2008; Haux, 1999). Altered number and positions of hydroxyl groups, methyl groups, and epoxide groups create variations in this backbone producing a wide array of different genins (Agrawal et al., 2012). Different sugar moieties also contribute to the variability in biological actions.

The structural details of a cardenolide are given in Figure 21.1 with the variations in genins, which are the basis of their nomenclature. Two types of genins are shown in Figure 21.1, they are cannogenin (Thevetin A and Peruvoside) and digitoxigenin (Thevetin B and Neriifolin). Cannogenin is a 3β-14-Dihroxy-19oxo-5βcard-20–22, β-Androstane 14-diol as defined by Repke (1985). It is a steroid aldehyde, a 5β-Cardenolide, MW 388.5, with β-Hydroxyl groups at carbon 3, 5, 14, an oxo-derivative at 19-Carbon, and a C-20(22) enolide in the lactone ring. Thevetin B and Neriifolin are digitoxigenin-derived cardenolide with a methyl group attached at C-10 instead of an aldehyde group (C-19) attached to C-10, which is a feature of cannogenin.

Neriifolin ($C_{30}H_{46}O_8$, MW 534.7) is a digitoxigenin derivative like thevetin B and C. Peruvoside ($C_{30}H_{44}O_9$, MW 548.7) is a monosaccharide with a single L-Thevetose and a cannogenin derivative like thevetin A. Thevetin, peruvoside, and neriifolin contain the butenolide lactone at C-17. Thevetin A ($C_{42}H_{64}O_{19}$, M.W. 872.94; Figure 21.2a) and Thevetin B ($C_{42}H_{66}O_{18}$, MW 859; Figure 21.2b) are trisaccharides with L-Thevetose and two glucose units attached to C3. The sugar moieties of cardenolides are unique for naturally occurring cardenolides and vary in number of saccharide units from 1 to 4 (Foye, 1989). Unique names of these saccharides are given according to the name of the cardiac glycosides like digitoxose for digitoxin and digoxin (Figure 21.2c); thevetose for thevetin. The variability in the glycones in the presence of unique sugars in different combinations

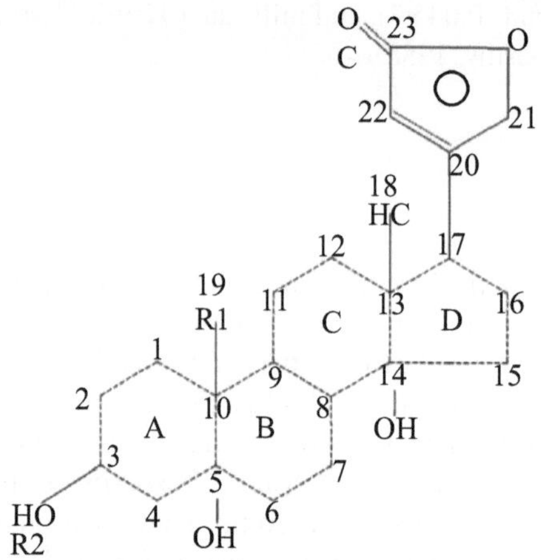

Digitoxi[genin]/Cannogenin

O Represents Lactone ring

R1 ----CH3 for Digitoxigenin
R1 ----CHO for Cannogenin
R2 ----Sugars like digitoxose/thevetose
Hydroxyl groups are linked at 3C, 5C, 14C

R1 ----CH3 for Digitoxigenin
R1 ----CHO for Cannogenin
R2 ----Sugars like digitoxose/thevetose
Hydroxyl groups are linked at 3C, 5C, 14C

Cannogenin is a steroid aldehyde and a 5(β-OH) Cardenolide

Steroid ring (denoted by dotted line) with the lactone ring at C17 forms the aglycone portion and is called the genin of the cardenolide, e.g., canno[genin], digitoxi[genin]
The cardenolides, e.g., thevetin, are is formed of genin with sugar moieties (glycone) linked by glycosidic bond at C3 of the steroid ring mentioned as R2 in the figure

The steroid and lactone rings are the genin and refer to the aglycone portion of the cardenolide

FIGURE 21.1 Cardenolides are 23 carbon steroids with glycosides at C3 and a lactone ring at C17.

affect the toxicity and biological activity of the compounds (Foye, 1989). Peruvoside and Neriifolin contain a single thevetose unit, whereas thevetins (A/B/C) are trisaccharides (thevetoside) shown in Figure 21.2. An 18,20-oxido bond and a 20,22-Dihydro in Thevetin C differentiate it from Thevetin B. 20,22-Dihydro makes the lactone ring saturated and renders thevetin C cardio-inactive (Cruz et al., 1979; Kohls et al., 2012).

A list of compounds isolated from the plant with structural variations at C-10 and C-3 of the steroid nucleus are presented in Table 21.2. All the compounds obtained from seed extracts of *T. peruviana* listed in Table 21.2 are structurally related CGs.

The cardenolides bind the ubiquitous animal enzyme Na^+/K^+-ATPase for their biologic actions (Schwartz and Whitmer, 1982; Demiryurek and Demiryurek, 2005). Minimum structure of the glycoside for binding to the enzymes are the genins comprised of the steroid nucleus and the lactone.

21.4.2 Other Phytochemical Constituents

Terpenes like pulegone, spathulenol, citronellol, carvacrol, nerolidol and fatty acids like palmitic, linoleic acid, benzoic, caprylic, and oleic acid were isolated from seed extracts and stem bark extract (Table 21.3; Gata-Goncalves et al., 2003). Leaves of *T. peruviana* contain several compounds like iridoid glycosides (Abe et al., 1995), flavonoids (Abe et al., 1995), triterpenes (Ali et al., 2000), and monoterpenes (Abe et al., 1996). Phenolic content of *T. peruviana* is highest (41 mg/g) in leaves compared to the other parts of the plant followed by fruit (9.75 mg/g), stem (7.3 mg/g), and flower (5.75 mg/g). Flowers (2.6 mg/g) and fruits (2.5 mg/g) contain nearly the same level of flavonoid contents, whereas its stem and leaves contain 1.37 and 0.75 mg/g flavonoid, respectively (Srivastava et al., 2012). The flowers have been found to contain quercetin, kaempferol, and quercetin 7-O-Galactoside. Phytochemical screening of *T. peruviana* leaves revealed isoprenoid compound like squalene, flavonoids like rutin, phenolic compound like rosmarinic acid, and secondary metabolite like di-Hydroxycinnamic or caffeic acid (El-Sawi et al., 2020).

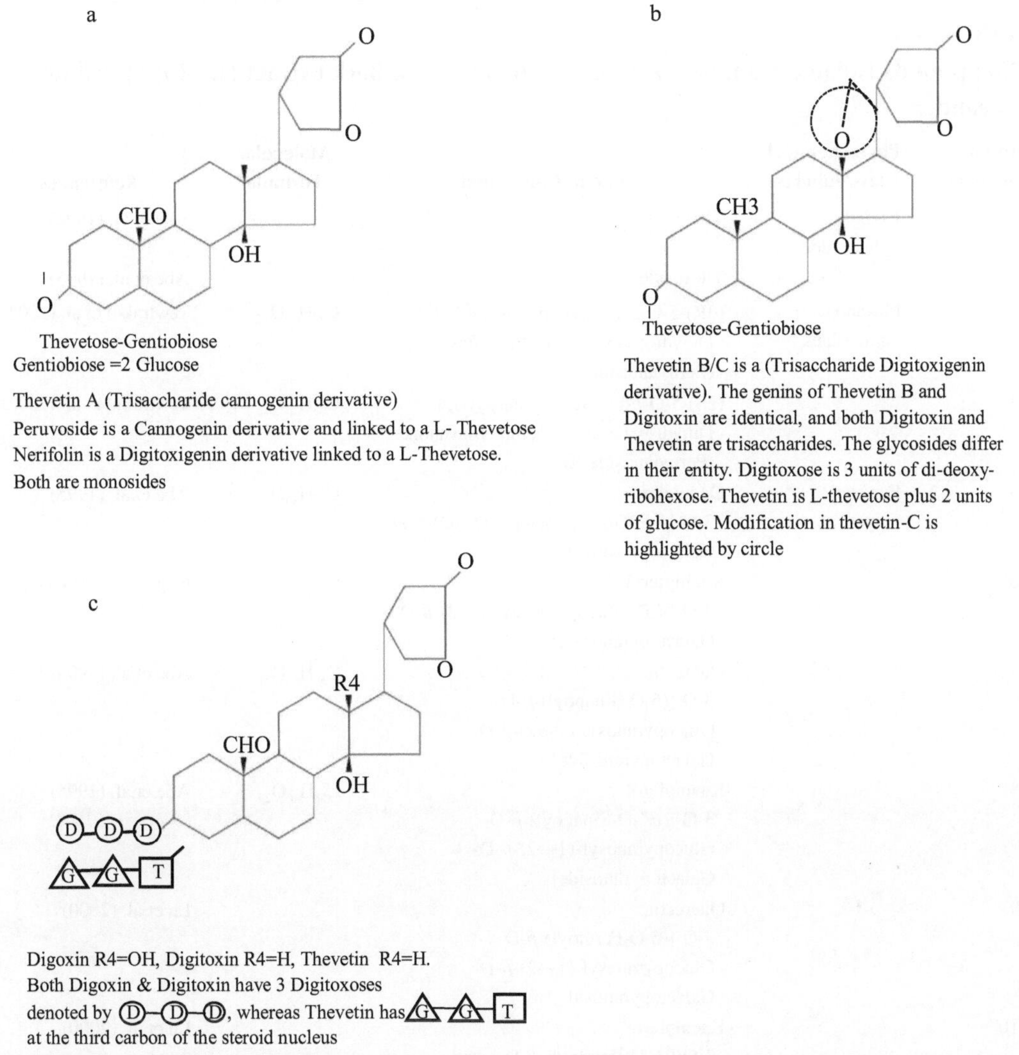

FIGURE 21.2 Thevetin A (a), Thevetin B (b), Digitoxin and Digoxin (c) are compared to highlight the structural differences between the molecules.

21.5 PHARMACOLOGICAL STUDIES

The toxicities produced in seed and leaf poisonings in human and animal studies are mentioned. As they resemble digitalis cardenolides in generating ionotropic actions on heart, their use in congestive heart failure is suggested in few studies. However, the therapeutic index of these compounds is very narrow and close to their toxic doses producing arrhythmia and GI side effects (Roberts et al., 2006). Hence, their use requires further clinical studies.

CGs from oleander exhibited greater toxic effects than digoxin in *in vitro* model of cardiomyocytes derived from pluripotent stem cells. The effects of CGs were graded according to their toxicity. Neriifolin exerted highest toxicity, followed by peruvoside, digoxin, and thevetin A. The calculated lethal dose in man hugely varies from 160 to 1.6 mg for CGs extracted from *T. peruviana* (Herrera, 1991; Eddleston et al., 1999; Roberts et al., 2016). Pharmacologically, thevetoxin has similar effects like thevetin but is less toxic (Thilagavathi et al., 2010). These cardenolides can cross the blood–brain barrier (Marx and Pretorius et al., 2006). Cardiotonic steroids inhibiting the Na^+/K^+-ATPase

TABLE 21.3
Compounds Isolated from Leaf Extract [1–16] and Stem Bark Extract [17–20] of Yellow Oleander

Serial Number	Phytochemical Class/Subclass	Name of Compound	Molecular Formula	References
1	Iridoid glucosides	Theviridoside		Abe et al. (1995)
2		Theveside		Abe et al. (1995)
3	Flavanone glucosides	(2R)-5-O-β-D-Glucopyranosyl-7,4′-Dihydroxy-3′,5′-Dimethoxyflavanone (Peruvianoside I)	$C_{23}H_{25}O_{12}$	Tewtrakul et al. (2002)
4		(2S)-5-O-β-D-Glucopyranosyl-7,4′-Dihydroxy-3′,5′-Dimethoxyflavanone (Peruvianoside II)	$C_{23}H_{25}O_{12}$	Tewtrakul et al. (2002)
5	Flavanol glycosides	Quercetin 3-O-[β-D-Glucopyranosyl-(1→2)-β-D-Galactopyranoside]	$C_{27}H_{30}O_{17}$	Abe et.al. (1995)
6		Kaempferol 3-O-[β-D-Glucopyranosyl(1→2)-β-D-Galactopyranoside]	$C_{27}H_{30}O_{16}$	Nagy et.al. (1984)
7		Quercetin 3-O-[(6-O-Sinapoyl)-β-D-Glucopyranosyl-(1→2)-β-D-Galactopyranoside]	$C_{38}H_{40}O_{21}$	Abe et.al. (1995)
8		Kaempferol 3-O-[(6‴-O-Sinapoyl)-β-D-Glucopyranosyl-(1→2)-β-D-Galactopyranoside]	$C_{38}H_{40}O_{20}$	Abe et.al. (1995)
9		Quercetin 3-O-[(6-O-Feruloyl)-β-D-Glucopyranosyl-(1→2)-β-D-Galactopyranoside]		Lu et al. (2000)
10		Kaempferol 3-O-[(6-O-Feruloyl)-β-D-Glucopyranosyl-(1→2)-β-D-Galactopyranoside]		Lu et al. (2000)
11		Quercetin 3-O-[β-D-Glucopyranosyl-(1→2)-β-D-Glucopyranoside]		Markham et al. (1978)
12		Kaempferol 3- O-[β-D-Glucopyranosyl-(1→2)-β-D-Glucopyranoside]		Williams et al. (1993)
13		Quercetin 3-O-{β-D-Glucopyranosyl-(1→2)-[α-L-Rhamonopyranosyl-(1→6)]-β-D-Galactopyranoside} (Peruvianoside III)	$C_{33}H_{40}O_{21}$	Tewtrakul et al. (2002)
14		Kaempferol 3-O-{β-D-Glucopyranosyl-(1→2)-[α-L-Rhamonopyranosyl-(1→6)]-β-D-Galactopyranoside}		Bashir et al. (1991)

(Continued)

TABLE 21.3 (*Continued*)
Compounds Isolated from Leaf Extract [1–16] and Stem Bark Extract [17–20] of Yellow Oleander

Serial Number	Phytochemical Class/Subclass	Name of Compound	Molecular Formula	References
15		Kaempferol 3-O-β-D-Glucopyranosyl (1→4) [6″-O-Sinapoyl-β-D-Glucopyranosyl] (1→2)-β-D-Galactopyranoside	$C_{44}H_{50}O_{25}$	Abe et al. (1995)
16		Kaempferol 3-O-[2‴′-O-Sinapoyl-β-D-Glucopyranosyl] (1 → 4) [6″-O-Sinapoyl-β-D-Glucopyranosyl] (1→2)-β-D-Galactopyranoside	$C_{55}H_{60}O_{29}$	Abe et al. (1995)
17	Cardenolides	Neriifolin	$C_{30}H_{46}O_8$	Jolad et al. (1981)
18		Peruvoside	$C_{30}H_{44}O_9$	Siddiqui et al. (1992)
19		Thevefolin	$C_{30}H_{46}O_8$	Ahmed et al. (2016)
20		(20S)-18,20 EpoxydigitoxigeninR- L-Thevetoside		Miyagawa et al. (2009)

affect a variety of systemic functions of the body which are a consequence of malfunctioning of the ATPase pump. Overall cellular functions are affected due to disturbances in electrolyte homeostasis, pH regulation, cell proliferation, and cell cycle (Schoner and Bobis, 2007). However, preliminary reports from the extracts of the plant suggest its use as antidiarrheal, antimicrobial, antiviral, anti-inflammatory, antioxidant, antispermatogenic, neuroprotective, haemodynamic, and apoptosis inducing actions (Amend and Worek et al., 2021).

21.5.1 Studies on Cardiac Glycoside Activities

21.5.1.1 Cytotoxic Activity/Anticancer Activity/Apoptotic Activity/Antitumour Activity

Tian et al. (2016) working with the ethanolic seed extract of *T. peruviana* containing glycosides of variable sugar chain lengths reported differential cytotoxic activities against cancer cell lines depending on sugar chain lengths. The study reported gradual decrease in cytotoxic activities. Monoglycosides were most toxic, followed by diglycosides and triglycosides. 19-Norcardenolides were less cytotoxic than CGs with formyl, methyl, or hydroxymethyl groups at C-10. Conversely, 19-Carboxycardenolides showed lesser cytotoxicity than 19-Norcardenolides. Among those compounds shown in Table 21.2, the CGs 19-nor-Neriifolin (1), Neriifolin (22), Thevefolin (26), and Ruvoside (28) were most cytotoxic against human gastric cancer cells by inducing apoptosis in dose-dependent manner. The extracts also inhibited proliferation of different cancer cell lines (Tian et al., 2016). Effects like altering membrane permeability, motility, and fragmentation of DNA inducing apoptosis in cells were seen (Ramos-Silva et al., 2017). Peruvoside is more effective cytotoxic agent than ouabain and digitoxin and induces cell death specifically in more mature myeloid leukemic cells without significant cytotoxicity on normal blood cells. Peruvoside has properties of arresting cell cycle and has apoptotic inducing effect by degrading caspase 3, 8 and PARP (Feng et al., 2016). Li et al. (2012) suggested that peruvoside may be a more potent anticancer candidate than other CGs. Peruvoside has an immediate haemodynamic effects like positive inotropic and negative chronotropic effects in congestive heart failure (Bhatia, 1970).

TNF-related apoptosis-inducing ligand (TRAIL) is employed in anticancer therapy because of its effect in triggering apoptosis in a variety of cancer cells by activating caspase signalling. Four cardenolides neriifolin, thevetosides, thevetin, and peruvoside showed significant activity in overcoming the TRAIL resistances (Miyagawa et al., 2009). Thevefolin and neriifolin showed potential as a TRAIL-resistance overcoming compound enhancing the levels of death receptors, DR4 and DR5, in human gastric adenocarcinoma cell lines (Miyagawa et al., 2009). Ethanolic extract of *T. peruviana* flowers enhances TNF-α induced and TRAIL-induced apoptosis in human cervical cancer cells (Managit et al., 2017). Improved targeted drug delivery through nanotechnology-based drug delivery is shown to decrease the adverse effects of Acetyl-thevetin B. There was a reduction in general cytotoxicity. Chitosan-Pluronic P123 micelle encapsulation was used to improve its anti-tumour efficacy against lung cancer (Zhu et al., 2017).

Squalene, rutin, rosmarinic acid, and di-Hydroxycinnamic or caffeic acid isolated from phytochemical screening of *T. peruviana* leaves showed cytoprotective activity. Specifically, squalene has potential action against chemotherapy-induced toxicity. Rutin has the capacity to improve radiation-induced harmful effects (Radwan and Kenawy, 2008) and also can act as a potential anticancer agent against colorectal and lung carcinogenesis (Ben Sghaier et al., 2016). Rosmarinic acid has the ability to reduce synthesis of reactive oxygen species and apoptosis along with anti-proliferative effect (Renzulli et al., 2004). It has been also reported that it is a potent chemotherapeutic agent with less cardiotoxicity (Kim et al., 2005).

21.5.1.2 Neuroprotective Action

Neriifolin has significant potential to rescue cortical pyramidal neurons from degeneration after transient oxygen and glucose deprivation in a dose-dependent manner. Specifically, neriifolin seemed to enhance neuroprotection in delaying fashion, i.e., after several hours of oxygen and glucose deprivation (Wang et al., 2006). Both flavanol glycosides, quercetin and kaempferol 3-O-[(6-Feruloyl)-Glucopyranosyl-Galactopyranoside], showed neuroprotection in glutamate-induced damage in primary cultures of rat cortical cells (Kim et al., 2001).

21.5.2 Studies on Other Activities

21.5.2.1 Antiviral Activity

There are numerous compounds isolated from an ethanolic extract of the leaves and stem barks of *T. peruviana* having inhibitory activities against HIV-1 integrase, reverse transcriptase associated RNA-dependent DNA polymerase, and DNA-dependent DNA polymerase (Tewtrakul et al., 2002). In this study, quercetin derivatives showed relatively higher DNA polymerase inhibitory activities than kaempferol derivatives. Among those isolated compounds, 7, 9, 11, 14, and quercetin compounds as shown in Table 21.3 displayed significantly higher inhibitory activities against RNA-dependent DNA polymerase of HIV-1. Compounds with sinapoyl moieties exhibited weaker activity than compounds with feruloyl moieties (Tewtrakul et al., 2002).

21.5.2.2 Wound Healing Activity

Extracts are reported to prevent oxidative damage in inflammation, lowering the oxidative stress markers. Lipid peroxidation (LPO) inflammatory markers like nitric oxide (NO) were lowered aiding wound healing process (Rahman et al., 2017).

21.5.2.3 Antispermatogenic Activity

Specific weight reduction in gonad of male rats seen by Gupta and Kachhawa et al. (2011) with no reduction in overall body weight after oral administration of stem bark methanolic extract. Grossly, the amount of total protein, glycogen and sialic acid content of the testes, epididymis, seminal vesicle, and ventral prostate were decreased. Overall spermatogenesis was depressed.

21.5.2.4 Antioxidant Activity

Dihydroquercetin, a flavanol found in the plant and a precursor of quercetin, has antioxidant activity and promising therapeutic properties in chronic inflammatory states, in cancer, cardiovascular and hepatic diseases (Tiukavkina et al., 1997; Weidmann, 2012; Oi et al., 2012). Cell suspension cultures of *T. peruviana* exhibit higher antioxidant capacity than the actual plant extract (Arias et al., 2016). Fruits exhibited maximum antioxidant capacity (Srivastava et al., 2012).

21.6 TOXIC RESPONSE

21.6.1 Studies on Yellow Oleander Seed Poisoning (YOP/YOSP) and Case Reports

Accidental and occasional cases of poisonings and suicidal attempts through consuming seeds are reported throughout the world (Ansford and Morris, 1981; Kakrani et al., 1981; Brewster, 1986; Zamani and Aslani, 2010; Mandal, 2012). Oleander resulted in several plant poisonings in children in Australia during 1972–78 reported by Shaw and Pearn (1979). Widespread use for suicides by consuming seeds of yellow oleander referred as YOSP occurred first in Jaffna, Sri Lanka between 1981 and 1983. (Saravanapavananthan, 1985). This happened after a publicity in newspaper reporting the consumption of seeds by two girls. After 15 years of that report, Eddleston reported that in some areas, especially in northern Sri Lanka, 40% of self-poisoning cases are linked to YOP particularly in youths, with an annual rate being 150 per 100,000 population (Eddleston et al., 2000b; Refer case reports presented in Table 4). This epidemic is a concern in India and Sri Lanka (Eddleston et al., 1999; Bose et al., 1999; Roberts et al., 2015).

Inhalation of smoke from burning plant parts (Senthilkumaran et al., 2011) or intake of milky juice or honey produced from oleanders are toxic (Osterloh et al., 1982). Seasonal variation results in different quantities of glycoside within various parts of the plant (Karawya et al., 1973). Tea produced from oleander leaves has also resulted in toxicity (Haynes et al., 1985). Prolonged high temperature treatments did not impair the toxicity of the plant extracts (Oji and Okafor, 2000). Flowering stages produced highest level of glycosides (Langford and Boor, 1996). The milky sap causes irritation to skin and may result in contact dermatitis in some individuals (Dorsey, 1962). In Sri Lanka, seeds are usually eaten uncrushed in chunks, whereas Indians are reported to crush the seeds before consuming. Crushing is found to increase cardiotoxicity (Bose et al., 1999; Aparna and Sharmila, 2017). Ingestion of dried mature seeds also increased the doses of glycosides found in body. In one case report, female patients with low BMI recorded significant increases in mortality compared to males (Gopalakrishnan, 2017). Insignificant correlation of the number of seeds ingested with severity and toxicity is reported but many authors have found positive correlation (Eddleston et al., 2000a). Anandhi et al. (2019) reported that even fractions of seed might result in increased severity of cardiotoxicity. Variability in cardiac glycoside concentration among seeds might be a factor. The seeds in the dry, hot season are associated with more instances of poisonings (Madhura et al., 2016). Time elapsed after ingestion, variable absorption, vomitus eliminating seeds after intake, physiological variation in cardiovascular responses, and comorbidities lead to unpredictable outcomes. These cardenolides recirculate in hepato-biliary circulation due to which patients were noted to develop significant cardiotoxicity after 2 days and even as late as after 92 hours (Bandara et al., 2010; Eddleston et al., 1999, 2000b). A smaller number of seeds fail to induce vomiting and get absorbed thoroughly with entero-hepatic recycling (Kanagasingam et al., 2019). Age appears to be a factor as people above 64 years have a 13.8 times risk of fatal outcome than people below 25 years (Eddleston et al., 2006).

Erratic pharmacokinetic behaviour is noted in blood contents of cardenolides observed through diagnosing anti-digoxin antibodies (Roberts et al., 2006). Glycosides are found in the blood beyond 50 hours post-ingestion. Treatment decisions are complicated due to several factors like nonuniformity in hospital setups, variation in toxic threshold, requirement of diagnostic tests, effective and specific treatment requirements, variable onset of toxicity, and long and delayed hospital transfers when required (Roberts et al., 2015). The case fatality for yellow oleander poisoning is around 10% in patients admitted in hospitals with limited resources or limited access to specific treatments and procedures, but as low as 3% in fully equipped tertiary care hospitals. A report from India had a mortality of 17.65% with significantly higher female mortality rate of 23.41% (Gopalkrishnan, 2017). Severity of cardiotoxicity varies unpredictably and frequent changes from mild to severe forms of toxicity occur (Anandhi et al., 2019). Definitive evaluation and staging are still not available. Monitoring and observing patients for 3–4 days after ingestion is recommended (Roberts et al., 2006; Anandhi et al., 2019).

21.6.2 Mechanism of Toxicity

The sodium and potassium gradients across the plasma membrane are used by animal cells for multitudes of physiological processes throughout the body like electrolyte homeostasis, maintaining membrane polarization, secretion, absorption, and pH regulation. Steroidal ring of the glycosides binds the alpha subunit of the Na^+/K^+-ATPase enzyme (Demiryurek and Demiryurek, 2005). Hydroxyl groups of the nucleus interact with the protein. Lactone ring is inserted within the transmembrane domain of the pump. The sugars face the extracellular side of the pump (Ogawa et al., 2009; Yatime et al., 2011). The pump operates to drive out Na^+ from cells which has entered in response to a depolarization wave. Inhibition of the pump leads to failure of Na^+ expulsion and K^+ import. (Refer pump 1 of Figure 21.3). This results in ionic imbalances in intracellular and extracellular compartments of our body. Na^+ gets overloaded in intracellular compartments, whereas K^+ levels outside the cell rise (Figure 21.3). The ionic load of Na^+ happens to affect the function of another ionic exchange mechanism (Refer pump 2 to the left of pump 1 in Figure 21.3). Na^+ overloaded cardiac cells cannot exchange Na^+ for calcium by the NA^+/Ca^{2+}-exchanger as it has to put in Na^+ inside cells where it is already overloaded. Now the exchanger failing to operate due to Na^+ overload Ca^{2+} ions cannot get out of the cells (Demiryurek and Demiryurek, 2005; Heard, 2004). This way calcium concentration in the cytosol increases in Na^+ overloaded conditions in cardiomyocytes. Calcium ions act as a second messenger and stimulate the Ryanodine channels (open configuration state of RYR2 isoform) in the sarcoplasmic reticulum (SR; left side of Figure 21.3) to release more calcium from the SR stores (Refer Figure 21.3; Sagawa et al., 2002). Ca^{2+} fluctuations inside the cell are responsible for transient inward currents. Two-third of these transient currents are attributed to the Na^+/Ca^{2+} exchanger mentioned above (Lederer and Tsien, 1976). Failure of another propagation of depolarization wave to contract the myocardium due to ionic imbalances mentioned affect the myocardium as well as the junctional tissues leading to bradycardia and conduction block.

21.6.2.1 Schematic Representation of Alterations in Ionic Distributions

Binding of cardenolides leads to inactivation of the pump-1, which affects Na^+/Ca^{2+} exchange pump-2.

Intracellular overload of Na^+ leads to loss of gradient; as a result, the exchanger pump extruding Ca^{2+} cannot operate, resulting in Ca^{2+} accumulation.

21.6.3 Symptoms and Diagnosis

The symptoms are similar to digitalis poisoning. A history of plant ingestion (with identification of plant parts if possible) presenting with features of toxicity, like nausea, vomiting, diarrhoea, dysrhythmias, and hyperkalaemia, dizziness, abdominal pain (Eddleston et al., 2000a; Bandara et al., 2010). All these features accompanied by having ECG abnormalities are likely due to oleander poisoning. Neurological tremor, drowsiness, ataxia, mydriasis, dysesthesias, numbness, burning sensation in mouth, buccal erythema, and mucus secretion are commonly reported symptoms (Barceloux, 2008; Haynes et al., 1985). Ventricular tachyarrhythmia and ventricular ectopic found in digoxin poisoning are relatively uncommon in YOP (Rajapakse, 2009).

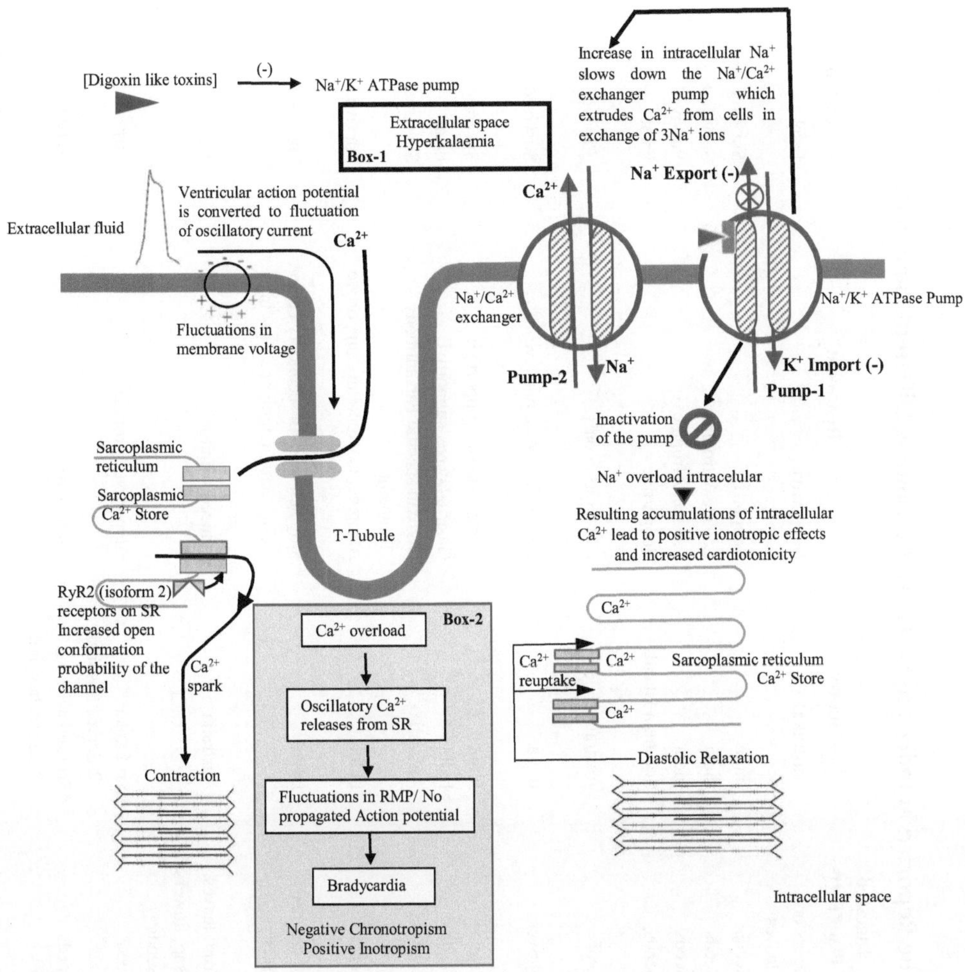

FIGURE 21.3 Mechanism of actions of oleander cardenolides on heart.

ECG remains a gold standard for detection and a primary diagnostic as well as predictor of outcome in YOP (Brewster, 1986; Saravanapavananthan and Ganeshamoorthy, 1988; Bose et al., 1999). Multiple and varying cardiac rhythm, frequent sinus pauses, with the absence of P waves, junctional escape (Shumaik et al., 1988), wide QRS complexes, tall peaked T waves, variable A-V blocks, i.e., first degree/second degree/complete atrioventricular block with marked bradycardia are characteristics of oleander poisoning. Flattening or inversion of the T-waves, ST depression over 2.5 mm sometimes deep and saucer shaped, Mobitz type-II AV conduction block that is rare in digoxin poisoning are seen to occur in oleander toxicity (Eddleston et al., 2000; Kelly and Smith, 1992). Another feature that is uncommon in oleander toxicity is atrial fibrillation and flutter. Ventricular arrhythmias are dangerous and do not respond to cardiopulmonary resuscitation, drugs, and cardioversion (Osterloh et al., 1982). Marked bradycardia and poor response to atropine are bad prognostic features (Saravanapavananthan, 1985). YOP findings in ECG resemble characteristic Brugada phenocopy ECG found in sudden death syndrome of young adults as reported by Gunaseelan et al. (2020).

21.6.4 Specific Detection

Cardenolides like thevetin and peruvoside cross react with digoxin and digitoxin antibodies which forms the basis of toxicological analysis. Measurement of the blood levels of these glycosides and

TABLE 21.4
Case Reports of Yellow Oleander Ingestion Continue to be Reported in India and Sri Lanka Consuming Different Plant Parts

	Key References	Case Description	Ingested Plant Parts	Country	Treatment	Outcome
#1	Ansford and Morris (1981)	3-year-old female	Seeds or leaves	Australia	Supportive	Fatal
*2	Kakrani et al. (1981)	12-year-old male	Seeds	India	Supportive	Survived
*3	Mallick et al. (1984)	26-year-old male	Seeds	India	Lavage, Atropine	Survived
*4	Mallick et al. (1984)	22-year-old male	Seeds	India	Lavage, Atropine	Survived
#5	Brewster (1986)	2 1/2-year-old male	Seeds	Solomon Islands	Supportive	Fatal
*6	Cheung et al. (1989)	17-year-old female	Leaf	Australia	Observation only	Survived
#7	Ahlawat et al. (1994)	32-year-old female	Flowers	India	Gastric lavage, supportive	Survived
		25-year-old male	Seeds	India	Gastric lavage, supportive	Survived
		28-year-old male	Seeds	India	Gastric lavage, supportive	Fatal
#8	Camphausen et al. (2005)	7-year-old female	Seeds (1)	Australia	Digoxin specific Fab antibody fragments	Survived
#9	Dwivedi et al. (2006)	30-year-old female	Seeds (5)	India	Atropine, isoprenaline, supportive	Survived
10	Gunaseelan et al. (2020)	17-year-old male	10 crushed seeds	India	Stomach wash in secondary centre, resuscitation, temporary pacing	Survived
	Case Series					
*11	Shaw and Pearn (1979)	10 male, 3 female ages	Fruit, leaves, sap, flowers, seeds	Australia	Induced vomiting	All survived
#12	Sreeharan et al. (1985)	20 cases	Seeds	Sri Lanka 2-degree hospital	Supportive, atropine	10%
#13	Saravanapavananthan and Ganeshamoorthy (1988)	170 cases	Seeds	Sri Lanka 2-degree hospital	Supportive	0.041

(*Continued*)

TABLE 21.4 (*Continued*)
Case Reports of Yellow Oleander Ingestion Continue to be Reported in India and Sri Lanka Consuming Different Plant Parts

	Key References	Case Description	Ingested Plant Parts	Country	Treatment	Outcome
*14	Saravanapavananthan and Ganeshamoorthy (1988)	114 female, 56 male	Seeds	Sri Lanka	Supportive	7 deaths
*15	Samal (1989)	7 cases	Seeds	India	Supportive	4 survived
#16	Samal (1990)	4 cases	Seeds	India	Supportive	All survived
*17	Saraswat (1992)	13 cases	Seeds	India	Fluids, dopamine, atropine	2 deaths
18	Zamani and Aslani (2010)	17 female, 4 male	Seeds	Iran	Supportive, transferred to coronary care unit, temporary pacemaker insertion and defibrillation	
#19	Bose et al. (1999)	300 cases	Seeds	India	Supportive	4.60%
#20	Eddleston et al. (1999)	79 cases	Seeds or leaves	Sri Lanka 2 degree hospital	Supportive	6.30%
#21	Eddleston et al. (2000b)	66 cases	Seeds	Sri Lanka, 2 and 3 degree hospital	Digoxin specific Fab antibody fragments and placebo	Selected sample, all survived
#22	Fonseka et al. (2002)	168 cases	Seeds	Sri Lanka 2 degree hospital	Atropine, isoprenaline, temporary cardiac pacing	3.60%
#23	de Silva et al. (2003)	401 cases	Seeds	Sri Lanka, 2 and 3 degree hospital	Activated charcoal, placebo, atropine	5.20%
#24	Eddleston et al. (2003)	194 cases	Seeds	Sri Lanka, 2 and 3 degree hospital	Digoxin specific Fab antibody fragments	3.10%
		381 cases	Seeds		Digoxin specific Fab antibody fragments and cardiac pacing	4.20%
		1,217 cases	Seeds		Supportive and cardiac pacing	8.60%
		279 cases	Seeds		Supportive and cardiac pacing	9.30%
		383 cases	Seeds		Supportive	10.90%
#25	Roberts et al. (2006)	104 cases	Seeds	Sri Lanka, 2 and 3 degree hospital	Activated charcoal (SD vs MD)	Selected sample, all survived

(*Continued*)

TABLE 21.4 (*Continued*)
Case Reports of Yellow Oleander Ingestion Continue to be Reported in India and Sri Lanka Consuming Different Plant Parts

	Key References	Case Description	Ingested Plant Parts	Country	Treatment	Outcome
#26	Eddleston (2008a, b)	549 cases	Seeds	Sri Lanka, 2 and 3 degree hospital	Single dose activated charcoal	4.70%
		541 cases	Seeds		Multiple does activated charcoal	4.20%
		555 cases	Seeds		No charcoal, supportive	4.30%
27	Lokesh and Arunkumar (April 2011–January 2013)	30 cases	Seeds	India	Supportive, significant cardiac arrhythmia patients shifted to Intensive Cardiac Care Unit	All survived
28	Mandal (2012)		Oleander poisoning	Nepal	Supportive, Atropine, Anti -digoxin Fab anti toxin	
29	Gopalakrishnan et al. (2017).	46 male and 55 female	Seeds/ Leaves/ Fruits	India	Resuscitation, Gastric lavage, supportive/Inotrope/orciprenaline/ atropine	18 deaths
30	Anandhi et al. (2019)	192 cases	Seeds	India	Gastric lavage, supportive	10 deaths
31	Karthik et al. (2020)	30 cases	Seeds	India	Gastric lavage, TPI, intravenous magnesium, oral activated charcoal	2 deaths
32	Sivakumar and Nithya (2018)	30 male 20 female	Crushed/ whole seeds/ leaves/Fruits	India	Drug/temporary pace making	2%
33	Aparna and Sharmila (2017)	22 males, 28 females.	Seeds	India	Gastric lavage, injection atropine, tablet Orciprenaline.	2 deaths

The case reports marked (*) are taken from Langford and Boor (1996); The case reports marked (#) are taken from Bandara et al. (2010). Other reports are taken from the respective authors.

detection through radioimmunoassay's and fluorescence immunoassays are achieved with the help of these antibodies (Smith et al., 1976; Osterloh et al., 1982; Dasgupta and Hart, 1997; Dasgupta et al., 2008). For sensitive medico-legal identification of criminal poisoning cases, more sensitive and direct methods like Liquid chromatography-electrospray tandem mass spectrometry (LC-MS/MS) methods can be used (Tor et al., 2005; Le Couteur and Fisher, 2002).

21.6.5 Treatment

History is to be recorded regarding time of intake, amount ingested, and time of appearance of symptoms in detail whenever possible. Treatment for YOP is based on symptoms and supportive care aimed specifically for poisoning patients. Bradyarrhythmia are a feature of YOP. Initial management should be based on the time elapsed since ingestion; as severity increases with time, immediate care should be on haemodynamic stability, hydration status, and checking for aspirations in severely vomiting patients. Fluid management with normal saline for dehydrated, hypotensive patients forms the first line of management with vomiting and diarrhoea (Rajapakse, 2009). Antiemetic drugs like metoclopramide or ondansetron are given. Gastric decontamination with lavage and induction of vomiting through syrups of ipecacuanha were practiced and are still followed in resource-constrained conditions, although hazards of these treatments may outweigh outcome (Eddleston et al., 2007). Similarly, activated charcoal is administered orally to make the glycosides adsorb on the surface of charcoals and get eliminated (de Silva et al., 2003; Eddleston et al., 2008a).

Commencement of continuous cardiac monitoring through ECG with rhythm strip is necessary. Marked electrolyte disturbances were found in patients who succumbed. Shock, hyperkalaemia (Osterloh et al., 1982), and prolonged vomiting with abdominal pain are features of severe toxicity and more likely to deteriorate. Extreme bradycardia (HR<40 bpm) and circulatory shock like manifestations with lowering of mean arterial pressure and more significantly lowering of diastolic pressure should be treated aggressively (Kanagasingam et al., 2019). Atropine boluses are used commonly to increase the heart rate. Bradycardia refractory to drug therapy is managed by Temporary cardiac pacing (TCP). TCP is required in severe bradycardia (<40 bpm) and if patients do not respond well to anticholinergics and sympathomimetics like isoprenaline. Hyperkalaemia is treated with insulin dextrose (Rajapakse, 2009; Roberts et al., 2016). Potassium levels should be checked at 6-hour interval. It is impossible to predict whether corrected hyperkalaemia would be beneficial. Hypokalaemia predisposes to more severe arrythmias. Marked cardiac arrythmias and electrolyte disbalances are reflective of severity of poisoning and can be correlated to higher glycoside concentration measure through Digoxin immunoassays. Higher levels of glycosides are found in severe cases compared to less severe cases. Patients who were discharged early have half the value of glycosidic concentration than that of the severely affected (Roberts et al., 2006). CG concentration is correlated with hyperkalaemia and conduction defects in SA and AV nodes (Roberts et al., 2016). Digoxin-specific antibody fragments effectively revert arrhythmias and hyperkalaemia (Shumaik et al., 1988), and they are considered as the specific antidote in oleander poisoning. Reduction in cardiac arrhythmias and correction of hyperkalaemia achieved result in a reduced mortality (Shumaik et al., 1988; Clark et al., 1991; Safadi et al., 1995; Eddleston et al., 2000b; Rajapakse, 2009). Digoxin-specific antibody fragment is very expensive and is still not affordable in developing countries like Sri Lanka and India (Eddleston et al., 2003).

Class I anti-arrhythmic agent like quinidine and calcium-channel blockers are contraindicated as they may increase digitalis concentrations. The use of beta-blockers may worsen heart block (Demiryurek, 2005; Rajapakse, 2009). Serum magnesium levels' disturbances are not correlated with intensity of poisoning and cardiac arrythmias (Eddleston et al., 2000a). However, both hypo- and hypermagnesemia should be checked and corrected. Intracellular calcium in YOP poisonings remains high due to the presence of cardiac glycosides which shift ECF calcium to the cell interior. Intravenous calcium is not generally administered to treat hyperkalaemia (Rajapakse, 2009).

Whether ventricular arrythmias can be treated with intravenous potassium is not answered yet. Further studies are needed to ascertain how magnesium affects toxicity and outcome. If pacing facility is not available, anticholinergics with other accelerator drugs are to be continued with increased vigilance for the development of fatal arrythmias. Debates persist on applications of multiple or single dose of activated charcoal (de Silva et al., 2003; Eddleston et al., 2008). The effects of atropine stay for a short time and repeated doses or infusion of atropine may be required. Atropine reduces gastric motility and theoretically increases absorption of ingested toxins. Features of atropine toxicity should be under careful observation. Higher dose of atropine and isoprenaline may predispose to sudden adrenergic activity and tachyarrhythmias which are very difficult to control. Titrating the dose of atropine for a maintenance heart rate of 60–90 bpm is safe.

More studies are required to ascertain whether TCP confers a survival benefit (Fonseka et al., 2002). TCP has its own hazards and high-class expertise levels are required to minimize errors. Direct myocardial stimulation of pacing wires in TCP may trigger fatal tachyarrhythmias (Eddleston et al., 2000; de Silva et al., 2003). The role of atropine doses contributing to reducing mortality or lowering use of TCP is also a question unresolved. Lidocaine is generally used to treat ventricular tachyarrhythmias. Electrical cardioversion is best avoided unless absolutely required in treating patients with resistant ventricular tachycardias. When cardioversions are used, low energy levels are better in digitalis and oleander toxicity. Large volume of distribution of cardiac glycoside poisoning renders haemodialysis ineffective.

Gastric lavage although hazardous should be started early along with activated charcoal ingestions (Rajapakse, 2009; de Silva et al., 2003; Roberts et al., 2006). Studies conducted to find efficacy of doses (single or multiple doses) of activated charcoal to be practiced are inconclusive (de Silva et al., 2003; Eddleston et al., 2008a). In clinical practice, single-dose activated charcoal is being used widely in Sri Lanka but sparsely in India. Gopalakrishnan et al. (2017) found that all patients who received gastric lavage within 1 hour of consumption suffered less with low fatal outcomes than those who received lavage after 2 hours. Early gastric lavage to all patients irrespective of number of seed ingestion and early initial dose of activated charcoal is important (Aparna and Sharmila, 2017; Kanagasingam et al., 2019). Considering hazards associated with lavage, further studies are needed to assess its benefit specially after a long delay post ingestion of seeds. However, in clinical settings in environments with limited resources, all types of decontamination through lavage appear to be required, as does the use of activated charcoal to prevent hepato-biliary recycling. Availability of activated charcoal, temporary pacing facilities, and anti-digoxin Fab are important for managing and treating serious cases of YOP.

21.6.6 Animal Toxicity

All parts of the oleander plant are readily accessible to animals and children for accidental poisoning (Langford and Boor, 1996) as these plants exhibit shrubby characteristics. Leaves, fruit, and flower are all accessible from ground (Kyerematen et al., 1985). Leaves, flowers, and seed-pods frequently litter the ground surrounding these plants leading to oleander contaminated feed to animals (Middleton and Chen, 1936; Szabuniewicz et al., 1972). Animal exposure has been extensively reviewed by Langford and Boor (1996). Contamination of fodder with leaves leads to cattle poisonings (Liu, 1957; Mahin et al., 1984; Rexakhani and Maham, 1992). Lehmann (1984) reported toxic responses in dogs and cats due to ingestion of leaves. Dogs have ATPase similar to humans and are vulnerable to oleanders. Monkeys were administered oleander leaf extract in an experimental study (Schwartz et al., 1974) and they were found to survive in low doses. Szabuniewicz et al. (1972) conducted experimental studies on monkeys, dog, cat, rat, mouse, and chicken and determined cumulative toxic doses for each species. Alfonso et al. (1994) and Bardosi (1963) reported deaths in geese by seed and leaf contamination respectively. Even toxic responses to bear ingesting leaves are reported in Ratigan (1921).

21.7 TRADITIONAL USES

Plants from Apocynaceae family are already documented in traditional books of ethnomedicinal practices like Unani, Ayurveda, Siddha, and Homeopathy for their broad spectrum formulations (Bhadane et al., 2018). *T. peruviana* has various ethnomedicinal, industrial, and pharmacological application. Extracts from every part (bark, leave, fruit, flower, and seed) of this plant have several ethnomedicinal uses. Reports of seed intake for suicidal and homicidal purposes in Bengal, India are found (Chen and Chen, 1934b).

In India, different parts of the plant were used as purgatives, as an emetic, and intermittent fever (febrifuge) remedies (Xabregas, 1950). Traditionally, extracts of yellow oleander were used as a household remedies to dissolve haemorrhoids and tumours, for ulcers and scabies, as an abortifacient, in treating congestive heart failure, malaria, leprosy, venereal disease, as a diuretic in patients with heart failure slowing ventricular response in patients with atrial fibrillation, for the treatment of GI and inflammatory diseases and skin tumours (El-Sawi et al., 2020; Osterloh et al., 1982; Bandara et al., 2010; Oji and Okafor, 2000). Seed kernels were used as an emetocathartic in Antilles (Weitz and Boulay, 1923), as treatment of nervous shock in Haiti, as a fish poison in Brazil (Chen and Chen, 1934b), as cattle poison in Bombay (Bhattacharya and Ayyar, 1927) and also extensively used in the preparation of the *ipos* or Malayan arrow poisons (Weitz and Boulay, 1923). In Mexico, the seeds are widely used as supplement to reduce body weight (González-Stuart and Rivera, 2017) and the milky juice is used to treat mange (itch mites; McLaughlin et al., 1980). In addition, the crushed kernels of yellow oleander act against insect infestations to protect stored seeds (Pandey et al., 1977). This plant commonly known as Indrajav is used in treating liver ailments in South India (Ramchandra et al., 1993). Extracts from various plant parts like bark, leaves, roots, and seeds have been applied to cure bladder stones, dropsy, oedema, insomnia, constipation, nausea, cardiac disorders, fever, measles, rheumatism, and ulcers (Joshi, 2000; Pullaiah, 2006). The seed oil is applied externally to treat skin infections and in healing acne. Root extract infusion is used to treat snakebites.

The bark and seeds are used to poison rats, and the use of plant for criminal purposes has historical records. In India, leaf extracts are used to kill lice (Atal and Kapur, 1977). In ethno-veterinary medicine, grinded leaves of *T. peruviana* have been used for treatment of infection and wounds, ring worms, and tumours (Rahman et al., 2017). It is held that that these cardiac glycosides were used for intentional contamination of food and water supplies in criminal poisonings (Krenzelok, 2003; Jurica et al., 2019; Le Couteur and Fisher, 2002).

21.8 OTHER POTENTIAL USES

The plant has other uses recorded. Some of these uses depicting the bioactivities that the plant has are pesticidal, rodenticidal, anti-termite, molluscicidal, larvicidal, piscicidal, antifungal, and antimicrobial activities against plant pathogens (Oji and Okafor, 2000; Gata-Goncalves et al., 2003). Flower extracts containing the flavonoids and anthocyanins that exhibited sharp colour changes in points of titrations at different pH can be used as a compound indicator (Singh et al., 2011).

21.8.1 Industrial Uses

Each fruit of yellow oleander contains one to four seeds in its kernel and these seeds are very rich in oil content (60%–65%; Oluwaniyi et al., 2007). Yellow oleander oil has been explored to be used in paints, surface coating industry, adhesive and binder for composite (Iodine value 71.4 g I_2/100 g; Obasi et al., 1990). It has the potential to synthesize alkyd resins which is a biodegradable, renewable, and low-cost polymer used in coating industry. Saponification value of the oil is 121 mg KOH/g which indicates its use to produce soaps. It has a peroxide value of 4.8 meq.O_2/kg, which indicates rancidity (Bora et al., 2014).

21.8.2 Use in Biodiesel

Different types of trials of transesterification of the yellow oleander seed oil with the help of heterogeneous catalyst (Deka and Basumatary, 2011) are underway to yield biodiesel. Fuel properties have been explored in comparison with Indian Automotive Diesel Fuels norms and a cetane number 61.5 of the oil is found indicating that it is a fuel with ignition and combustion quality. Blending of yellow oleander methyl and ethyl esters is effective as a cetane and lubricity improver (Sarmah and Deka, 2019). The seed oil has the properties like renewability, biodegradability, nontoxicity, carbon neutrality, low kinematic viscosity, good density, acid value, calorific value and makes it a viable feedstock for biodiesel industries (Yarkasuwa et al., 2013; Deka and Basumatary, 2011).

21.8.3 Anti-Termite Activity

The anti-termite activity of yellow oleander might be a significant application in the wood and furniture industry. An oil-based paint extracted from yellow oleander kernel seeds displayed the repellent action against *Microtermes* spp. in a concentration dependent manner. Pure oleander oil exhibited repellent action in protecting timber. Oleander paint can be used against microbes (Kareru and Keriko et al., 2010).

21.8.4 Molluscicidal and Pesticidal Activity

It is also used in the field to protect crops from insects and microbes. The slug *Laevicaulis alte (Férussac)* and the snail *Achatina fulica Bowdich* are known as most serious Agri-horticultural pests in Indo-Pacific countries. Panigrahi and Raut (1994) reported that extracts from kernels can be used as naturally occurring molluscicides which are relatively safer for environment. 20% concentration of the extract was effective for killing with 100% mortality within a short period of time. The slugs and snails became immobile and secrete mucus until their death (Panigrahi and Raut, 1994).

21.8.5 Antimicrobial Activity

Benzoic acid and oxo-Benzoate, phytochemicals present in active fraction of leaf extract, had antimicrobial activity (Meena et al., 2021). The secondary metabolite thevetin acts against herbivory due to its lethality (cardiac arrest of insects) so it is used as insecticidal agent and also shows antibacterial activity against pathogenic bacteria (Basile et al., 1993).

21.8.6 Antifungal Activity

Terpenes isolated from seed extracts of *T. peruviana* showed a strong fungicidal activity against *Phytophthora megakarya*. It has the properties of biocide mainly acting against black pod disease of *Theobroma cacao* (Ambang et al., 2010). Growth of different strains of *Phytophthora megakarya* was effectively inhibited by crude extracts of ethyl acetate. Seed methanolic crude extract exerted greater inhibitory effect against particular strains (Ibiyemi et al., 2002). Pulegone had antifungal photoactivity against *Cladosporium cucumerinum* (Gata-Goncalves et al., 2003). Leaf extract of *T. peruviana* inhibits cell wall synthesis. Withdrawal of cytoplasm led to death of the hyphae in mycelium (Sasidharan et al., 2011).

21.8.7 Use in Nanoparticles

Amphiphilic drug molecules, including CGs, can be incorporated in nanosized micelles with essential circulatory properties making it biocompatible to specifically get delivered in tumour tissues

(Biswas et al., 2016). These micelles have enhanced permeability and retention (Wang et al., 2014; Liu et al., 2009). Leaf extracts of yellow oleander were shown to act as a capping agent for Cadmium Telluride nanoparticles synthesis which has applications in measurement of power conversion efficiency in dye-sensitized solar cell (Bera et al., 2018).

21.9 FUTURE PERSPECTIVES

To beautify the road side views and maintain ecological balance due to its tolerance to extreme environmental conditions, yellow oleander is preferred to be grown along roadsides in many countries. State initiatives about protecting the plant as well the awareness of its potential toxicity will help further to understand the mysteries that this plant holds. The uses that hold promise are further industrial applications regarding its anti-termite properties in treatment and processing of plyboards in wood industries and its use as an anti-insecticidal, bio-pesticidal, and biodiesel.

Yellow oleander has been reported to be used all around the world despite its toxicity in all traditional medicines. These plants constituted the earliest healing options when there were no purified drugs or purified compounds. Mostly, the use of herbal products comes from the belief that all natural products are safe. Different herbal products with aggressive marketing are found in the internet and are popular along the U.S. Mexico border where they are sold as herbal food supplements for weight loss targeting the increasing obese population, and such herbal food supplements do not require a prescription. The seed is known as India seed (Semilla de la India). Mexican Herbal Pharmacopoeia states that yellow oleander is a highly toxic plant. Rampant use as a weight losing agent under commercial names like Easy figure and Capslim, which contain a misnomer compound named troncomin, is reported (González-Stuart and Rivera, 2017). Mexican authorities have issued warning against the use of such products. However, the use of seeds cannot be controlled due to its use in ethnic dances and are available in the market. These herbal products are mixed with various other compounds like antibiotics, hormones, steroids, and diuretics to add efficacy in formulations. Mislabelling and wrong identification of plant species that are traditionally used also add to the risks of use of herbal formulations.

Suicidal reports consuming parts of the plant are a matter of concern in India and Sri Lanka. Although anti-digoxin antibodies have the potential to neutralize the glycosides and thereby reduce toxicological burdens, they are prohibitively expensive. Activated charcoal should be made available in areas where suicides are rampant. A cost effective alternative or an initiative to reduce the price of anti-digoxin antibody (like other antibodies used to treat snake venom) may reduce the expenses associated with transfer and the infrastructural requirements associated with treatment referrals in severe cases. Fructose-6-Disphosphate (FDP) was used as antidote in canine model and effectively prevented hyperkalaemia with improvement in haemodynamic and reduced arrythmias (Markov et al., 1997). Gawarammana et al. (2010) reported a small phase II study of FDP in YOP patients in 2006 where the results were encouraging. Reports of FDP's use in clinical trial are there but results are awaiting. Anticalins, a non-biologic alternative with high binding affinity for digoxin, are effective in reducing free plasma digoxin in rats but human data are currently lacking (Roberts et al., 2015; Eyer et al., 2012). Traditional medicinal uses of this plant mentioned here in herbal formulations should be validated after clinical trials proving the efficacy of formulations as herbal drugs.

NOTES

Subhrajit Banerjee, Department of Physiology, Surendranath College, Kolkata, India
Senjuti Banerjee, Department of Microbiology, St. Xavier's College, Kolkata, India
Kasturi Sarkar, Department of Microbiology, St. Xavier's College, Kolkata, India
Parames C. Sil, Department of Molecular Medicine, Bose Institute, Kolkata, India

REFERENCES

Abe, F., Chen, R. F., & Yamauchi, T. (1996). Dinormonoterpenoids and their apiosylglucosides from *Thevetia peruviana. Phytochemistry*, 43, 161–163.

Abe, F., Iwase, Y., Yamauchi, T., Yahara, S., & Nohara T. (1995). Flavonol sinapoyl glycosides from leaves of *Thevetia peruviana. Phytochemistry*, 40, 577–581.

Abe, F., Yamauchi, T., Yahara, S., & Nohara, T. (1994). Glycosides of 19- formylthevetiogenin and 5a-thevetiogenin from Thevetia neriifolia. *Phytochemistry*, 37, 1429–1432.

Abe, F., Yamauchi, T., Yahara, S., & Nohara, T. (1995). Minor iridoids from *Thevetia peruviana. Phytochemistry*, 38, 793–794.

Abe, F., Yamauchi, T., & Wan, A. S. C. (1992). Cardiac glycosides from the leaves of Thevetia neriifolia. Phytochemistry. *The International Journal of Plant Biochemistry*, x; 31(9), 3189–3193.

Agrawal, A., Petschenka, G., Bingham, R., Weber, M., & Rasmann, S. (2012). Toxic cardenolides: Chemical ecology and coevolution of specialized plant-herbivore interactions. *The New Phytologist*, 194, 28–45. DOI: 10.1111/j.1469–8137.2011.04049.x.

Ahlawat, S.K., Agarwal, A.K., & Wadhwa, S. (1994). Rare poisoning with Cerebra thevetia (yellow oleander): a report of three cases. *Tropical Doctor*, 37–38.

Ahmed F., & Ishibashi M. (2016). Bio-active natural products with TRAIL-resistance overcoming activity. *Chemical and Pharmaceutical Bulletin*, 64(2), 119–127.

Alfonso, H. A., Sanchez, L. M., Merino, N., & Gomez, B. C. (1994). Intoxication due to Nerium oleander in geese. *Veterinary and Human Toxicology*, 36, 47.

Ali, M., Ravinder, E., & Ramachandram, R. (2000). New ursane-type triterpenic esters from the stem bark of Thevetia peruviana. *Die Pharmazie*, 55(5), 385–389.

Alvarado-Cárdenas, L. O., & Ochoterena, H. (2007). A phylogenetic analysis of the cascabela–Thevetia species complex (plumerieae, apocynaceae) based on morphology. *Annals of the Missouri Botanical Garden*, 94(2), 298–323.

Alvarado-Cárdenas, L. O., Villaseñor, J. L., López-Mata, L., Cadena, J., & Ortiz, E. (2017). Systematics, distribution and conservation of Cascabela (Apocynaceae: Rauvolfioideae: Plumerieae) in Mexico. *Plant Systematics and Evolution*, 303(3), 337–369. DOI: 10.1007/s00606-016-1375-6.

Ambang, Z., Ngoh Dooh, J. P., Essono, G., Bekolo, N., Chewachong, G., & Asseng, C. C. (2010). Effect of Thevetia peruviana seeds extract on in vitro growth of four strains of Phytophthora megakarya. *Plant Omics*, 3(3), 70–76.

Amend, N., Worek, F., Thiermann, H., & Wille, T. (2021). Investigation of cardiac glycosides from oleander in a human induced pluripotent stem cells derived cardiomyocyte model. *Toxicology Letters*, 350, 261–266. DOI: 10.1016/j.toxlet.2021.07.020.

Anandhi, D., Vinay, R. P, Kadhiravan, T., Soundaravally, R., & Prakash Raju, K. N. J. (2019). Cardiac arrhythmias, electrolyte abnormalities and serum cardiac glycoside concentrations in yellow oleander (Cascabelathevetia) poisoning—A prospective study. *Clinical Toxicology*, 57(2), 104–111. DOI: 10.1080/15563650.2018.1499930.

Ansford, A. J., & Morris, H. (1981). Fatal oleander poisoning. *The Medical Journal of Australia*, 1(7), 360–361.

Aparna, S., & Sharmila, M. (2017). Yellow oleander seed poisoning—A profile. *IOSR Journal of Dental and Medical Sciences*, 16, 64–71.

Arias, J. P., Zapata, K., Rojano, B., & Arias, M. (2016). Effect of light wavelength on cell growth, content of phenolic compounds and antioxidant activity in cell suspension cultures of Thevetia peruviana. *Journal of Photochemistry & Photobiology, B: Biology*, 163, 87–91.

Atal, C. K., & Kapur, B. N. (1977). *Cultivation and utilization of aromatic plants*. Council of Scientific and Industrial Research, Jammu-Tawi: Regional Research Laboratory. Edited by C.K. Atal & B.M. Kapur. Companion volume to Cultivation and utilization of medicinal plants. New Delhi, India, p. 568.

Bandara, V., Weinstein, S. A., White, J., & Eddleston, M. (2010). A review of natural history, toxinology, diagnosis and clinical management of Nerium oleander (common oleander) and Thevetia peruviana (yellow oleander) poisoning. *Toxicon,* 56, 273–281.

Barceloux, D. G. (2008). *Medical Toxicology of Natural Substances: Foods, Fungi, Medicinal Herbs, Plants, and Venomous Animals*. John Wiley & Sons Inc. Hoboken, NJ.

Bardosi, Z. (1963). Intoxications caused by oleander leaves. *Magy Allatorv Lap*, 18, 361.

Bashir, A., Hamburger, M., Gupta, M. P., Solis, P. N., & Hostettmann, K. (1991). Flavonol glycosides from Monnina sylvatica. *Phytochemistry*, 30, 3781–3784.

Basile, A., Giordano, S., & Castaldo-Cobianchi, R. (1993). Antibiotic activity in Thevetia Netifolia Juss and Thevetia Peruviana K. Schum (Apocinaceae). *Pharmacolgical Research*, 27(1).

Ben Sghaier, M., Pagano, A., Mousslim, M., Ammari, Y., Kovacic, H., & Luis, J. (2016). Rutin inhibits proliferation, attenuates superoxide production and decreases adhesion and migration of human cancerous cells. *Biomedicine & Pharmacotherapy,* 84, 1972–1978.

Bera, S. R., & Saha, S. (2018). Biosynthesis and characterization of Thevetia peruviana leaf extract capped CdTe nanoparticles in photoconductive and photovoltaic applications. *Materials Today: Proceedings*, 5, 3476–3485.

Bhadane, B. S., Patil, M. P., Maheshwari, V. L., & Patil, R. H. (2018). Ethnopharmacology, phytochemistry, and biotechnological advances of family Apocynaceae: A review. *Phytotherapy Research*, 32, 1–30.

Bhatia, M. L., Manchanda, S. C., & Roy, S. B. (1970). Haemodynamic Studies With Peruvoside In Human Congestive Heart Failure. *The British Medical Journal*, 3(5725), 740–743. http://www.jstor.org/stable/20385306

Bhattacharya, R., & Ayyar, R. P. (1927). Oils and fats from the seeds of Indian forest plants. Part VIII. The oil from the seeds of Thevetia Nerifolia (Juss.). *Journal of the Indian Institute of Science*, 10A, 15.

Biswas, S., Kumari, P., Lakhani, P. M., & Ghosh, B. (2016). Recent advances in polymeric micelles for anticancer drug delivery. *European Journal of Pharmaceutical Sciences*, 83, 184–202.

Blas, M. C., Bull. Acad. ray. m&d. Belgique, 2, 745 (1868).

Bora, M. M., Gogoi, P., Deka, D. C., & Kakati, D. K. (2014). Synthesis and characterization of yellow oleander (Thevetia peruviana) seed oil-based alkyd resin. *Industrial Crops and Products*, 52, 721–728. DOI: 10.1016/j.indcrop.2013.11.012.

Bose, T. K., Basu, R. K., Biswas, B., De, J. N., Majumdar, B. C., & Datta, S. (1999). Cardiovascular effects of yellow oleander ingestion. *Journal of Indian Medical Association*, 97, 407–410.

Brewster, D. (1986). Herbal poisoning: A case report of a fatal yellow oleander poisoning from the Solomon Islands. *Annals of Tropical Paediatrics,* 6, 289–291.

Camphausen, C., Haas, N.A., & Mattke, A.C. (2005). Successful treatment of oleander intoxication (cardiac glycosides) with digoxin-specific Fab antibody fragments in a 7-year-old child: case report and review of literature. *Zeitschrift für Kardiologie* 94, 817–823.

Chen, K. K., & Chen, A. L. (1934a). The constituents of be-still nuts, thevetia Neriifolia. *Journal of Biological Chemistry*, 105, 2.

Chen, K. K., & Chen, A. L. (1934b). The action of crystalline the vetin, a cardiac Glucoside of thevetia neriifolia. *Journal of Pharmacology and Experimental Therapeutics*, 51, 23.

Cheung, K., Hinds, J. A., & Duffy, P. (1989). Detection of poisoning by plant-origin cardiac glycoside with the Abbott TDx analyzer. *Clinical Chemistry* 25, 295–297.

Clark, R. F., Selden, B. S., & Curry, S. C. (1991). Digoxinspecific Fab fragments in the treatment of oleander toxicity in a canine model. *Annals of Emergency Medicine,* 20, 1073–1077.

Cruz, A., Guzmán, A., Iriarte, J., Medina, R., Muchowski, J. M., & Maddox, M. L. (1979). 18,20-Oxido-20,22-dihydroneriifolin, an unusual oxygenated cardenolide. *The Journal of Organic Chemistry,* 44, 3511–3515.

Dasgupta, A., & Hart, A. P. (1997). Rapid detection of oleander poisoning using fluorescence polarization immunoassay for digitoxin. Effect of treatment with digoxin-specific Fab antibody fragment (ovine). *American Journal of Clinical Pathology,* 108, 411–416.

Dasgupta, A., Risin, S. A., Reyes, M., & Actor, J. K. (2008). Rapid detection of oleander poisoning by Digoxin III, a new Digoxin assay: Impact on serum Digoxin measurement. *American Journal of Clinical Pathology,* 129, 548–553.

de Silva, H. A., Fonseka, M. M. D., Pathmeswaran, A., Alahakone, D. G. S., Ratnatilake, G. A., Gunatilake, S. B., Ranasinha, C. D., Lalloo, D. G., Aronson, J. K., & de Silva, H. J. (2003). Multiple-dose activated charcoal for treatment of yellow oleander poisoning: A single-blind, randomised, placebo-controlled trial. *Lancet*, 361, 1935–1938.

Deka, D. C., & Basumatary, S. (2011). High quality biodiesel from yellow oleander (Thevetia peruviana) seed oil. *Biomass Bioenergy*, 35, 1797–1803.

Demiryurek, A. T., & Demiryurek, S. (2005). Cardiotoxicity of digitalis glycosides: Roles of autonomic pathways, autacoids and ion channels. *Autonomic & Autacoid Pharmacology,* 25, 35–52.

Dorsey, C. S. (1962). Plant dermatitis in California. *California Medicine,* 96, 412–413.

Dwivedi, S., Rajpal, S., Narang, S., (2006). Cardiotoxic manifestations of yellow oleander (Thevetia nerifolia) poisoning and its treatment: a case report. *Indian Heart Journal*, 58, 450–451.

Eddleston, M., Ariaratnam, C. A., Meyer, W. P., Perera, G., Kularatne, A. M., Attapattu, S., Sheriff, M. H., & Warrell, D. A. (1999). Epidemic of self poisoning with seeds of the yellow oleander tree (Thevetia peruviana) in northern Sri Lanka. *Tropical Medicine & International Health*, 4, 266–273.

Eddleston, M., Ariaratnam, C. A., Sjostrom, L., Jayalath, S., Rajakanthan, K., Rajapakse, S., Colbert, D., Meyer, W. P., Perera, G., Attapattu, S., Kularatne, S. A., Sheriff, M. R., & Warrell, D. A. (2000a). Acute yellow oleander (Thevetia peruviana) poisoning: Cardiac arrhythmias, electrolyte disturbances, and serum cardiac glycoside concentrations on presentations to hospital. *Heart,* 83, 301–306.

Eddleston, M., Dissanayake, M., Sheriff, M. H. R., Warrell, D. A., & Gunnell, D. (2006). Physical vulnerability and fatal self-harm in the elderly. *The British Journal of Psychiatry,* 189, 278–279.

Eddleston, M., Juszczak, E., Buckley, N. A., Senarathna, L., Mohamed, F., Dissanayake, W., Hittarage, A., Azher, S., Jeganathan, K., Jayamanne, S., Sheriff, M. H. R., & Warrell, D. A. (2008a). Multiple-dose activated charcoal in acute self-poisoning: A randomised controlled trial. *Lancet*, 371, 579–587.

Eddleston, M., Juszczak, E., & Buckley, N. A. (2008b). Multiple-dose activated charcoal in yellow oleander poisoning. *Lancet*, 371, 2171–2172.

Eddleston, M., Rajapakse, S., Rajakanthan, K., Jayalath, S., Sjostrom, L., Santharaj, W., Thenabadu, P. N., Sheriff, M. H., & Warrell, D. A. (2000b). Anti-digoxin Fab fragments in cardiotoxicity induced by ingestion of yellow oleander: A randomised controlled trial. *Lancet*, 355, 967–972.

Eddleston, M., Senarathna, L., Mohamed, F., Buckley, N., Juszczak, E., Sheriff, M. H. R., Ariaratnam, A., Rajapakse, S., Warrell, D., & Rajakanthan, K. (2003). Deaths due to absence of an affordable antitoxin for plant poisoning. *Lancet*, 362, 1041–1044.

Eddleston, M., & Warrell, D. A. (1999). Management of acute yellow oleander poisoning. *Quarterly Journal of Medicine,* 82, 483–485.

Eddleston, M., Haggalla, S., Reginald, K., Sudarshan, K., Senthilkumaran, M., Karalliedde, L., Ariaratnam, A., Sheriff, M. H. R., Warrell, D. A., & Buckley, N. (2007). The hazards of gastric lavage for intentional self-poisoning in a resource poor location. *Clinical Toxicology*, 45, 136–43.Eddleston, M., Juszczak, E., & Buckley, N. A. (2008b). Multiple-dose activated charcoal in yellow oleander poisoning. *Lancet*, 371, 2171–2172.

El-Sawi, A. S., Maamoun, A. A., Salama, H. A., & Farghaly, A. A. (2020). Chemical profiling of Thevetia peruviana leaves cytotoxic active extracts enhanced by microemulsion formulation. *Bulletin of the National Research Centre*, 44, 93. DOI: 10.1186/s42269-020-00339-3.

Eyer, F., Steimer, W., Nitzsche, T., Jung, N., Neuberger, H., Muller, C., Schlapschy, M., Zilker, T., & Skerra, A. (2012). Intravenous application of an anticalin dramatically lowers plasma digoxin levels and reduces its toxic effects in rats. *Toxicology and Applied Pharmacology*, 263, 352–359.

Feng, Q., Leong, W. S., Liu, L., & Chan, W. (2016). Peruvoside, a cardiac glycoside, induces primitive myeloid leukemia cell death. *Molecules*, 21, 534. DOI: 10.3390/molecules21040534.

Fonseka, M. M., Seneviratne, S. L., de Silva, C. E., Gunatilake, S. B., & de Silva, H. J. (2002). Yellow oleander poisoning in Sri Lanka: Outcome in a secondary care hospital. *Human & Experimental Toxicology*, 21(6), 293–295.

Foye, W. O. (1989). *Principles of Medicinal Chemistry*, 3rd ed. Lea & Febiger. Philidelphia, PA.

Gata-Goncalves, L., Nogueira, J. M. F., Matos, O., & Bruno de Sousa, R. (2003). Photoactive extracts from Thevetia peruviana with antifungal properties against Cladosporium cucumerinum. *Journal of Photochemistry and Photobiology B: Biology*, 70, 51–54.

Gawarammana, I., Mohamed, F., Bowe, S. J., Rathnathilake, A., Narangoda, S. K., Azher, S., Dawson, A. H., & Buckley, N. A. (2010). Fructose-1, 6-diphosphate (FDP) as a novel antidote for yellow oleander-induced cardiac toxicity: A randomized controlled double blind study. *BMC Emergency Medicine*, 10, 15. DOI: 10.1186/1471-227X-10-15.

Ghatak, N., & Pendse, G. P., Bull. Acad. SC. United Provinces Agra, Oudh, India, 2, 259 (1933).

González-Stuart, A., & Rivera, J. O. (2017). Yellow oleander seed, or "Codo de Fraile" (Thevetia spp.): A review of its potential toxicity as a purported weight-loss supplement. *Journal of Dietary Supplements*, 15, 352–364. DOI: 10.1080/19390211.2017.1353565.

Gopalakrishnan, S. K., Kandasamy, S., Isaac, B., Jayasankar C., & Chandru, C. (2017). Oleander toxicity—The clinical spectrum and mortality predictors: An observational study. *Internet Journal of Medical Update*, 12(1), 4–8. DOI: 10.4314/ijmu.v12i1.2.

Gunaseelan, R., Sasikumar, M., Aswin, K., Dhar, S., Balamurugan, N., & Pillai, V. (2020). Brugada phenocopy induced by consumption of yellow oleander seeds—A case report. *Journal of Electrocardiology*, 62, 107–109. DOI: 10.1016/j.jelectrocard.2020.07.014.

Gupta, R., Kachhawa, J. B., Gupta, R. S., Sharma, A. K., Sharma, M. C., & Dobhal, M. P. (2011). Phytochemical evaluation and antispermatogenic activity of Thevetia peruviana methanol extract in male albino rats. *Human Fertility (Cambridge, England),* 14(1), 53–59. DOI: 10.3109/14647273.2010.542230.

Haux, J. (1999). Digitoxin is a potential anticancer agent for several types of cancer. *Medical Hypotheses*, 53, 543–548.

Haynes, B. E., Bessen, H. A., & Wightman, W. D. (1985). Oleander tea: Herbal draught of death. *Annals of Emergency Medicine,* 14, 350–353.

Heard, K. 2004. Digoxin and therapeutic cardiac glycosides. In: Dart, R. C. (Ed.), *Medical Toxicology*, 3rd ed. Lippincott Williams & Wilkins. Amsterdam, The Netherlands.

Herrera, J. (1991). The reproductive biology of a riparian Mediterranean shrub, Nerium oleander L. (Apocynaceae). *Botanical Journal of the Linnean Society,* 106(2), 147–172. DOI: 10.1111/j.1095-8339.1991.tb02289.x.

Husemann, T., & König, A. Arch. ezp. Path. u. Pharmakol., 6, 228 (1876).

Ibiyemi, S. A., Fadipe, V. O., Akinremi, O. O., & Bako, S. S. (2002). Variation in oil composition of Thevetia peruviana Juss "Yellow oleander" fruit seed. *Journal of Applied Sciences and Environment Management,* 6, 61–66.

Jacobs, W. A. (1933). The chemistry of the cardiac glucosides. *Physiological Reviews*, 13, 222–245.

Jolad, S. D., Hoffmann, J. J., Cole, J. R., Tempesta, M. S., & Bates, R. B. (1981). 3'-O-Methylevomonoside: a new cytotoxic cardiac glycoside from Thevetia ahouia A. DC (Apocynaceae). *Journal of Organic Chemistry*, 46, 1946–1947.

Joshi, S. G. (2000). *Medicinal Plants*. Oxford and IBH Publishing Company Pvt. Ltd. New Delhi, India.

Jurica, K., Vrdoljak, J., & Karaconji, I. B. (2019). Food defence systems as an answer to food terrorism. *Arhiv za higijenu rada i toksikologiju,* 70(4), 232–255.

Kakrani, A. L., Rajput, C. S., Khandare, S. K., & Redkar, V. E. (1981). Yellow oleander seed poisoning with cardiotoxicity; a case report. *Indian Heart Journal*, 33, 31–33.

Kanagasingam, A., Francis, G. R., Komagarajah, B., Ladchumanan, D., & Sivapramyan, A. (2019). Impact of severe yellow oleander poisoning on cardiac function and hemodynamics. *Journal of Clinical Toxicology,* 9, 423.

Karawya, M. S., Balbaa, S. I., & Khayyal, S. E. (1973). Estimation of cardenolides in Nerium oleander. *Planta Medica,* 23, 70–73.

Kareru, P. G., Keriko, J. M., Kenji, G. M., & Gachanja, A. N. (2010). Anti-termite and antimicrobial properties of paint made from Thevetia peruviana (Pers.) Schum. oil extract. *African Journal of Pharmacy and Pharmacology,* 4(2), 87–89.

Karthik, G., Iyadurai, R., Ralph, R., Prakash, V., Abhilash, K. P. P., Sathyendra, S., Abraham, O. C., Truman, C., & Reginald, A. (2020). Acute oleander poisoning: A study of clinical profile from a tertiary care center in South India. *Journal of Family Medicine and Primary Care*, 9, 136–140.

Kelly, R. A., & Smith, T. W. (1992). Recognition and management of digitalis toxicity. *American College of Cardiology,* 69, 108G–119G.

Kim, D.-S., Kim, H.-R., Woo, E.-R., Hong, S.-T., Chae, H.-J., & Chae, S.-W. (2005). Inhibitory effects of rosmarinic acid on adriamycin-induced apoptosis in H9c2 cardiac muscle cells by inhibiting reactive oxygen species and the activations of cJun N-terminal kinase and extracellular signal-regulated kinase. *Biochemical Pharmacology,* 70(7), 1066–1078.

Kim, Y., Park, E. J., Kim, J., Kim, Y. B., Kim, S. R., & Kim, Y. C., (2001). Neuroprotective constituents from Hedyotis diffusa. *Journal of Natural Products*, 64, 75–78.

Kohls, S., Scholz-Böttcher, B. M., Teske, J., Zark, P., & Rullkötter, J. (2012). Cardiac glycosides from Yellow Oleander (Thevetia peruviana) seeds. *Phytochemistry*, 75, 114–127. DOI: 10.1016/j.phytochem.2011.11.019.

Krenzelok, E. P. (Ed.), (2003). *Biological and Chemical Terrorism: A Pharmacy Preparedness Guide*. American Society of Health-System Pharmacists. Bethesda, MD, p. 232.

Kyerematen, G., Hagos, M., Weeratunga, G., & Sandbrg, F. (1985). The cardiac glycosides of Thevetia ovata A. Dc. and Thevetia neriifolia Juss. Ex Stend. *Acta Pharmaceutica Suecica*, 22, 37–44.

Langford, S. D., & Boor, P.J. (1996). Oleander toxicity: An examination of human and animal toxic exposures. *Toxicology*, 109, 1–13.

Le Couteur, D. G., & Fisher, A. A. (2002). Chronic and criminal administration of Nerium oleander. *Journal of Toxicology. Clinical Toxicology,* 40, 523–524.

Lederer, W. J., & Tsien, R. W. (1976). Transient inward current underlying arrhythmogenic effects of cardiotonic steroids in Purkinje fibres. *Journal of Physiology*, 263, 73–100.

Lehmann, V. H. D. (1984). Zur Wirkung pflanzlicher Glycoside auf Widerstandsgeflifie und KapazitBtsgefal3e. *Arzneim Forsch,* 34, 423–429.

Li, H., Zhou, H., Wang, D., Qiu, J., Zhou, Y., Li, X., Rosenfeld, M. G., Ding, S., & Fu, X. D. (2012). Versatile pathway-centric approach based on high-throughput sequencing to anticancer drug discovery. *Proceedings of the National Academy of Sciences of USA*, 109, 4609–4614.

Liu, J., Jiang, Z. Z., Zhang, S. M., & Saltzman, W. M. (2009). Poly(omega-pentadecalactone-co-butylene-co-succinate) nanoparticles as biodegradable carriers for camptothecin delivery. *Biomaterials*, 30, 5707–5719.

Liu, S. K. (1957). The pathology of oleander poisoning in cattle. *Memoirs of the College of Agriculture; National Taiwan University*, 5, 75–82.

Lokesh, S., & Arunkumar, R. (2013). A clinical study of 30 cases of Acute Yellow Oleander poisoning (Thevetia nerifolia). *Journal of Current Trends in Clinical Medicine & Laboratory Biochemistry*, 1(2).

Lu, C. M., Yang, J. J., Wang, P. Y., & Lin, C. C. (2000). A new acylated flavonol glycoside and antioxidant effects of Hedyotis diffusa, *Planta Medica*, 66, 374–376. https://doi.org/10.1055/s-2000-8544

Madhura, J., Arulnithy, K., Mathiventhan, U., & Mathiventhan, T. (2016). Yellow oleander (Thevetia peruviana) seed poisoning (YOSP) in the Batticaloa District, Sri Lanka: Is related with Fruiting Season? *International Journal of Research Studies in* Biosciences, 4, 8–13. DOI: 10.20431/2349-0365.0408002.

Mahin, L., Marzou, A., & Huart, A. (1984). A case of Neriwn oleander poisoning in cattle. *Veterinary and Human Toxicology,* 26, 303–304.

Mallick, B. K. (1984). Cardiotoxicity in yellow oleander seed poisoning. *Journal of the Indian Medical Association*, 82, 296–297.

Managit, C., Sakurai, H., & Saiki, I. (2017). Ethanolic extract of Thevetia peruviana flowers enhances TNF-α and TRAIL-induced apoptosis of human cervical cancer cells via intrinsic and extrinsic pathways. *Oncology Letters,* 13, 2791–2798. DOI: 10.3892/ol.2017.5748.

Mandal, L. (2012). Yellow oleander-Theveti a peruviana-poisoning. *PMJNM Postgraduate Medical Journal of NAMS*, 12(2).

Markham, K. R., Ternai, B., Stanley, R., Geiger, H., & Mabry, T. J. (1978). Carbon-13 NMR studies of fjlavonoids-III naturally occurrinci flavonoid glycosides and their acylated derivatives. *Tetrahedron*, 34, 1389–1397.

Markov, A. K., Brumley, M. A., Figueroa, A., Skelton, T. N., & Lehan, P. H. (1997). Hemodynamic effects of fructose 1,6-diphosphate in patients with normal and impaired left ventricular function. *American Heart Journal*, 133(5), 541–549.

Marx, J., Pretorius, E., & Bornman, M. S. (2006). The neurotoxic effects of prenatal cardiac glycoside exposure: A hypothesis. *Neurotoxicology and Teratology*, 28(1), 135–143.

McLaughlin, J. L., Freedman, B., Powel, R. G., & Smith, C. R. (1980). Neriifolin and 2'acetylneriifolin. Insecticidal and cytotoxic agents of Thevetia theveotides seeds. *Journal of Economic Entomology,* 73, 398–402.

Meena, B. R., Meena, S., Chittora, D., & Sharma, K. (2021). Antifungal efficacy of Thevetia peruviana leaf extract against Alternaria solani and characterization of novel inhibitory compounds by Gas Chromatography-Mass Spectrometry analysis. *Biochemistry and Biophysics Reports,* 25, 100914.

Middleton, W. S., & Chen, K. K. (1936). Clinical results from oral administration of thevetin. a cardiac glycoside. *American Heart Journal*, 11, 75–88.

Miyagawa, T., Ohtsuki, T., Koyano, T., Kowithayakorn, T., & Ishibashi, M. (2009). Cardenolide glycosides of Thevetia peruviana and triterpenoid saponins of Sapindus emarginatus as TRAIL resistance-overcoming compounds. *Journal of Natural Products*, 72(8), 1507–1511.

Nagy, E., Seres, I., Verzar-Petri, G., & Neszmelyi, A. (1984). Kaempferol 3-O-[β-D-Glucopyranosyl (1-2) β-D-galactopyranoside], a New Flavonoid from Lilium candidum L. *Zeitschrift für Naturforschung B*, 39(12), 1813–1815.

Obasi, N. B. B., Igboechi, A. C., & Bejamin, T. V. (1990). Seasonal variations in the seed oil of Thevetia peruviana (Pers.) K. Schum. *Journal of the American Oil Chemists' Society*, 67(10), 624–625.

Oi, N., Chen, H., Ok Kim, M., Lubet, R. A., Bode, A. M., & Dong, Z. (2012). Taxifolin suppresses UV-induced skin carcinogenesis by targeting EGFR and PI3-K. *Cancer Prevention Research*, 5(9), 1103–1114. DOI: 10.1158/1940-6207.CAPR-11-0397.

Oji, O., & Okafor, Q. E. (2000). Toxicological studies on stem bark, leaf and seed kernel of yellow oleander (Thevetia peruviana). *Phytotherapy Research,* 14, 133–135.

Oluwaniyi, O. O., Ibiyemi, A. S., & Usman, A. L. 2007. Effect of detoxification on the nutrient content of Thevetia peruviana seed cake. *Research Journal of Applied Sciences,* 2, 188–191.

Osterloh, J., Herold, S., & Pond, S. (1982). Oleander interference in the digoxin radioimmunoassay in a fatal ingestion. *JAMA*, 247, 1596–1597.

Pandey, N. D., Singh, S. R., & Tewari, G. C. (1977). Use of some plant powders, oils, and extracts as protectants against pulse beetle, Callosobruchus chinensis Linn. *Indian Journal of Entomology,* 38, 110–113.

Panigrahi, A., & Raut, S. K. (1994). Thevetia peruviana (Family: Apocynaceae) in the control of slug and snail pests. *Memórias do Instituto Oswaldo Cruz*, 89(2), 247–250.

Prassas, I., & Diamandis, E. P. (2008). Novel therapeutic applications of cardiac glycosides. *Nature Reviews Drug Discovery*, 7, 926–935.

Pullaiah, T. (2006). *Encyclopedia of World Medicinal Plants. Volume IV.* Regency Publications. New Delhi, India.

Radwan, R. R. S. E., & Kenawy, S. A. (2008). Hepatoprotective efficiency of combined administration of natural antioxidants (rutin and vitamin E) and cysteine in hyperthermic irradiated rats. *Egyptian Journal of Hospital Medicine,* 32, 441–454.

Rahman, N., Rahman, H., Haris, M., & Mahmood, R. (2017). Wound healing potentials of Thevetia peruviana: Antioxidants and inflammatory markers criteria. *Journal of Traditional and Complementary Medicine*, 7, 519–525. DOI: 10.1016/j.jtcme.2017.01.005v.

Rajapakse, S. (2009). Management of yellow oleander poisoning. *Journal of Toxicology. Clinical Toxicology,* 47, 206–212.

Ramchandra, P., Basheermiya, M., Krupadanam, G. L. D., & Srimannarayana, G. (1993). Wrightial, a new terpene from Wrightia tinctoria. *Journal of Natural Products*, 56(10), 1811–1812.

Ramos-Silva, A., Tavares-Carreón, F., Figueroa, M., De la Torre-Zavala, S., GastelumArellanez, A., Rodríguez-García, A., Galán-Wong, L. J., & Avilés-Arnaut, H. (2017). Anticancer potential of Thevetia peruviana fruit methanolic extract. *BMC Complementary Medicine and Therapies,* 17(1), 241.

Ratigan, W. J. (1921). Oleander poisoning in a bear. *Journal of the American Veterinary Medical Association,* 60, 96–98.

Renzulli, C., Galvano, F., Pierdomenico, L., Speroni, E., & Guerra, M. (2004). Effects of rosmarinic acid against aflatoxin B1 and ochratoxin-A-induced cell damage in a human hepatoma cell line (Hep G2). *Journal of Applied Toxicology,* 24(4), 289–296.

Repke, K. R. H. (1985). New developments in cardiac glycoside structure-activity relationships. *Trends in Pharmacological Sciences*, 6, 275–278.

Rexakhani, A., & Maham, M. (1992). Oleander poisoning in cattle of the Fars province, Iran. *Veterinary and Human Toxicology,* 34, 549.

Roberts, D. M., Gallapatthy, G., Dunuwille, A., & Chan, B. S. (2015). Pharmacological treatment of cardiac glycoside poisoning. *British Journal of Clinical Pharmacology*, 81(3), 488–495.

Roberts, D. M., Gallapatthy, G., Dunuwille, A., & Chan, B. S. (2016). Pharmacological treatment of cardiac glycoside poisoning. *British Journal of Clinical Pharmacology*, 81(3), 488–495. DOI: 10.1111/bcp.12814.

Roberts, D. M., Southcott, E., Potter, J. M., Roberts, M. S., Eddleston, M., & Buckley, N. A. (2006). Pharmacokinetics of digoxin cross-reacting substances in patients with acute yellow Oleander (Thevetia peruviana) poisoning, including the effect of activated charcoal. *Therapeutic Drug Monitoring*, 28, 784–792.

Romano, G. A., & Mombelli, G. (1990). Poisoning with oleander leaves. *Schweizerische Medizinische Wochenschrift*, 120(16), 596–597.

Safadi, R., Levy, I., Amitai, Y., & Caraco, Y. (1995). Beneficial effects of digoxin-specific Fab antibody fragments in oleander intoxication. *Archives of Internal Medicine,* 155, 2121–2125.

Sagawa, T., Sagawa, K., Kelly, J. E., Tsushima, R. G., & Wasserstrom, J. A. (2002). Activation of cardiac ryanodine receptors by cardiac glycosides. *American Journal of Physiology-Heart and Circulatory Physiology,* 282, H1118–H1126.

Samal, K. K. (1989). Yellow oleander poisoning with jaundice and renal failure. *Journal of the Association of Physicians of India,* 37, 232–233.

Samal, K.K. (1990). Yellow oleander poisoning with jaundice and renal failure. *JAPI* 38, 821–822.

Saravanapavananthan, N., & Ganeshamoorthy, J. (1988). Yellow oleander poisoning–A study of 170 cases. *Forensic Science International,* 36, 247–250.

Saraswat, D. K., Garg, P. K., & Saraswat, M. (1992). Rare poisoning with cerebra thevetia (yellow oleander). Review of 13 cases of suicidal attempt. *Journal of the Association of Physicians of India* 40, 628–629.

Saravanapavananthan, T. (1985). Plant poisoning in Sri Lanka. *Jaffna Medical Journal,* 20, 17–21.

Sarmah, D., & Deka, D. C. (2019). Use of yellow oleander (Thevetia peruviana) seed oil biodiesel as cetane and lubricity improver for petrodiesel. *Rasayan Journal of Chemistry*, 12, 1547–1556.

Sasidharan, S., Chen, Y., Saravanan, D., Sundram, K. M., & Latha L. Y. (2011). Extraction, isolation and characterization of bioactive compounds from plants' extracts. *African Journal of Traditional, Complementary, and Alternative Medicines,* 8(1).

Schoner, W., & Bobis, G. S. (2007). Endogenous and exogenous cardiac glycosides: Their roles in hypertension, salt metabolism, and cell growth. *American Journal of Physiology-Cell Physiology*, 293(2), C509–C536.

Schwartz, A., & Whitmer, K. (1982). Mechanism of action of digitalis: Is the Na, K-ATPase the pharmacological receptor? *Annals of the New York Academy of Sciences*, 402, 253–271.

Schwartz, W. L., Bay, W. W., Dollahite, J. W., Strorts, R. W., & Russell, L. H. (1974). Toxicity of Nerium oleander in the monkey (Cebus ape/la). *Veterinary Pathology,* 11, 259–277.

Senthilkumaran, S., Meenakshisundaram, R., Michaels, A. D., & Thirumalaikolundusubramanian, P. (2011). Electrocardiographic changes during inhalational oleander toxicity. *Journal of Electrocardiology*, 44, 470–472.

Shaw, D., & Pearn, J. (1979). Oleander poisoning. *Medical Journal of Australia*, 8, 267–269.

Shumaik, G. M., Wu, A. W., & Ping, A. C. (1988). Oleander S.D. Lmglbrd, P.J. Boor/ Toxicology IO9 (19%) l-13 13 poisoning: Treatment with digoxin-specific Fab antibody fragments. *Annals of Emergency Medicine*, 17, 732–735.

Siddiqui, S., Siddiqui, B. S., Adil, Q., & Begum, S. (1992). Cardenolides and triterpenoids of the leaves of Thevetia neriifolia. Phytochemistry. *The International Journal of Plant Biochemistry*, 31(10), 3541–3546.

Singh, S., Bothara, S. B., & Singh, S. (2011). Acid-base indicator properties of dyes from local flowers: Cassia aungostifolia Linn., Thevetia peruviana (Pers.) K. Schum and Thevetia thvetiodes (Kunth) K. Schum. *Pharmacognosy Journal*, 3(19), 35–39.

Sivakumar, D. K., & Nithya, G. (2018). Study on clinical profile, electrolyte and electrocardiographic abnormalities in patients with yellow oleander poisoning. *International Journal of Advances in Medicine*, 5(2), 299–306. http://www.ijmedicine.com

Smith, T. W., Haber, E., Yeatman, L., & Bulter, V. P., Jr. (1976). Reversal of advanced digoxin intoxication with Fab fragments of digoxin specific antibodies. *New England Journal of Medicine*, 294, 797–800.

Sreeharan, N., Puthrasingam, S., Ranjadayalan, K., Satkurunathan, K., Ganeshamoorthy, J., 1985. Yellow oleander poisoning – clinical manifestations and prognostic criteria. *Jaffna Medical Journal*, 20, 100–101.

Srivastava, N., Chauhan, A. S., & Sharma, B. (2012). Isolation and characterization of some phytochemicals from Indian traditional plants. *Hindawi Publishing Corporation Biotechnology Research International*, 2012, 8. DOI: 10.1155/2012/549850.

Srivastava, V., Negi, A. S., Kumar, J. K., Gupta, M. M., & Khanuja, S. P. S. (2005). Plant-based anticancer molecules: A chemical and biological profile of some important leads. *Bioorganic & Medicinal Chemistry,* 13, 5892–5908.

Swaminathan, M., & Kochhar, S. (2019). *Major Flowering Trees of Tropical Gardens.* Cambridge University Press. Cambridge, England. DOI: 10.1017/9781108680646.

Szabuniewicz, M., Schwartz, W. L., & McCrady, J. D. (1972). Experimental oleander poisoning and treatment. *Southwestern Veterinarian*, 25, 105–114.

Szabuniewicz, M., Schwartz, W. L., McCrady, J. D., Russell, L. H., & Camp, B. J. (1971). Treatment of experimentally induced oleander poisoning. *Archives internationales de pharmacodynamie et de thérapie,* 189, 12–21.

Tewtrakul, S., Nakamura, N., Hattori, M., Fujiwara, T., & Supavita, T. (2002). Flavanone and flavonol glycosides from the leaves of Thevetia peruviana and their HIV-1 reverse transcriptase and HIV-1 integrase inhibitory activities. *Chemical and Pharmaceutical Bulletin*, 50(5), 630–635.

Thilagavathi, R., Kavitha, H., and Venkatraman, B. (2010). Isolation, characterization and anti-inflammatory property of *Thevetia Peruviana. Journal of Chemistry*, 7, 7. DOI: 10.1155/2010/759497.

Tian, D. M., Cheng, H. Y., Jiang, M. M., Shen, W. Z., Tang, J. S., & Yao, X. S. (2016). Cardiac glycosides from the seeds of Thevetia peruviana. *Journal of Natural Products*, 79, 38–50.

Tiukavkina, N. A., Rulenko, I. A., & Kolesnik, I. A. (1997). Dihydroquercetina new antioxidant and biologically active food additive. *Voprosy Pitaniia,* 6, 12–15.

Tor, E. R., Filigenzi, M. S., & Puschner, B. (2005). Determination of oleandrin in tissues and biological fluids by liquid chromatography-electrospray tandem mass spectrometry. *Journal of Agricultural and Food Chemistry,* 53, 4322–4325.

Vijayababu, M. R., Arunkumar, A., Kanagaraj, P., Venkataraman, P., Krishnamoorthy, G., & Arunakaran, J. (2006). Quercetin downregulates matrix metalloproteinases 2 and 9 proteins expression in prostate cancer cells (PC-3). *Molecular and Cellular Biochemistry,* 287, 109–116.

Voigtländer, H.-W., Balsam, G., Herbst, G., (1969). Über acetyl-cardenolide aus Thevetia peruviana. *Archiv der Pharmazie* 302, 538–544.

Wang, J., Portbury, S., Thomas, M. B., Barney, S., Ricca, D. J., Morris, D. L., Warner, D. S., & Lo, D. C. (2006). Cardiac glycosides provide neuroprotection against ischemic stroke: discovery by a brain slice-based compound screening platform. *Proceedings of the National Academy of Sciences of the United States of America*, 103(27), 10461–10466. https://doi.org/10.1073/pnas.0600930103

Wang, X., Yang, C., Zhang, Y., Zhen, X., Wu, W., & Jiang, X. (2014). Delivery of platinum (IV) drug to subcutaneous tumor and lung metastasis using bradykinin-potentiating peptide-decorated chitosan nanoparticles. *Biomaterials*, 35, 6439–6453.

Weidmann, A. E. (2012). Dihydroquercetin: More than just an impurity? *European Journal of Pharmacology*, 684(1–3), 19–26. DOI: 10.1016/j.ejphar.2012.03.035.

Weitz and Boulay: Bull. Sci. Pharmacol., 30, 81,1923. www.gbif.org. Thevetia neriifolia Juss. ex Steud. https://doi.org/10.15468/rjarmt. accessed via GBIF.org on 2022-05-08.

Williams, C.A., Richardson, J., Greenham, J., & Eagles, J. (1993) Correlation between leave flavonoids, taxonomy and plant geography in the genus Disporum. *Phytochemistry*, 34, 197–203.

Xabregas, J. A. (1950). Thevetia neriifolia JUSS como oleaginosa, Agron. light exposure. *Angolana*, 3, 121–132.

Yarkasuwa, C. I., Wilson, D., & Michael, E. (2013). Production of biodiesel from yellow oleander (Thevetia Peruviana) oil and its biodegradability. *Journal of Korean Chemical Society*, 2013(57), 1–6.

Yatime, L., Laursen, M., Morth, J. P., Esmann, M., Nissen, P., & Fedosova, N. U. (2011). Structural insights into the high affinity binding of cardiotonic steroids to the Na+, K+ -ATPase. *Journal of Structural Biology*, 174, 296–306.

Zamani, J., & Aslani, A. (2010). Cardiac findings in acute yellow oleander poisoning. *Journal of Cardiovascular Disease Research*, 1, 27–29.

Zhu, J. J., Zhang, X. X., Miao, Y. Q., He, S. F., Tian, D. M., Yao, X. S., Tang, J. S., & Gan, Y. (2017). Delivery of acetylthevetin B, an antitumor cardiac glycoside, using polymeric micelles for enhanced therapeutic efficacy against lung cancer cells. *Acta Pharmacologica Sinica*, 38(2), 290–300. https://doi.org/10.1038/aps.2016.113

22 *Toxicodendron succedaneum* (Rhus or Wax Tree)

Godwin Anywar and Patience Tugume

CONTENTS

22.1 INTRODUCTION

Toxicodendron succedaneum is a vascular flowering plant that belongs to family Anacardiaceae. Its synonyms include *Rhus erosus* Radlk., *Rhus succedanea* var. *Japonica* Engl. and *Toxicodendron succedaneum* var. *succedaneum* (WFO, 2022). It is commonly referred to as the wax tree (Slaughter et al., 2017) but also known by different local names in different localities. For instance, in Japan, it is known as the Japanese Hazenoki tree (Sumac tree), Japanese lacquer tree, Japanese tallow tree, Japanese wax, Japanese wax tree, poison ivy, poison sumac, rhus, rhus tree, scarlet rhus, sumac, varnish tree and wax tree. It is called sơn in Vietnam or charão in Portugual and *jy bbo* in China (Addi et al., 2022).

The plant produces an oily resin in leaves that contain active toxins called urushiols (Derraik, 2007). These toxins may also be found within the entire plant including dead tissue. Toxicity ensues in the autumn coinciding with the leaves changing colour from green to red (Kim et al., 2019). Thus, it is very important that people are aware of this fact when handling the plant. *Toxicodendron succedaneum* has beautiful foliage which attracted its cultivation as an ornamental by gardeners who were unaware of the allergic reactions it causes (Derraik, 2007). *Toxicodendron succedaneum* is as allergenic as poison ivy (*T. radicans*), (Derraik, 2007) and the most common cause of allergic contact dermatitis worldwide (Allen, 2006).

Toxicodendron succedaneum has long been grown for commercial purposes. However, following its declaration as a noxious weed, it could no longer be sold in Australia (Monaghan & McMaugh, 2002), Japan (Rademaker & Duffill, 1995a, b). Public education regarding its toxicity resulted in reduction in the number of trees grown in New South Wales (Monaghan & McMaugh, 2002). Despite the toxicity of the plant, it has been traditionally used by different communities for medicine and a source of lacquer.

DOI: 10.1201/b23017-24

22.2 BOTANICAL DESCRIPTION AND ECOLOGY OF *TOXICODENDRON SUCCEDANEUM*

T. succedaneum is a large shrub or tree whose crown grows up to 8 m tall on a single erect trunk. It is deciduous and loses its leaves in autumn. The leaves are hairless on both surfaces, compound pinnate (0–35 cm long) and alternately arranged along the branches consisting of 4–7 pairs of leaflets that are 4–10 cm long and 1.5–3 cm wide and a single terminal leaflet (Johnson & Johnson, 2006). Each leaflet is borne on a short petiolule 1–4 mm long. The leaflets (4–10 cm long and 1.5–3 cm wide) are lanceolate or elliptic with entire margins and acuminate apices. The leaves are green but turn bright red, scarlet or crimson before they are shed in autumn (Flora of China Editorial Committee, 2016; Figure 22.1).

Flowers are white to yellowish green 5–6 mm wide borne in clusters near the tips of the branches. Separate male and female flowers are present in these clusters. Each flower has five tiny greenish sepals (about 1 mm long) and five small greenish-yellow or whitish petals (about 2 mm long), and it is borne on a short stalk (i.e., pedicel) that is about 2 mm long. They also have five stamens and an ovary topped with a style and stigma. Flowering occurs mostly during spring and early summer. The fruit is a drupe, papery (5–11 mm wide) that turns from green to pale brown to black when ripe and contains a single hard, dark brown round seed (3–5 mm wide). These fruits (5–10 mm long and 7–11 mm across) are borne in large groups that hang from the branches. The trunk is greyish-brown coloured bark. The younger branches are hairless (i.e., glabrous) and smooth in texture, apart from some small raised bumps (i.e., lenticels; Flora of China Editorial Committee, 2016).

22.3 DISTRIBUTION OF *T. SUCCEDANEUM*

T. succedaneum is widely distributed in temperate zones across China, Japan, Indochina, Malaysia and India and grows on a wide range of soil types (Hiraoka et al., 2018). It is native to India, Myanmar, Thailand, Vietnam, China, Taiwan, Japan and Malaysia (Aggarwal, 2001; Flora of China

FIGURE 22.1 *Toxicodendron succedaneum.*

Editorial Committee, 2016; Plants of the World Online, 2022). It has been introduced to and is naturalized in Australia, New Zealand, South Africa, Brazil and Cuba (NZPCN, 2016; Weeds of Australia, 2022).

The tree grows in several habitats in different countries. For instance, in Asia, *T. succedaneum* grows in lowlands, hill forests, thickets and along streams in montane forests at an altitude of 100–2,500 m (Aggarwal, 2001; Flora of China Editorial Committee, 2016). In New Zealand, it grows among indigenous vegetation along coasts, urban gardens and disturbed sites (NZPCN, 2016). In Australia, it grows as a weed of disturbed sites, forests, open woodlands, urban bushland, roadsides, gardens and waste areas (Weeds of Australia, 2022). The varied habitats are attributed to the invasive nature of the species with the ability to spread from domestic gardens (NZPCN, 2016; Weeds of Australia, 2022). It is thus a potential weed of disturbed sites, forests, open woodlands, urban bushland, roadsides, gardens and waste areas in temperate and sub-tropical regions. The plant is regarded as a weed in Australia (Weeds of Australia, 2022).

22.4 ETHNOBOTANY OF *T. SUCCEDANEUM*

Regardless of its toxicity, *T. succedaneum* is used traditionally or used as raw material in small scale industries in different areas. In Japan, *T. succedaneum* is grown for the production of sumac wax, which is extracted from the mesocarp (Hiraoka et al., 2018). In Ryukyu kingdom, southeast Asia, the sap from *T. succedaneum* was used to produce lacquer an excellent adhesive and painting material used to make lacquerware. The wax is used for making anticorrosive and decorative paints and dyes (Nie et al., 2009), wooden coffins (Nakagawa et al., 2019) and fine art paintings in Vietnam and Taiwan (Leipe et al., 2022). Additionally, the wax is used in varnishes, polishes, ointments and plasters (Flora of China Editorial Committee, 2016). A wax obtained from the fruit is used as an ingredient in cosmetic preparations as an emollient, binding agent and viscosity controller (https://atamankimya.com/sayfalaralfabe.asp?LanguageID=2&cid=3&id=2868&id2=3610).

Traditionally, the Japanese used vegetable oil from *T. succedaneum* to produce traditional candles (Hiraoka et al., 2018; Katsumi et al., 2016). Recently, however, traditional candles have been substituted with imported candles made of paraffin due to their affordability. Another reason for the preference of imported paraffin candles is that they release vaporized paraffin compared to traditional candles which emit soot that has earlier been shown to cause lipoid pneumonia (Glynn & Gale, 1990). *T. succedaneum* is traditionally used in West Garo hills of Meghalaya in India for construction works, firewood and charcoal (Das et al., 2018). In Azad Jammu and Kashmir district of Neelum valley, *T. succedaneum* is used as a source of fodder and fuel (Ajaib et al., 2020). The resin of *T. succedaneum* is used in the production of insulation materials (Yun-Yang et al., 2006).

T. succedaneum has been used in native medicine to treat asthma, cough and colicky pains (Do Hyung Kim et al., 2018) and Korean folk medicine (Oh et al., 2003). It has also been used to treat diarrhea, nose and gum bleedings, vomiting, dysentery, tuberculosis, fever, liver ailments and ear infections (Kottakkal, 1995; Sathasivampillai et al., 2007).

22.5 PHYTOCHEMISTRY OF *T. SUCCEDANEUM*

T. succedaneum contains urushiols (Evans & Schmidt, 1980) and bioflavonoids (Wang et al., 1998) such as amentoflavone, agathisflavone, robustaflavone, hinokiflavone, volkensiflavone, morelloflavone, rhusflavanone and succedaneaflavanone (Lin et al., 1997). Urushiols are the allergenic principles of poison ivy and related species of the Anacardiaceae. Urushiol has a variable chemical structure but is mainly comprised of catechol with a long hydrocarbon chain. They are mixtures of homologous long chain phenolic compounds (Dawson, 1956; Evans & Schmidt, 1980; Symes & Dawson, 1953). *T. succedaneum* also contains the cytotoxic hinokiflavone (Lin et al., 1989; Wang et al., 1998).

22.6 BIOLOGICAL ACTIVITY OF *T. SUCCEDANEUM*

Extracts from *T. succedaneum* possess anticancer activities (Wang et al., 1998), anti-HIV activity (Lin et al., 1997), antioxidant (Wu et al., 2002) and cytotoxic activities (Wang et al., 1998). The bioflavonoids from *T. succedaneum* had varying anti-HIV-1 activity against HIV-1 reverse transcriptase. Their activities ranged from inactive to significant in HIV-1 in human lymphocytes. For instance, morelloflavone demonstrated significant antiviral activity against HIV-1 (strain LAV-1) in phytohemagglutinin-stimulated primary human peripheral blood mononuclear cells ($EC_{50}=6.9\,\mu M$) and a selectivity index of approximately 10 (Lin et al., 1997).

22.7 TOXICITY OF *T. SUCCEDANEUM*

T. succedaneum is highly toxic and causes severe allergies when its sap comes in contact with the skin. The allergies manifest as a rash, redness, itching and blisters (Johnson & Johnson, 2006; McGovern & Barkley, 1998). Exposure to urushiol leads to the production of a type IV hypersensitivity reaction (Fisher, 1996). This causes the release of local cytokines and inflammatory mediators that initiate the process of sensitization. Urushiol is rapidly absorbed due to its lipophilic nature. As it is picked up by Langerhans cells in the epidermis, efforts to wash or remove the compound after exposure are largely ineffective after absorption (McGovern & Barkley, 1998). It is estimated that 50%–75% of adults are allergic to urushiol (Kim et al., 2019).

Ingestion of *Toxicodendron* can cause systemic contact dermatitis. This manifested as four main types of clinical reaction, viz., generalized erythematous maculopapular eruptions, generalized erythroderma, vesiculobullous lesions and erythema multiform-like lesions (Oh et al., 2003). Initial exposure results in pruritus and erythema, followed by a papulovesicular eruption, oedema, and oozing within 10–14 days. Symptoms of re-exposure are more acute and appear in 24–72 hours (Williams et al., 1999). Secondary exposure can occur when the skin is exposed to oxidized urushiol or animal fur that came in contact with the Toxicodendron species. Aerosolized urushiol from forest fires or burning plant debris has been known to cause airway inflammation or generalized dermatitis in severe cases (Lofgran & Mahabal, 2022).

22.8 CONCLUSION

T. succedaneum is a multipurpose tree with a long history of commercial cultivation as an ornamental. The plant has an equally long history of use in folk medicine in different parts of the world, even though it is highly toxic. *T. succedaneum* causes severe allergies when its sap comes in contact with the skin and many gardeners were unaware of the allergic reactions it causes. *T. succedaneum* is the most common cause of allergic contact dermatitis worldwide. The plant produces an oily resin in leaves that contain active toxins called urushiols. Regardless of its toxicity, *T. succedaneum* is used traditionally or used as raw material in small scale industries in different areas such as in the production of sumac wax cosmetics, adhesives like lacquer varnishes, polishes, ointments and plasters, painting materials used to make lacquerware preparations, as an emollient, binding agent and viscosity controller as well as traditional candles.

NOTES

Godwin Anywar, Department of Plant Sciences, Microbiology & Biotechnology, Makerere University, Kampala, Uganda

Patience Tugume, Department of Plant Sciences, Microbiology & Biotechnology, Makerere University, Kampala, Uganda

REFERENCES

Addi, Y. W., Zhang, Y., Ding, X. Y., Guo, C. A., & Wang, Y. H. (2022). A study of the plant folk nomenclature of the Yi people in Xiaoliangshan, Yunnan Province, China, and the implications for protecting biodiversity. *Journal of Ethnobiology and Ethnomedicine*, *18*(1), 1–29.

Aggarwal, S. (2001). Rhus succedanea L. Record from Proseabase. Bogor, Indonesia: PROSEA (Plant Resources of South-East Asia) Foundation. http://www.proseanet.org

Ajaib, M., Maqbool, M., Hussain, T., Bhatti, K. H., Khan, R., Ghani, A., … & Khatoon, S. (2020). Exploration of ethnobotanical and ethnomedicinal importance of naturally growing plants of District Neelum from areas of Dawarian to Ratti Gali, Azad Jammu and Kashmir. *Research Square.* doi: 10.21203/rs.3.rs-103983/v1

Allen, P. L. J. (2006). Leaves of three, let them be: If it were only that easy! *Dermatology Nursing*, *18*, 236–242.

Das, G., Sharma, R. K., & Borborah, M. (2018). Ethnomedicinal use of different medicinal plants used by the traditional practitioners in north East India for the treatment of jaundice. *World Journal of Pharmacy & Pharmaceutical Sciences*, *8*(2), 1133–1138.

Dawson, C. R. (1956). The chemistry of poison ivy. *Transactions of the New York Academy of Sciences, 18*(5), 427–443. doi: 10.1111/j.2164-0947.1956.tb00465.x.

Derraik, J. G. (2007). *Heracleum mantegazzianum* and *Toxicodendron succedaneum*: Plants of human health significance in New Zealand and the National Pest Plant Accord. *New Zealand Medical Journal, 120*, U2657.

Do Hyung Kim, S. J. L., Da Som Oh, I. D. L., Jun Sik Eom, H. Y. P., & Seong Ho Choi, S. S. L. (2018). In vitro evaluation of Rhus succedanea extracts for ruminants. *Asian-Australasian Journal of Animal Sciences, 31*(10), 1635.

Evans, F. J., & Schmidt, R. J. (1980). Plants and plant products that induce contact dermatitis. *Planta Medica*, 38(4), 289–316. doi: 10.1055/s-2008-1074883.

Fisher, A. A. (1996). Poison Ivy/Oak/Sumac. Part II: Specific features. *Cutis*, *58*(1), 22–24.

Flora of China Editorial Committee. (2016). Flora of China. St. Louis, Missouri and Cambridge, Massachusetts, USA: Missouri Botanical Garden and Harvard University Herbaria. http://www.efloras.org/flora_page.aspx?flora_id=2

Glynn, K. P., & Gale, N. (1990). Exogenous lipoid pneumonia due to inhalation of spray lubricant (DW-40 lung). *Chest*, *97*, 1265–1266.

Hiraoka, Y., Tamaki, I., & Watanabe, A. (2018). The origin of wild populations of Toxicodendron succedaneum on mainland Japan revealed by genetic variation in chloroplast and nuclear DNA. *Journal of Plant Research, 131*(2), 225–238.

Johnson, A., & Johnson, S. (2006). Garden plants poisonous to people, Primefact 359, NSW Department of Primary Industries, Orange.

Katsumi, H., Tominaga, M., Tajiri, M., Shimizu, S., Sakazaki, Y., Kinoshita, T., … & Hoshino, T. (2016). A case of lipoid pneumonia caused by inhalation of vaporized paraffin from burning candles. *Respiratory Medicine Case Reports, 19*, 166–168.

Kim, Y., Flamm, A., ElSohly, M. A., Kaplan, D. H., Hage, R. J., Jr., Hamann, C. P., & Marks, J. G., Jr. (2019). Poison ivy, oak, and sumac dermatitis: What is known and what is new? *Dermatitis*, *30*(3), 183–190.

Kottakkal, A. V. S. (1995). *Indian Medicinal Plants*. Orient Longman Ltd., Vijayawada, Andhra Pradesh.

Leipe, C., Aquaro, A., & Tarasov, P. E. (2022). Scanning electron microscopy for differentiating charred endocarps of Rhus/Toxicodendron species and tracking the use of the lacquer tree and Asian poison ivy in Japanese prehistory. *Journal of Archaeological Science: Reports*, *41*, 103335.

Lin, Y. M., Anderson, H., Flavin, M. T., Pai, Y. H., Mata-Greenwood, E., Pengsuparp, T., Pezzuto, J. M., Schinazi, R. F., Hughes, S. H., & Chen, F. C. (1997). *In vitro* Anti-HIV activity of biflavonoids isolated from *Rhus succedanea* and *Garcinia multiflora*. *Journal of Natural Products*, *60*(9), 884–888. doi: 10.1021/np9700275.

Lin, Y. M., Chen, F. C., & Lee, K. H. (1989). Hinokiflavone, a cytotoxic principle from *Rhus Succedanea* and the cytotoxicity of the related biflavonoids. *Planta Medica*, *55*(2), 166–168. doi: 10.1055/s-2006-961914.

Lofgran, T., & Mahabal, G. D. (2022). Toxicodendron toxicity. Treasure Island (FL).

McGovern, T. W., & Barkley, T. M. (1998). Botanical dermatology. *International Journal of Dermatology*, *37*(5), 321–334. doi: 10.1046/j.1365-4362.1998.00385.x.

Monaghan, N., & McMaugh, J. Rhus – an urban weed. Agfact P7.6.41 (2nd ed), 21 March 2002. New South Wales Department of Primary Industries. https://assets-global.website-files.com/5e332a62c703f653182faf47/5e332a62c703f684402fc8fd_Vol-120-No-1259-10-August-2007.pdf

Nakagawa, R., Nakai, S. I., Miyakoshi, T., & Honda, T. (2019). Materials and provenance determination of lacquerware from the Ryukyu Kingdom period by pyrolysis-gas chromatography/mass spectrometry and 87Sr/86Sr isotope ratio. *Journal of Archaeological Science: Reports, 25*, 72–76.

Nie, Z., Sun, H., Meng, Y., & Wen, J. (2009). Phylogenetic analysis of Toxicodendron (Anacardiaceae) and its biogeographic implications on the evolution of north temperate and tropical intercontinental disjunctions. *Journal of Systematics and Evolution, 47*(5), 416–430.

NZPCN. (2016). *Toxicodendron Succedaneum.* New Zealand Plant Conservation Network, Wellington, New Zealand. http://www.nzpcn.org.nz/flora_details.aspx?ID=4412

Oh, S.-H., Haw, R.-C., & Lee, M.-H. (2003). Clinical and immunologic features of systemic contact dermatitis from ingestion of Rhus (Toxicodendron). *Contact Dermatitis, 48*(5), 251–254.

Plants of the World Online (2022) https://powo.science.kew.org/results?q=Toxicodendron%20succedaneum-Rademaker, M., & Duffill, M. B. (1995a). *Toxicodendron succedaneum* (Rhus tree), New Zealand's poison ivy. *Contact Dermatitis, 33*, 357–358.

Rademaker, M., & Duffill, M. B. (1995b). Allergic contact dermatitis to *Toxicodendron succedaneum* (rhus tree): An autumn epidemic. *New Zealand Medical Journal, 108*, 121–123.

Sathasivampillai, S. V., Rajesh, A., Sebastian, P. R., & Varatharasan, S. (2007). Bioactivities of Toxicodendron succedaneum (L.) Kuntze extracts and isolated compounds. *Anatolian Journal of Biology, 1*(2), 69–73.

Slaughter, R., Beasley, M., & Schep, L. (2017). Dermatitis due to Toxicodendron plants: A common occurrence during autumn. *Dermatitis, 130*, 1451.

Symes, W. F., & Dawson, C. R. (1953). Separation and structural determination of the olefinic components of poison ivy urushiol, cardanol and cardol. *Nature, 171*(4358), 841–842. doi: 10.1038/171841b0.

Wang, H. K., Xia, Y., Yang, Z. Y., Morris Natschke, S. L., & Lee, K. H. (1998). Recent advances in the discovery and development of flavonoids and their analogues as antitumor and anti-HIV agents. *Flavonoids in the Living System, 439*, 191–225.

Weeds of Australia (2022) *Toxicodendron succedaneum* (L.) Kuntze https://profiles.ala.org.au/opus/weeds-australia/profile/Toxicodendron%20succedaneum

WFO. (2022). Toxicodendron succedaneum (L.) Kuntze. Published on the Internet. http://www.worldfloraonline.org/taxon/wfo-0001052331. Accessed on: 05 May 2022.

Williams, J. V., Light, J., & Marks, J. G., Jr. (1999). Individual variations in allergic contact dermatitis from urushiol. *Archives of Dermatology, 135*(8), 1002–1003. doi: 10.1001/archderm.135.8.1002.

Wu, P.-L., Lin, S.-B., Huang, C.-P., & Chiou, R. Y. Y. (2002). Antioxidative and cytotoxic compounds extracted from the sap of *Rhus Succedanea. Journal of Natural Products, 65*(11), 1719–1721. doi: 10.1021/np0201467.

Yun-Yang, W., Du, Y., Fang-Xing, Y., Ying, X., Rong-Zhi, C., & Kennedy, J. F. (2006). Purification and characterization of hydrosoluble components from the sap of Chinese Lacquer tree *Rhus vernicifera. International Journal of Biological Macromolecules, 38*, 232–240. doi:10.1016/j.ijbiomac.2006.02.019.

Index

Note: **Bold** page numbers refer to tables and *italic* page numbers refer to figures.